Quality and Reliability of Large-Eddy Simulations II

ERCOFTAC SERIES

VOLUME 16

Series Editors

Aims and Scope of the Series

ERCOFTAC (European Research Community on Flow, Turbulence and Combustion) was founded as an international association with scientific objectives in 1988. ERCOFTAC strongly promotes joint efforts of European research institutes and industries that are active in the field of flow, turbulence and combustion, in order to enhance the exchange of technical and scientific information on fundamental and applied research and design. Each year, ERCOFTAC organizes several meetings in the form of workshops, conferences and summerschools, where ERCOFTAC members and other researchers meet and exchange information.

The ERCOFTAC Series will publish the proceedings of ERCOFTAC meetings, which cover all aspects of fluid mechanics. The series will comprise proceedings of conferences and workshops, and of textbooks presenting the material taught at summerschools.

The series covers the entire domain of fluid mechanics, which includes physical modelling, computational fluid dynamics including grid generation and turbulence modelling, measuring-techniques, flow visualization as applied to industrial flows, aerodynamics, combustion, geophysical and environmental flows, hydraulics, multiphase flows, non-Newtonian flows, astrophysical flows, laminar, turbulent and transitional flows.

For other titles published in this series, go to
www.springer.com/series/5934

Quality and Reliability of Large-Eddy Simulations II

Edited by

Maria Vittoria Salvetti
University of Pisa, Italy

Bernard Geurts
University of Twente, Netherlands

Johan Meyers
Katholieke Universiteit Leuven, Belgium

and

Pierre Sagaut
Université Pierre et Marie Curie 6, France

Springer

Maria Vittoria Salvetti
University of Pisa
Aerospace Engineering Department
Via G. Caruso 8
56122 Pisa
Italy
mv.salvetti@ing.unipi.it

Bernard Geurts
Multiscale Modeling and Simulation
Faculty EEMCS
University of Twente
P.O. Box 217
7500 AE Enschede
The Netherlands
b.j.geurts@utwente.nl
Anisotropic Turbulence
Faculty Applied Physics
Eindhoven University of Technology
P.O. Box 513
5600 MB Eindhoven
The Netherlands

Johan Meyers
Katholieke Universiteit Leuven
Div. Appl. Mechanics & Energy Conversion
Celestijnenlaan 300A
3001 Leuven
Belgium
johan.meyers@mech.KULeuven.be

Pierre Sagaut
Université Pierre et Marie Curie 6
Institut Jean le Rond d'Alembert
4 place Jussieu - case 162
75252 Paris cedex 5
France
pierre.sagaut@upmc.fr

ISSN 1382-4309
ISBN 978-94-007-0230-1 e-ISBN 978-94-007-0231-8
DOI 10.1007/978-94-007-0231-8
Springer Dordrecht Heidelberg London New York

Cover design: VTEX, Vilnius

Printed on acid-free paper

Springer is part of Springer Science+Business Media (www.springer.com)

Preface

The second Workshop on "Quality and Reliability of Large-Eddy Simulations", QLES2009, was held at the University of Pisa from September 9 to September 11, 2009. Its predecessor, QLES2007, was organized in 2007 in Leuven (Belgium). The focus of QLES2009 was on issues related to predicting, assessing and assuring the quality of LES. The development of computational resources and the corresponding tendency to apply LES-methodologies to turbulent flow problems of significant complexity, such as arise in various applications in technology and in many natural flows, makes the issue of assessing and optimizing the quality of LES predictions a timely challenge. Different error sources are present in LES, which are mainly related to physical modeling (especially as regards subgrid scales), to numerical discretization techniques, to boundary-condition treatment, and to grid resolution and design. These errors may interact in a complex non-linear manner, eventually leading to unpredictable and unexpected effects on LES results.

To establish the credibility of LES as a tool for innovation in industrial flow applications and for the study of complex-physics problems, clear standards and criteria to assess and predict the quality and the reliability of the simulation results should be devised. To this aim, an understanding of the non-linear accumulation and interaction of the different errors arising in large-eddy simulations, and of their dependence on the different simulation parameters, is required. This is also crucial for the development of methodologies and techniques aimed at controlling the different errors and, hence, at optimizing the quality of LES results.

The main goal of QLES2009 was to enhance the knowledge on error sources and on their interaction in LES and to devise criteria for the prediction and optimization of simulation quality, by bringing together mathematicians, physicists and engineers and providing a platform specifically addressing these aspects for LES.

In total 64 participants from 12 countries registered for the workshop. The majority of participants was from academia and research institutes. In addition, several companies and consultancy agencies were represented.

QLES2009 gathered 7 invited lectures, held by speakers from different scientific fields: Johan Meyers (Katholieke Universiteit Leuven, Belgium), Thierry Poinsot (Institut de Mécanique des Fluides de Toulouse, CNRS, France), Philippe Spalart

(Boeing Commercial Airplanes, USA), Marc Parlange and Chad Higgins (Ecole Polythecnique Fédérale de Lausanne, Switzerland), Andreas M. Kempf (Imperial College London, United Kingdom), Lars Davidson (Chalmers University, Sweden), Jean-Luc Guermond (Texas A&M University, USA).

Johan Meyers presented an overview of recent developments of the "error-landscape" methodology, aimed at investigating the quality and reliability of large-eddy simulations by constructing a full response surface of the LES error behavior. Thierry Poinsot illustrated the application of LES to the simulation of complex reacting flows and discussed issues related to the reliability and the repeatability of LES results for such applications. Philippe Spalart drew the attention to a careful grid generation and optimization as a key issue to obtain accurate and reliable LES predictions for external flows. Marc Parlange and Chad Higgins described recent a-priori tests of models for subgrid-scale processes in stable and unstable atmospheric boundary-layers, carried out by using data from field experiments. Jean-Luc Guermond reviewed the mathematical properties of the 3D incompressible Navier-Stokes equations and their relation to LES. He also illustrated how the notion of suitable weak solutions can be used to devise LES closure models. Andreas Kempf discussed quality issues and the possibility of using quality indicators and error-charts in combustion LES, for which the presence of a wide range of chemical and mixing scales makes the assessment of simulation quality and reliability even more challenging than for "fluid-flow only" LES. Finally, Lars Davidson addressed the issue of how to estimate the resolution of LES simulations of recirculating flows.

Next to the invited lectures, 33 contributed presentations were selected by a Scientific Committee of experts.

From the presentations and the discussions held during the workshop, it was clear that the tendency to apply LES to various, very complex, industrial and environmental problems, already observed during the previous QLES workshop, further enhanced in the last two years. Several examples of such complex applications, comprising atmospheric and geophysical problems, particle-laden flows, combustion, aeronautical engines or conductive fluids and plasmas, were shown during the workshop, and LES was generally found to be able to give satisfactory results. It was also made clear, however, that this requires a profound knowledge of the problem and a careful combination of physical modeling, numerics, grid resolution and quality. Although SGS modeling is still felt as the most critical issue in LES, several contributions were given at the workshop on the sensitivity of the quality of LES results also to numerical methods, boundary conditions treatment and grid resolution. Recent developments and applications of methodologies aimed at understanding, predicting and minimizing error dynamics in LES were also presented.

In the spirit of the QLES series, this workshop gave a stimulating contribution to the development of higher standards for the assurance of quality and reliability of large-eddy simulations. Critical and open issues remain in order to increase the accessibility of LES to non-specialist users. As also highlighted at QLES2007, the development of a fully consistent theory on errors in LES, comprising the definition of mathematically sound quantitative error measurements and the simulation or modeling of error dynamics, is certainly needed. A related crucial issue is the sensitivity of the LES results to different parameters. This sensitivity analysis is complicated for LES by some peculiar features: for instance, grid independence can only be reached in the DNS limit and even the complete repeatability of a LES simulation may be questionable due to round-off errors, which may be a source of random disturbances, especially in massively parallel simulations. However, from a practical viewpoint, the main difficulty is that it is not affordable to carry out a large number of LES simulations, especially for complex applications, due to the large computational costs of each simulation. Thus, the development of tools aimed at estimating LES sensitivity from a limited number of simulations is required to obtain significant achievements in this direction. As previously mentioned, progress in these fields has been reported at QLES2009, but there is still room for development.

The present book contains the written contributions to QLES2009 and is divided in three parts, which reflect the main topics addressed at the workshop: (i) SGS modeling and discretization errors; (ii) Assessment and reduction of computational errors; (iii) Mathematical analysis and foundation for SGS modeling.

Financial support was provided, on a European scale, by COST Action P20 'LESAID' (LES-Advanced Industrial Design) and ERCOFTAC (European Research Community on Flow, Turbulence and Combustion), and, locally, by the University of Pisa. This support was essential for the organization of this event and is gratefully acknowledged.

Finally, we gratefully acknowledge the help of the members of the scientific committee: N.A. Adams (Technische Universität München, Germany), V. Armenio (Università di Trieste, Italy), A. Boguslawski (Politechnika Czestochowska, Poland), D. Carati (Université Libre de Bruxelles, Belgium), D. Drikakis (Cranfield University, United Kingdom), J. Fröhlich (Technical University of Dresden, Germany), C. Fureby (FOI, and Chalmers, Sweden), J.G.M. Kuerten (Eindhoven University of Technology, The Netherlands), A. Soldati (Università di Udine, Italy) and G. S. Winckelmans (Université Catholique de Louvain, Belgium).

Pisa,
February 2010

Maria-Vittoria Salvetti
Bernard J. Geurts
Johan Meyers
Pierre Sagaut

Contents

List of Contributors

Orhan Aybay
Helmut-Schmidt-Universität Hamburg,
22043 Hamburg, Germany

Peter Baas
Meteorology and Air Quality Group,
Wageningen University, PO Box 47,
Wageningen, The Netherlands;
Royal Netherlands Meteorological
Institute KNMI, De Bilt,
The Netherlands

Christophe Bailly
Institut Universitaire de France and
LMFA, UMR CNRS 5509, Ecole
Centrale de Lyon, 69134 Ecully, France

Guillaume Balarac
MOST/LEGI, BP53, 38041, Grenoble
Cedex 09, France

Sara Barsotti
Istituto Nazionale di Geofisica e
Vulcanologia, Sezione di Pisa, Via della
Faggiola 32, 56126 Pisa, Italy

Sukanta Basu
Atmospheric Science Group, Department of Geosciences, Texas Tech
University, Lubbock, Texas, USA

Luigi C. Berselli
Dipartimento di Matematica Applicata
"U. Dini", Università di Pisa, Via F.
Buonarroti 1/c, I-56127, Pisa, Italy

Christophe Bogey
LMFA, UMR CNRS 5509, Ecole
Centrale de Lyon, 69134 Ecully, France

Andrzej Boguslawski
Institute of Thermal Machinery,
Czestochowa University of Technology, Al. Armii Krajowej 21, 42-200
Czestochowa, Poland

Fred C. Bosveld
Royal Netherlands Meteorological Institute KNMI, De Bilt,
The Netherlands

Michael Breuer
Helmut-Schmidt-Universität Hamburg,
22043 Hamburg, Germany

Bryan A. Burkholder
School of Meteorology, University of
Oklahoma, 120 David L. Boren Blvd.,
Norman, OK, USA 73072

Daniele Carati
Statistical and Plasma Physics Laboratory, Université Libre de Bruxelles,
CP 231, Boulevard du Triomphe, 1050
Brussels, Belgium

Andrea Cimarelli
DIEM, II Facoltà di Ingegneria, Università di Bologna, Viale Risorgimento, 40136 Bologna, Italy

Lars Davidson
Division of Fluid Dynamics, Department of Applied Mechanics, Chalmers University of Technology, SE-412 96 Göteborg, Sweden

Elisabetta De Angelis
DIEM, II Facoltà di Ingegneria, Università di Bologna, Viale Risorgimento, 40136 Bologna, Italy

James R. DeBonis
NASA Glenn Research Center, Cleveland, Ohio 44135, USA

Chris De Langhe
Department of Flow, Heat and Combustion Mechanics, Faculty of Engineering, Ghent University, St. Pietersnieuwstraat 41, 9000 Ghent, Belgium

Filippo M. Denaro
Dip. Ingegneria Aerospaziale e Meccanica, Seconda Università di Napoli, Via Roma 29, 81031 Aversa (CE), Italy

Jordan A. Denev
Institut für Technische Chemie und Polymerchemie, Universität Karlsruhe, 76128 Karlsruhe, Germany

Alain Dervieux
INRIA, 2004 Route des lucioles, BP 93, 06902 Sophia Antipolis, France

Giuliano De Stefano
Dip. Ingegneria Aerospaziale e Meccanica, Seconda Università di Napoli, Via Roma 29, 81031 Aversa (CE), Italy

Erik Dick
Department of Flow, Heat and Combustion Mechanics, Faculty of Engineering, Ghent University, St. Pietersnieuwstraat 41, 9000 Ghent, Belgium

F. Dinkelacker
Institut für Technische Verbrennung, Leibniz Universität Hannover, Hannover, Germany

Pascale Domingo
CORIA - CNRS and INSA de Rouen, Technopole du Madrillet, BP 8, 76801 Saint-Etienne-du-Rouvray, France

Tomaso Esposti Ongaro
Istituto Nazionale di Geofisica e Vulcanologia, Sezione di Pisa, Via della Faggiola 32, 56126 Pisa, Italy

Carlos J. Falconi
Institut für Technische Chemie und Polymerchemie, Universität Karlsruhe, 76128 Karlsruhe, Germany

Dieter Fauconnier
Department of Flow, Heat and Combustion Mechanics, Faculty of Engineering, Ghent University, St. Pietersnieuwstraat 41, 9000 Ghent, Belgium

Evgeni Fedorovich
School of Meteorology, University of Oklahoma, 120 David L. Boren Blvd., Norman, OK, USA 73072

J. Franke
Institut für Fluid- und Thermodynamik, Universität Siegen, Germany

Jochen Fröhlich
Institut für Strömungsmechanik, TU Dresden, 01062 Dresden, Germany

Andrey Garbaruk
New Technologies and Services, St.-Petersburg 197198, Russia

Massimo Germano
Dip. di Ing. Aeronautica e Spaziale, Politecnico di Torino, C.so degli Abruzzi 24, 10129 Torino, Italy

Bernard J. Geurts
Multiscale Modeling and Simulation, Faculty EEMCS, University of Twente, P.O. Box 217, 7500 AE Enschede, The Netherlands;
Anisotropic Turbulence, Faculty Applied Physics, Eindhoven University of Technology, P.O. Box 513, 5600 MB Eindhoven, The Netherlands

Laurent Y.M. Gicquel
CERFACS, 42 Avenue G. Coriolis 31057 Toulouse Cedex, France

Mikhael Gorokhovski
LMFA UMR 5509 CNRS Ecole Centrale de Lyon Université Claude Bernard Lyon 1, 36 avenue Guy de Collongue, 69134 Ecully Cedex, France

T. Görler
Max-Planck-Institut für Plasmaphysik, Boltzmannstr. 2, D-85748 Garching, Germany

Jean-Luc Guermond
Department of Mathematics, Texas A&M University, College Station, Texas 77843, USA

Claudia Hertel
Institut für Strömungsmechanik, TU Dresden, 01062 Dresden, Germany

Johan Hoffman
Computational Technology Laboratory, Computer Science and Communication, KTH, SE-10044 Stockholm, Sweden

Darryl Holm
Department of Mathematics, Imperial College London, London, UK;
Computer and Computational Science Division, Los Alamos NationalLaboratory, New Mexico, USA

Traian Iliescu
Department of Mathematics, Virginia Tech, Blacksburg, VA, United States

Niclas Jansson
omputational Technology Laboratory, Computer Science and Communication, KTH, SE-10044 Stockholm, Sweden

Marek Jaszczur
AGH - University of Science and Technology 30-059 Krakow, Al.Mickiewicza 30, Poland

F. Jenko
Max-Planck-Institut für Plasmaphysik, Boltzmannstr. 2, D-85748 Garching, Germany

Colleen M. Kaul
Dept. of Aerospace Engineering and Engineering Mechanics, The University of Texas at Austin, Austin, TX 78712, USA

Andreas M. Kempf
Department of Mechanical Engineering, Imperial College London South Kensington Campus, SW7 2AZ London, UK

Bruno Koobus
Département de Mathématiques, Université de Montpellier 2, Case Courrier 051, Place Eugène Bataillon, 34095 Montpellier, France

J. G. M. Kuerten
Department of Mechanical Engineering, Eindhoven University of Technology, P.O. Box 513, 5600 MB Eindhoven, The Netherlands

Guido Lodato
CORIA - CNRS and INSA de Rouen, Technopole du Madrillet, BP 8, 76801 Saint-Etienne-du-Rouvray, France

Mirosław Łuniewski
Institute of Fluid-Flow Machinery, Polish Academy of Sciences, Fiszera 14, 80952 Gdańsk, Poland

Michael Manhart
Fachgebiet Hydromechanik, Technische Universität München, 80333 München, Germany

Bhuvaneswaran Manickam
Institut für Technische Verbrennung, Leibniz Universität Hannover, Hannover, Germany

Cristian Marchioli
Dipartimento di Energetica e Macchine, Università degli Studi di Udine, Via delle Scienze 208, 33100 Udine, Italy

Marcello Meldi
Dipartimento di Ingegneria Meccanica e Civile, Università di Modena, 41125 Modena, Italy

Suresh Menon
School of Aerospace Engineering, Georgia Institute of Technology, 270 Ferst Drive, Atlanta, GA, 30332-0150, USA

F. Merz
Max-Planck-Institut für Plasmaphysik, Boltzmannstr. 2, D-85748 Garching, Germany

Johan Meyers
Department of Mechanical Engineering, Katholieke Universiteit Leuven, Celestijnenlaan 300A, B3001 Leuven, Belgium

Pablo Mininni
National Center for Atmospheric Research Boulder, Colorado, USA; Departamento de Física, Facultad de Ciencias Exactas y Naturales, Universidad de Buenos Aires, Buenos Aires, Argentina

Arnold F. Moene
Meteorology and Air Quality Group, Wageningen University, PO Box 47, Wageningen, The Netherlands

P. Morel
Statistical and Plasma Physics Laboratory, Université Libre de Bruxelles, CP 231, Boulevard du Triomphe, 1050 Brussels, Belgium

S. P. R. Muppala
Faculty of Engineering, Kingston University, London, U.K

Augusto Neri
Istituto Nazionale di Geofisica e Vulcanologia, Sezione di Pisa, Via della Faggiola 32, 56126 Pisa, Italy

Hilde Ouvrard
Département de Mathématiques, Université de Montpellier 2, Case Courrier 051, Place Eugène Bataillon, 34095 Montpellier, France

Tamay Özgökmen
RSMAS/MPO, University of Miami, Miami, Florida, USA

Richard Pasquetti
Lab. J.A. Dieudonné, UMR CNRS 6621, Université de Nice-Sophia Antipolis, Nice, France

Alessandro Pecenko
Department of Mechanical Engineering, Eindhoven University of Technology, P.O. Box 513, 5600 MB Eindhoven, The Netherlands

Nikolaus Peller
Fachgebiet Hydromechanik, Technische Universität München, 80333 München, Germany

Federico Perini
Dipartimento di Ingegneria Meccanica e Civile, Università di Modena, 41125 Modena, Italy

Mike Pettit
Department of Mechanical Engineering, Imperial College London, South Kensington Campus, SW7 2AZ London, UK

Jonathan Pietarila Graham
Max-Planck-Institut für Sonnensystemforschung, Katlenburg-Lindau, Germany

Heinz Pitsch
Center for Turbulence Research, Stanford University, Stanford, CA 94305, USA

Thierry Poinsot
Institut de Mécanique des Fluides de Toulouse, CNRS UMR 5502, France

Bojan Popov
Department of Mathematics, Texas A&M University, College Station, Texas 77843, USA

Annick Pouquet
National Center for Atmospheric Research, Boulder, Colorado, USA

Jacek Pozorski
Institute of Fluid-Flow Machinery, Polish Academy of Sciences, Fiszera 14, 80952 Gdańsk, Poland

Venkat Raman
Dept. of Aerospace Engineering and Engineering Mechanics, The University of Texas at Austin, Austin, TX 78712, USA

Christoph Rapp
Fachgebiet Hydromechanik, Technische Universität München, 80333 München, Germany

Maria Vittoria Salvetti
Dipartimento di Ingegneria Aerospaziale, Università di Pisa, Via G. Caruso 8, 56122 Pisa, Italy

Alan Shapiro
School of Meteorology, University of Oklahoma, 120 David L. Boren Blvd., Norman, OK, USA 73072

Andrew G. Smith
School of Aerospace Engineering, Georgia Institute of Technology, 270 Ferst Drive, Atlanta, GA, 30332-0150, USA

Alfredo Soldati
Dipartimento di Energetica e Macchine, Università degli Studi di Udine, Via delle Scienze 208, 33100 Udine, Italy

Philippe R. Spalart
Boeing Commercial Airplanes, Seattle, WA 98124, USA

Srikant Srinivasan
School of Aerospace Engineering, Georgia Institute of Technology, 270 Ferst Drive, Atlanta, GA, 30332-0150, USA

Gabriel Staffelbach
CERFACS, 42 Avenue G. Coriolis 31057 Toulouse Cedex, France

Mikhail Kh. Strelets
New Technologies and Services, St.-Petersburg 197198, Russia

Artur Tyliszczak
Institute of Thermal Machinery, Czestochowa University of Technology, Al. Armii Krajowej 21, 42-200 Czestochowa, Poland

Oleg V. Vasilyev
Department of Mechanical Engineering, University of Colorado, Engineering Center, ECME 126, CO 80309 Boulder, USA

Roel Verstappen
Johann Bernoulli Institute for Mathematics and Computer Science, University of Groningen, P.O. Box 407, 9700 AK Groningen, The Netherlands

Luc Vervisch
CORIA - CNRS and INSA de Rouen, Technopole du Madrillet, BP 8, 76801 Saint-Etienne-du-Rouvray, France

Denis Veynante
EM2C - CNRS and Ecole Centrale Paris, Grande Voie des Vignes, 92295 Châtenay-Malabry, France

Ivana Vinkovic
LMFA UMR 5509 CNRS Ecole Centrale de Lyon Université Claude Bernard Lyon 1, 36 avenue Guy de Collongue, 69134 Ecully Cedex, France

Pierre Wolf
CERFACS, 42 Avenue G. Coriolis 31057 Toulouse Cedex, France

Hongyi Xu
Rotary-Wing Aerodynamics, Aerodynamics Laboratory, Institute for Aerospace Research, National Research Council Canada, Canada

Remi Zamansky
LMFA UMR 5509 CNRS Ecole Centrale de Lyon Université Claude Bernard Lyon 1, 36 avenue Guy de Collongue, 69134 Ecully Cedex, France

Part I

SGS modeling and discretization errors

Error-landscape assessment of large-eddy simulations: a review

Johan Meyers

Department of Mechanical Engineering, Katholieke Universiteit Leuven Celestijnenlaan 300A, B3001 Leuven, Belgium. johan.meyers@mech.kuleuven.be

Summary. A review is presented of the error-landscape methodology. This approach evaluates the error-response surface of large-eddy simulations (LES) to essential model and numerical parameters by a systematic variation of these parameters. Using an error landscape constructed for LES of decaying homogeneous isotropic turbulence, it is shown that the determination of LES quality based on one error measure alone, can lead to misleading results, related to underlying error-balancing mechanisms. This problem can be avoided by considering a range of errors simultaneously, emphasizing different scales in the solution. Subsequently, the error-landscape method is further illustrated by comparing different numerical discretizations for Smagorinsky LES. Finally, a more complex case, i.e. a high (infinite) Reynolds number boundary layer, is considered.

Key words: Quality, reliability, large-eddy simulation, homogeneous isotropic turbulence, atmospheric boundary layer, Smagorinsky model

1 Introduction

Several LES (large-eddy simulation) studies have been published in the past, demonstrating the high level of accuracy with which turbulent flow predictions can be attained, without having to resort to the excessive requirements on computational resources imposed by direct numerical simulations. However, the setup and use of large-eddy simulations requires a profound knowledge of fluid mechanics, numerical techniques, and the application under consideration. The susceptibility of LES to errors in modeling, in numerics, and in the treatment of boundary conditions, can be quite large and the nonlinear accumulation of these different errors over time, often leads to intricate and unpredictable situations. Therefore, error-behavior of large-eddy simulation, and the assurance of quality in LES has recently gained considerable attention [1, 2, 3, 4]. These studies aim to formulate a rigorous standard for the assessment of accuracy and reliability in LES.

One of the tracks explored for the study of quality and reliability of large-eddy simulations, is based on a systematic variation of influencing parameters (e.g., model constants and grid resolution), such that a full response surface of the LES error

M.V. Salvetti et al. (eds.), *Quality and Reliability of Large-Eddy Simulations II*,
ERCOFTAC Series 16, DOI 10.1007/978-94-007-0231-8_1,

behavior is obtained [5, 6, 2, 7]. The current paper gives a review of this 'error-landscape' methodology. First of all, it is demonstrated that the evaluation of LES quality is best performed using a series of flow properties that are sensitive to various scales of the flow. Examples are given where a careless selection of error definitions may lead to an incorrect evaluation of simulation quality.

The error-landscape methodology can also facilitate the formulation of new insights in the behavior of subgrid-scale models, numerical discretization, and their interaction. In this context, some changes have been proposed to the Smagorinsky model in Refs. [8, 6], leading to better low-Reynolds-number behavior of the model. In the current work we will demonstrate the error-landscape approach by using it as a platform for the comparison of different discretization methods in Smagorinsky LES (cf. also Ref. [2]). Results presented are based on LES of decaying homogeneous isotropic turbulence with reference data obtained from direct numerical simulations [5]. The error-landscape can also be a very useful tool when experimental reference data are used: the reader is referred to Ref. [7] for a full discussion, where the high-Reynolds number grid-turbulence data of Kang et al. [9] were used as experimental reference in combination with the error-landscape approach.

Finally, an extensive variation of all influencing parameters in large-eddy simulations of more complex cases can become infeasible when too much parameters are involved. Nevertheless, a systematic variation of some parameters can still be very useful. An example is given on a calibration of the Smagorinsky model for high-Reynolds-number boundary-layer flows, where it is demonstrated that good logarithmic velocity profiles and turbulence spectra can be obtained when the Smagorinsky coefficient and near wall damping are appropriately selected.

The paper is further organized as follows. First, in Section 2 the error-landscape methodology is reviewed. In §2.1 the basic principles are explained. Subsequently, in §2.2 the error-balancing mechanisms are highlighted, and the importance of including a range of errors in the analysis is discussed. In §2.3 the method is used to compare different numerical discretizations of Smagorinsky LES. Next, in Section 3, a systematic variation of model parameters in LES of a high-Reynolds-number boundary-layer is presented, identifying model-parameter regions where both the velocity profile and the velocity spectra are well predicted. Finally, a summary is presented in Section 4.

2 Error landscapes and multi-objective refinement trajectories

2.1 The basic principles

In the current section, the error-landscape methodology, first introduced in Ref. [5], and later refined in Ref. [6, 2], is reviewed. The methodology is based on a systematic variation of relevant LES setup parameters such that a full response surface, a so-called error landscape, of the LES error behavior can be constructed. Next to a large set of LES runs, the methodology requires a reference case, for which a direct numerical simulation has been used in Ref. [5, 6, 2] (the use of experimental reference data is addressed in Ref. [7]).

The error-landscape approach is here demonstrated based on decaying homogeneous isotropic turbulence, and a DNS reference at $Re_\lambda = 100$ is used [5]. The large-eddy simulations employ a standard Smagorinsky model, i.e.

$$\tau_{sgs}^M = 2(C_s\Delta)^2(\mathbf{S}:\mathbf{S})\mathbf{S}, \tag{1}$$

with $\mathbf{S}$ the strain-rate tensor ($S_{ij} = (\partial\bar{u}_i/\partial x_j + \partial\bar{u}_j/\partial x_i)/2$), $\Delta = h$ the grid-spacing, and C_s the Smagorinsky constant.

A systematic variation of the Smagorinsky constant C_s and the grid resolution $N \sim 1/h$ is carried out, such that an error landscape can be constructed. Central to this approach is the reduction of the results of a simulation to one single number, quantifying the 'error' of that simulation with respect to the reference. Initially, in [5] this error was defined as the L_2 norm of the difference between the kinetic energy of the LES with the filtered DNS integrated in time, i.e.

$$\delta_E = \left(\int_0^T \left[E_{LES}(t) - E_{\overline{ref}}(t) \right]^2 \mathrm{d}t \Big/ \int_0^T E_{\overline{ref}}(t)^2 \mathrm{d}t \right)^{1/2}, \tag{2}$$

with E_{LES} the kinetic energy of the DNS data filtered (projected) onto the LES grid. An example of the resulting error landscape is presented in Figure 1(a). It is clear from this figure that the error δ_E is very sensitive to a good selection of C_s at a given resolution N. However, once such an error landscape is constructed, an optimal $\widehat{C}_s(N)$ trajectory for that given LES code (and for the given error definition) emerges.

When evaluating the error landscape based on resolved enstrophy $\mathscr{E}$ instead of resolved kinetic energy E in Figure 1(b), it is apparent that the 'optimal refinement trajectory' $\widehat{C}_s(N)$ depends on the error definition. In order to investigate this in a more systematic way, the error definition is formalized, such that a range of definitions can be easily included. Recognizing that $E(t) = \int E(k,t)\mathrm{d}k$ (with $E(k)$ the 3D energy spectrum at time t, and k the wavenumber), and $\mathscr{E}(t) = \int k^2 E(k,t)\mathrm{d}k$ (cf., e.g., Pope 2000 [10]), we introduce a set of errors errors D_p ($p = -1$ to 2) which are based on weighted integrals of the energy spectrum, i.e.,

$$D_p(N, C_s) = \left[\frac{\int_0^T \left\{ \int_0^{k_c} k^p \left(E_{LES}(k) - E_{EXP}(k) \right) \mathrm{d}k \right\}^2 \mathrm{d}t}{\int_0^T \left\{ \int_0^{k_c} k^p E_{EXP}(k)\, \mathrm{d}k \right\}^2 \mathrm{d}t} \right]^{1/2}, \tag{3}$$

with $k_c = \pi/\Delta$ the grid cut-off frequency. Depending on the values of p, this approach defines errors on large-scale properties ($p = -1, 0$) as well as resolved fine-scale properties ($p = 1, 2$). For $p = -1$, 0, and 2, it is readily shown that D_p respectively represents the relative error on the integral length scale $\delta_{\mathscr{L}}$, the resolved turbulent kinetic energy δ_E, and the resolved enstrophy $\delta_{\mathscr{E}}$ (cf. Ref. [2] for details).

In order to include a series of errors D_p in a systematic approach, the notion of near-optimal regions and a multi-objective refinement trajectory was introduced in [6]. The near optimal region Ω_p related to $D_p(N, C_s)$ is defined as

$$\Omega_p(a) = \left\{ N \in \mathbb{N}; C_s \in \mathbb{R}^+ \;\middle|\; \frac{D_p(N, C_s)}{D_p(N, \widehat{C}_s^{(p)}(N))} \leq a \right\}, \tag{4}$$

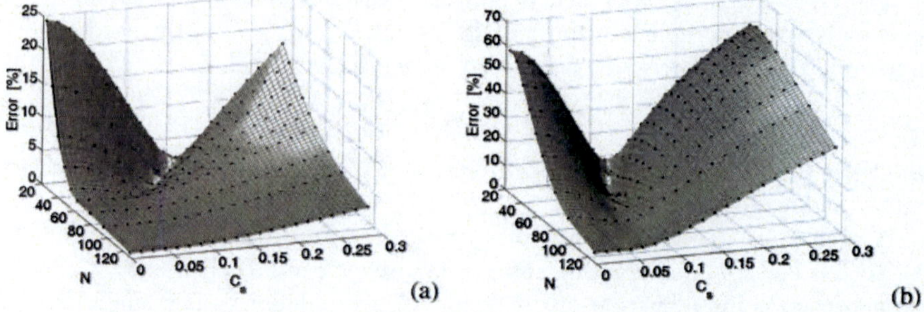

Fig. 1. Error landscapes of LES employing the Smagorinsky model. Errors are shown for δ_E (a) and $\delta_{\mathscr{E}}$ (b). The different simulations that were conducted are indicated by (•).

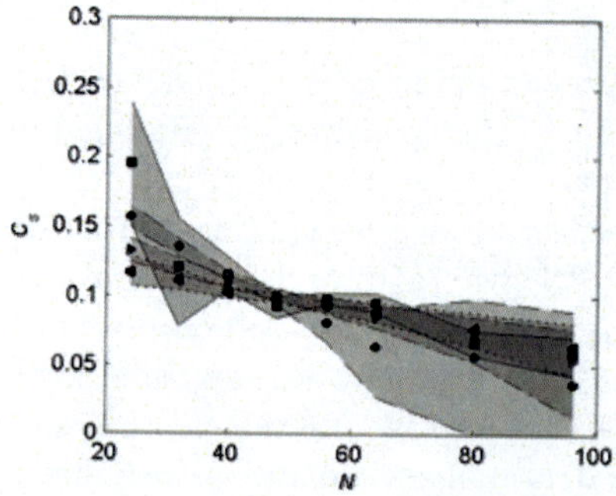

Fig. 2. 'Near optimal' regions based on D_p ($p = -1$ to 2) for Smagorinsky LES. Different 'near optimal' regions are displayed in gray and semitransparent, such that regions with overlap appear with darker shades of gray. The curves (—), (− −), (−·), and (···) respectively mark the boundaries of the $D_{-1}(=\delta_{\mathscr{L}})$, $D_0(=\delta_E)$, D_1 and $D_2(=\delta_{\mathscr{E}})$ 'near optimal' regions. Symbols (■, •, ▶, ◀) correspond respectively to the optimal refinement strategies for the different error definitions.

with $\widehat{C}_s^{(p)}(N))$ the optimal refinement trajectory related to the error measure D_p, and where we select $a = 1.2$ [2]. Hence, Ω_p defines a area around $\widehat{C}_s^{(p)}(N)$ where the error is close to optimal with respect to D_p.

In Figure 2, the near optimal regions are displayed for $p = -1$ to 2, based on Smagorinsky LES using a second order discretization (for more on the discretizatoin, see Section 2.3). The figure shows that near-optimal regions can overlap, hence leading to a multi-objective optimal region, where several errors are simultaneously (near) optimal. It is now further useful to define a multi-objective optimal refinement trajectory, by an appropriate weighting of different errors. To this end, a global weighted error is defined as [2]

$$\widetilde{D}(N,C_s) = \frac{\Sigma_p \left[D_p(N,C_s) / D_p\left(N, \widehat{C}_s^{(p)}(N)\right)\right]}{\Sigma_p \left[1 \Big/ D_p\left(N, \widehat{C}_s^{(p)}(N)\right)\right]}, \tag{5}$$

Based on $\widetilde{D}(N,C_s)$, a multi-objective optimal refinement trajectory $\widetilde{C}_s(N)$ is defined.

One of the ideas introduced in the context of the error-landscape methodology is that a fair comparison of LES codes (i.e. the combination of discretization and model) should be based on comparing the LES codes along there respective multi-objective refinement trajectories, such that every LES method is known to be operating at its best [6, 2]. Before pursuing this idea further with an example in Section 2.3, the existence of multi-objective optimal regions, and the relevance of a multi-objective optimal refinement trajectory in case a multi-objective optimal region does not exist, are discussed in the next subsection.

2.2 Error balancing and error definitions

In Figure 2 a multi-objective optimal region was found for $N > 40$. For lower resolutions, no overlap exists between the different near-optimal regions. While exploring different numerical discretizations, more extreme cases (without overlap) were found [2], and an example is given in Figure 3(a) using a 4th order 2nd order mixed discretization (cf. next subsection, and Ref. [2] for details). Now, a multi-objective optimal region only exists for $N > 64$; at lower resolutions optimal settings for one flow property are not necessary optimal for other flow properties. The reasons for this are further explored in Figure 3.

In order to understand the lack of a multi-objective optimal region in Figure 3(a), a series of three-dimensional energy spectra is plotted in Figure 3(c) for a range of C_s values, and for $N = 64$, and the DNS reference is also displayed. Two LES spectra are highlighted, one resulting from a C_s selection near the optimal setting for a minimal error on the kinetic energy, and the other similarly for the resolved enstrophy. It is appreciated that the 'optimal' result for the kinetic energy is a quite poor prediction of the DNS spectrum. However, the level of energy (resulting from integrating this spectrum over wavenumbers) is close to the total resolved DNS energy, since the strong under prediction of the LES spectrum in the low-wavenumber range is compensated with a pile-up of energy in the high-wavenumber range of the spectrum. Hence, in this example, the low error on the total resolved energy is due to a balancing of errors in the energy spectrum. Similar balancing mechanisms have been identified in Ref. [11] for poorly resolved direct numerical simulations of turbulent channel flow.

In order to avoid this type of balancing in wavenumber space, a more robust error definition may be used, which penalizes incorrect spectral distributions, and corresponds to [2]

$$d_p(N,C_s) = \left[\frac{\Sigma_{x/M} \int_0^{k_c} k^{2p} \left(E_{LES}(k) - E_{EXP}(k)\right)^2 \, \mathrm{d}k}{\Sigma_{x/M} \int_0^{k_c} k^{2p} (E_{EXP}(k))^2 \, \mathrm{d}k} \right]^{1/2}. \tag{6}$$

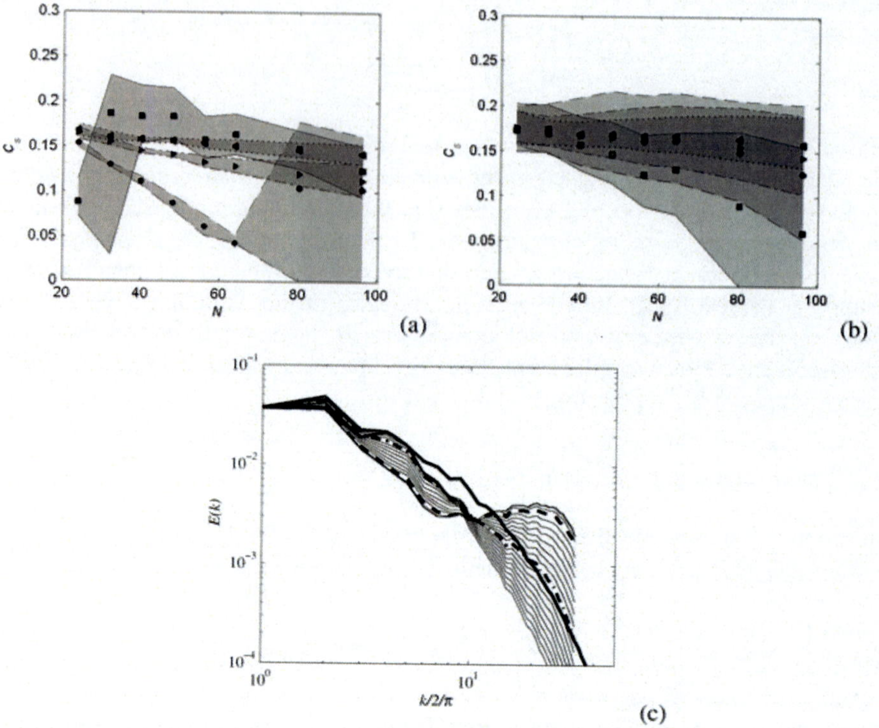

Fig. 3. (a,b) 'Near optimal' regions based on D_p, and d_p respectively ($p = -1$ to 2) for Smagorinsky LES using a 4–2 discretization (cf. next section for details). Lines and symbels for (a) and (b) as in Figure 2. (c) C_s dependence of LES spectra (at the end of the simulation) for $N = 64$. $C_s = \cdots$. Bold lines – (—): DNS reference; (– –) optimal coefficient $\widehat{C}_s^{(0)}$ ($N = 64$) for kinetic energy; (–·): optimal coefficient $\widehat{C}_s^{(2)}$ ($N = 64$) for resolved enstrophy.

In Figure 3(b), the near-optimal regions based on the new definition d_p are displayed. Now it is appreciated that a multi-objective optimal region exist over the whole range of simulation resolutions.

2.3 A comparison of numerical methods

The effect of the numerical discretization on Smagorinsky LES of homogeneous isotropic turbulence is now assessed using the error-landscape framework (cf. Ref. [2] for more details). Four different discretizations are considered: the convective terms are discretized using either a second-order or a fourth-order scheme, and similarly for the viscous (and subgrid-scale) terms. Hence, four different combinations are possible, i.e. a 2–2, 2–4, 4–2, and 4–4 scheme, with the first number the order of the convective discretization, and the second the order of the viscous and subgrid-scale-model discretization.

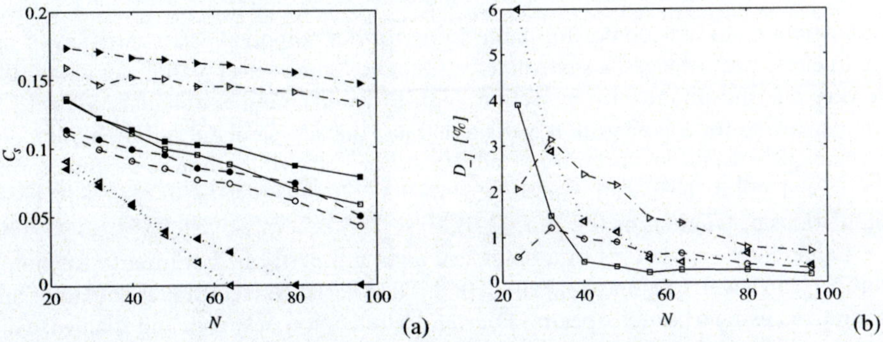

Fig. 4. (a) Multi-objective refinement strategies $\widetilde{C}_s$ as function of the resolution N for the 2–2 scheme (–□–), the 4–2 scheme (-▷·), the 2–4 scheme (◁·), and the 4–4 scheme (–○–). (Opens symbols): using $\{D_p\}$; (Closed symbols): using $\{d_p\}$. (b) Comparisons of relative errors D_{-1}, along the respective multi-objective optimal refinement strategies $\widetilde{C}_D$ for the 2–2 scheme (–□–), the 4–2 scheme (-▷·), the 2–4 scheme (◁·), and the 4–4 scheme (–○–).

In a first step, the near-optimal regions and multi-objective optimal refinement trajectories are determined either using the error definition D_p or d_p. In Figure 4(a) the multi-objective optimal trajectories are displayed for both error definitions, and for the four different discretization schemes (near-optimal regions are not shown here, except for the 2–2 scheme in Figure 2, and for the 4–2 scheme in Figure 3). These results show that multi-objective optimal refinement trajectories based on D_p and d_p are close together, irrespective of the fact that some of the D_p errors may be subject to 'error balancing'. Hence, as long as a series of error measures is included in the determination of a multi-objective optimal trajectory, the results are independent of the selected error definition (D_p or d_p).

Following one of the key ideas behind the error-landscape methodology (cf. §2.1), a 'fair' comparison between the different numerical methods (for Smagorinsky LES) is now performed by evaluating their errors along their respective multi-objective optimal refinement trajectories. Results are shown in Figure 4(b) (similar results are obtained based on other error measures in Ref. [2]). It is now striking that the 2–2 method is giving the lowest errors closely followed by the 4–4 discretization. Further, the use of 'equal-order' finite volume methods (2–2 and 4–4) appears to be advisable over the 'mixed-order' methods.

3 Error-landscape results for a high-Reynolds-number boundary layer

In the previous section, the error-landscape methodology was demonstrated for decaying homogeneous isotropic turbulence and Smagorinsky LES, based on a systematic variation of two set-up parameters, i.e. the model constant and the resolution. In case the model or the test case becomes more complex, the number of relevant parameters quickly becomes too large to perform a comprehensive variation of all parameters. Nevertheless, a systematic variation of some set-up parameters may still be very relevant. In the current section, high-Reynolds number boundary-layer LES are considered for a fixed grid. It is demonstrated that a systematic variation of model parameters is useful: in the current study, this provided a calibration of the Smagorinsky model and its near-wall damping for further use in atmospheric-boundary-layer simulations of wind farms (cf. Ref. [12]).

Large-eddy simulations of a boundary layer are performed at infinite Reynolds number, and a wall roughness $z_0/H = 10^{-4}$. The filtered Navier–Stokes equations are discretized using a pseudo-spectral Fourier discretization in stream and span wise directions x, and y, and a 4th order energy-conservative finite-volume discretization[13] is used in the normal direction z. A standard Smagorinksy model is used. The computational box size is $2\pi \times \pi \times 1$ (taking $H = 1$), and all simulations are performed on a $64 \times 64 \times 48$ mesh.

Since a high Reynolds number boundary layer is considered (i.e. the viscosity is set to zero in the simulations), the near-wall layer cannot be resolved, and a high-Reynolds-number boundary condition is required. At the bottom surface, we use a classic ‘imposed wall stress’ boundary condition relating the wall stress to the velocity at the first grid-point using the standard log (Monin-Obukhov) similarity law [14]:

$$\tau_{w1} = -\left(\frac{\kappa}{\ln z/z_{0,lo}}\right)^2 \left(\widehat{\tilde{u}}^2 + \widehat{\tilde{v}}^2\right)^{0.5} \widehat{\tilde{u}} \tag{7}$$

$$\tau_{w2} = -\left(\frac{\kappa}{\ln z/z_{0,lo}}\right)^2 \left(\widehat{\tilde{u}}^2 + \widehat{\tilde{v}}^2\right)^{0.5} \widehat{\tilde{v}}, \tag{8}$$

where the hat on $\widehat{\tilde{u}}$ and $\widehat{\tilde{v}}$ represents a local average obtained by filtering the LES velocity field with filter width 4Δ (see Bou-Zeid et al. [15] for more details about such filtering). On the top boundary, a symmetry condition is used.

A constant-coefficient Smagorinsky model is known to be too dissipative near the wall, where a length scale $\lambda \sim z$ is expected instead of a constant length scale $C_s\Delta$. Consequently, Mason and Thomson [16] proposed a adaptation of the Smagorinsky length $C_s\Delta$ near the wall corresponding to

$$\lambda^{-n} = [C_s\Delta]^{-n} + [\kappa(z+z_0)]^{-n}. \tag{9}$$

In the current work, a systematic variation of C_s and n is considered, keeping the resolution constant. In Figure 5, the error on the mean-velocity profile is presented, defined by

$$\delta_U = \frac{1}{u_\tau}\left(\int \left[U_{LES(z)} - \frac{1}{\kappa}\log(z/z_0)\right]^2 \mathrm{d}z/H\right)^{1/2}, \tag{10}$$

with $U_{LES(z)}$ the time-averaged mean-velocity profile predicted by the large-eddy simulations. The Smagorinsky constant C_s is varied between 0.09 and 0.15, and n between 1 and 3. From this Figure, it is appreciated that an optimal setting of C_s very much depends on n.

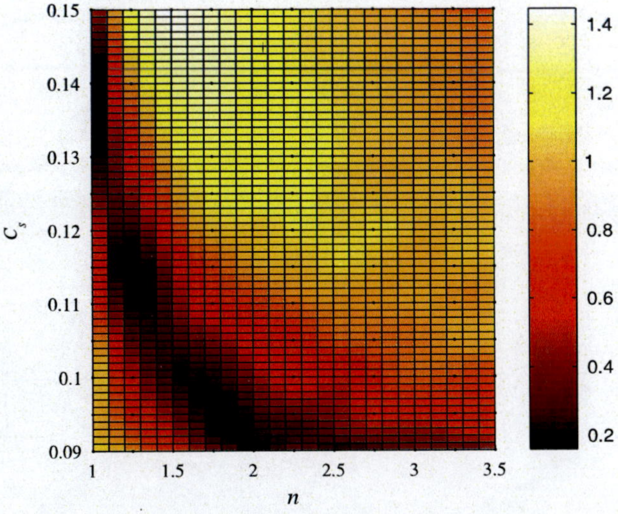

Fig. 5. Smagorinsky LES of a high-Reynolds-number boundary layer: error on the mean velocity profile as function of C_s and n. The different simulations that were conducted are indicated by (•).

Three different C_s–n combinations are now investigated in more detail in Figure 6. First of all, consider $C_s = 0.15$, and $n = 2$ (A Smagorinsky constant $C_s = 0.15$ roughly corresponds to the Lilly value for the current grid—i.e., the well-known $C_s = 0.17$ value should be corrected for the shape of the grid cut-off filter, cf. Ref. [17, 8, 6]). It is appreciated in Figure 5 that this combination does not result in a low error on the mean velocity profile, and this is further confirmed in Figure 6(a), where a large log-layer mismatch is visible for this case.

In Figure 6(b) the streamwise energy spectra are displayed at different heights for the $C_s = 0.15$, $n = 2$ case. Close to the wall, the tails of the spectra should collapse (in log–log scaling) onto a -1 slope, while further away from the wall (where the turbulence becomes homogeneous isotropic) a collapse onto a -5/3 slope is expected (see, e.g. Ref [18]). As clear from the figure, the -5/3 slope far from the wall is well recovered, but close to the wall, the tails of the spectra drop too fast, indicating that the Smagorinsky model is too dissipative in this region.

A second large-eddy simulation displayed in Figure 6 uses $C_s = 0.09$, and keeps $n = 2$. Lowering the Smagorinsky constant is a solution often used to rectify too high dissipation near the wall, and Figure 5 confirms that $C_s = 0.09$, and $n = 2$ is situated in a 'near optimal' zone. Figure 6(a) further illustrates this, showing the improvement of the mean velocity profile over the $C_s = 0.15$ case. When the stream-wise energy spectra are studied in more detail in Figure 6(c), it is clear that the tails of the spectra near the wall collapse better. However, now the -5/3 slope of the spectra for from the wall is not apparent anymore. This is a clear indication of a Smagorinsky constant

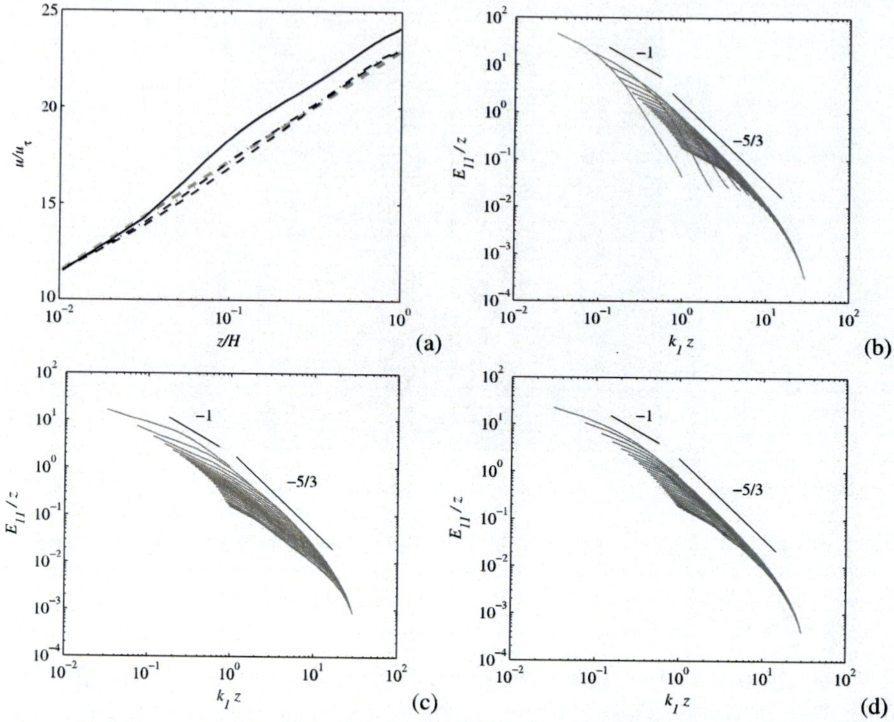

Fig. 6. Smagorinsky LES of the high-Reynolds-number boundary layer. (a) Mean velocity profiles; (—): $C_s = 0.15$, $n = 2$; (−−): $C_s = 0.09$, $n = 2$; (−·): $C_s = 0.14$, $n = 1$; (−−, in gray): log law. (b,c,d) Stream-wise energy spectra respectively for $C_s = 0.15$, $n = 2$; $C_s = 0.09$, $n = 2$; and $C_s = 0.14$, $n = 1$.

which is not well selected for the (nearly) homogeneous isotropic turbulence far from the wall.

Consequently, for a good 'far-wall' behavior a Smagorinsky constant closer to the Lilly value is required. The error-landscape in Figure 5 is useful for selecting such a value: it is appreciated that a combination $C_s = 0.14$ and $n = 1$ is also situated in an optimal C_s–n region. Figure 6(a) also shows the velocity profile for this case, which matches the theoretical log-profile quite well. The stream-wise velocity spectra for $C_s = 0.14$ and $n = 1$ are plotted in Figure 6(d). Now, good spectra are obtained both close to the wall and in the 'far-wall' field.

4 Summary

A review of the error-landscape methodology has been presented. This method relies on a systematic variation of model and numerical parameters in large-eddy simulations, and helps gaining insight in the complex interaction between numerics and modeling. First of all, LES of decaying homogeneous isotropic turbulence was used to illustrate the basic principles. It was shown that error definitions matter, and that a reliable quality analysis of LES should rely on a range of errors emphasizing both large and small scales in the resolved solution: an analysis based on one error only can be misleading due to possible error-balancing mechanisms. As an illustration, the error-landscape method was used to compare different discretization variants for Smagorinsky LES. Finally, a tuned-down version of the methodology was employed to elaborate an error landscape for Smagorinsky LES of high (infinite) Reynolds number boundary layers, varying the Smagorinsky constant and the wall damping exponent in Mason and Thomson's [16] wall damping function. This analysis highlighted the importance of a sufficiently high Smagorinsky constant for correct $-5/3$ spectral slopes in the 'far-wall' region of the simulation. Using the constructed error response surface, a combination of $C_s = 0.14$ and $n = 1$ was found to give quite good velocity profiles and stream-wise energy spectra.

Acknowledgements

JM thanks Charles Meneveau for useful discussions on the atmospheric boundary layer. JM acknowledges funding from the Research Foundation – Flanders (FWO – Vlaanderen).

References

1. Gualtieri P, Casciola CM, Benzi R, Piva R (2007) Preservation of statistical properties in large-eddy simulation of shear turbulence. J Fluid Mech 592:471–494
2. Meyers J, Geurts BJ, Sagaut P (2007) A computational error assessment of central finite-volume discretizations in large-eddy simulation using a Smagorinsky model. J Comput Phys 227:156–173
3. Park N, Mahesh K (2007) Analysis of numerical errors in large eddy simulation using statistical closure theory. J Comput Phys 222:194–216
4. Viré A, Knaepan B (2009) On discretization errors and subgrid scale model implementations in large eddy simulations. J Comput Phys 228:8203–8213
5. Meyers J, Geurts BJ, Baelmans M (2003) Database analysis of errors in large-eddy simulations. Phys Fluids 15:2740–2755
6. Meyers J, Sagaut P, Geurts BJ (2006) Optimal model parameters for multi-objective large-eddy simulations. Phys Fluids 18: Art no 095103
7. Meyers J, Meneveau C, Geurts BJ (2009) Error-landscape assessment of LES accuracy using experimental data. In: Armenio V, Geurts BJ, Fröhlich J (Eds) Direct and large-eddy simulation VII. Springer, Berlin Heidelberg New York
8. Meyers J, Sagaut P (2006) On the model coefficients for the standard and the variational multi-scale Smagorinsky model. J Fluid Mech 569:287–319
9. Kang HS, Chester S, Meneveau C (2003). Decaying turbulence in an active-grid-generated flow and comparisons with large-eddy simulation. J Fluid Mech 480:129–160
10. Pope SB (2000) Turbulent Flows. Cambridge University Press
11. Meyers J, Sagaut P (2007) Is plane-channel flow a friendly case for the testing of LES subgrid-scale models? Phys Fluids 19: Art no 048105

12. Meyers J, Meneveau C (2010) Large Eddy Simulations of large wind-turbine arrays in the atmospheric boundary layer. 48th AIAA Aerospace Sciences Meeting Including the New Horizons Forum and Aerospace Exposition, Orlando, FL
13. Verstappen RWCP, Veldman AEP (2003) Symmetry-preserving discretization of turbulent flow. J Comput Phys 187:343–368
14. Moeng C-H (1984) A large-eddy simulation model for the study of planetary boundary-layer turbulence. J Atmos Sci 6:2311–2330
15. Bou-Zeid E, Meneveau C, Parlange MB (2005) A scale-dependent Lagrangian dynamic model for large eddy simulation of complex turbulent flows. Phys Fluids 17: Art no 025105
16. Mason PJ, Thomson DJ (1992) Stochastic backscatter in large-eddy simulations of boundary layers. J Fluid Mech 242:51–78
17. Scotti A, Meneveau C, Lilly DK (1993) Generalized Smagorinsky model for anisotropic grids. Phys Fluids A 5:2306–2308
18. Porté-Agel F, Meneveau C, Parlange MB (2000) A scale-dependent dynamic model for large-eddy simulation: applications to a neutral atmospheric boundary layer. J Fluid Mech 415:261–284

Numerical and physical instabilities in massively parallel LES of reacting flows

Dr Thierry Poinsot[1]

Institut de Mecanique des Fluides de Toulouse
CNRS UMR 5502, FRANCE poinsot@imft.fr

Summary. LES of reacting flows is rapidly becoming mature and providing levels of precision which can not be reached with any RANS (Reynolds Averaged) technique. In addition to the multiple subgrid scale models required for such LES and to the questions raised by the required numerical accurcay of LES solvers, various issues related the reliability, mesh independence and repetitivity of LES must still be addressed, especially when LES is used on massively parallel machines. This talk discusses some of these issues: (1) the existence of non physical waves (known as 'wiggles' by most LES practitioners) in LES, (2) the effects of mesh size on LES of reacting flows, (3) the growth of rounding errors in LES on massively parallel machines and more generally (4) the ability to qualify a LES code as 'bug free' and 'accurate'. Examples range from academic cases (minimum non-reacting turbulent channel) to applied configurations (a sector of an helicopter combustion chamber).

Key words: Combustion, Large-Eddy Simulation

1 Introduction

This paper focuses on the quality of LES in terms of mesh dependency and repetitivity, especially for reacting flows. LES codes are now routinely used to simulate combustion and more specifically to predict instabilities (mostly due to acoustics) in reacting flows [43, 23]. The fact that flames can couple with acoustics has been known for a long time [27], even though it is still not fully understood. Since acoustic waves can propagate in any direction in subsonic flows (which is the case in most combustors), they can create a feedback from any point of the flow to any other point, thereby creating multiple paths for absolute instabilities. More importantly, combustion instabilities are difficult to predict and are usually discovered at a late stage during the development of engine programmes so that they represent a significant industrial risk. These instabilities take various forms.

(1) In steady combustors like gas turbines, instabilities can lead to oscillations of all flow parameters, reaching levels which are incompatible with the normal operation of the chamber. They have been the source of multiple failures in rocket engines, as early as the Saturne or the Ariane 4 project, in aircraft engines (main chamber of

M.V. Salvetti et al. (eds.), *Quality and Reliability of Large-Eddy Simulations II*,
ERCOFTAC Series 16, DOI 10.1007/978-94-007-0231-8_2,

post combustion chamber), in industrial gas turbines and furnaces, etc. Fig. 1 shows an example of simulation of 'mild' oscillation in a gas turbine [10] where the flame position (visualized by an isosurface of temperature colored by axial velocity) pulsates at four instants of a cycle occuring at 120 Hz).

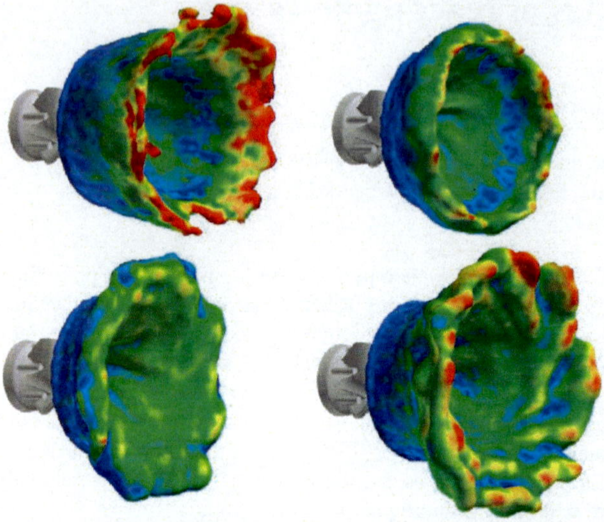

Fig. 1. Snapshots of flame position (isosurface of temperature) during one oscillation cycle at 120 Hz in an industrial gas turbine [10].

(2) In piston engines or in pulse combustors (such as the one used in German V 1 rocket during the second world war) where an external periodic motion or timing is imposed, instabilities take other forms: the most famous one is cycle-to-cycle variations. Fig. 2 shows LES results [28] in a four-valve four-stroke engine where all phases are explicitly computed (intake, compression, combustion, exhaust). Fig. 2 displays the reaction rate at exactly the same crank angle for various cycles: none of these cycles is similar. In extreme cases, certain cycles can actually not ignite or not burn at all. The source of these differences is obviously some type of instability. It can be an intrinsic bifurcation of the flow within the chamber or a coupling with the acoustic waves in the intake and exhaust pipes of the engine.

Predicting and controlling combustion instabilities is a major challenge for combustion research. Today, the most promising path is LES [23, 19, 32] something which was impossible 10 years ago with classical Reynolds Averaged Navier -Stokes methods. Evaluating the error margins associated to such LES is becoming a critical question. Multiple 'instabilities' are found in combustion LES. Instabilities exist in individual flame fronts and lead to the formation of cells and of various unstable modes depending on molecular transport of chemical species and heat [43, 4]. Like any shear flow, reacting flows are submitted to hydrodynamic modes [7, 13] and to vortex formation. But acoustics play the first role in reacting flows: by coupling

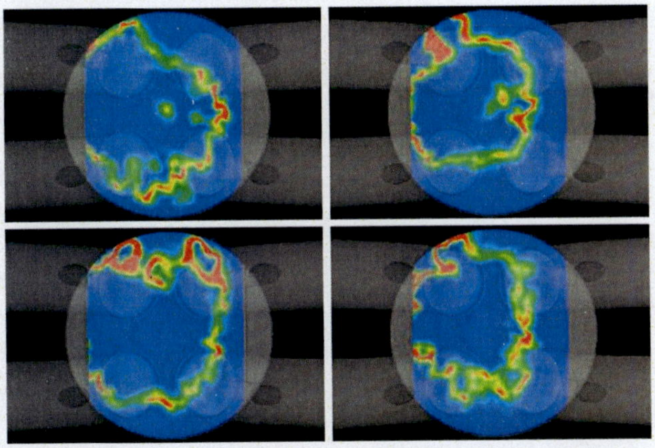

Fig. 2. Snapshots of reaction rate in a simulation of piston engine at four successive cycles for the same crank angle. None of them is the same.

with heat release, they are the source of most combustion instabilities [43, 23]. Instabilities are present in the physical problem studied but they are also present in the numerical methods used to simulate these mechanisms. Most high-fidelity numerical schemes required for Computational Fluid Dynamics exhibit low dissipation and therefore multiple non-physical instabilities (wiggles) arise which can require significant efforts to be kept under control [40, 33, 23]. Finally, CFD for reacting flowss are performed today on massively parallel machines: these architectures coupled with centered schemes for turbulent flows lead to an additional type of instability linked to the growth of rounding errors and to a new type of instability where the solution depends on unexpected parameters such as the commutativity errors of addition, the initial condition or the number of processors.

All these phenomena are 'instabilities' even though they correspond to different mechanisms. In some cases, they can couple: for example, in LES of combustion instabilities, the first issue is to be able to control the non-physical waves due to the high-order spatial scheme as well as the rounding errors due to massively parallel computing. This paper describes numerical instabilities found in LES (Section 2), proposes an analysis of the repetitivity of LES especially on parallel machines (Section 3) and ends with a study of the influence of mesh size on the LES of a sector of a gas turbine combustion chamber (Section 4).

2 Numerical waves in LES of reacting flows

Predicting combustion instabilities requires an accurate computation of all waves mentioned in the previous section. Unfortunately, LES (like DNS and unlike RANS) can propagate other waves: these numerical waves (called Q waves [40, 33, 23]) are a significant difficulty for most high-fidelity simulations. Q waves are produced by sharp gradients, approximate initial conditions, boundary conditions, etc. They interact with physical waves, making LES difficult and sensitive to unexpected behaviors. Knowing how Q waves interact with physics is a necessary exercice but one which is not often discussed because studying wiggles is not an exciting topic and also because most RANS codes use excessive artificial viscosity and large turbulent viscosity levels (due to turbulence models) which kill all numerical waves (they also kill all acoustic waves and hydrodynamic modes...) Methods which can compute accurately waves must use centered schemes and low turbulent viscosity levels. A convenient way to illustrate this point is to compare the various viscosities playing a role in a CFD code:

- The laminar viscosity ν is the only true flow characteristic and it defines the true Reynolds number of the flow: $Re_{real} = UL/\nu$ where U and L are a reference velocity and length of the flow respectively.
- In CFD codes, a turbulent viscosity ν_t is added through the model for turbulent fluctuations.
- Many CFD codes[1] also add an artificial viscosity ν_a or upwind (dissipative) schemes. An important dissipation is also introduced by large time steps and implicit schemes which are commonly used in RANS.

Adding two viscosities ν_t and ν_a to the true viscosity ν leads to a lower Reynolds number really seen by the code:

$$Re_{code} = UL/(\nu + \nu_t + \nu_a) \tag{1}$$

which is much smaller than Re_{real}. In RANS formulations, the turbulent viscosity introduced by models such as the k-ε model can reach 1000 times the physical viscosity while very high levels of artificial viscosity ν_a are used for robustness and speed. As a result, RANS results are steady: in other words, laminar. Of course, the local viscosity is tuned to match the mean characteristics of the mean turbulent flow but the turbulent character of the flow is completely lost. At the other end of the spectrum, DNS methods strive to use $\nu_a = \nu_t = 0$ so that $Re_{code} = Re_{real}$. LES formulations have to use non-zero values for ν_a and ν_t but these must be kept to very small values.

Method	Physical viscosity	Numerical viscosity	Turbulent viscosity	Capacity to propagate waves
DNS	1	0	0	excellent
LES	1	1 to 10	5 to 50	good
RANS	1	10 to 500	100 to 1000	none

Table 1. Orders of magnitude of physical, numerical and turbulent viscosity in DNS, LES and RANS codes. All values are scaled by the laminar viscosity.

[1] Certain numerical methods introduce dissipation terms which are not second order and are more difficult to compare to other viscosities: the well-known Jameson approach for example uses a fourth-order dissipation. They are not discussed here.

Table 1 shows typical levels of physical viscosity, turbulent viscosity ν_t and artificial viscosity ν_a (scaled by the laminar viscosity) reached in a combustion chamber for a standard regime. Note that viscosity affects all scales and not only the small scales. For example, acoustic waves are very strongly dissipated in a RANS code because the turbulent viscosity acts on them too. A less pleasant implication of Table 1 is that, as soon as high-fidelity methods such as DNS or LES are developed, they have to avoid large values of turbulent and artificial viscosities. This requires small mesh sizes, high-order schemes, small time steps [5, 33, 23]. These improvements unfortunately, also make these methods sensitive to numerical Q waves [34, 40, 23].

3 The growth of rounding errors in LES

The main strength of LES compared to classical Reynolds Averaged (RANS) methods is that compressible LES explicitly captures large scale unsteady motions due to turbulence and the instability modes found in reacting flows. An often ignored aspect of this feature is that LES is also submitted to a well-known feature of turbulent flows: the exponential separation of trajectories [39] implies that the flow solution exhibited by LES is very sensitive to any "small perturbations":

- Rounding errors are an unavoidable forcing for the Navier-Stokes equations and may lead to LES variability. The study of error growth in finite precision computations is an important topic in applied mathematics [37, 3] but has found few applications in multidimensional fluid mechanics because of the complexity of the codes used in CFD.
- Initial conditions are a second source of LES results variability: these conditions are often unknown and any small change in initial conditions may trigger significant changes in the LES solution. Boundary conditions, in particular the unsteady velocity profiles imposed at inlets and outlets, can have the same effect as initial conditions but are not studied here.
- Due to its large computational resource requirements, modern LES heavily relies on parallel computing. However, in codes using domain decomposition, it is also an additional "noise" source in the Navier-Stokes equations especially at partition interfaces. Even in explicit codes, where the algorithm is independent of the number of processors, the different summation orders with which a nodal value is reconstructed at partition interfaces, may induce non-associativity errors. For example, in explicit codes on unstructured meshes using cell vertex methods [31], the residual at one node is obtained by adding the weighted residuals of the surrounding cells. In some cases, summation may yield distinct results for floating-point accumulation: the rounding errors in $(a+b)+c$ and in $a+(b+c)$ may be different [11]. After thousands of iterations, the LES result may be affected. Since these rounding errors are induced by non deterministic message arrival at partition interfaces, such behaviour may occur for any CFD code, re-

gardless of the numerical scheme and lead to results which depend on the number of processors used for the LES.[2]

- Even on a single processor, internal parameters of the partitioning algorithm may couple with rounding errors. For example, a different reordering of nodes using the Cuthill-McKee (CM) or the reverse Cuthill-McKee (RCM) algorithm [6, 16] may produce the same solution divergence.

LES solutions are known to have a meaning only in a statistical manner [25]: observing that the solution of a given LES/DNS at a given instant changes when the rounding errors or the initial conditions change is not really surprising. It is however a difficulty in the practical use of LES: running the same simulation on two different machines or one machine with a different number of processors or slightly different initial conditions can lead to totally different instantaneous results. For steady flows in the mean, statistics should not depend on these changes and mean profiles must be identical. However, when the objective of the LES is unsteady phenomena such as ignition or quenching in a combustor [36], knowing that results depend on these parameters is certainly a sobering thought. This section tries to address these issues and answer a simple question [35]: how does the solution produced by LES depend on the number of processors used to run the simulation? On the initial condition? On internal details of the algorithm?

The configuration is a rectangular channel computed with a fully explicit LES code [21]. The following section then gives a systematic description of the effects of rounding errors in two flows: a turbulent channel and a laminar Poiseuille flow.

3.1 Effects of the number of processors on LES

This first example is the LES of a rectangular fully developed turbulent channel of dimensions: 75x25x50 mm (Fig. 3). An homogeneous force is applied to a periodic channel flow to provide momentum; random disturbances are added to trigger transition to turbulence. There are no boundary conditions except for the walls in y direction. The Reynolds number is $Re_\tau = \delta u_\tau / \nu = 1500$, where δ is half the channel height and u_τ the friction velocity at the wall: $u_\tau = (\tau_{wall}/\rho)^{1/2}$ with τ_{wall} being the wall stress. The mesh contains 30^3 hexahedral elements, it is not refined at walls. The first grid point is at a reduced distance $y^+ = y u_\tau / \nu \approx 100$ of the wall. The subgrid model is the Smagorinsky model and a law-of-the-wall is used at the walls [30]. The initial condition corresponds to a snapshot of the flow at a given instant, long after turbulence was initialized so that it is fully established. The computation is performed with an explicit code where domain decomposition is such that the method is perfectly equivalent on any number of processors. The Recursive Inertial Bisection (RIB) [38] algorithm is used to partition the grid and the Cuthill-McKee algorithm is the default graph reordering strategy. The scheme is the Lax-Wendroff scheme [12]. Additional tests using a third-order Taylor-Galerkin scheme [5] led to the same conclusions.

[2] The case of implicit codes [18, 17, 8] or in space (such as compact schemes) [15, 1, 34] is not considered here.

Table 2. Summary of turbulent LES runs (fully developed turbulent channel).

Run Id	Nbr proc	Init. cond.	Precision	Graph ordering	CFL λ
TC1	4	Fixed	Double	CM	0.7
TC2	8	Fixed	Double	CM	0.7
TC3	1	Fixed	Double	CM	0.7
TC4	1	Modif.	Double	CM	0.7
TC5	1	Fixed	Double	RCM	0.7
TC6	4	Fixed	Double	CM	0.35
TC7	8	Fixed	Double	CM	0.35
TC8	4	Fixed	Simple	CM	0.7
TC9	8	Fixed	Simple	CM	0.7
TC10	28	Fixed	Quadr.	CM	0.7
TC11	32	Fixed	Quadr.	CM	0.7

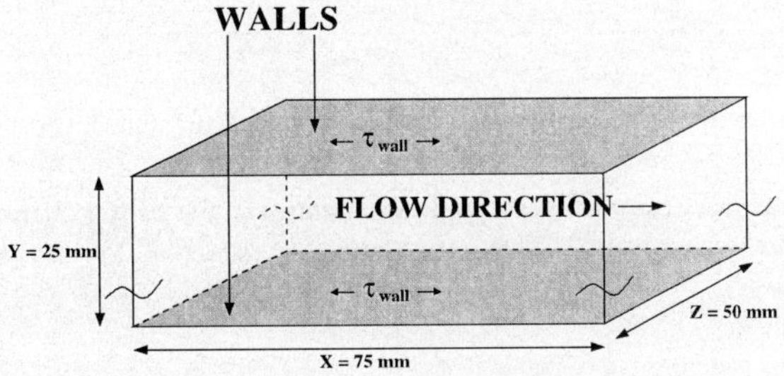

Fig. 3. Schematic of a periodic channel. The upper and lower boundaries consist of walls, all other boundaries are pairwise periodic.

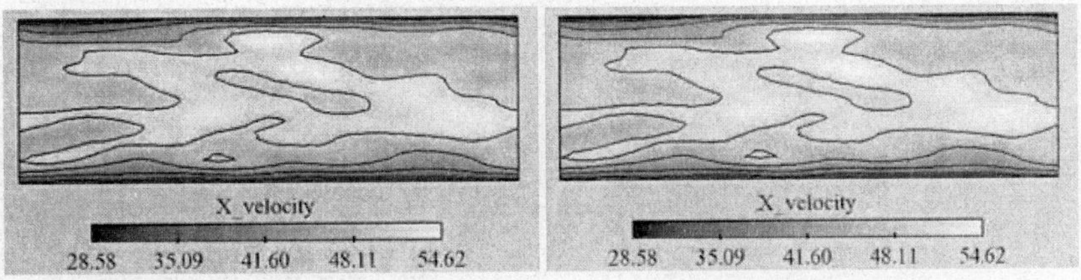

(a) Run with 4 processors (b) Run with 8 processors

Fig. 4. Instantaneous field of axial velocity in the central plane of the channel at $t+ = 7.68$. a) run TC1 (4 processors), b) run TC2 (8 processors).

Figs. 4 and 5 show fields of axial velocity in the central plane of the channel at two instants after the initialization. Two simulations performed on respectively 4 (TC1) and 8 processors (TC2) with identical initial conditions and meshes are compared (Table 2 and 3). The instants correspond to (in wall units) $t^+ = 7.68$ and $t^+ = 26.11$ respectively where $t^+ = u_\tau t/\delta$. Obviously, the two flow fields observed at $t^+ = 7.68$ are identical. Flow fields start to diverge after $t^+ = 15$ and at $t^+ = 26.11$, the instantaneous flow fields obtained in TC1 and TC2 are totally different (Figs. 5). Even though the instantaneous flow fields are different, statistics remain the same: mean and root mean square axial velocity profiles averaged over $t^+ \approx 60$ are identical for both simulations (Figs. 6).

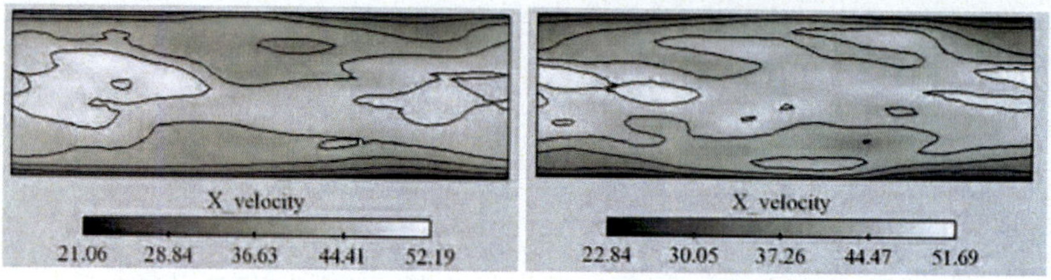

(a) Run with 4 processors (b) Run with 8 processors

Fig. 5. Instantaneous field of axial velocity in the central plane of the channel at $t+ = 26.11$. a) run TC1 (4 processors), b) run TC2 (8 processors).

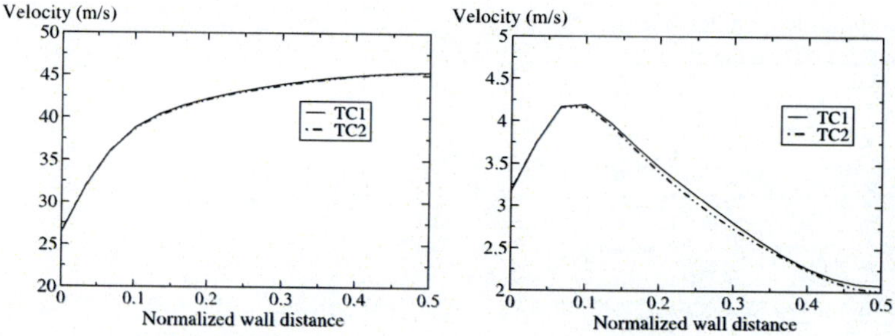

(a) Mean velocity (b) RMS velocity

Fig. 6. Comparison of the mean (left) root mean square (right) velocity profiles for TC1 (4 processors) and TC2 (8 processors) simulations over half channel height.

3.2 Sensitivity of LES in laminar and turbulent flows

To understand how LES can produce diverging instantaneous results such as those shown above, tests were performed to investigate the effects of various aspects of the methodology: laminar/turbulent baseline flow, number of processors, initial condition, graph ordering, time step and machine precision. The objective is to quantify differences between two LES solutions produced by a couple of simulations in Table 2 and 3. Let u_1 and u_2 be the scalar fields of two given instantaneous solutions at the same instant. A proper method to compare them is to use the iteration evolution of the following norms: N_{max} provides the maximum local velocity difference in the field between two solutions while N_{mean} yields a volumetrically averaged difference between the two solutions. Note that performing any of the LES of Table 2 twice on the same machine with the same number of processors, the same initial conditions and the same partition algorithm leads to exactly the same solution, N_{max} and N_{mean} being zero to machine accuracy. In that sense, the LES remains fully deterministic. However, this is true only if the order of operations at interfaces is not determined by the order of message arrival so that summations are always carried out in the same order. Otherwise, the randomness induced by the non-deterministic order of message arrival is enough to induce diverging solutions. The following test is to compare a turbulent channel flow studied in the previous section and a laminar flow. A three-dimensional Poiseuille flow in a pipe geometry was used as test case. The flow is laminar and the Reynolds number based on the bulk velocity and diameter is approximately 500. The boundary conditions are periodic at the inlet/outlet and no-slip at the duct walls. A constant axial pressure gradient is imposed in the entire domain.

Table 3. Summary of laminar runs (Poiseuille flow).

Run Id	Nbr proc	Init. cond.	Precision	Graph ordering	CFL λ
LP1	4	Fixed	Double	CM	0.7
LP2	8	Fixed	Double	CM	0.7

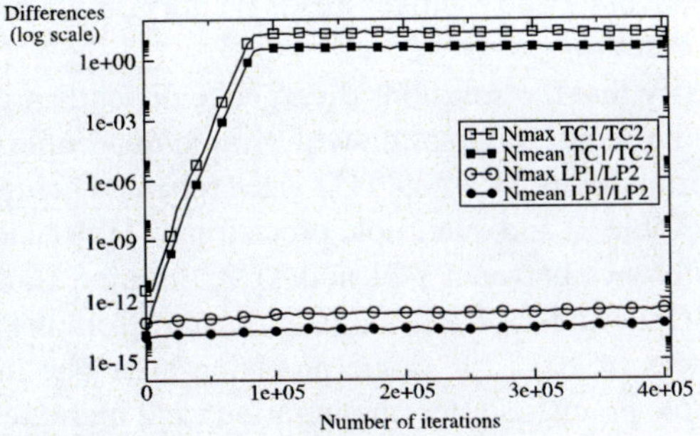

Fig. 7. Effects of turbulence. Differences between solutions: N_{max} (open symbols) and N_{mean} (closed symbols) vs iteration. Squares: differences between TC1 and TC2 (turbulent channel). Circles: differences between LP1 and LP2 (laminar flow).

Figure 7 shows the evolutions of N_{max} and N_{mean} versus iteration for runs TC1/TC2 and LP1/LP2. The only parameter tested here is a change of the number of processors. As expected from the snapshots of Figs. 4–5, the simulations are sensitive to a change in the number of processors and the solutions of TC1 and TC2 diverge rapidly leading to a maximum difference of 20 m/s and a mean difference of 3-4 m/s after 90,000 iterations. The stagnation of absolute and mean differences between TC1/TC2 simply implies that after 90,000 iterations solutions have become fully uncorrelated. On the other hand, the difference between the laminar simulations LP1 and LP2 hardly increases and levels off when reaching values of the order or 10^{-12}. This is expected since there is obviously only one stable solution for the Poiseuille flow: laminar flows do not induce exponential divergence of trajectories. This test case confirms that the turbulent character of the flow is the source of the divergence of solutions.

The basic mechanism leading to Figs. 4–5 is that the turbulent flow acts as an amplifier for rounding errors generated by the fact that the mesh is decomposed differently in TC1 and TC2. This implies a different ordering when adding the contributions to a cell residual for nodes at partition interfaces. This random noise roughly starts at machine accuracy (Fig. 7) at a few points in the flow and grows continuously if the flow is turbulent.

The previous results show that turbulence combined with a different domain decomposition is sufficient to lead to different instantaneous flow realizations. A perturbation in initial conditions has the same effect as domain decomposition as verified in runs TC3 and TC4 which are run on one processor only, thereby eliminating issues linked to parallel implementation. The only difference between TC3 and TC4 is that in TC4, the initial solution is identical to TC3 except at one random point where a 10^{-16} perturbation is applied to the streamwise velocity component. Solutions exhibited the same divergence [35]. Finally, the results of Senoner et al [35] show that the numerical scheme and the time step do not influence the growth rate of the solutions difference: for example, simulations TC6 and TC7 are performed with a time step reduced by a factor 2 compared to simulations TC1 and TC2. TC6 and TC7 are carried out on respectively 4 and 8 processors. The norms between TC6 and TC7 are similar to the other cases.

A last test to verify that the solutions divergence depends primarily on rounding errors is to perform the same computation with simple/quadruple precision instead of double precision. Simulations TC1 and TC2 were repeated using single precision in runs TC8 and TC9 (Table 2) and quadruple precision in TC10 and TC11. Results are compared to the difference between TC1 and TC2. Figure 8 shows that the solution differences for TC8/TC9 and TC10/TC11 roughly start from the respective machine accuracies (differences of 10^{-6} for single precision after one iteration, differences of 10^{-30} for quadruple precision after one iteration) and increase exponentially with the same growth rate before reaching the same difference levels for all three cases. Higher precision computations cannot prevent the exponentional divergence of trajectories but only delay it.

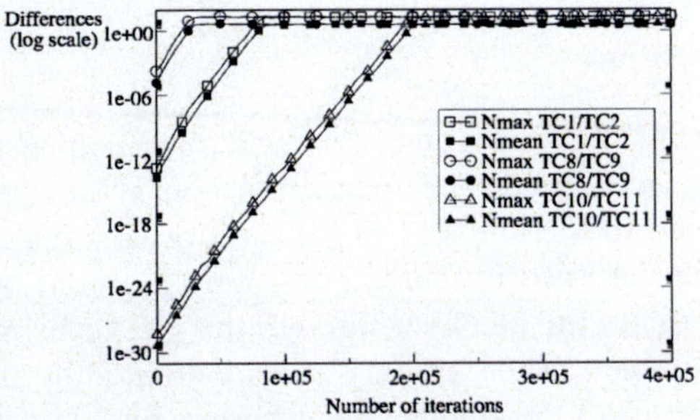

Fig. 8. Effects of machine accuracy. Differences between solutions: N_{max} (open symbols) and N_{mean} (closed symbols) vs iteration. Squares: differences between TC1 and TC2 (double precision). Circles: differences between TC8 and TC9 (single precision). Triangles: differences between TC10 and TC11 (quadruple precision).

4 Mesh effects in LES of gas turbine chamber

4.1 Objectives

The last topic discussed here is related to mesh dependency effects in LES of reacting flows. Multiple authors have underlined the importance of this point for LES [29, 25]. LES depend on mesh resolution (unlike RANS) and produce grid independent results only if the mesh is sufficiently refined. LES must then satisfy multiple properties: time-averaged values must converge, Root Mean Square (RMS) resolved values must increase when the mesh cell size decreases and the SGS turbulence level diminishes, the resolved velocity spectra must fill towards larger wave-numbers. In practice, these behaviors are expected to be controlled by the LES models, the flow Reynolds number, the grid resolution as well as the accuracy of the numerical solver (in the context of implicit filtering [29, 9, 24]). Mesh dependency analysis of non-reacting LES predictions has recently been addressed [42, 20, 26] and quality criteria have been proposed for *a posteriori* evaluation of the LES flow predictions [25, 41, 14].

For reacting flows, the computer power needed to simulate realistic geometries is so large that the grids used for LES are are usually still too coarse to resolve all flow zones: multiplying the number of grid points by a significant factor to verify the effects of grid resolution on the LES results was impossible until very recent times. Very few LES of reacting flows have been devoted to mesh dependency in simple configurations [42, 20, 26] and none of them has addressed this issue in complex geometry combustors. The situation has changed in the last two years: porting LES codes on massively parallel machines in the Top 20 list has allowed a sudden increase of power for combustion computations. Speed-ups of nearly 95 percent obtained on 2 to 10 000 processors allow to address the problem of mesh dependency by performing one 'coarse grid' simulation with a reasonable mesh (typically 1.5 million cells) and then comparing it with an 'intermediate grid' simulation (8 times more cells) and finally with a 'fine grid' simulation (32 times more cells). In the present work, the mesh dependency of the LES predictions is studied for a helicopter combustion chamber [2].

4.2 Target configuration

The configuration (Fig. 9) corresponds to a helicopter combustion chamber where fuel is injected using an inverted cane injection system, also called pre-vaporizer. The computational domain focuses on a 36 degree section of a full annular reverse-flow combustion chamber designed by Turbomeca (Safran group). A premixed gaseous mixture of $C_{10}H_{16}$ enters the chamber through the pre-vaporizer, Fig. 9 (a) & (b). Fresh gases are consumed in the primary zone, delimited by the chamber dilution holes and the liner dome of the combustion chamber, Fig. 9 (a). To ensure full combustion, this region of the chamber is fed with air by primary jets located on the inner liner, Fig. 9 (b). Burnt gases are then cooled by dilution jets or cooling films located on the inner and outer liners as well as on the return bend of the combustion chamber. Multi-perforated plates also ensure local wall cooling in areas of the chamber shown on Fig. 9 (b).

(a)
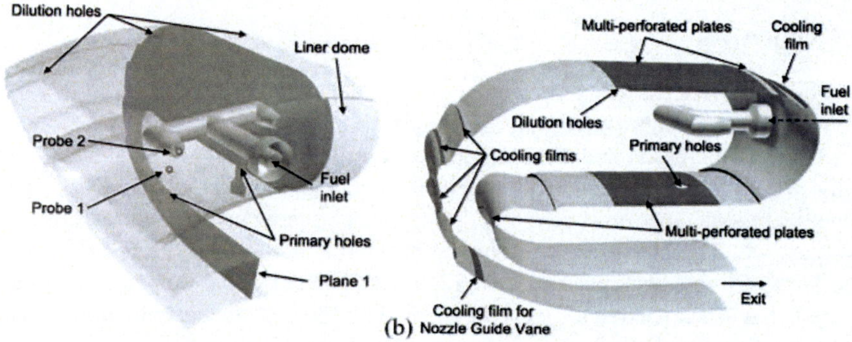

Fig. 9. Combustion chamber: (a) 3*D* view and (b) side view.

	'Coarse'	'Intermediate'	'Fine'
Total number of points	230,118	1,875,835	7,661,005
Total number of cells	1,242,086	10,620,245	43,949,682
Max. cell volume [m^3]	$3.12\,10^{-8}$	$8.97\,10^{-9}$	$4.05\,10^{-9}$
Min. cell volume [m^3]	$1.81\,10^{-11}$	$8.29\,10^{-12}$	$1.18\,10^{-12}$
Time step [s]	$1.52\,10^{-7}$	$0.88\,10^{-7}$	$0.49\,10^{-7}$
Averaging time [ms]	10	10	10
Number of iterations for averaging	65,790	108,695	204,082
CPU time(hours) for a 10ms LES	315	4,550	30,200

Table 4. Mesh characteristics.

The combustion regime expected in such burners mixes rich partially premixed flames in the chamber primary zone (the gases injected in the canes can be considered as premixed gases at an equivalence ratio of 3.17) and diffusion flames in the dilution region: *i.e.* RQL concept. A turbulent combustion model able to handle both regimes is therefore needed and the DTF model offers this capacity [30, 19, 36]. The meshes are refined in the primary zone, particularly in the lower part where combustion occurs, and in the regions of cooling films (Table 4). The Navier-Stokes Characteristic Boundary Conditions (NSCBC) [21, 22] are applied on boundaries to control the acoustic behavior of the system. Walls are adiabatic and are treated with a turbulent law-of-the-wall. Side boundaries of the computational domain are axi-periodic. The operating point corresponds to cruising conditions and is the same for the three grids. The cost of the three computations for the same physical time goes from 315 hours on the coarse mesh to 30,200 on the fine one.

4.3 Instantaneous flow topology and flame structure

A crucial requirement for the LES method when applied in such complex configurations is the right prediction of the combustion phenomenon. Modelling, which is needed to supply proper combustion enhancement due to lack of interactions at the unresolved scales, is paramount in that context. If improperly parameterized, a turbulent combustion model can yield different flame positions for LES computed with different mesh resolutions.

Figure 10 compares instantaneous fields of temperature for the three resolutions. The cutting plane goes through one of the pre-vaporizer outlets (Plane 1, Fig. 9) and is colored by the instantaneous field of temperature scaled by the inlet mean temperature. The observations drawn for the axial component of the velocity field also apply to Fig. 10: the temperature fields are clearly enriched with increasing grid resolution and the impact on the temperature levels seems reduced.

4.4 Mean flow results

Figure 11 shows the mean temperature field scaled by the inlet temperature. Grid resolution has an impact in various highly localized regions and some discrepancies are detected, especially in the mean temperature fields of Fig. 11. The improved mesh quality of the intermediate and fine grids makes the flow behave differently in near-wall regions where the chamber flow interacts with the flow issuing from cooling devices. For example, the thermal boundary layer created by the multi-perforated plates (Zone 1 and 3 on Fig. 11 (a)) on the intermediate mesh is thinner than the one on the coarse grid. Likewise, the penetration of the cooling film located in the upper part of the liner dome is different on the coarse grid. Despite these discrepancies, the agreement between these sets of predictions underlines mesh independence of the first moments for these calculations.

The paper of Boudier et al [2] provides more comparisons showing that velocity and temperature fields are well converged on the medium and fine grids while the reaction rate field continues to evolve when the grid is refined. This is not surprising as reaction rates are essentially subgrid scale quantities: increasing the mesh resolution allows to capture more wrinkling. However, the flame position does not change

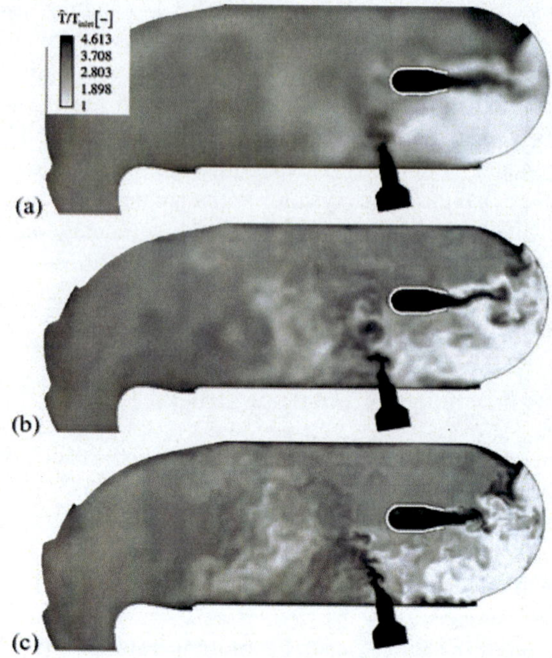

Fig. 10. Instantaneous temperature field scaled by the mean inlet temperature and obtained on (a) the coarse grid, (b) the intermediate one and (c) the fine one.

and even the acoustic activity in the chamber remains reasonably independent of the mesh.

5 Conclusions

This paper has shown that any turbulent flow computed by LES exhibits significant sensitivity to parameters such as initial solution, number of processors, message ordering. This sensitivity leads to instantaneous turbulent solutions which can be totally different while laminar flows are almost insensitive to these parameters. The divergence of solutions is due to the well known- exponential separation of trajectories in turbulent flows and to the non-deterministic rounding errors induced by different domain decompositions or different ordering of operations. More generally any change in the code lines affecting rounding errors will have the same effects. These results confirm the expected nature of LES [25] in which solutions are meaningful only in a statistical sense and instantaneous values can not be used for analysis. However, on a more practical level, they point out various difficulties to develop LES codes:

- Repeating the results of a given LES after modifying the code and verifying that instantaneous solutions have not changed is not always possible. Since any programming error will also lead to a change in instantaneous solutions, identifying

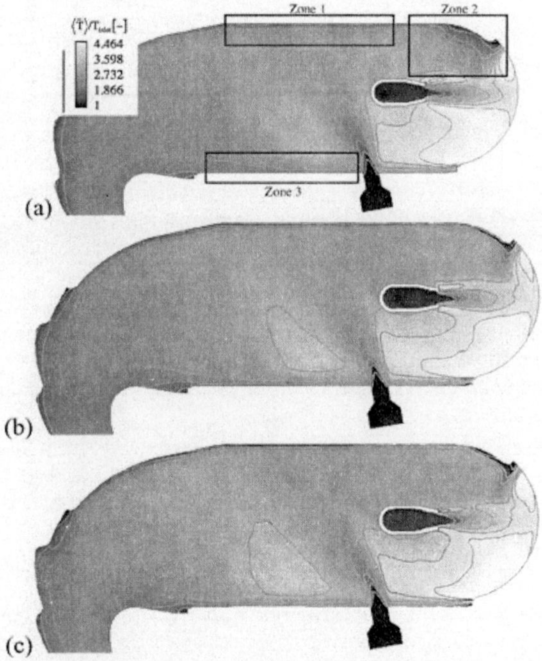

Fig. 11. Mean temperature field (scaled by the mean temperature at the pre-vaporizer inlet) obtained on (a) the coarse grid, on (b) the intermediate grid and on (c) the fine grid.

errors introduced by new lines will require a detailed analysis based on average fields (and not on instantaneous fields) and a significant loss of time.

- Verifying an LES code on a parallel machine is a difficult task: running the code on different numbers of processors will lead to different solutions and make comparisons impossible.
- Porting a LES code from one machine to another will also produce different solutions for turbulent runs, making comparison and validations of new architectures difficult.

When used on a complete reacting simulation of a sector of combustion chambers, the same conclusions are reached: instantaneous solutions can differ but the mean flow can be consistently captured with LES. The quality of the models themselves and of the numerical method accuracy was evaluated by repeating the same LES on three different meshes ranging from 1.4 million to 40 million cells and a reasonable mesh independency was observed even though more resolved wrinkling is observed when the mesh resolution is increased. More generally, these results demonstrate that the concept of "quality" in LES will require much more detailed studies and tools than what has been used up to now in Reynolds Averaged simulations. Instabilities appearing in a given LES on a given computer can have sources which were not expected at first sight (like the number of processors). Mastering these instabilities will be an important task to get the full power of LES techniques.

References

1. S. S. Abarbanel and A. E. Chertock. Strict stability of high-order compact implicit finite-difference schemes: the role of boundary conditions for hyperbolic PDEs, I. *J. Comput. Phys.*, 160:42–66, 2000.
2. G. Boudier, L. Y. M. Gicquel, T. Poinsot, D. Bissières, and C. Bérat. Effect of mesh resolution on large eddy simulation of reacting flows in complex geometry combustors. *Combust. Flame*, 155(1-2):196–214, 2008.
3. F. Chaitin-Chatelin and V. Frayssé. *Lectures on Finite Precision Computations*. SIAM, Philadelphia, 1996.
4. P. Clavin. Dynamics of combustion fronts in premixed gases: from flames to detonations. *Proc. Combust. Inst.*, 28:569–586, 2000.
5. O. Colin and M. Rudgyard. Development of high-order taylor-galerkin schemes for unsteady calculations. *J. Comput. Phys.*, 162(2):338–371, 2000.
6. E. Cuthill and J. McKee. Reducing the bandwidth of sparse symmetric matrices. In *Proceedings of the 24th National Conference of the ACM*, pages 157–172, 1969.
7. P. G. Drazin and W. H. Reid. *Hydrodynamic stability*. Cambridge University Press, London, 1981.
8. M. Freitag and J. Janicka. Investigation of a strongly swirled premixed flame using LES. *Proc. Combust. Inst.*, 31:1477–1485, 2007.
9. S. Ghosal and P. Moin. The basic equations for the large eddy simulation of turbulent flows in complex geometry. *J. Comput. Phys.*, 118:24 – 37, 1995.
10. A. Giauque, L. Selle, T. Poinsot, H. Buechner, P. Kaufmann, and W. Krebs. System identification of a large-scale swirled partially premixed combustor using LES and measurements. *J. Turb.*, 6(21):1–20, 2005.
11. G. Hanrot, V. Lefèvre, D. Stehlé, and P. Zimmermann. Worst cases for a periodic function with large arguments. In Peter Kornerup and Jean-Michel Muller, editors, *Proceedings of the 18th IEEE Symposium on Computer Arithmetic*, pages 133–140, Los Alamitos, CA, 2007. IEEE Computer Society Press.
12. C. Hirsch. *Numerical Computation of Internal and External Flows*. John Wiley, New York, 1988.
13. C. M. Ho and P. Huerre. Perturbed free shear layers. *J. Fluid Mech.*, 16:365, 1984.
14. M. Klein. An attempt to assess the quality of large eddy simulations in the context of implicit filtering. *Flow, Turb. and Combustion*, 75(1-4):131–147, 2005.
15. S.K. Lele. Compact finite difference schemes with spectral like resolution. *J. Comput. Phys.*, 103:16–42, 1992.
16. W.-H. Liu and A. H. Sherman. Comparative analysis of the Cuthill-McKee and the reverse Cuthill-McKee ordering algorithms for sparse matrices. *SIAM Journal of Numerical Analysis*, 13(2):198–213, 1976.
17. K. Mahesh, G. Constantinescu, and P. Moin. A numerical method for large-eddy simulation in complex geometries. *J. Comput. Phys.*, 197(1):215–240, 2004.
18. F. Di Mare, W. P. Jones, and K. Menzies. Large eddy simulation of a model gas turbine combustor. *Combust. Flame*, 137:278–295, 2004.
19. C. Martin, L. Benoit, Y. Sommerer, F. Nicoud, and T. Poinsot. LES and acoustic analysis of combustion instability in a staged turbulent swirled combustor. *AIAA Journal*, 44(4):741–750, 2006.

20. J. Meyers, B. J. Geurts, and M. Baelmans. Database analysis of errors in large-eddy simulation. *Phys. Fluids*, 15(9):2740–2755, September 2003.
21. V. Moureau, G. Lartigue, Y. Sommerer, C. Angelberger, O. Colin, and T. Poinsot. Numerical methods for unsteady compressible multi-component reacting flows on fixed and moving grids. *J. Comput. Phys.*, 202(2):710–736, 2005.
22. T. Poinsot, T. Echekki, and M. G. Mungal. A study of the laminar flame tip and implications for premixed turbulent combustion. *Combust. Sci. Tech.*, 81(1-3):45–73, 1992.
23. T. Poinsot and D. Veynante. *Theoretical and Numerical Combustion*. R.T. Edwards, 2nd edition, 2005.
24. S. B. Pope. *Turbulent flows*. Cambridge University Press, 2000.
25. S. B. Pope. Ten questions concerning the large-eddy simulation of turbulent flows. *New Journal of Physics*, 6:35, 2004.
26. C. Prière, L. Y. M. Gicquel, P. Gajan, A. Strzelecki, T. Poinsot, and C. Bérat. Experimental and numerical studies of dilution systems for low emission combustors. *Am. Inst. Aeronaut. Astronaut. J.*, 43(8):1753–1766, 2005.
27. L. Rayleigh. The explanation of certain acoustic phenomena. *Nature*, July 18:319–321, 1878.
28. S. Richard, O. Colin, O. Vermorel, A. Benkenida, C. Angelberger, and D. Veynante. Towards large eddy simulation of combustion in spark ignition engines. *Proc. Combust. Inst.*, 31(3059-3066), 2007.
29. P. Sagaut. *Large Eddy Simulation for incompressible flows*. Scientific computation series. Springer-Verlag, 2000.
30. P. Schmitt, T. J. Poinsot, B. Schuermans, and K. Geigle. Large-eddy simulation and experimental study of heat transfer, nitric oxide emissions and combustion instability in a swirled turbulent high pressure burner. *J. Fluid Mech.*, 570:17–46, 2007.
31. T. Schønfeld and M. Rudgyard. Steady and unsteady flows simulations using the hybrid flow solver AVBP. *AIAA Journal*, 37(11):1378–1385, 1999.
32. L. Selle, L. Benoit, T. Poinsot, F. Nicoud, and W. Krebs. Joint use of compressible large-eddy simulation and Helmoltz solvers for the analysis of rotating modes in an industrial swirled burner. *Combust. Flame*, 145(1-2):194–205, 2006.
33. T. K. Sengupta. *Fundamentals of Computational Fluid Dynamics*. Universities Press, Hyderabad (India), 2004.
34. T. K. Sengupta, G. Ganerwal, and A. Dipankar. High accuracy compact schemes and Gibbs' phenomenon. *J. Sci. Comput.*, 21(3):253–268, 2004.
35. J.-M. Senoner, M. García, S. Mendez, G. Staffelbach, O. Vermorel, and T. Poinsot. Growth of rounding errors and repetitivity of large eddy simulations growth of rounding errors and repetitivity of large eddy simulations growth of rounding errors and repetitivity of large eddy simulations. *AIAA Journal*, 46(7):1773–1781, 2008.
36. Y. Sommerer, D. Galley, T. Poinsot, S. Ducruix, F. Lacas, and D. Veynante. Large eddy simulation and experimental study of flashback and blow-off in a lean partially premixed swirled burner. *J. Turb.*, 5, 2004.
37. J. S. Stoer and R. Bulirsch. *An Introduction to Numerical Analysis*. Springer, Berlin, 1980.
38. V. E. Taylor and B. Nour-Omid. A study of the factorization fill-in for a parallel implementation of the finite element method. *Int. J. Numer. Meth. Eng.*, 37:3809–3823, 1994.
39. H. Tennekes and J. L. Lumley. *A first course in turbulence*. M.I.T. Press, Cambridge, 1972.
40. R. Vichnevetsky and J. B. Bowles. *Fourier analysis of numerical approximations of hyperbolic equations*. SIAM Studies in Applied Mechanics, Philadelphia, 1982.

41. A. W. Vreman, B. J. Geurts, and J. G. M. Kuerten. A priori tests of large eddy simulation of the compressible plane mixing layer. *J. Eng. Math.*, 29:299–327, 1995.
42. B. Vreman, B. Geurts, and H. Kuerten. Comparison of numerical schemes in large-eddy simulation of the temporal mixing layer. *Int. J. Numer. Meth. Fluids*, 22:297–311, 1996.
43. F. A. Williams. *Combustion Theory*. Benjamin Cummings, Menlo Park, CA, 1985.

Quality Issues of Combustion LES

A. M. Kempf, M. Pettit

Department of Mechanical Engineering, Imperial College London
South Kensington Campus, SW7 2AZ London, UK
a.kempf@imperial.ac.uk

Summary. Combustion LES requires additional modelling of physics beyond the flow-field only. These additional models lead to further quality issues and an even stronger need to quantify the errors. The present paper illustrates stability problems, the need for consistent modelling in premixed and non-premixed combustion, and shows how RANS models that have frequently been applied to LES can lead to strong conceptual errors. The paper then outlines the application of Meyer's error landscape approach to a complex non-premixed flame, and mentions several error-indicators that have been developed for situations where no experimental reference data is available.

Key words: Combustion, Large-Eddy Simulation, Quality, Reliability

1 Introduction

Fossil fuels are the major source of primary energy today, but alternatives must be sought due to the growing need for cheap energy, the predicted depletion of oil reserves, and the need to reduce pollution and combat global warming. Pollutant emissions can be lowered dramatically through newly developed combustion modes like lean premixed, stratified, and mild combustion. To counter global warming, fossil fuels can be replaced through carbon-neutral bio-mass; the oxidiser can be substituted through pure oxygen, so that carbon dioxide is the only major combustion product, which can be captured and sequestered (CCS) under ground. The new combustion modes, oxidisers and fuels require more expensive, complicated, and heavy combustors, and often lead to flame instability that did not occur in classical non-premixed flames. The successful deployment of the new technologies can only be achieved through a detailed understanding generated from experiments and simulation. However, measurements in flames are notoriously difficult due to the aggressive environment and the generally poor optical access; in the case of pulverised coal or bio-mass firing, flame diagnostics are almost impossible. Overall, the difficulty in measuring the data and the need for safe, clean, and affordable energy are strong incentives for detailed, quality-managed combustion simulations.

M.V. Salvetti et al. (eds.), *Quality and Reliability of Large-Eddy Simulations II*, ERCOFTAC Series 16, DOI 10.1007/978-94-007-0231-8_3,

1.1 Modes of Combustion

Most existing large-eddy simulations have been performed for single-phase combustion, where air is used to oxidise gaseous fuels like hydrogen, methane, or natural gas. One normally distinguishes the premixed and the non-premixed combustion modes, although most engineering applications burn a partially premixed fuel/air mixture. In non-premixed combustion, a homogeneous fuel air mixture enters the combustion chamber, where a thin premixed flame front propagates through the reactants. The conditions of premixed combustion are 'well-defined' since the fuel and oxidiser are perfectly mixed on a molecular level, so a flame can be tuned to minimise the emission of pollutants like carbon monoxide (CO) or nitric oxides (NOx). As a result, premixed flames are often used in advanced low-emission combustors, although the stabilisation of premixed flames is often difficult. In some cases, strong accoustic instabilities have been observed in premixed flames, which can destroy the combustion device.

In the non-premixed mode, fuel and oxidiser are only mixed in the combustion chamber, where the fast chemical reaction is limited by the rate of molecular mixing. Such non-premixed flames are safe and stable, as only a very small amount of combustible mixture is available in the mixing layer at any time. However, pollutant formation is much harder to control as the fuel-air mixture is no longer well-defined. In recent years, non-premixed flames were often replaced with premixed flames to meet newly introduced pollution legislation.

In real engineering applications such as gas turbines or internal combustion engines, one usually encounters some level of premixing and inhomogenity, so that the flames will burn in a partially premixed way. However, the mode of combustion determines which combustion models can be used and hence what level of error will result.

2 Combustion LES

The major quality problem with Combustion LES is the introduction of a wide range of chemical and mixing time-scales, that increase the computational cost and require additional closure modelling.

However, most combustion devices are relatively 'LES friendly', as the hard-to-model near-wall flow is usually of little interest near the flames: a burner is usually designed to minimise the heat-load on the flame's confinement and to keep the flame from extinguishing near walls, which would otherwise lead to the emission of unburned hydrocarbons. This means that in many combustion devices, the flow near walls can be ignored, or at least does not need to be described accurately. However, walls can be critical for certain burner geometries, for flame stabilisation, and for the occurance of flash-backs, and very accurate wall-modelling may be required if such phenomena are to be investigated. A further reason for the 'LES friendliness' of many combustion devices is that classical non-premixed combustors are usually designed to burn stably for a wide range of operating conditions in terms of flow rate, pressure, and fuel composition. The result is a very robust flame for which small errors in the simulation should not alter the results too much.

However, Combustion LES often suffers from numerical instability due to the large density gradients between the burned and unburned gases, and the rapid expansion due to heat release. Instabilities are also related to the description of convective transport, where non-oscillating schemes must be applied for reactive scalars to avoid excessive fluctuations in dependent quantities. An example would be the density's dependence on mixture fraction f, a scalar that describes the mixing of fuel ($f = 1$) and oxidiser ($f = 0$). For many fuels, the stoichiometric mixture fraction is so low (e.g. $f_{st} \approx 0.055$ for methane) that an absolute error of only 0.05 could reduce the density from that of ambient air to that at adiabatic flame temperature.

Extending large-eddy simulation for chemically reacting flows requires changes to the flow-solver, which must cope with variable density and viscosity. For air flames at atmospheric pressure, density ratios of eight between the unburned fuel or oxidiser and the hot burned gas are common. In many cases, the reactants and products will only be separated by a thin premixed flame front, so that density 'jumps' can occur that are similar to shocks in super-sonic flows. The resulting gradients and curvatures require that scalar convection is discretised by non-oscillating schemes such as ENO or TVD. Although the density field of a flame is very inhomogeneous, most simulations still apply a low-Mach assumption to achieve a less restrictive CFL criterion for the time-step width. In such cases, 'incompressible' is interpreted as $\partial\rho/\partial p = 0$ with $\partial\rho/\partial T < 0$. The high temperature resulting from combustion will significantly increase viscosity, which can impose severe limits for the time-step width, in particular when cylindrical grids are used. (Such grids are very common as most laboratory flames are stabilised on axi-symmetric burners.) The density and viscosity depend on the chemical state and must be provided from the combustion model. In general, only combustion models are suitable for LES where the density does not depend too sensitively on other variables, as strong spurious expansion may induce velocities much higher than what is physically realistic, leading to serious stability problems.

2.1 Modelling in the limit of very fine grids

Normally, combustion and diffusive transport occur on scales that are too small to be resolved, leading to a closure problem that is often as hard as for RANS methods. However, one will often achieve sufficient resolution to locally resolve transport and reaction, and the models must remain consistent even for very fine grid resolutions. Ensuring this consistency is particulary hard for classical RANS models that are now applied with LES.

Convergence to DNS for premixed combustion

The need for a model to stay consistent in the limit of very fine grids (i.e. approaching DNS) can easily be outlined for premixed flames. For premixed flames, a classical approach is based on the Bray-Moss Libby (BML) [6] concept that a premixed flame is sufficiently thin to treat it as an interface between reactants and combustion products, which means that the probability of finding species that undergo reaction becomes zero. This assumption holds well for RANS simulations and for LES on coarse grids, but many simulations will actually resolve the typical flame thickness of the order of 0.1 mm. In such highly resolved large-eddy simulations, combustion models that are based on the BML analysis will lead to large errors near the flame-front. Most flame surface density (FSD) models (e.g. [1, 4, 8]) as well as the G-equation level-set approach [22, 24] fall in this category, and while they have proven to be elegant and successful in RANS simulation, great care is required in their application to highly resolved LES.

Convergence to DNS for non-premixed combustion

In non-premixed combustion, a flame is established at the mixing-layer between fuel and oxidiser, where stoichiometric conditions can be found. Non-premixed combustion is limited by the rate of mixing, which means that fuel and oxidiser are assumed to react as soon as they meet, since the mixing takes far longer than the reaction. In many cases, non-premixed combustion is modelled using a flamelet approach according to Williams [28] and Peters [21] or the CMC technique of Bilger [2] and Klimenko [17], using Bilger's mixture fraction [3]. The mixture fraction f gives the local ratio of atomic mass from the fuel nozzle to the atomic mass from the air-coflow, for which a transport equation must be solved.

In LES, the subgrid distribution of the mixture fraction is not known. However, the thermo-chemistry depends on the mixture fraction in a non-linear way so the a subgrid distribution must be known to calculate the filtered chemical state. (For example, the temperature in an LES cell of mixture fraction 0.5 would be different if the 'left' half was filled with fuel and the 'right' half was filled with air, or if the entire cell was filled with a homogeneous and hence burned mixture). It is common practise to use the mixture fraction mean and variance to parameterise an assumed distribution based on β-functions, Gaussian fuctions and top-hat functions. (A Dirac function would correspond to a fully resolved DNS.) Most large-eddy simulations (including those by the author) of non-premixed combustion have so far used the β-function in the same way as in RANS.

However, Oefelein and Frank [12] have recently performed a very detailed experimental and computational analysis of instantaneous scalar fields, in which they confirm that very small scalar dissipation structures exist, which are very hard to model. However, they have also observed that many of these scalar dissipation structures correspond to mixing between very similar mixture fractions, so that the mixture fraction distribution within an LES cell can be relatively narrow. Such a narrow distribution is somewhat inconsistent with the β-function that would normally assume a mixture fraction distribution between zero and one.

Floyd et al. [11] have further analysed the implications of using a β-function in LES. They found that the β-function is much less compatible with LES than with RANS, due to the conceptual differences between the two approaches. The β-function leads to strong contradictions in the LES context, whereas it is a very powerful and elegant model for RANS. Floyd et al. [11] have shown that in the LES context, a β-function would imply infinite scalar gradients and hence infinite rates of scalar dissipation and mixing. It is also well known that a β-function is not suitable for multi-stream mixing unless an additional equation is solved, but this constraint is less severe for RANS than for LES, where areas of relatively homogeneous mixture can be found throughout the computational domain. Such locally homogeneous areas cannot be described by a β-function that always spans over the entire mixture fraction range, from zero to one. Floyd et al. [11] have tried to analyse how severe this constraint is by applying the assumed β-function model and the Smagorinsky model to a synthetic scalar field based on a sine wave of wave-length λ as shown in Fig. 1. For this field, an FDF that spans the entire mixture fraction range (like the β-function) can only be justified if the filter-width Δ is *larger* than one half the wave-length ($\Delta > 1/2\lambda$). However, Pope's recommendation that an LES should resolve at least 80% of energy can only be achieved for filters smaller than a certain threshold. For a top-hat filtered sine-wave, the energy content of the filtered wave decays with filter width Δ according to eq. 1:

$$\frac{E_{\overline{sin(x)}}}{E_{sin(x)}} = \left(\frac{2}{\Delta}\sin\frac{\Delta}{2}\right)^2 \tag{1}$$

According to Eq. 1, approximately 80 % of energy is left for a filter-width of one quarter of the wave-length ($\pi/2$), so that the filter must be *finer* than this ($\Delta < 1/4\lambda$). There is therefore no filter width Δ for which the β-function could be consistent with an LES – as illustrated in Fig. 2.

Interestingly, with LES one can circumvent the before-mentioned problems of the β FDF by assuming a much simpler top-hat FDF, which also offers computational advantages. Firstly, a top-hat FDF is much easier to accurately integrate than a β-function for situations where the variance approaches its minimum or maximum values. Secondly, pre-integrating a lookup table with a top-hat FDF does not add a new dimension to the original table as would be the case with a β-function, so that effectively the dimensionality of the lookup table will be one half that of a table for the β-function. This greatly reduces the memory requirements in situations where more than one parameter affect the thermo-chemical state [11].

3 Quantifying Errors

To manage the quality of combustion LES, similar methods are available as for non-reactive LES. With LES becoming available in many commercial CFD codes, industry becomes interested in applying LES and as a result, the quality of these simulations must be managed.

3.1 Quality Estimators and Error Estimators

In recent years, several error-indicators have been developed for LES, which attempt to provide insight into whether a given simulation should be trusted. Such error indicators would ideally be available in a truely predictive way, without requiring any further reference data.

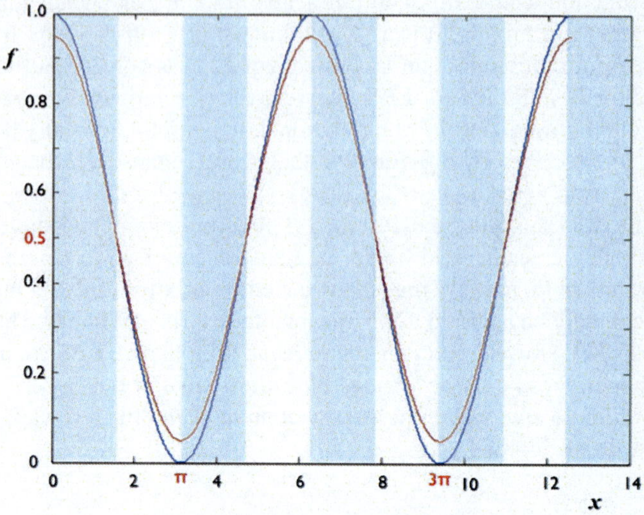

Fig. 1. A filtered and an unfiltered sine wave. A filter-width of at least $\Delta = \lambda/2 = \pi$ is required to cover the entire mixture fraction range $0 \leq f \leq 1$. The (tophat-) filter width Δ must not be larger than $\Delta \approx \lambda/4 = \pi/2$ to maintain 80% of scalar energy. The shaded bands have a width of $\pi/2 = \lambda/4$.

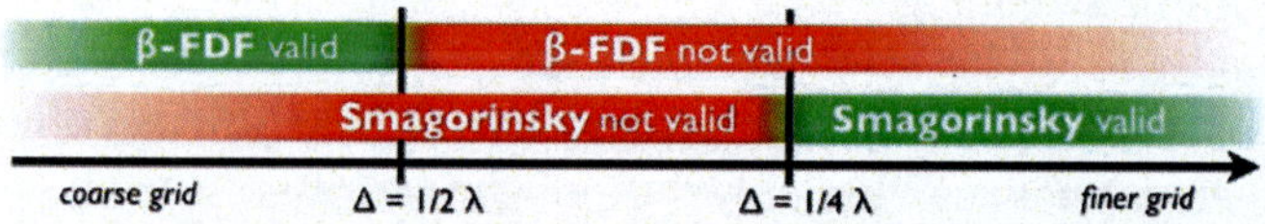

Fig. 2. Ranges of validity for the Smagorinsky model and for the β FDF model for a wave of wave-length λ according to [11]. There is no filter-size Δ for which both models can be applied at the same time.

The currently existing indicators (e.g. [7, 16, 19]) are useful and can help to appreciate the problems in a given LES. A positive assessment from any of these indices is possibly a *necessary requirement* for an accurate LES, but a high quality index is not a *sufficient criterion*: most error-estimators are necessarily based on modelled quantities, which cannot be relied upon if the grid is coarse. Hence, an error-indicator may imply that a solution is 'good' where in reality, the grid was under-resolved.

For combustion LES, there are some special issues with these error-indicators. Firstly, combustion LES normally relies on stabilised numerical descriptions, for example using time-implicit schemes or total variation diminishing schemes to avoid numerical oscillations in the scalar fields. These 'stabilised schemes' can induce artificial dissipation beyond the physical dissipation due to molecular viscosity and turbulent viscosity. As a result, an error-estimator that is (even implicitly) based on viscosities will be biased by the effect of 'numerical viscosity' and numerical dissipation, so that the error-estimator would over-estimate a simulation's quality.

The heat release due to combustion induces expansion of the gases, and very often excessive expansion results from numerical errors in both the mixing and combustion models. These increased velocity fluctuations will contribute to the resolved velocity fluctuations while hardly contributing to the turbulent viscosity, so that an error-estimator would be likely to over-estimate the level of quality.

Overall, error estimators are a very useful step towards quality managed LES, and they can help to identify insufficient grid resolution. However, one must not rely on such quality estimates; they are not sufficient. In particular, such indicators must not be used to compare or tune LES subgrid models, as the 'highest' quality would falsely be observed for the (eddy-viscosity) model yielding the smallest turbulent viscosity: if an error-estimator was abused to tune the Smagorinsky constant C_s, the resulting optimal value would be determined as $C_s = 0$, at least if the simulation remains stable.

3.2 The Error Landscape

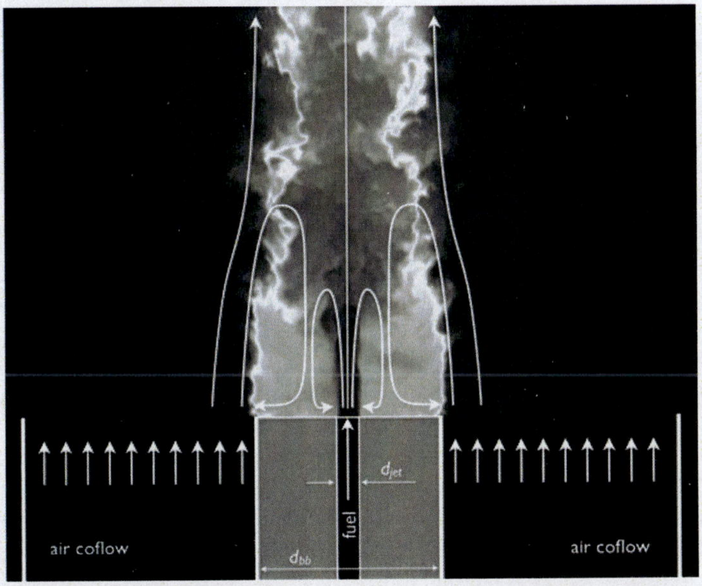

Fig. 3. Configuration of the Sydney Bluff-Body Burner [9, 10]. A jet in the centreline provides fuel that is burned in the recirculation zone downstream of the cylindrical bluff-body. A coflow-provides the oxidiser to the recirculation zone.

A very interesting approach for the analysis of computational errors is based on the concept of an error landscape, as demonstrated by Meyers et al. [19] for decaying homogeneous isotropic turbulence. The approach is currently extended [15] for the complex variable density flow in the Sydney Bluff-Body flame [9, 10, 13] illustrated in Fig. 3. The error landscape is normally based on deviations from DNS results, but it can also be calculated in comparison to experimental data [5, 14]. The resulting error landscape can help analyse computational and modelling errors, but as it needs a set of reference data, it is not a complete tool for the quality management of predictive simulations.

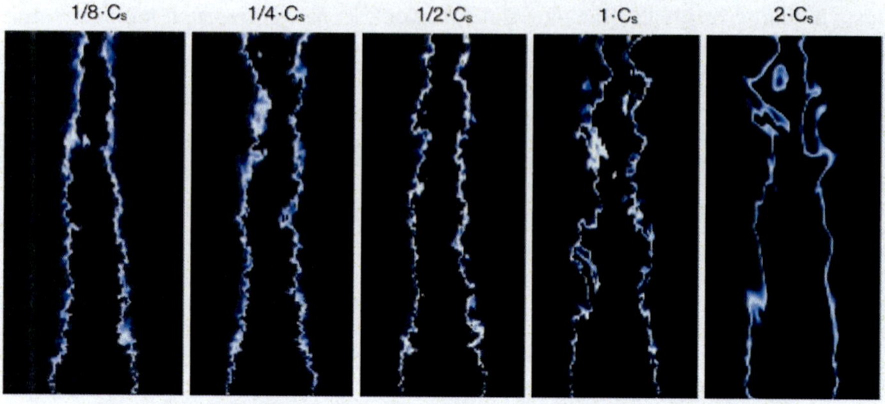

Fig. 4. Concentration of the hydroxyl (OH) radical for different values of the Smagorinsky constant ($C_s = 0.173$ as suggested by Lilly [18]). In its excited state (OH*), hydroxyl emits UV light.

For the Sydney bluff body flame, error landscapes were constructed for seven different values of the Smagorinsky parameter $[0.25, 0.5, 0.75, 1.0, 1.25, 1.5, 2.0] \cdot C_s = 0.173$ [18] and five different sets of grids with $[0.3, 0.625, 1.48, 5, 40]$ million cells and resolutions of $[2.5, 2.0, 1.5, 1.0, 0.5]$ mm. Constructing the error landscape required a total of 35 simulations and 2.7 years of CPU time, which is approximately half the computational cost that would be required for the 'next' grid refinement to 0.25 mm resolution with 320 million cells. The simulations were performed using the efficient parallel PsiPhi LES program, which permitted simulations with 40 million cells on 12 CPUs in only 3200 CPU hours. An overview of all simulations is given in table 3.2.

Figure 4 illustrates the effect of the Smagorinsky parameter C_s on the concentration of the hydroxyl radical OH, that only exists close to the flame-front. For large C_s, the flame is hardly wrinkled as the eddy viscosity has dampened most of the small scale turbulence.

Figures 5 and 6 show a comparison of the radial velocity fluctuations predicted at different axial positions x/D for different grid resolutions and for two extrem values of the Smagorinsky constant $C_s = 0.043$ (Fig. 5) and $C_s = 0.346$ (Fig. 6). As expected, the fluctuation levels are largely over-predicted for the small model-constant that results in insufficient dissipation (Fig. 5), whereas the effect of too much dissipation can be clearly seen in Fig. 6. For both (extreme) values of C_s, the results converge with grid-refinement towards the experimental data.

Figures 7 and 8 show a comparison similar to that in figs. 5 and 6, albeit for the mean temperature. These plots clearly show that the mean temperatures are significantly less affected by the Smagorinksy constant than the velocity fluctuations, although the differences between the grid resolutions are still significant.

The error landscapes were calculated for the means and fluctuations of the three velocity components, of mixture fraction, and of the temperatures. Figure 9 shows an overall error landscape that was calculated based on an equal weighting of the different (normalised) quantities [15]. The error landscapes are consistent with that presented by Meyers et al. [2], showing the expected large errors on coarse grids and for extreme values of the Smagorinsky parameter, with a minimum for the optimal Smagorinsky value. However, the landscape is no longer smooth for coarse grids, implying that the error only converges in a predictable monotonic way for fine grid-resolutions. In turn, this non-monotonic convergence implies that quality-indices calculated from coarse-grid simulations cannot be expected to be accurate, stressing that a positive result from any error-estimator is not sufficient.

Table 1. A total of 2.7 CPU years were used for the following simulations. ($C_s = 0.173$)

# cells	Δ mm	$\frac{1}{4}C_s$ t/h	$\frac{1}{2}C_s$ t/h	$\frac{3}{4}C_s$ t/h	C_s t/h	$\frac{5}{4}C_s$ t/h	$\frac{6}{4}C_s$ t/h	$2C_s$ t/h	CPU cores
320k	2.5	1.5	1.5	1.5	1.5	1.5	1.5	1.5	1
625k	2.0	6	6	6	6	6	6	6	1
1.48M	1.5	28	28	28	28	28	28	28	4
5.00M	1.0	150	150	150	150	150	150	150	4
40.0M	0.5	3.2k	3.2k	3.2k	3.2k	3.2k	3.2k	3.2k	3×4
	t/h	3.4k	3.4k	3.4k	3.4k	3.4k	3.4k	3.4k	

4 Conclusions

This paper has outlined the need for combustion LES, and the major modes of non-premixed, partially premixed, and premixed combustion were described. Some special features of combustion LES that affect the simulation error were highlighted, in particular the need for stabilising numerical discretisation schemes. The paper illustrated that even though combustion typically occurs on very small scales that require modelling just like RANS techniques, combustion models developed for the RANS of high Reynolds number flow can often not be used in LES, as these models do not converge properly with grid refinement. These convergence problems were discussed in great detail for the classical assumed β FDF, which has been demonstrated to only be appropriate in situations where the filter-width is coarser than it should be for an accurate LES. The paper also discussed the applicability of error-indicators, and concludes that the resulting error-indices are not a sufficient indication for an accurate

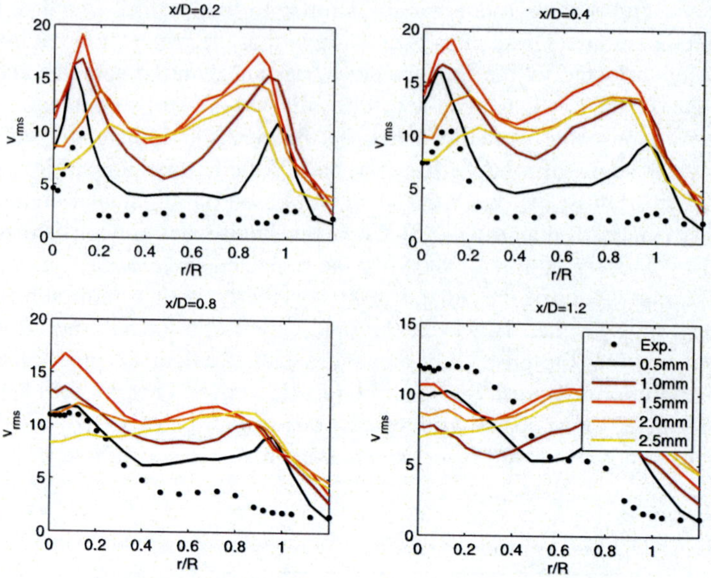

Fig. 5. Fluctuations in the radial velocity for a very low Smagorinksy constant of $C_s = 0.043$; showing the expected over-estimation of the fluctuation level. The computational data was sampled at the same points as the experiments.

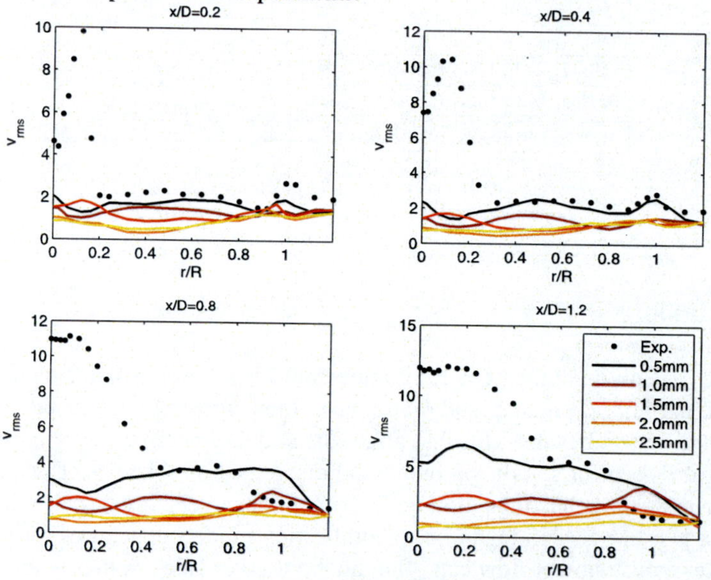

Fig. 6. Fluctuations in the radial velocity for a very high Smagorinksy constant of $C_s = 0.346$; showing the expected under-estimation of the fluctuation level. The computational data was sampled at the same points as the experiments.

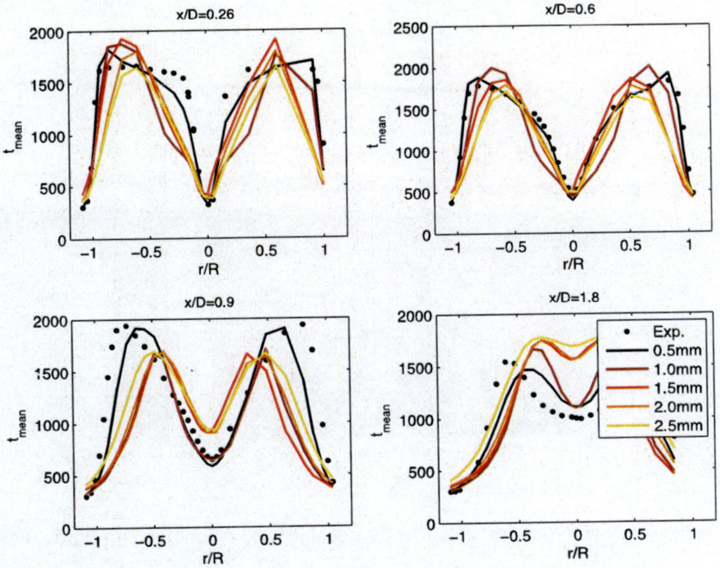

Fig. 7. Mean temperature fields resulting from a very low Smagorinksy constant of $C_s = 0.043$. The computational data was sampled at the same points as the experiments.

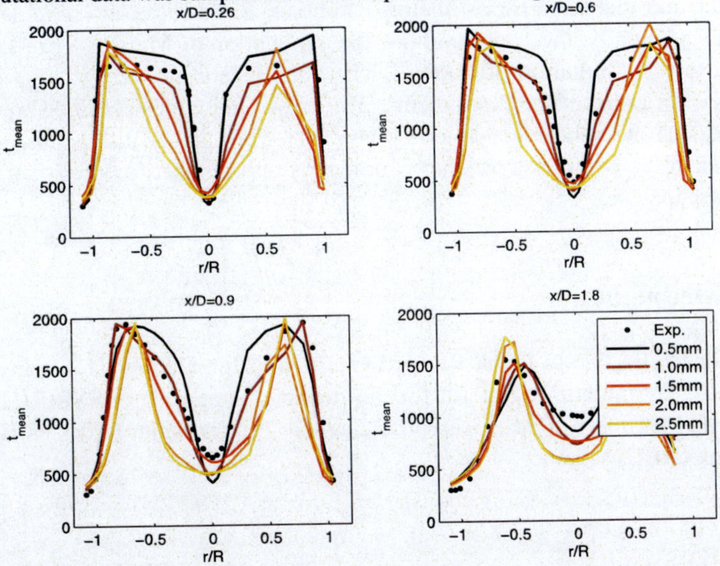

Fig. 8. Mean temperature fields resulting from a very high Smagorinksy constant of $C_s = 0.346$. The computational data was sampled at the same points as the experiments.

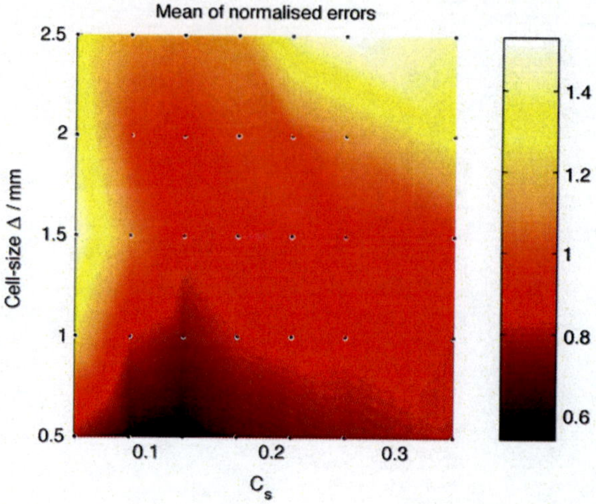

Fig. 9. Global error landscape calculated as the mean of the normalised individual error landscapes.

simulation, and that the error-estimators can be expected to be less reliable in combustion LES. Finally, the paper outlines the application of Meyers' error landscape approach to the simulation of a flame, where the reference data was not obtained from DNS but rather from experiments. The error landscapes obtained were found to be consistent with those generated by Meyers for turbulence dissipation; albeit the landscapes in the complex flows have somewhat less distinct features than Meyers' error landscape.

Acknowledgements

The author would like to thank Bernard Geurts and Joe Oefelein for many helpful discussions. The author is grateful for the financial support by EPSRC (UK Engineering and Physical Sciences Research Council), in particular under its High End Computing (HEC) scheme.

References

1. Angelberger, C., Veynante, D., Egolfopoulos, F., Poinsot, T., A flame surface density model for large eddy simulations of turbulent premixed flames. Proceedings of the Summer Program, Center for Turbulence Research, Stanford (1998) 66-82.
2. Bilger, R.W., Conditional moment closure for turbulent reacting flows. Phys. Fluids A 5 (1993) 436-444.
3. Bilger, R., Starner, S., Kee, R., On Reduced Mechanisms for Methane-Air Combustion in Nonpremixed Flames. Combust. Flame, 80 (1990) 135-149.

4. Boger, M., Veynante, D., Boughanem, H., Trouve, A., Direct numerical simulation analysis of flame surface density concept for large eddy simulation of turbulent premixed combustion. Proc. Combust. Inst. 27 (1998) 917.
5. Böhm, B., Brübach, J., Ertem, C., Dreizler, A., Experiments for LES validation. Flow Turb. Combust. 80 (2008) 507-529.
6. Bray, K.N.C., Moss, J.B., Libby, P.A., Unified modeling approach for premixed turbulent combustion Part I: General formulation. Combust. Flame 61 (1985) 87-102.
7. Celik, I., Cehreli, Z., Yavuz, I., Index of resolution quality for large eddy simulations. J. fluids eng. 127 (2005) 949-958.
8. Chakraborty, N., Klein, M., A priori direct numerical simulation assessment of algebraic flame surface density models for turbulent premixed flames in the context of large eddy simulation. Phys. Fluids 20 (2008), 085108.
9. Dally, B., Masri, A., Barlow, S., Fiechtner, G., Instantaneous Mean Compositional Structure of Bluff-Body Stabilized Non-premixed Flames. Combust. Flame 114 (1998) 119-148.
10. Dally, B., Fletcher, D., Masri, A., Modelling of Turbulent Flames Stabilised on a Bluff-Body. Combust. Theory Modeling 2 (1998) 193-219.
11. Floyd, J., Kempf, A., Kronenburg, A., Ram, R.H., A Simple Model for the Filtered Density Function for Passive Scalar Combustion LES. Combust. Theory Modelling 13 (2009) 559-588.
12. Frank, J.H., Kaiser, S.A., Oefelein, J.C., Coupling Imaging Diagnostics and Large Eddy Simulation in Turbulent Non-reacting Jets and Non-premixed Jet Flames, Proc. European Combustion Meeting (ECM), Vienna, Austria, 2009.
13. Kempf, A., Lindstedt, R.P., Janicka, J., Large-eddy simulation of a bluff-body stabilized non-premixed flame. Combust. Flame 144 (2006) 170-189.
14. Kempf A.M., LES Validation from Experiments. Flow Turb. Combust. 80 (2008) 351-373.
15. Kempf, A.M., Geurts, B.J., Oefelein, J.C., Error Analysis of the Large-Eddy Simulation of the Turbulent Non-Premixed Sydney Bluff-Body Flame, in preparation.
16. Klein, M., An attempt to assess the quality of LES in the context of implicit filtering. Flow Turb. Combust. 75 (2005) 131-147.
17. Klimenko, A.Y., Multicomponent diffusion of various scalars in turbulent flows. Fluid Dyn. 25 (1990) 327-334.
18. Lilly, D.K., The representation of small-scale turbulence in numerical simulation experiments. Proc. IBM Scientific Computing Symp. Environmental Sciences (1967) 195-210.
19. Meyers, J., Geurts, B.J., Baelmans, M., Database-analysis of errors in large-eddy simulations. Phys. Fluids 15 (2003) 2740-2755.
20. Navarro-Martinez, S., Kronenburg, A., LES-CMC simulations of a turbulent bluff-body flame. Proc. Combust. Inst. 31 (2007) 1721-1728.
21. Peters, N., Laminar Diffusion Flamelet Models in Non-Premixed Turbulent Combustion. Prog. Energy Combust. Sci. 10 (1984) 319-339.
22. Peters, N., Turbulent Combustion. Cambridge Monographs on Mechanics, Cambridge University Press, Cambrdige, U.K., 2000.
23. Raman, V., Pitsch, H., Large-Eddy Simulation of a Bluff-Body Stabilized Flame Using a Recursive-Refinement Procedure. Combust. Flame 142 (2005) 329-347.
24. Sethian, F.A., Level set methods. Cambridge Monographs on Applied and Computational Mathematics, Cambridge University Press, Cambridge, U.K., 1996.
25. Sick, V., Hildenbrand, F., Lindstedt, R.P., Quantitative Laser-Based Measurements and Detailed Chemical Kinetic Modelling of Nitric Oxide Concentrations in Methane-Air Counterflow Diffusion Flames. Proc. Combust. Inst. 27 (1998) 1401-1409.

26. Smagorinsky, J., General circulation experiments with the primitive equations. Mon. Weather Rev., 91 (1963) 99.
27. The International Workshop on Experiments and Simulations of Turbulent Non-Premixed Flames, http://www.ca.sandia.gov/TNF/, 11/10/2009.
28. Williams, F.A., Recent Advances in Theoretical Descriptions of Turbulent Diffusion Flames, in S.N.B. Murthy (ed.), Turbulent Mixing in Nonreactive and Reactive Flows, Plenum Press, New York, US, 1975.

Energy cascade and spatial fluxes of filtered wall-turbulent flows

A. Cimarelli and E. De Angelis

DIEM, II Facoltà di Ingegneria, Università di Bologna, Viale Risorgimento, 40136 Bologna, Italy andrea.cimarelli2@unibo.it, e.deangelis@unibo.it

Summary. The analysis of the subgrid energy transfer which is traditionally accomplished with the turbulent kinetic energy [1], and with the energy spectrum [2], is performed by a scale energy balance. A generalized Kolmogorov equation for filtered velocity field will be applied and discussed. The results will show which are the effects of subgrid scales on the resolved motion in both physical and scale space, singling out the prominent role of the filter scale compared to the cross-over scale between production dominated scales and inertial range, l_c, and the reverse energy cascade region Ω_B.

Key words: Filtered Kolmogorov equation, Subgrid-energy transfer

1 Introduction

In the Large Eddy Simulation the energy carrying structures that are directly affected by the boundary conditions are computed explicitly, while the small scales are modeled. Since the small scales tend to be more homogeneous and isotropic than the large ones, it is thought that relatively simple and universal models can be used to describe them. The most important feature of such models should be their ability to reproduce accurately the energy transfer between resolved and unresolved scales. Most used LES models assume that the main role of the subgrid scales is to remove energy from the large scales and dissipate it via viscous forces accordingly to the idea of an inertial range in the spectrum. Indeed, as asserted by the 4/5 law, in the inertial range the energy flux is independent of the scale under consideration, is from large to small scales and it is equal to the energy dissipation. This picture is claimed to be highly universal and most of the LES models attempt to reproduce it.

This approach has given good results in homogeneous and in unbounded shear flows but less in wall flows where difficulties arise. Indeed, in this case an energy cascade through scales and spatial momentum transfer occur simultaneously and a model should be able to capture subgrid scale effects on both these energy processes. Furthermore, as recently quantified in Marati et al. [3], in wall-turbulence the picture of the Richardson energy cascade is strongly modified and a reverse energy cascade is observed in the buffer layer.

M.V. Salvetti et al. (eds.), *Quality and Reliability of Large-Eddy Simulations II*,
ERCOFTAC Series 16, DOI 10.1007/978-94-007-0231-8_4,

A tool for the study of these phenomena is the equation for the filtered second order structure function. This generalized Kolmogorov equation allows to describe the scale-dependent dynamics when spatial fluxes and inhomogeneity are present. Furthermore, it is possible to quantify how the framework of the classical energy cascade is modified by anisotropy and inhomogeneity. This is a crucial point in a context of subgrid energy transfer modeling.

2 Analysis of the Direct Numerical Simulation data base

The present data set is a turbulent channel flow at $Re_\tau = u_\tau h/\nu = 300$, where u_τ is the friction velocity, ν is the kinematic viscosity and h is half the channel height. The simulations have been carried out with a pseudo-spectral code. Details of the numerical scheme can be found in [4]. The computational domain is $2\pi h \times 2h \times \pi h$ with $512 \times 193 \times 256$ grid points respectively, corresponding to a resolution in wall units in the homogeneous directions of $\Delta x^+ = \Delta z^+ = 3.64$, suited to investigate the velocity field up to the dissipative scale.

Energy transfer, production and dissipation are the relevant processes in turbulence. They take place in various ranges of scales and, for inhomogeneous flows, may change appreciably in different flow regions. As a consequence, a full understanding of these interacting phenomena requires a detailed description of the processes occurring simultaneously in physical and scale space. To analyze the energy content of a given scale as function of the spatial position, we study the second order structure function $\langle \delta u^2 \rangle$, where $\delta u^2 = \delta u_i \delta u_i$ and $\delta u_i = u_i(x_s + r_s) - u_i(x_s)$ is the fluctuating velocity increment. It can be thought as a scale energy at $r = \sqrt{r_s r_s}$. It is a function of the separation vector r_i and of the mid-point $X_{ci} = 1/2(x_i' + x_i)$, allowing to describe the scale-dependent energy processes in the presence of inhomogeneity. The governing equation of $\langle \delta u^2 \rangle$ in wall flows is the generalized Kolmogorov equation [5], which for a flow with longitudinal mean velocity $U(y)$ reads

$$\frac{\partial \langle \delta u^2 \delta u_i \rangle}{\partial r_i} + \frac{\partial \langle \delta u^2 \delta U \rangle}{\partial r_x} + 2\langle \delta u \delta v \rangle \left(\frac{dU}{dy} \right)^* + \frac{\partial \langle v^* \delta u^2 \rangle}{\partial Y_c} =$$
$$-4\langle \varepsilon^* \rangle + 2\nu \frac{\partial^2 \langle \delta u^2 \rangle}{\partial r_i \partial r_i} - \frac{2}{\rho} \frac{\partial \langle \delta p \delta v \rangle}{\partial Y_c} + \frac{\nu}{2} \frac{\partial^2 \langle \delta u^2 \rangle}{\partial {Y_c}^2} \qquad (1)$$

where an asterisk denotes a mid-point average, i.e. $u_i^* = (u_i(x_s') + u_i(x_s))/2$ and the Y_c dependence is associated with inhomogeneity. A numerical analysis of equation 1 was performed in [3] and in the following paragraphs for the sake of clarity we will report a similar analysis performed on this new dataset. The generalized equation 1 can also be rewritten as

$$\nabla_r \cdot \Phi_r(\mathbf{r}, Y_c) + \frac{d}{dY_c} \Phi_c(\mathbf{r}, Y_c) = s(\mathbf{r}, Y_c) \qquad (2)$$

where $\Phi_r = \langle \delta u^2 \delta \mathbf{u} \rangle + \langle \delta u^2 \delta U \rangle - 2\nu \nabla_r \langle \delta u^2 \rangle$, $\Phi_c = \langle v^* \delta u^2 \rangle + 2\langle \delta p \delta v \rangle/\rho - \nu d\langle \delta u^2 \rangle / 2dY_c$ and $s = 2\langle \delta u \delta v \rangle (dU/dy)^* - 4\langle \varepsilon^* \rangle$. Two kind of scale-energy fluxes

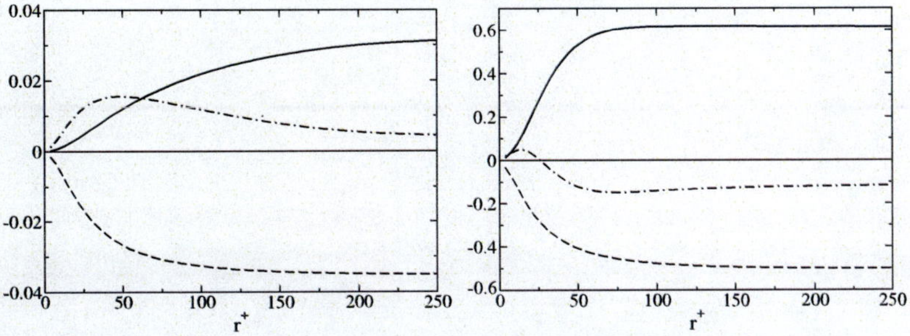

Fig. 1. Scale energy balance 3 in the log-layer $y^+ = 160$ (left) and in the buffer layer $y^+ = 20$ (right). $-\Pi_e$ (solid line), E_e (dashed line) and $-T_r$ (dashed-dotted line).

appear, namely Φ_r which identifies the transfer through the scales and Φ_c which is the flux of scale energy in physical space. These fluxes balance with a source term s which accounts for energy production and dissipation.

To highlight the scale processes , it is useful to consider the r-averaged form of 1. In a channel flow however the integration over a ball of radius r is unfeasible, hence the r-average is performed on a square domain of side r belonging to wall parallel planes. The r-averaged form of 1 follows as

$$T_r + \Pi + T_c = E + D_r + P + D_c \tag{3}$$

where T_r, D_r are the contributions to the scale transfer due to the inertial fluctuations and viscous diffusion. T_c, D_c and P are the inhomogeneous contributions to the spatial flux related to the inertial fluctuations, viscous diffusion and pressure-velocity correlation. Π is the energy production by mean shear and E is the dissipation. It is useful to group together some terms of 3 in a sort of effective production, $\Pi_e = \Pi + T_c - P$ and modified dissipation rate, $E_e = E + D_r + D_c$. Therefore, the r-averaged balance can be rewritten as $\Pi_e(r,Y_c) + T_r(r,Y_c) = E_e(r,Y_c)$, whose analysis permits us to characterize the different regions of the channel in term of scale by scale dynamics.

In the logarthmic layer, the large-scale production range is followed by a range dominated by the inertial energy cascade which is closed by dissipation, see left plot of Fig. 1. A cross-over scale l_c which splits scales into an inertial range at small r from a production dominated range at large r is identified as $\Pi(l_c,Y_c) = T_r(l_c,Y_c)$. l_c is dimensionally related to the shear scale $L_s = \sqrt{\varepsilon/|S^3|}$ which is found crucial for the small scale isotropy recovery [6] and for the subgrid scale stresses modelization [7]. Instead, in the buffer layer a direct scale-energy cascade at small scales and an inverse cascade at large scales exists see also right plot of Fig. 2. More details about the reverse energy cascade can be found in [3] [2]. Such process can be related to the dynamics of the coherent structures which live in this region and are responsible to build up Reynolds stresses with regeneration cycle of large structures and breakdown to small scales.

In a context of Large Eddy Simulation, the so-called *subgrid dissipation* is interpreted as the rate of energy flux between scales in the inertial range. Therefore,

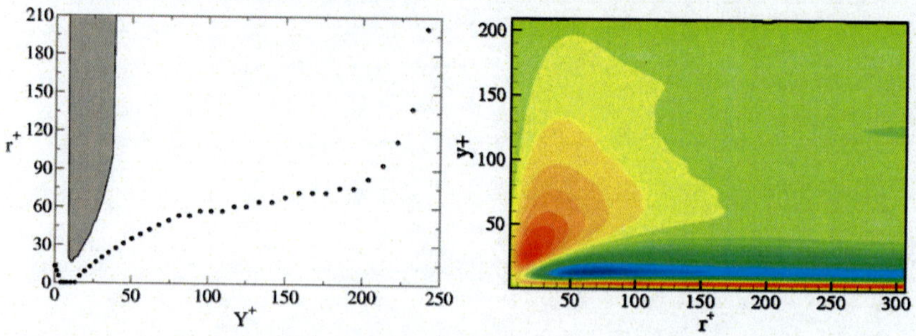

Fig. 2. Left: characteristics lenght scales in the channel; the cross-over scale l_c (circle) and the reverse energy cascade plane Ω_B (grey region). Right: energy cascade, $T_r(r^+, y^+)$.

most of the LES models attempt to reproduce the energy cascade assuming isotropy and homogeneity at small scales of any flows. In wall-turbulence the scale dynamics is dominated by the spatial energy flux and production by mean shear and these processes strongly modify the energy cascade up to a reverse energy cascade in the buffer layer. Therefore, the assumption of a inertial range without production and spatial fluxes drops out when dealing with wall flows and should be taken into account in LES models.

In order to rationalize the present scenario it is suitable to identify the various regions of the (r^+, Y_c^+)-plane where the relevant processes take place. Firstly the curve $l_c(y^+)$ identifies a production dominated region at large r^+. Second, the region Ω_B of the plane (r, y) where the energy cascade term changes sign leading to a reverse energy cascade from small to large scales. The reverse energy cascade will be shown responsible of the bakward energy transfer observed in Large Eddy Simulation. This phenomena leads to the opposite energy exchange between subgrid and resolved scales usually reproduced by the LES models which are based on the classical Richarson energy cascade. Both l_c and Ω_B are shown in the left plot of Fig. 2 as function of the wall distance.

3 The Kolmogorov equation for filtered velocity field

The analysed data are filtered with respect to the wall-parallel directions, using a sharp cutoff filter in wavenumber space, whose corresponding scales in the homogeneous directions are equal to $l_F^+ = 20, 30, 60$. This filter allows to divide the turbulent kinetic energy into the sum of the resolved energy and subgrid energy without the cross-product, i.e. $\langle u^2 \rangle = \langle \bar{u}^2 \rangle + \langle u_{sgs}^2 \rangle$ and $2\langle \bar{u} u_{sgs} \rangle = 0$ where $\langle \cdot \rangle$ stands for average in the homogeneous directions. The evolution equation of the resolved turbulent kinetic energy, $\langle \bar{u}_i^2 \rangle$, is

$$\frac{1}{2}\frac{d}{dy}\langle \bar{u}_i^2 \bar{v} \rangle + \langle \bar{u}\bar{v} \rangle \frac{dU}{dy} = -\frac{1}{\rho}\frac{d}{dy}\langle \bar{p}\bar{v} \rangle + \frac{\nu}{2}\frac{d^2}{dy^2}\langle \bar{u}_i^2 \rangle - \langle \bar{\varepsilon} \rangle - \frac{d}{dy}\langle \bar{u}_i \tau_{i,2} \rangle - \langle \varepsilon_{sgs} \rangle \quad (4)$$

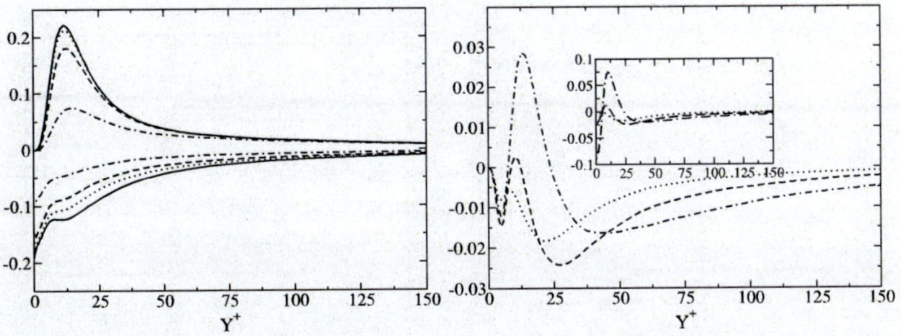

Fig. 3. Left: y-behavior of production and dissipation. Right: subgrid dissipation $\langle \varepsilon_{sgs} \rangle$ and in the inset $[(P-\bar{P})-(D-\bar{D})]$. DNS (solid line), $l_F^+ = 20$ (dotted line), $l_F^+ = 30$ (dashed line) and $l_F^+ = 60$ (dashed-dotted line).

where $\tau_{ij} = \overline{u_i u_j} - \bar{u}_i \bar{u}_j$ are the subgrid stresses, $\bar{\varepsilon} = \nu(\partial \bar{u}_i/\partial x_j)(\partial \bar{u}_i/\partial x_j)$ is the resolved viscous pseudo-dissipation, $\varepsilon_{sgs} = -\tau_{ij}\bar{S}_{ij}$ is the *subgrid dissipation* with $\bar{S}_{ij} = 1/2(\partial \bar{u}_i/\partial x_j + \partial \bar{u}_j/\partial x_i)$ the resolved strain-rate tensor.

In analogy with 1, the evolution equation of filtered second order structure function, $\langle \delta \bar{u}^2 \rangle = \delta \bar{u}_i \delta \bar{u}_i$, for a turbulent channel flow reads as

$$\frac{\partial \langle \delta \bar{u}^2 \delta \bar{u}_i \rangle}{\partial r_i} + \frac{\partial \langle \delta \bar{u}^2 \delta U \rangle}{\partial r_x} + 2\langle \delta \bar{u} \delta \bar{v} \rangle \left(\frac{dU}{dy} \right)^* + \frac{\partial \langle \bar{v}^* \delta \bar{u}^2 \rangle}{\partial Y_c} = -4\langle \varepsilon^* \rangle$$

$$+2\nu \frac{\partial \langle \delta \bar{u}^2 \rangle}{\partial r_i \partial r_i} - \frac{2}{\rho} \frac{\partial \langle \delta \bar{p} \delta \bar{v} \rangle}{\partial Y_c} + \frac{\nu}{2} \frac{\partial^2 \langle \delta \bar{u}^2 \rangle}{\partial {Y_c}^2} - 4\langle \varepsilon^*_{sgs} \rangle - 4 \frac{\partial \langle \tau^*_{ij} \delta \bar{u}_i \rangle}{\partial r_j} - \frac{\partial \langle \delta \tau_{i2} \delta \bar{u}_i \rangle}{\partial Y_c}. \quad (5)$$

Equation 5 allows to analyze how the resolved processes change as function of l_F in different regions and scale range and to appreciate the subgrid stresses both in physical and scale space. The three new terms represent the exchange of energy between grid and subgrid scales, a redistribution of resolved scale energy in the spectrum of scales and in physical space, respectively.

The r-averaged form of equation 5 follows as,

$$\bar{T}_r + T_r^{sgs} + \bar{\Pi} + \bar{T}_c + T_c^{sgs} = \bar{E} + \bar{D}_r + \bar{P} + \bar{D}_c + E_{sgs} \quad (6)$$

where together with the processes of Eq. 3 the effects of subgrid stresses are the redistribution of energy in scales, T_r^{sgs}, and physical space, T_c^{sgs} and the draining or sourcing of resolved energy, E_{sgs}. It is useful again to group together some terms and the r-averaged balance 6 follows as

$$\left(\bar{\Pi}_e + T_c^{sgs}\right) + \left(\bar{T}_r + T_r^{sgs}\right) = \left(\bar{E}_e + E_{sgs}\right). \quad (7)$$

In the following, a detailed analysis of the scale balance 6 is performed. Energy transfer in the resolved motion will be discussed and emphasis will be spend on the action of the subgrid stresses on the resolved motion in both physical and scale space, underlining the prominent role of the filter scale compared to the cross-over scale l_c and the region of reverse energy cascade $\Omega_B(r, y)$.

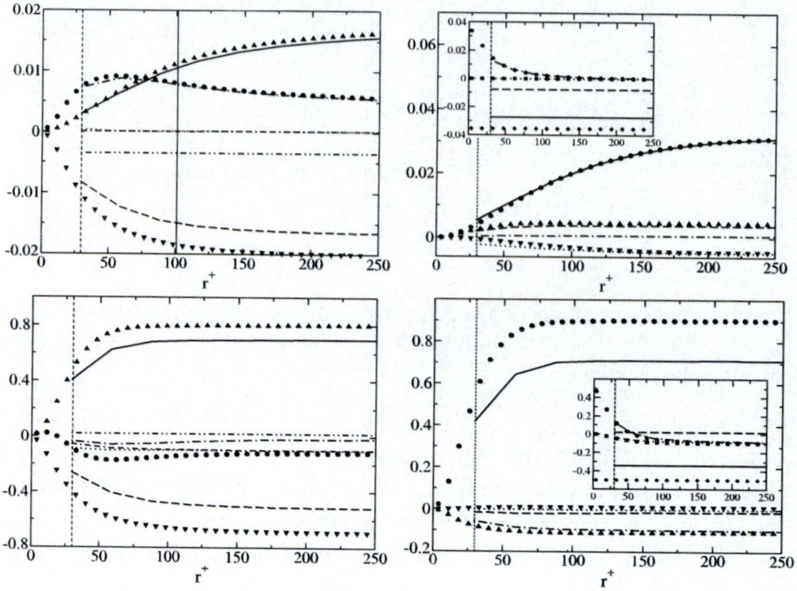

Fig. 4. Scale-energy balance at $y^+ = 160$ (top) and $y^+ = 12$ (bottom), for $l_F^+ = 30$. Left: terms of 7. Filtered field: $\bar{E}_e$ dashed line, $-\bar{\Pi}_e$ solid line, $-\bar{T}_r$ dashed dotted line, $-\bar{T}_r^{sgs}$ dotted line, $-\bar{T}_c^{sgs}$ dashed-dashed-dotted line and $\bar{E}_{sgs}$ dashed-dotted-dotted line. Unfiltered field: E_e triangle down, $-\Pi_e$ triangle up and $-T_r$ circle. Right: effective production. Filtered field: $-\bar{\Pi}$ solid line, $-\bar{P}$ dotted line, $-\bar{T}_c$ dashed line and $-T_c^{sgs}$ dashed-dotted line. Unfiltered field: $-\Pi$ circle, $-P$ triangle down and $-T_c$ triangle up. Inset: effective dissipation. Filtered field: $\bar{E}$ solid line, E_{sgs} dashed line, $\bar{D}_r$ dashed-dotted line and $\bar{D}_c$ dotted line. Unfiltered field: E diamonds, D_r circle and D_c square. The vertical solid and dashed line are l_c and l_F.

In the left panel of Fig. 3 the resolved energy production by mean shear $\langle \bar{u}\bar{v} \rangle dU/dy$ and the resolved viscous dissipation $\langle \bar{\varepsilon} \rangle$ are shown for the three values of the filter length l_F^+. In the core flow, the former remains unaltered and a depletion of the latter at increasing filter scale is observed. Whereas, approaching the wall both these quantities decrease with l_F^+, indeed, as the shear scale diminishes, $L_s \approx ky^+$, the production involves more and more the subgrid scales. The right panel of Fig. 3 shows $\langle \varepsilon_{sgs} \rangle$ at the same values of l_F^+. From the inspection of the plot the role of the subgrid scale energy dissipation is deduced. Resolved turbulent kinetic energy is drained in the core flow and in the viscous sublayer while in the buffer layer, for the larger filter scales, it becomes opposite in sign in an increasing region, implying that the subgrid scales feed the large scale of motion, i.e. a backward energy transfer occurs.

The backward energy transfer can be related to the excess of turbulent energy in the subgrid scales. Let us consider, $[(P - \bar{P}) - (D - \bar{D})]$, where $\bar{P} = \langle \bar{u}\bar{v} \rangle dU/dy$, $\bar{D} = \bar{\varepsilon}$ and $P = \langle uv \rangle dU/dy$, $D = \varepsilon$. This quantity represents somehow the net balance of energy production and dissipation acting in the subgrid scales. From the inset in the right panel of Fig.3, it is observed that dissipation is dominant in the subgrid scales in the core flow and the viscous sublayer, whereas, in the buffer layer, production overcomes the viscous effects. The corrispondance between the region where $[(P-\bar{P})-(D-\bar{D})]>0$ and the region where $\varepsilon_{sgs}>0$ is remarkable.

The scale energy balance 6 in the log-layer is shown in the top part of Fig. 4. At this location the filter scale is always smaller than the cross-over scale, $l_F<l_c$, see 2, and the subgrid scale dynamic reduces to the Richardson energy cascade. Such condition leads to a preservation of the statistical properties of the filtered field, i.e. $E_e \approx \bar{E}_e + E_{sgs}$, $\Pi_e \approx \bar{\Pi}_e + T_c^{sgs}$, $T_r \approx \bar{T}_r + T_r^{sgs}$. Indeed, for $l_F<l_c$ the energy production in the subgrid scales is negligible and the resolved energy production equals the unfiltered process, $\Pi \approx \bar{\Pi}$. Therefore, the rate of energy exchange between resolved and subgrid motion, E_{sgs}, is exclusively determined by the viscous dissipation in the subgrid scales, implying that $E \approx \bar{E} + E_{sgs}$. The role of the filter position in the log-layer is exploited in the top panels of Fig. 5. The same phenomenology is observed also in the bulk of the flow, not shown here, where $l_c \to \infty$.

The budget in the buffer layer is reported for $y^+=12$, corresponding to the peak of the production. In this region the cross-over scale is very small and the filter scale is always larger than l_c. Furthermore the filter scale lies in a range of scales where a reverse energy transfer occurs, see Ω_B in 2.

From the bottom of Fig. 4, it is shown that a large part of the production acts in the subgrid scales, $\bar{\Pi} \ll \Pi$. Indeed, approaching the wall l_c reaches the Kolmogorov scale η, as shown in Fig.2 and the rate of energy exchange between grid/subgrid motion, E_{sgs}, is not determined by the viscous dissipation, but by the balance between production and dissipation in the subgrid scales, therefore $\bar{E}+E_{sgs} \ll E$. The resolved field cannot reproduce the unfiltered dynamics because most of the processes act in the subgrid scales and, in the buffer layer, the role of E_{sgs} is strongly modified up to a backward energy transfer, $E_{sgs}>0$. The position of filter scale for $y^+=12$ lies in the reverse energy cascade region, Ω_B, leading to an energy flux from subgrid to resolved scales. In this range the nonlinear interactions in the resolved scales do not reproduce the energy flux across scales, T_r, and the spatial flux T_c. Whereas, the subgrid stresses significantly act in the energy redistribution with T_r^{sgs} and T_c^{sgs} meaning that most of the resolved processes depends on the scales below l_F^+ that, in the wall region, are those responsible for the coherent structures. Infact, from the bottom plots of Fig. 5, it is interesting to observe that the role of T_r^{sgs} and T_c^{sgs} for $l_F^+=60$ is very significant to capture the energy redistribution mechanisms, while for $l_F^+=30$ is negligible. Indeed, only for $l_F^+=60$ the subgrid scales lies in the reverse energy cascade region, Ω_B. The effects of nonlinear interactions T_r^{sgs} are shown in the left plots of Fig. 6 as function of the wall-distance, where also T_r and $\bar{T}_r$ appear. In the core flow the transfer is only due to the nonlinear interactions in the resolved scales, $\bar{T}_r$, while T_r^{sgs} is negligible, also for other scales and filters considered. Whereas approaching the wall, the subgrid interactions, T_r^{sgs}, becomes more and more significant. This analysis performed for the spatial fluxes T_c, $\bar{T}_c$ and T_c^{sgs} lead to similar conclusions.

The common interpretation of $E_{sgs}=-\tau_{ij}\bar{S}_{ij}$ is a measure of the rate of energy transfer between scales larger and smaller than l_F^+. Indeed, from the right plot of Fig. 6 showing $\langle E_{sgs}\rangle$ for different filter lengths and wall distances, it is noticeable that as l_F approaches η, $\langle E_{sgs}\rangle$ decreases since a larger fraction of dissipation occurs

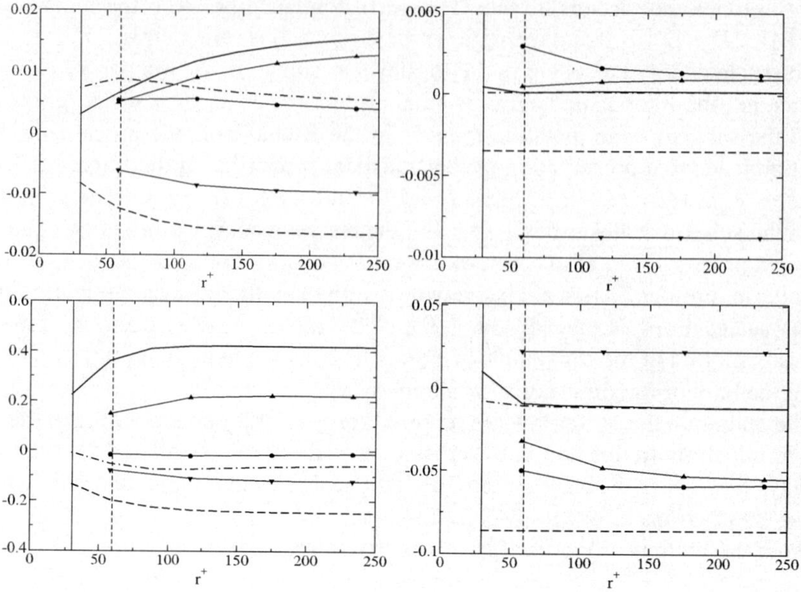

Fig. 5. Filter scale effects at $y^+ = 220$ (top) and $y^+ = 25$ (bottom). Left: $l_F^+ = 30$: $-\bar{\Pi}_e$ solid line, $\bar{E}_e$ dashed line and $-\bar{T}_r$ dashed-dotted line. $l_F^+ = 60$: $-\bar{\Pi}_e$ triangle up, $\bar{E}_e$ triangle down and $-\bar{T}_r$ circle. Right: $l_F^+ = 30$: $-T_r^{sgs}$ solid line, $-T_c^{sgs}$ dashed-dotted line and E_{sgs} dashed line. $l_F^+ = 60$: $-T_r^{sgs}$ circle, $-T_c^{sgs}$ triangle up and E_{sgs} triangle down.

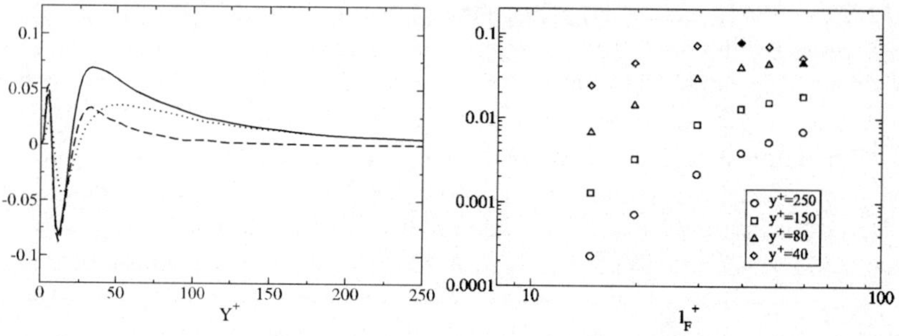

Fig. 6. Left: y-behaviour of T_r (solid line), $\bar{T}_r$ (dotted line) and T_r^{sgs} (dashed line) for $r^+ = 30$ and $l_F^+ = 30$. Right: $E_{sgs}(l_F)$ at different wall distances (right). The filled symbols correspond to $l_F/l_c = 1$.

due to the resolved flow. When l_F/η is large, $\langle E_{sgs} \rangle$ increases monotonically up to approach the crossover scale, $l_F/l_c \approx 1$ where a maximum is observed. At higher Reynolds number a plateau is expected whose extension measures the amplitude of the inertial range.

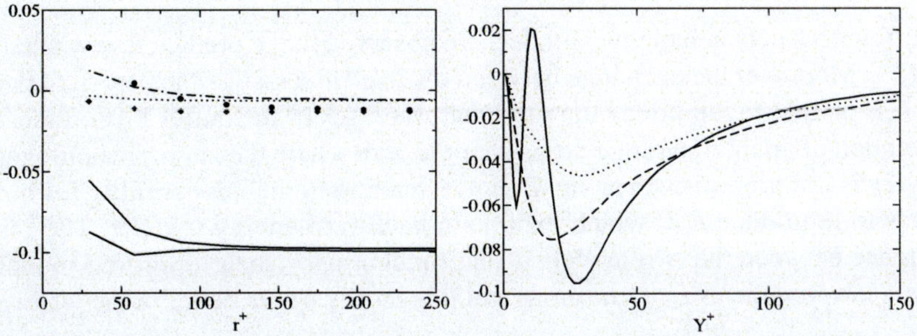

Fig. 7. Right: y-behavior of E_{sgs} estimated with exact subgrid stresses (solid line) and with Smagorinsky (dashed line) and Smagorinsky shear-improved (dotted line). Left: $\bar{T}_r$ and T_r^{sgs} evaluated with the Smagorinsky (symbols), Smagorinsky shear-improved (lines) and with the filtered DNS (solid lines). $l_F^+ = 30$

4 Assessment of the LES models

In this section, the filtered Kolmogorov equation 5 will be presented as a new tool for the assessment of LES models. According to the previous analysis, LES models should reproduce a backward energy transfer in the wall region if l_F^+ lies in Ω_B. Hence, it is expected that dissipative models are not able to capture the correct subgrid stresses. Nonetheless, well-known isotropic eddy-viscosity models will be analysed. Moreover, a sharp Fourier cutoff in the homogeneous directions will be used even if this is a very anisotropic filter.

Accordingly to the idea of isotropic recovery at small scales of all type of flow for sufficiently large Reynolds number, the eddy viscosity models assume that the energy exchange between resolved and subgrid scales is similar to the viscous dissipation leading to subgrid stresses in the form, $\tau_{ij} - 1/3\tau_{kk}\delta_{ij} = -2\nu_T\bar{S}_{ij}$. Infact, when isotropic conditions are recovered, equation 1 leads to an energy exchange between resolved and subgrid scales,

$$\left.\frac{\partial\langle\delta u^2\delta u_i\rangle}{\partial r_i}\right|_{l_F} = -4\nu\langle S_{ij}S_{ij}\rangle \tag{8}$$

which is similar to E_{sgs} evaluated with eddy-viscosity, i.e. $E_{sgs} = -8\nu_T\langle\bar{S}_{ij}\bar{S}_{ij}\rangle$.

In this context, the classical Smagorinsky model assumes that $\nu_T = (C_s l_F)^2|\bar{S}|$ producing a non-zero eddy-viscosity in laminar flow regions. To alleviate these deficiencies for wall-bounded flows, the Smagorinsky constant C_s is often multiplied by a damping factor depending on the wall-normal distance. In other models like Smagorinsky shear-improved [8], the eddy-viscosity takes into account mean-shear effects and naturally decreases to zero at the wall, i.e. $\nu_T = (C_s l_F)^2(|\bar{S}| - |\langle\bar{S}\rangle|)$. These two models will be here considered.

The predicted subgrid stresses for $l_F^+ = 30$ lead to a redistribution effect in physical and scale space, T_r^{sgs} and T_c^{sgs}, which are always negligible also in the buffer when l_F is in the region of the reverse energy transfer Ω_B, see left panel of Fig. 7.

No differences between the two models are observed in the predicted very small T_r^{sgs} and T_c^{sgs}. Moreover the exchange between resolved and subgrid scales, right panel of Fig. 7, is unable to reproduce the physical processes in the buffer layer. Indeed, the assumption of an inertial range drops near the wall where the isotropic homogeneous recovery is not present even at small scales, leading to the impossibility for both the models to reproduce E_{sgs} which exhibits a backward energy transfer. The relevant difference between the two models is that Smagorinsky shear-improved model estimates a lower value of E_{sgs}, meaning that this model works better in the wall region.

5 Conclusions

A generalized Kolmogorov equation specialized for filtered velocity field has been applied to filtered DNS data and to LES models. The results single out the prominent role of the filter scale compared to the cross-over scale, l_c, and the reverse cascade region, Ω_B, and the incapability of the eddy-viscosity models to reproduce the filtered near-wall physics. The present tool appears to be appropriate to test the validity of LES models to capture the relevant processes occurring in wall-turbulence both in physical and scale space.

References

1. Härtel C, Kleiser L, Unger Fi, Friedrich R (1994) Phys. Fluids 6:3130–3143
2. Domaradzki JA, Liu W, Härtel C, Kleiser L (1994) Phys. Fluids 6:1583–1599
3. Marati N, Casciola CM, Piva R (2004) J. Fluid Mech. 521:191–215
4. Lundbladh A, Henningson DS, Johansson AV (1992) FFA-TN-28 - Sweden
5. Hill RJ (2002) J. Fluid Mech. 468:317–326
6. Casciola CM, Gualtieri P, Jacob B, Piva R (2005) Phys. Rev. Lett. 95:024503(4)
7. Gualtieri P, Casciola CM, Benzi R, Piva R (2007) J. Fluid Mech. 592:471–494
8. Lévêque E, Toschi F, Shao L, Bertoglio JP (2007) J. Fluid Mech. 570:491–502

A computational study of turbulent flow separation for a circular cylinder using skin friction boundary conditions

Johan Hoffman[1] and Niclas Jansson[2]

[1] Computational Technology Laboratory, Computer Science and Communication, KTH, SE-10044 Stockholm, Sweden. jhoffman@kth.se
[2] Computational Technology Laboratory, Computer Science and Communication, KTH, SE-10044 Stockholm, Sweden. njansson@kth.se

Summary. In this paper we present a computational study of turbulent flow separation for a circular cylinder at high Reynolds numbers. We use a stabilized finite element method together with skin friction boundary conditions, where we study flow separation with respect to the decrease of a friction parameter. In particular, we consider the case of zero friction corresponding to pure slip boundary conditions, for which we observe an inviscid separation mechanism of large scale streamwise vortices, identified in our earlier work. We compare our computational results to experiments for very high Reynolds numbers. In particular, we connect the pattern of streamwise vorticity in our computations to experimental findings of spanwise 3d cell structures reported in the literature.

Key words: Turbulent boundary layer, flow separation, General Galerkin method, a posteriori error estimation, adaptive finite element method, skin friction boundary conditions

1 Introduction

The choice of boundary conditions at a solid wall is critical for accurate modeling of turbulent flow, in particular to capture flow separation phenomena. Since full resolution of a turbulent boundary layer is out of reach, the standard way to handle the problem is to divide the computational domain into: (i) an interior part, and (ii) a boundary layer. In the boundary layer a simplified model of the flow is used to provide boundary conditions to the equations to be solved in the interior part. Boundary conditions may be in the form of velocities or stresses, and the coupling between (i) and (ii) may be one-way from (ii) to (i), or more closely coupled. Boundary layer models are developed based on experimental data (e.g. velocity profiles or skin friction stresses), theory or computation (in a multiscale framework). Here we note in particular Detached-eddy simulation [21] used for high Reynolds number separated flow. There are also approaches relying on the numerical method to handle boundary layer modeling, e.g. by a weak implementation of the no slip velocity boundary conditions [16, 12]. For an overview of boundary layer modeling, see [1, 17].

M.V. Salvetti et al. (eds.), *Quality and Reliability of Large-Eddy Simulations II*,
ERCOFTAC Series 16, DOI 10.1007/978-94-007-0231-8_5,

Viscous effects in the boundary layer are traditionally used as a mechanism to explain flow separation, not only for low Reynolds numbers *Re* but also for high *Re* where otherwise inertial effects dominate [19]. In particular, viscous effects in the boundary layer are often presented as the resolution of the d'Alembert paradox [20], seemingly disqualifying the inviscid Euler equations with slip boundary conditions as a model for high *Re* flow. The significance of the boundary layer for explaining turbulent flow separation was questioned in [11], where instead a mechanism for inviscid separation was suggested based on exponential growth of streamwise vorticity at separation. In particular, a new resolution of the d'Alembert paradox was presented based on this instability of potential solutions at separation.

In [6, 11] preliminary computations for a circular cylinder showed that using a stabilized finite element method, which we refer to as a General Galerkin (G2) method, with slip boundary conditions, thus without the boundary layer, flow separation was observed with significant drag, in qualitative agreement with the high *Re* experimental data available in the literature [22]. In this paper we follow up the predictions of [11] for a circular cylinder at high *Re*, in the form of a computational study where the goal is to (i) verify the results in [11] using two different adaptive G2 methods, and (ii) to quantitatively determine drag for a circular cylinder with slip boundary conditions. We use adaptive mesh refinement based on quantitative a posteriori error estimation, taking into consideration both numerical errors from discretization and modeling errors from numerical stabilization. We note that using slip boundary conditions corresponds to an approximation of small skin friction by zero skin friction, and thus in particular this small skin friction is not part of the model and cannot be determined.

If indeed a slip boundary condition, without the boundary layer, is a good model for high *Re* flow separation, this would mean a breakthrough for turbulence simulation, which would open for new advanced simulations in aero- and hydrodynamics. We conclude the paper by comparing the computational results using slip boundary conditions with experimental results for very high *Re*, where in particular we connect the pattern of streamwise vorticity in our computations to experimental findings of spanwise 3d cell structures.

2 Computational model

As the basic model for incompressible Newtonian fluid flow we consider the Navier–Stokes equations (NSE) with constant kinematic viscosity $\nu > 0$, enclosed in $\Omega \subset \mathbb{R}^3$ over a time interval $I = (0,T]$:

$$\begin{aligned} \dot{u} + (u\cdot\nabla)u + \nabla p - \nu\Delta u = f, &\quad (x,t) \in \Omega \times I, \\ \nabla \cdot u = 0, &\quad (x,t) \in \Omega \times I, \\ u(x,0) = u^0(x), &\quad x \in \Omega, \end{aligned} \tag{1}$$

with $u(x,t)$ the velocity vector and $p(x,t)$ the pressure, $u^0(x)$ initial data and $f(x,t)$ a body force. The stress tensor $\sigma_{ij} = -\nu\varepsilon_{ij}(u) + p\delta_{ij}$, with strain rate tensor $\varepsilon_{ij}(u) =$

$1/2(\partial u_i/\partial x_j + \partial u_j/\partial x_i)$ and δ_{ij} the Kronecker delta function, and the relative importance of viscous and inertial effects in the flow is determined by the Reynolds number $Re = UL/\nu$, where U and L are characteristic velocity and length scales.

The mathematical theory of NSE for small ν does not guarantee existence of classical solutions, and in particular there are many open problems regarding the limit $\nu \to 0$. For $\nu = 0$ the system of equations (1) is referred to as the Euler equations, for which particular solutions can be constructed under certain assumptions (e.g. potential solutions), but no general existence result is known.

For the turbulence simulations in this paper we use cG(1)cG(1), a G2 method with linear interpolation in space and time described e.g. in [9] and demonstrated for turbulence benchmark problems in [4, 10, 5, 8]. cG(1)cG(1) is an adaptive finite element method with a posteriori error control, without any large eddy simulation filtering of NSE or subgrid models, but with a numerical stabilization dissipating kinetic energy proportional to the NSE residual.

For $\nu = 0$, cG(1)cG(1) solutions connects to the mathematical concepts of a suitable weak solution [18, 3], or a dissipative weak Euler solution [2], with dissipation from local non-smoothness of the solution, here in terms of residuals which are large for non-smooth solutions.

3 Turbulent boundary layers

In our work [6, 7, 11] on boundary layer modeling we have chosen to apply a skin friction stress as boundary condition, where we either use a weak implementation [14, 16, 6], or we apply the no penetration (slip) boundary condition strongly together with a weak implementation of the friction boundary condition. That is, we append the Navier-Stokes equations (1) with the following boundary conditions:

$$u \cdot n = 0, \tag{2}$$
$$u \cdot \tau_k + \beta^{-1} n^T \sigma \tau_k = 0, \quad k = 1,2, \tag{3}$$

for $(x,t) \in \Gamma_{solid} \times I$, with $n = n(x)$ an outward unit normal vector, and $\tau_k = \tau_k(x)$ orthogonal unit tangent vectors of the solid boundary Γ_{solid}. We use matrix notation with all vectors v being column vectors and the corresponding row vector is denoted v^T. The equation (3) corresponds to Navier's slip law where the friction parameter β can be related to a slip length, and if the tangent velocity $u \cdot \tau_k \sim 1$ then $\beta \sim F_f^h$, with F_f^h the skin friction stress of a computed solution on a mesh of size h.

With skin friction boundary conditions, the rate of kinetic energy dissipation in cG(1)cG(1) has a contribution of the form

$$\sum_{k=1}^{2} \int_0^T \int_{\Gamma_{solid}} |\beta^{1/2} u \cdot \tau_k|^2 \, ds \, dt,$$

from the kinetic energy which is dissipated as friction in the boundary layer. We note that since experimental studies indicate that skin friction vanish with increasing Re

(normalizing the velocity and length scales), we model $Re \to \infty$ by $\beta \sim F_f^h \to 0$, so that the dissipative effect of the boundary layer vanish with *Re*. This connects to the original idea of Euler of modeling high *Re* flow by an inviscid model with slip boundary conditions. Using a G2 method for high *Re* where numerical dissipation dominates the dissipation from the viscous term, we may as well drop the viscous term altogether so that we effectively have the Euler equations. The question is now if we for high *Re* similarly can drop the skin friction term at the boundary, resulting in the slip boundary condition of the Euler equations?

4 Experimental results

As the Reynolds number increases different regions of the flow past a cylinder undergoes transition to turbulence; the wake, shear layers and finally boundary layers, causing a delayed separation and the so called drag crisis. In his treatise of flow around circular cylinders [22], Zdravkovich refers to this high *Re* regime as the fully turbulent state, and notes an "almost total lack of data" for this regime. The reason is that to perform a high *Re* experiment either the length scale or the velocity need to be increased, and whereas the size of wind tunnels sets a limit for the length scale, compressibility effects limit the velocity that can be used while still considering the flow to be incompressible.

But even though the use of wind tunnel experiments is limited for flow with very high *Re*, this regime is the relevant one for a range of important applications; from aerodynamics of cars and airplanes, to civil, offshore and wind engineering, and geophysical flow.

Although limited, there are some experimental results available, from which conclusions can be drawn about the flow at these high *Re*, of particular interest here are the following observations [22]:

1. Transition to turbulence in the boundary layers causes a delayed separation, with associated decrease in mean drag of almost a factor 3; the so called drag crisis.
2. Associated with drag crisis is that the eddy formation in the wake is delayed about 2-3 diameters downstream.
3. Large scale three dimensional wake structures are formed, of the same size as the diameter, referred to as a spanwise cell pattern [15, 13] (with some pictures also reproduced at p.182 in [22]).

In the next section we present computational results where an increasing *Re* is modeled by decreasing the friction parameter β, with the goal of simulating drag crisis by reproducing the characteristics of 1-3 above, without resolving the turbulent boundary layers.

5 Computational results

We now report results from a computational study of the turbulent flow past a circular cylinder at high *Re*. In particular we assume *Re* to be high enough for the numerical dissipation in cG(1)cG(1) to dominate the dissipation from the viscous term in NSE, which is neglected in the computations. This assumption is consistent with a law of finite dissipation with kinetic energy dissipation independent of *Re* beyond a certain *Re*.

The computational domain is shown in Fig. 1, with $D = 0.1$, where we use an inflow boundary condition of constant unit inflow velocity from left to right, and a standard zero stress outflow boundary condition.

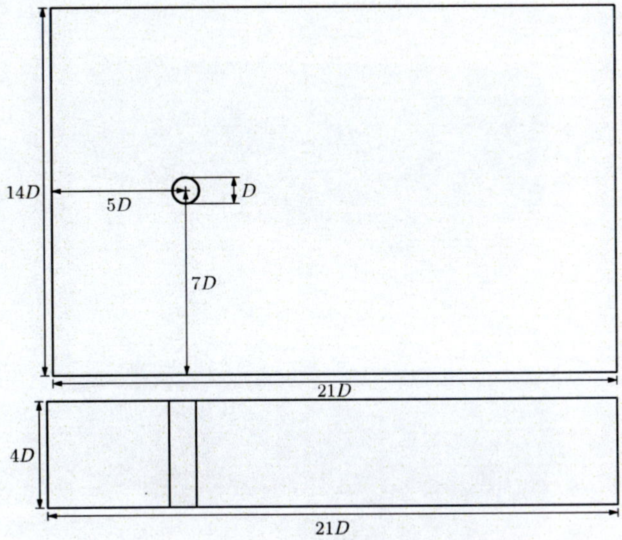

Fig. 1. The geometry of the test problem.

We compute over a time interval $I = (0, 50D)$, and to minimize the mesh dependency of the computational results, we employ two different mesh refinement strategies where we in each adaptive iteration refine 10% of the cells in the mesh based on the size of: (A) an error indicator for drag from duality based a posteriori error estimation, see e.g. [5], or (B) the L_2-norm of the mesh size times the residual of NSE for each cell.

In Fig. 2-4 we show the solution at final time $T = 50D$ for the finest mesh using mesh refinement criterion (A), in Fig. 5-6 we show the drag coefficient averaged over the time interval $[40D, 50D]$ for (A) and (B), and in Fig. 7 we show a long time simulation of drag (normalized as the drag coefficient) for the finest mesh corresponding to (A).

6 Discussion and summary

From the computational results we find that as the friction parameter β is reduced, corresponding to increasing *Re*, the separation of the flow is changing from a laminar

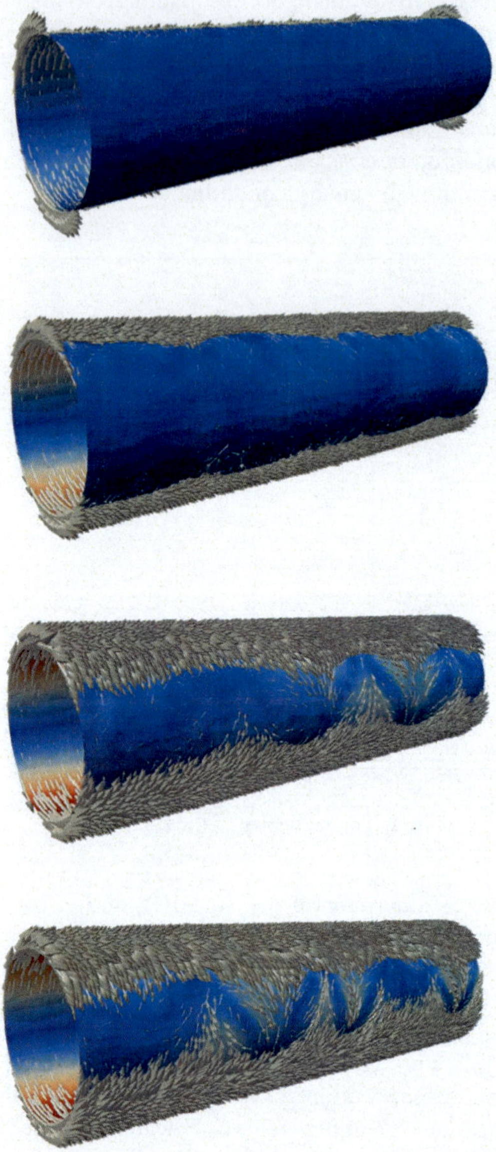

Fig. 2. Velocity vectors at surface of cylinder; for $\beta = 10^{-1}$, $\beta = 10^{-2}$, $\beta = 10^{-3}$ and $\beta = 0$ (from top to bottom).

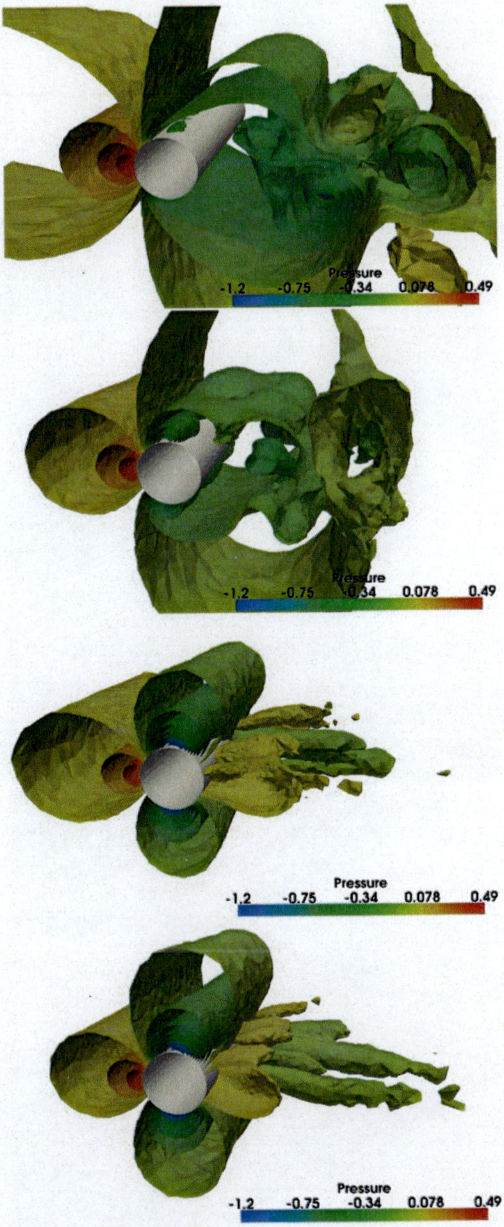

Fig. 3. Pressure isosurfaces; for $\beta = 10^{-1}$, $\beta = 10^{-2}$, $\beta = 10^{-3}$ and $\beta = 0$ (from top to bottom).

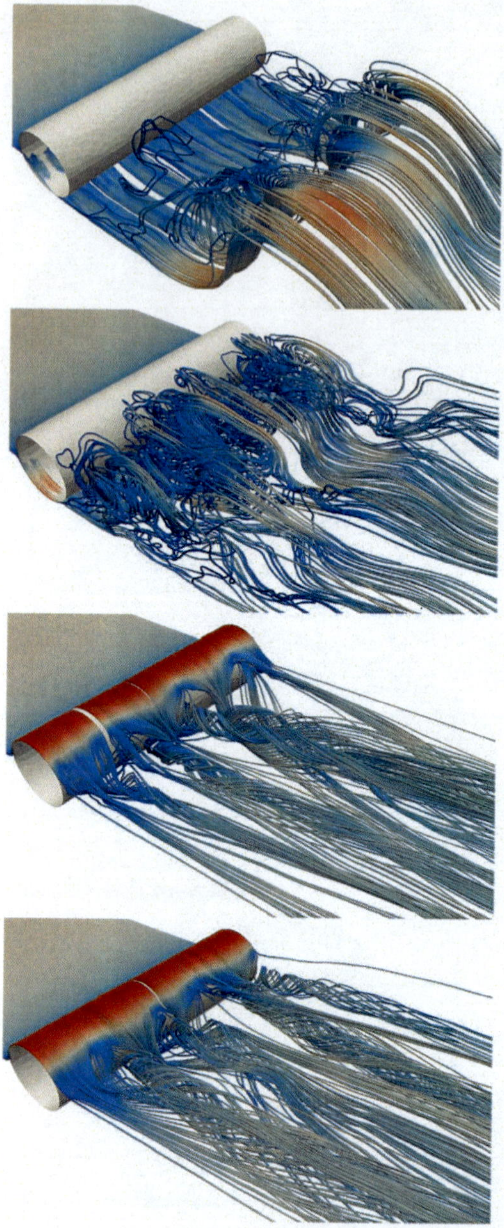

Fig. 4. Velocity streamlines; for $\beta = 10^{-1}$, $\beta = 10^{-2}$, $\beta = 10^{-3}$ and $\beta = 0$ (from top to bottom).

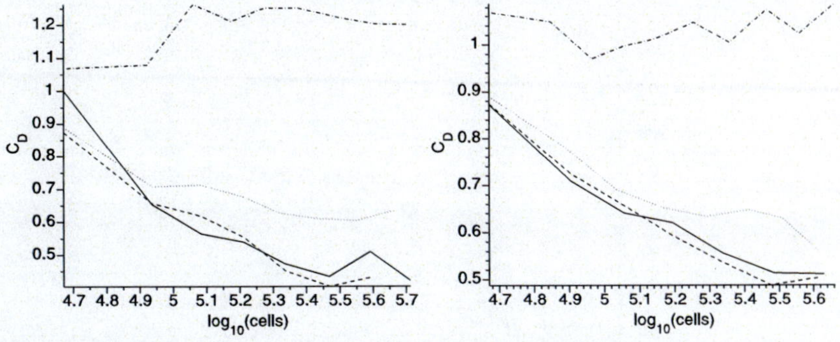

Fig. 5. Drag coefficient c_D for adaptively refined meshes; (A) duality based refinement (left) and (B) residual based refinement (right), with $\beta = 10^{-1}$ ('-.'), $\beta = 10^{-2}$ ('..'), $\beta = 10^{-3}$ ('–') and $\beta = 0$ ('solid').

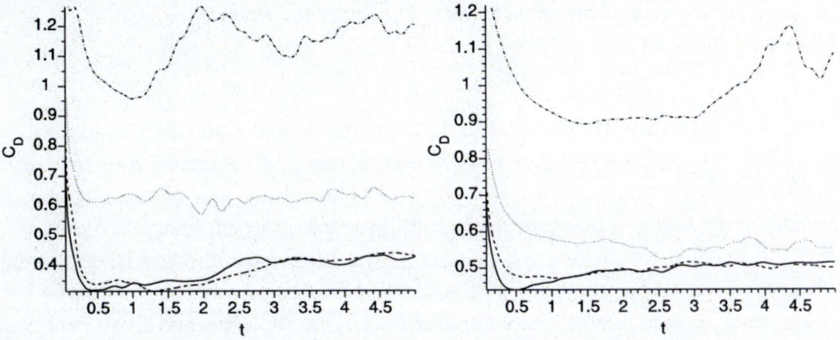

Fig. 6. Time series of drag (normalized as c_D) for the finest computational meshes; (A) duality based refinement (left) and (B) residual based refinement (right), with $\beta = 10^{-1}$ ('-.'), $\beta = 10^{-2}$ ('..'), $\beta = 10^{-3}$ ('–') and $\beta = 0$ ('solid').

separation at top and bottom of the cylinder, into an inviscid separation in streamwise vorticity described in [11], with drag from suction in the center of the streamwise vortices. We note that the streamwise vorticity rolls are arranged in a zig-zag pattern at the back of the cylinder, where the number of of vortex rolls appears to be independent of the mesh resolution, with a fixed distance of about the diameter of the cylinder. Further, we note little difference between the solutions for $\beta < 10^{-3}$. There is a striking resemblance of these vortex structures to the spanwise cell patterns reported from experiments [15, 13, 22]. The streamwise vorticity is also consistent with the experimental observation that eddy formation in the wake is delayed 2-3 diameters downstream [22].

Studying convergence of drag in Fig. 5 we find that for the case of $\beta = 10^{-1}$ with laminar separation; drag converges fast towards the expected value of $c_D \approx 1.1$, but for smaller β we cannot observe a clear convergence, even if the duality based refinement criterion (A) appears to show faster convergence than the residual based

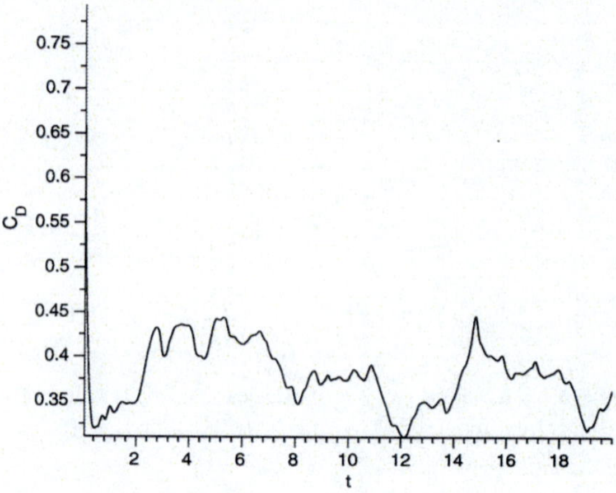

Fig. 7. Long time series of (normalized as c_D) drag for the finest computational mesh with duality based refinement (A) and $\beta = 0$.

refinement (B). To improve convergence, there are two obvious actions to take: (i) to further refine the mesh a number of iterations, and (ii) to increase the time interval over which drag is averaged. We are currently continuing this computational study on an IBM Blue Gene/L machine with 2048 processors to address (i)-(ii).

For the finest mesh of (A) we compute for a longer time to get better statistics, which gives a value for drag of $c_D \approx 0.4 \pm 10\%$, consistent with experimental results for drag crisis. This value of c_D is smaller than what we obtained from preliminary computations in [6, 11]. The computations in [6, 11] differs from the current computations in that one and the same mesh was used for all different β, whereas in this study the mesh was refined adaptively with respect to (A) and (B) for each β separately.

Although this is an ongoing investigation, with in particular more computational studies needed, the results of the computational study in this paper together with the linear stability analysis in [11], consistent with available experimental data, suggests a reinterpretation of the role of the boundary layer for turbulent flow separation. We are lead to the following conjectures:

- The boundary layer is not needed to explain flow separation for high Re, where instead inviscid separation in streamwise vorticity is dominating.
- We can model flow separation corresponding to the transition from a laminar to a turbulent boundary layer (drag crisis) by varying the friction parameter β in the simple skin friction boundary model (2)-(3).
- For many important applications in aero- and hydrodynamics, pure slip boundary conditions ($\beta = 0$) can give an accurate model of flow separation.

The boundary layer model (2)-(3) is not capable of predicting skin friction, but instead takes skin friction as data. The skin friction data can come from experiments, theory, or computation, but a key point is that the sensitivity of flow separation with respect to skin friction data appears to be small for high *Re*. The exact point of transition to turbulence in the boundary layer for high *Re* appears to be of minor importance for prediction of flow separation, as separation of a laminar boundary layer for high *Re* leads to an immediate reattachment as a turbulent boundary layer.

7 Acknowledgements

The authors would like to acknowledge the financial support from the Swedish Foundation for Strategic Research, and the Starting Grant of the European Research Council.

References

1. S. Deck, P. Sagaut, and M. Terracol, *Multiscale and multiresolution approaches in turbulence*, Imperial College Press, 2006.
2. J. Duchon and R. Robert, *Inertial energy dissipation for weak solutions of incompressible euler and navier-stokes solutions*, Nonlinearity, (2000), pp. 249–255.
3. J. L. Guermond and S. Prudhomme, *On the construction of suitable solutions to the navier-stokes equations and questions regarding the definition of large eddy simulation*, Physica D, 207(1-2) (2005), pp. 64–78.
4. J. Hoffman, *Computation of mean drag for bluff body problems using adaptive DNS/LES*, SIAM J. Sci. Comput., 27(1) (2005), pp. 184–207.
5. J. Hoffman, *Adaptive simulation of the turbulent flow past a sphere*, J. Fluid Mech., 568 (2006), pp. 77–88.
6. J. Hoffman, *Computation of turbulent flow past bluff bodies using adaptive general galerkin methods: drag crisis and turbulent euler solutions*, Comput. Mech., 38 (2006), pp. 390–402.
7. J. Hoffman, *Simulation drag crisis for a sphere using skin friction boundary conditions*, in proceedings for ECCOMAS CFD 2006, Egmond aan Zee, 2006.
8. J. Hoffman, *Efficient computation of mean drag for the subcritical flow past a circular cylinder using general galerkin g2*, Int. J. Numer. Meth. Fluids, 59(11) (2009), pp. 1241–1258.
9. J. Hoffman and C. Johnson, *Computational Turbulent Incompressible Flow: Applied Mathematics Body and Soul Vol 4*, Springer-Verlag Publishing, 2006.
10. J. Hoffman and C. Johnson, *A new approach to computational turbulence modeling*, Comput. Methods Appl. Mech. Engrg., 195 (2006), pp. 2865–2880.
11. J. Hoffman and C. Johnson, *Resolution of d'Alembert's paradox*, J. Math. Fluid Mech., (Published Online First at www.springerlink.com: 10 December 2008).
12. T. Hughes, Y. Bazilevs, C. Michler, and V. M. Cao, *Weak dirichlet boundary conditions for wall-bounded turbulent flows*, Comp. Meth. Appl. Mech. Engrng., 196 (2007), pp. 4853–4862.
13. J. S. Humphreys, *On a circular cylinder in a steady wind at transition Reynolds numbers*, Journal of Fluid Mechanics, 9(4) (1960), pp. 603–612.
14. V. John, *Slip with friction and penetration with resistance boundary conditions for the Navier–Stokes equations - numerical tests and aspects of the implementation*, J. Comp. Appl. Math., 147 (2002), pp. 287–300.

15. A. I. KOROTKIN, *The three dimensionality of the flow transverse to a circular cylinder*, Fluid Mechanics - Soviet Research, 5 (1976), p. 2.
16. W. LAYTON, *Weak imposition of no-slip boundary conditions in finite element methods*, Computers and Mathematics with Applications, 38 (1999), pp. 129–142.
17. P. SAGAUT, *Large Eddy Simulation for Incompressible Flows*, Springer-Verlag, Berlin, Heidelberg, New York, 2001.
18. V. SCHEFFER, *Hausdorff measure and the Navier-Stokes equations*, Comm. Math. Phys., 55(2) (1977), pp. 97–112.
19. H. SCHLICHTING, *Boundary Layer Theory*, McGraw-Hill, 1955.
20. K. STEWARTSON, *D'alembert's paradox*, SIAM Review, 23(3) (1981), pp. 308–343.
21. A. TRAVIN, M. SHUR, M. STRELETS, AND P. SPALART, *Detached-eddy simulation pas a circular cylinder*, Flow, Turbulence and Combustion, 63 (1999), pp. 293–313.
22. M. M. ZDRAVKOVICH, *Flow around circular cylinders: a comprehensive guide through flow phenomena, experiments, applications, mathematical models, and simulations. Vol.1 [Fundamentals]*, Oxford Univ. Press, Oxford, 1997.

LES-SSAM for a high Reynolds number turbulent channel flow

Remi Zamansky, Ivana Vinkovic and Mikhael Gorokhovski

LMFA UMR 5509 CNRS Ecole Centrale de Lyon Université Claude Bernard Lyon 1, 36 avenue Guy de Collongue, 69134 Ecully Cedex, France remi.zamansky@ec-lyon.fr, ivana.vinkovic@ec-lyon.fr, mikhael.gorokhovski@ec-lyon.fr

Summary. In [15] the large eddy simulation with stochastic modeling of the sub-grid acceleration (LES-SSAM) for homogeneous turbulence was proposed. The main motivation of this approach is to account for intermittency of the flow at sub-grid scales, by emphasizing the role of sub-grid acceleration. In this paper, we develop further this approach in order to simulate a high Reynolds number channel flow. We proposed a new sub-grid acceleration model for channel flow. This model introduces explicitly the cross-channel correlation of subgrid velocity gradients and includes two parameters: the Reynolds number based on the friction velocity, and the channel half-width. The objective is to assess the capability of this model in comparison to the standard large-eddy simulation (LES) and to direct numerical simulation (DNS).

Key words: large-eddy simulation, sub-grid acceleration, stochastic model, intermittency, turbulent channel flow

1 Introduction

The structure of well-developed turbulent wall layer in the channel flow is highly intermittent. Close to the wall, the low-speed regions are interleaved with tiny zones of high-speed motion. The main role in this intermittency is attributed to quasi-streamwise vortices formed in the near-wall layer [1, 8, 19].

Their anisotropic dynamics are Reynolds-number dependent. Sweeps from the outer layer toward the wall induce strong variations of the wall-normal velocity. The cross-channel correlation in the turbulent velocity field is amplified by merging of near-wall small-scale structures and their eruptions towards the outer region [6, 7, 18].

For a high Reynolds flow, the LES at moderate resolution has to be combined with a SGS model for the non-resolved turbulent motion. The majority of SGS models are focused on simulation of turbulent stresses generated by the non-resolved velocity field [3, 10, 14]. In these models the structure of subgrid flow is supposed to be independent of the Reynolds number, *i.e.* to be not intermittent. Therefore the

M.V. Salvetti et al. (eds.), *Quality and Reliability of Large-Eddy Simulations II*, ERCOFTAC Series 16, DOI 10.1007/978-94-007-0231-8_6,

approach recently proposed by Sabelnikov [15] is focused directly on the stochastic modeling of the subgrid acceleration (LES-SSAM).

It was shown, by Kolmogorov's scaling, that, for a given filter width Δ, the non-resolved acceleration may be substantially greater than the resolved acceleration: $(\overline{a}_k\overline{a}_k)/(a'_i a'_i) \approx (\eta/\Delta)^{2/3}$, where $\overline{a}_k$ and a'_i represent resolved and non-resolved accelerations and $\eta = L/Re_L^{3/4}$ is the Kolmogorov's length scale. This implies that in any SGS model, which is aimed to introduce the intermittency effects, the non-resolved acceleration must be a key variable. This motivated us to set up a new stochastic model for the subgrid acceleration of wall bounded flow. The aim of this paper is to assess the capability of the new model to reproduce the near-wall behavior compared to a standard LES and DNS.

2 LES-SSAM approach and model formulation

In the LES-SSAM framework of Sabelnikov [15], it is consider that the total instantaneous acceleration, governed by the Navier-Stokes equations, can be represented by the sum of two parts: $a_i = \overline{a}_i + a'_i$. The first part represents the spatially filtered total acceleration: $\overline{a}_i = \frac{\partial \overline{u}_i}{\partial t} + \frac{\partial \overline{u_k u_i}}{\partial x_k}$, and is equivalent, with spatial filtering of the Navier-Stokes equations, to:

$$\overline{a}_i \equiv \overline{\frac{du_i}{dt}} = -\frac{1}{\rho}\frac{\partial \overline{p}}{\partial x_i} + \nu\Delta\overline{u}_i \; ; \qquad \frac{\partial \overline{u}_k}{\partial x_k} = 0 \tag{1}$$

with ν the kinematic viscosity. The second part is associated with the total acceleration in the residual field and is considered as a stochastic variable. In the LES-SSAM approach, eq. 1 is modeled in the framework of the classical LES approach. The resulting model-equation, which reconstructs an approximation for the non-filtered velocity field, writes then as:

$$\frac{\partial \hat{u}_i}{\partial t} + \hat{u}_k\frac{\partial \hat{u}_i}{\partial x_k} = -\frac{1}{\rho}\frac{\partial \hat{p}}{\partial x_i} + \frac{\partial}{\partial x_k}(\nu+\nu_t)\left(\frac{\partial \hat{u}_i}{\partial x_k} + \frac{\partial \hat{u}_k}{\partial x_i}\right) + \hat{a}'_i \; ; \qquad \frac{\partial \hat{u}_k}{\partial x_k} = 0 \tag{2}$$

where $\hat{\bullet}$ represents a synthetic field and ν_{turb} is given by an eddy viscosity model (*e.g.* the Smagorinsky subgrid model).

In [15] a model have been introduced for subgrid scale acceleration in isotropic and homogeneous turbulent flow. For further development of the LES-SSAM approach, we propose a new model for turbulent channel flow. We introduce the separation of variables for the non-resolved acceleration $\hat{a}'_i$. On the basis of our DNS for turbulent channel flow (see table 1) and experiences [9, 11, 12], $|a|$, the modulus of the subgrid acceleration and e_i its orientation, are two independent random variables, characterized by long memory and rapid decorrelation, respectively. Then the non-resolved acceleration is written as:

$$\hat{a}'_i = |a| e_i \tag{3}$$

For $|a|$, our proposal is to emulate the modulus of the non-resolved acceleration in the following form:

$$|a| = f\delta_y u_*^2/\nu \tag{4}$$

where δ_y is the cell size in the normal to the wall direction and u_* the friction velocity, $u_*^2/\nu \equiv \frac{\partial u}{\partial y}|_{\text{wall}}$; so $\delta_y u_*^2/\nu$ will be considered as a typical normal to wall velocity increment in the near to wall region. f is the subgrid frequency, considered as a stochastic variable. The frequency f is supposed to have a stochastic evolution from the wall to the outer flow driven by the non-dimensional parameter τ defined as follows:

$$\tau = -ln\left(\frac{h-y}{h}\right) \tag{5}$$

where h is the channel half-width, and y is the wall distance ($y = 0 : \tau = 0$ and $y \to h : \tau \to \infty$). The near-wall region is characterized by strong velocity gradients (high values of f), which decrease in mean toward the outer flow through the highly intermittent boundary layer. Thereby we assumed that with increasing of the normal distance from the wall, the frequency f is changing by a random independent multiplier α ($0 < \alpha < 1$), governed by distribution $q(\alpha)$, $\int_0^1 q(\alpha)d\alpha = 1$, which is in principle unknown. In other words, we apply the fragmentation stochastic process under scaling symmetry for the frequency f. From [4], we derive the following stochastic equation corresponding to this process:

$$df = (\langle ln\alpha\rangle + \langle ln^2\alpha\rangle/2)\, f d\tau + \sqrt{\langle ln^2\alpha\rangle/2} f dW(\tau) \tag{6}$$

where $\langle ln^k\alpha\rangle = \int_0^1 q(\alpha) ln^k\alpha d\alpha$; $k = 1,2$, and $dW(\tau)$ is the Wiener process ($\langle dW(\tau)\rangle = 0$ and $\langle dW(\tau)^2\rangle = d\tau$, with τ given by eq. 5). In the present study, parameters are chosen in the following form:

$$-\langle ln\alpha\rangle = \langle ln^2\alpha\rangle = Re_+^{1/3} \tag{7}$$

where Re_+ is the Reynolds number, based on the friction velocity and the channel half-width. The starting condition, $\tau = 0$, for this stochastic process (the first grid cell on the wall) is given as follows. We introduce the characteristic value of the frequency f at the wall $f_+ = \lambda/u_*$, where λ is determined, as a Taylor-like scale, which can be estimated by the Kolmogorov's scaling in the framework of definitions of wall parameters. The Reynolds number, based on friction velocity, is $Re_+ = u_* h/\nu = h/y_0 \approx Re_h^{3/4}$ where y_0 is the thickness of the viscous layer, and Re_h is the Reynolds number based on the center-line velocity. One then yields: $\lambda \approx hRe_h^{-1/2} \approx hRe_+^{-2/3}$. Similar to Kolmogorov-Oboukhov 62, the starting condition for the random path given by eq. 6 is sampled from the stationary log-normal distribution of f/f_+:

$$P_0(f/f_+) = \frac{f_+}{f\sqrt{2\pi\sigma^2}} e^{-\frac{(ln(f/f_+)-\mu)^2}{2\sigma^2}} \tag{8}$$

with parameters $\sigma^2 = \ln 2$ and $\mu = -\frac{1}{2}\sigma^2$, such that $\langle f \rangle = (\langle f^2 \rangle - \langle f \rangle^2)^{1/2} = f_+$. The stochastic process giving by eq. 6 with initial condition from eq. 8 will relax f from a log-normal distribution at the wall ($\tau = 0$) to the power distribution as the distance to the wall increases ($\tau \to \infty$). The evolution through the channel, for distributions of the frequency predicted by the stochastic equation can be compared with the evolution of the frequency computed from DNS, via eq. 4. According to fig. 1a SSAM ensures a good relaxation of the frequency, as the distance to the wall increases.

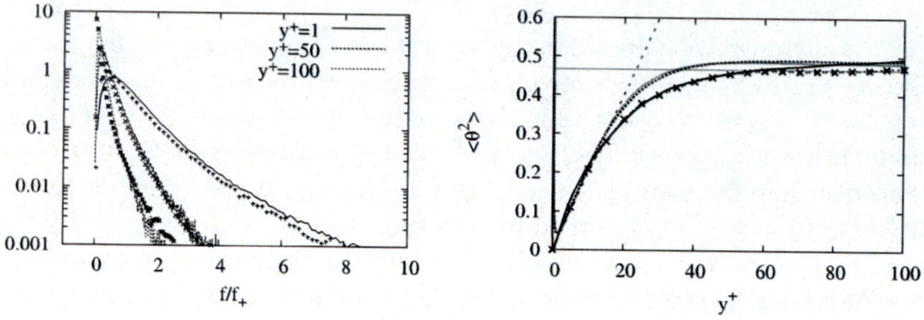

Fig. 1. (a) Distribution of f/f_+ from SSAM (cross) and comparison with DNS (line) at $Re_+ = 590$, for several distances from the wall. (b) Variance of θ for small scale acceleration from DNS (line) for $Re_+ = 180$, 590 and 1000 and from SSAM (cross). Straight line denote the variance of θ for an isotropic orientation.

In order to emulate the orientation vector of the subgrid scale acceleration, e_i, we consider a random walk evolving on the surface of a sphere of unity radius. The orientation vector may be defined by two angles which are longitude ϕ and latitude θ:

$$\begin{cases} e_x = cos(\theta)\cos(\phi) \\ e_y = \sin(\theta) \\ e_z = cos(\theta)\sin(\phi) \end{cases} \tag{9}$$

The ϕ angle characterizes the direction in the streamwise-spanwise (x,z) plan, and the other one, θ, defines the orientation in relation to the normal to wall direction ($\theta = 0$ means acceleration is parallel to the wall, and $\theta = \pm\pi/2$ means acceleration is normal to the wall), as can be seen on fig 2. First the computation of e_i from DNS was performed. The result are shown in fig. 1b. It is seen that e_i relaxes toward isotropy with increasing distance from the wall. Note that in the case of full isotropy, the distributions of θ and ϕ are respectively given by $P_{isotropic}(\theta) = \cos(\theta)/2$ and $P_{isotropic}(\phi) = \pi/2$.

In order to represent this tendency towards isotropy, we implement a Brownian motion on the sphere. This motion will defines the evolution of the unit vector e_i. Each increment $d\mathbf{e}$ of the random walk is given by

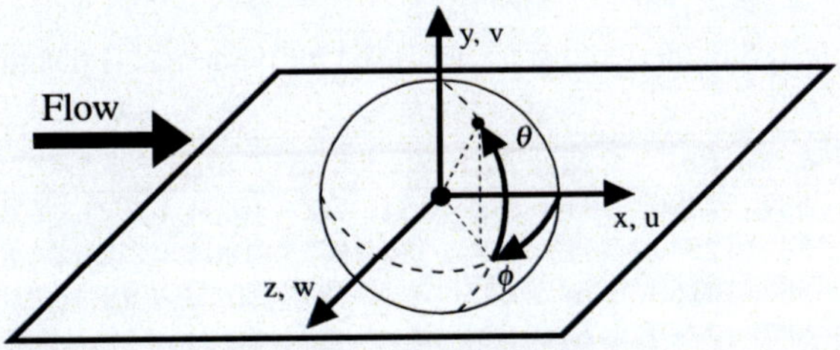

Fig. 2. Coordinate system.

$$d\mathbf{e} = \zeta \mathbf{e} \times d\mathbf{W}(dy^+) \tag{10}$$

where $d\mathbf{W}(y^+)$ is a 3-D Wiener process ($\langle d\mathbf{W}(y^+)\rangle = 0$ and $\langle d\mathbf{W}(y^+)^2\rangle = dy^+$, with y^+ the distance to the wall in wall unit) and $\times$ denotes the vector product. The constant ζ is given to be consitant with the evolution of orientation vector computed from DNS. We choose $\zeta = 25$ which is the characterisic thickness (in wall unit) of the layer near the wall where the orientation is strongly anisotropic. As the distance to the wall increases the random walk covers all the surface of the sphere, ensuring the relaxation towards isotropy for the subgrid scale acceleration orientation. This process is initialized on the wall with:

$$\begin{cases} P_\theta(\theta, y^+ = 0) = \delta(\theta) \\ P_\phi(\phi, y^+ = 0) = 1/2\pi \quad \text{if } 0 \le \phi < 2\pi \end{cases} \tag{11}$$

where P_θ and P_ϕ are the distribution of θ and ϕ respectively, and δ is the Dirac distribution, *i.e.* the orientation vector at the wall is parallel to the wall. The initial condition 11 is coherent with DNS. On fig. 1b we present the evolution of the variance of θ given by eq. 9, eq. 10 and eq. 11 and the one computed from the small-scale acceleration of DNS. Good agreement with the DNS is achieved.

3 Numerical Results and discussion

In order to make *a posteriori* tests of this subgrid scale model for acceleration we ran simulations of a pressure driven turbulent channel flow for three Reynolds numbers: $Re_+ = 590$, 1000 et 2000. We used a pseudo-spectral method with integration in time by the explicit Adam-Basforth algorithm for convective terms, and by semi-implicit algorithm for diffusion terms. Further details about the computational code can be found in [2].

The results of LES-SSAM tests have been compared with standard LES and DNS. We used our own DNS data as well as the DNS data from Moser et al. [13] and from Hoyas and Jiménez [5]. For LES and LES-SSAM simulations the classical Smagorinsky model with a wall damping function for the turbulent viscosity has been applied [16]:

Table 1. Summery of parameters used for numerical simulations

Name	Re_+	Re_c	$N_x \times N_y \times N_z$	$\Delta x^+ \times \Delta y^+ \times \Delta z^+$	C_s	A/h
DNS	587	12490	$384 \times 257 \times 384$	$7.2 \times (0.04 \sim 7.2) \times 3.6$	-	-
DNS [13]	587	12547	$384 \times 257 \times 384$	$9.7 \times (0.04 \sim 7.2) \times 4.8$	-	-
LES	587	14160	$64 \times 65 \times 64$	$87 \times (0.71 \sim 29) \times 29$	0.16	0.015
LES-SSAM	587	12760	$64 \times 65 \times 64$	$87 \times (0.71 \sim 29) \times 29$	0.16	0.015
DNS	1000	22250	$512 \times 385 \times 512$	$8.2 \times (0.03 \sim 8.3) \times 4.1$	-	-
DNS [5]	934	20960	$3072 \times 385 \times 2304$	$7.6 \times (0.06 \sim 7.6) \times 3.8$	-	-
LES	1000	25430	$96 \times 97 \times 96$	$99 \times (0.53 \sim 33) \times 33$	0.16	0.009
LES-SSAM	1000	23380	$96 \times 97 \times 96$	$99 \times (0.53 \sim 33) \times 33$	0.16	0.009
LES	1000	25500	$64 \times 65 \times 64$	$147 \times (1.2 \sim 49) \times 49$	0.2	0.015
LES-SSAM	1000	23700	$64 \times 65 \times 64$	$147 \times (1.2 \sim 49) \times 49$	0.2	0.015
DNS [5]	2003	48680	$6144 \times 633 \times 4608$	$8.2 \times 8.9 \times 4.1$	-	-
LES	2000	49350	$128 \times 129 \times 128$	$147 \times (0.60 \sim 49) \times 49$	0.16	0.006
LES-SSAM	2000	48950	$128 \times 129 \times 128$	$147 \times (0.60 \sim 49) \times 49$	0.16	0.006
LES	2000	52640	$64 \times 65 \times 64$	$295 \times (2.4 \sim 98) \times 98$	0.2	0.015
LES-SSAM	2000	49050	$64 \times 65 \times 64$	$295 \times (2.4 \sim 98) \times 98$	0.2	0.015

$$\begin{aligned} \nu_{turb} &= (C_s \Delta f_{VD})^2 |S| \\ |S| &= (2 S_{ij} S_{ij})^{1/2} \\ f_{VD} &= 1 - e^{-y/A} \end{aligned} \tag{12}$$

with C_s the Smagorinsky constant, $\Delta = (\Delta x \times \Delta y \times \Delta z)^{1/3}$ the typical cell size, $S_{ij} = \frac{1}{2}\left(\frac{\partial \bar{u}_i}{\partial x_j} + \frac{\partial \bar{u}_j}{\partial x_i}\right)$ the resolved rate of strain tensor, f_{VD} the Van Driest function and A the constant controlling the damping of f_{VD}. The constant A is computed in order to fulfill the suggestion of Shur et al. [17] for the subgrid length-scale ℓ definition: $\ell = \min(y, \Delta)$, y is the distance to the nearest wall. We choose A such that $\Delta f_{VD} \sim \min(y, \Delta)$ by least square regression. The parameters used for these simulations are summarized in table 1.

Note that in this code, Reynolds number are imposed via the setting of ν and $-\frac{1}{\rho}\frac{\partial P}{\partial x_i}$ (the mean pressure gradient). One may use Dean's correlation (Dean, 1978): $\nu = 0.110 U_c h Re_+^{-1.1296}$ and $-\frac{1}{\rho}\frac{\partial P}{\partial x_i} = Re_+^2 \nu^2 / h^3$, with U_c the center-line velocity, to choose suitable values. As shown in table 1, the Reynolds numbers computed from LES-SSAM are closer to the DNS than the ones computed by LES. For a given set of parameters (ν and $-\frac{1}{\rho}\frac{\partial P}{\partial x_i}$), LES-SSAM improves both center-line velocity and mass flow rate estimations.

For simplicity reasons, in the following, we only present results from LES-SSAM and LES with a $64 \times 65 \times 64$ grid for the three Reynolds numbers. It should be noted that for finer resolutions the differences between LES-SSAM and standard LES are less pronounced, but still present.

Fig. 3a shows evolution of the mean velocity across the channel. As pointed out in table 1 it is clear that LES-SSAM improves mean flow rate estimation as well as center-line velocity prediction. Moreover the mean velocity profile follows the logarithmic law contrary to LES.

On fig. 3 profiles of the standard deviation for streamwise, spanwise and normal to wall velocities are also presented. Standard deviations of streamwise velocity are notably improved. The peak position obtained by LES-SSAM is closer to the DNS than the one obtained with LES. For the spanwise velocity standard deviation the improvement is less visible. However, note that the shape of the profile obtained by LES-SSAM is closer to the DNS one, even if it is overestimated. Finally, the normal to the wall velocity standard deviation is slightly improved in comparison with LES.

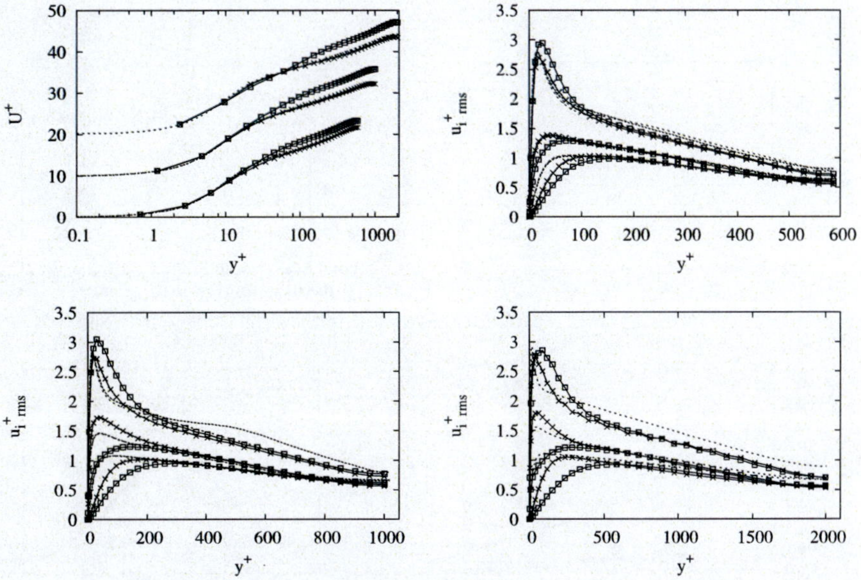

Fig. 3. (a) Streamwise mean velocity for $Re_+ = 590$, $Re_+ = 1000$ and $Re_+ = 2000$ from bottom to top, respectively, shifted by 10 wall units upward. Standard deviation of streamwise (u), spanwise (w) and normal to the wall (v) velocity, for (b) $Re_+ = 590$, (c) $Re_+ = 1000$ and (d) $Re_+ = 2000$ Square: LES; cross: LES-SSAM; dash: DNS (only for $Re_+ = 590$ and $Re_+ = 1000$); dots: DNS from [13] for $Re_+ = 590$ and from [5] for $Re_+ = 1000$ and $Re_+ = 2000$.

Fig. 4a illustrates the computation of turbulent and viscous stresses, $\tau_{turb} = -\rho\langle u'v'\rangle$ and $\tau_{visc} = -\rho\nu\langle\frac{\partial u}{\partial y}\rangle$, respectively ($\langle . \rangle$ denotes ensemble average). The results are presented as ratios $\tau_{turb}/(\tau_{turb}+\tau_{visc})$ and $\tau_{visc}/(\tau_{turb}+\tau_{visc})$. Here again the advantage of the LES-SSAM approach versus the classical LES is explicitly seen, and can be interpreted as a better estimation of momentum fluxes in the normal to the wall direction. In addition to this, from fig. 4a it can be seen that close to the wall the viscous stress is dominant, whereas as we move away from the wall the turbulent stress becomes preponderant. The LES-SSAM approach enables improvement of both the viscous and the turbulent stress away and close to the wall.

Velocity spectra are shown on fig. 4b. From this figure we can see that the anomalous small scale (high wave number) damping inherent to LES can be reduced by LES-SSAM. Fig. 4c represents the evolution of the longitudinal autocorrelation coefficient for the streamwise velocity component along the channel. Improvement of the decorrelation length can be seen, indicating that integral length scales, and so the size of strucutres in the near wall, computed by LES-SSAM are closer to DNS than in the classical LES.

This result is mainly due to the fact that decorrelation is ensured by small scale fluctuations modelled with LESS-SSAM as seen by the spectra on fig. 4b.

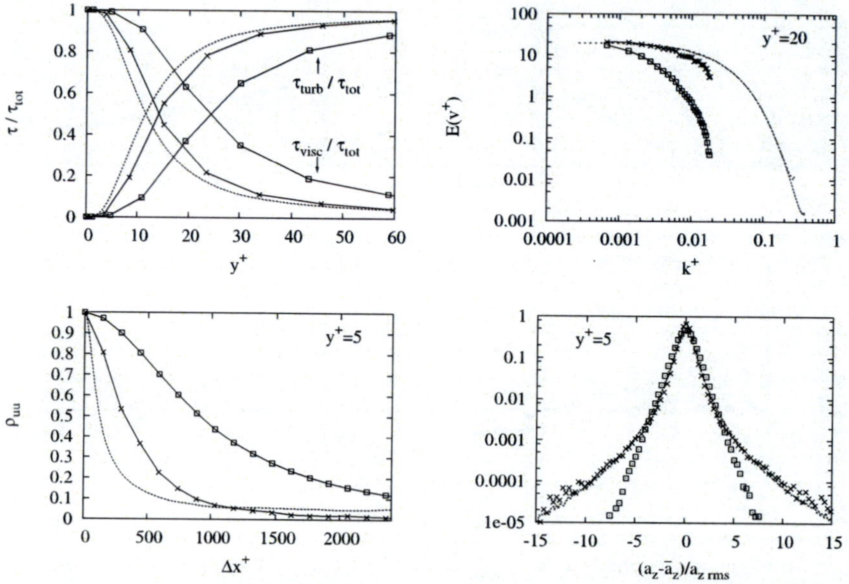

Fig. 4. (a) Fractions of turbulent $\tau_{turb} = -\rho\langle u'v'\rangle$ and viscous $\tau_{visc} = -\rho\nu\langle\frac{\partial u}{\partial y}\rangle$ stresses compared to the total one $\tau_{tot} = \tau_{visc} + \tau_{turb}$. (b) Normalized longitudinal 1-D spectra of normal to wall velocity for $y^+ = 20$. (c) Longitudinal autocorrelation of streamwise velocity at $y^+ = 5$. (d) Distribution of spanwise acceleration at $y^+ = 5$. $Re_+ = 1000$. Square: LES; cross: LES-SSAM; dash: DNS; dots: DNS from [5] (uniquely for (b)).

Fig. 4d illustrates the distribution of the spanwise acceleration. The distributions obtained by DNS present stretched tails, as a manifestation of intermittency. From fig. 4d it can be seen that in agreement with the DNS, the distributions of the a_z, obtained by LES-SSAM, present the stretched tails, while these distributions computed by LES stay close to the Gaussian distribution. By introducing the LES-SSAM model, small scale intermittency is included in the modelling of the acceleration.

4 Conclusion

In the framework of the LES-SSAM approach, a new SGS model is proposed for the subfilter scale acceleration in order to include the intermittency effects in the near-wall region of a high-Reynolds number channel flow. In this model, the modulus and the orientation of the non resolved acceleration are considered as two independent stochastic variables. The modulus is a function of a subgrid frequency given by a fragmentation stochastic process under the scaling symmetry. The mean and standard deviation of the process are functions of the Reynolds number. For the orientation of the acceleration a random walk evolving on the surface of a sphere is proposed. The orientation of the acceleration is such that away from the wall the process relaxes towarss isotropy. This was chosen in agreement with the DNS.

The results for the model proposed here are compared with DNS data ($Re_+ =$ 590, 1000 and 2000) and standard LES. The comparisons show that mass flow rate estimations and mean velocity profiles are improved by introducing the LES-SSAM model. In addition to this a better prediction near the wall of the streamwise standard deviation, the viscous and turbulent stresses are obtained with LES-SSAM. Small scale dumping inherent to LES is avoided with LES-SSAM. Finally, improved longitudinal velocity decorelation is obtained as well as the prediction of the stretched tails in the acceleration distribution.

Acknowledgement. M. Buffat is acknowledged for the development of the computational code. The authors express their gratitude to F. Laadhari who kindly provided his DNS fields. This work was granted access to the HPC resources of CINES under the allocation 2009-c200902560 made by GENCI (Grand Equipement National de Calcul Intensif)

References

1. R. J. Adrian, C. D. Meinhart, and C. D. Tomkins. Vortex organization in the outer region of the turbulent boundary layer. *Journal of Fluid Mechanics*, 422:1–54, 2000.
2. M. Buffat, L. Le Penven, and A. Cadiou. An efficient spectral projection method on divergence-free subspaces for transition analysis in wall bounded flow. *Journal of computational physics*, submitted, 2009.
3. J. A. Domaradzki and N. A. Adams. Direct modelling of subgrid scales of turbulence in large eddy simulations. *Journal of Turbulence*, 3:0–24, 2002.
4. M. Gorokhovski and V. L. Saveliev. Statistical universalities in fragmentation under scaling symmetry with a constant frequency of fragmentation. *Journal of Physics D: Applied Physics*, 41:085405, 2008.
5. S. Hoyas and J. Jiménez. Reynolds number effects on the reynolds-stress budgets in turbulent channels. *Physics of Fluids*, 20(10):101511, 2008.
6. N. Hutchins and I. Marusic. Evidence of very long meandering features in the logarithmic region of turbulent boundary layers. *Journal of Fluid Mechanics*, 579:1–28, 2007.
7. J. Jiménez, J. C. Del Alamo, and O. Flores. The large-scale dynamics of near-wall turbulence. *Journal of Fluid Mechanics*, 505:179–199, 2004.
8. D. Kaftori, G. Hetsroni, and S. Banerjee. Funnel-shaped vortical structures inwall structures. *Physics of Fluids*, 6(9):3035–3050, 1994.

9. C. Lee, K. Yeo, and J.-I. Choi. Intermittent nature of acceleration in near-wall turbulence. *Physical Review Letters*, 92(14), 2004.
10. C. Meneveau and J. Katz. Scale-invariance and turbulence models for large-eddy simulation. *Annual Reviews of Fluid Mechanics*, 32:1–32, 2000.
11. N. Mordant, J. Delour, E. Léveque, A. Arnéodo, and J.-F. Pinton. Long time correlations in lagrangian dynamics: a key to intermittency in turbulence. *Pysical Review Letters*, 89(25), 2002.
12. N. Mordant, E. Lévêque, and J.-F. Pinton. Experimental and numerical study of the Lagrangian dynamics of high Reynolds turbulence. *New Journal of Physics*, 6:116, 2004.
13. R. D. Moser, J. Kim, and N. N. Mansour. Direct numerical simulation of turbulent channel flow up to $Re_\tau = 590$. *Physics of Fluids*, 11(4):943–945, 1999.
14. N. Park and K. Mahesh. A velocity-estimation subgrid model constrained by subgrid scale dissipation. *Journal of Computational Physics*, 227:4190–4206, 2008.
15. V. Sabelnikov, A. Chtab, and M. Gorokhovski. The coupled LES - sub-grid stochastic acceleration model (LES-SSAM) of a high Reynolds number flows. In *Advances in Turbulence XI*, volume 117, pages 209–211, 11th EUROMECH European Turbulence Conference, June 25-28, 2007, Porto, Portugal, 2007. Springer Proceedings in Physics.
16. P. Sagault. *Large Eddy Simulation for Incompressible Flows: An introduction*. Springer Verlag, second edition, 2002.
17. M. L. Shur, P. R. Spalart, M. Kh. Strelets, and A. K. Travin. A hybrid RANS-LES approach with delayed-DES and wall-modelled LES capabilities. *International Journal of Heat and Fluid Flow*, 29(6):1638–1649, 2008.
18. S. Toh and T. Itano. Interaction between a large-scale structure and near-wall structures in channel flow. *Journal of Fluid Mechanics*, 524:249–262, 2005.
19. C. D. Tomkins and R. J. Adrian. Spanwise structure and scale growth in turbulent boundary layers. *Journal of Fluid Mechanics*, 490:37–74, 2003.

A new development of the dynamic procedure for the integral-based implicit filtering in large-eddy simulation

Filippo M. Denaro[1] and Giuliano De Stefano[2]

Department of Aerospace and Mechanical Engineering, Second University of Naples, Aversa (CE), Italy

[1]denaro@unina.it, [2]giuliano.destefano@unina2.it

Summary. This paper is focused on the role of integral-based Finite Volume (FV) discretizations in Large Eddy Simulation of turbulence. The integral-based form implicitly induces the top-hat filtering on the balanced variable. This leads us to rewrite also a different decomposition of the fluxes. As a consequence, the development of a new Germano identity can be achieved having some advantages over the classical differential-based form. However, the dynamic procedure requires an explicit test-filtering on a computational grid that, to be optimal, requires an evaluation of the shape of the numerical filter induced by the FV-based discretization. Therefore, the goal of this paper is the theoretical study of the effective filter shape induced by some 3D Finite Volume reconstructions. The induced shape and width are analyzed by means of a *modified wavenumber-like* analysis that is applied in the 3D Fourier space. Some schemes are considered and the differences in terms of either velocity or flux interpolations on either staggered or non-staggered grids are derived and analyzed.

Key words: Implicit Filtering, Finite Volume Methods, Germano identity

1 Introduction

Large Eddy Simulation (LES) of turbulence is based on the separation between large (resolved) and Sub-Grid Scale (unresolved) flow contributions, obtained by means of the application of a low-pass filtering operator on the Navier-Stokes equations. As a matter of fact, often the filtering procedure remains nothing but a formalism in writing the LES equations in continuous form. Indeed, in the so-called *implicit filtering* approach, the discretization of both domain and differential operators is practically used as built-in filtering, see Refs. [1-4]. Thus, numerical representation of the filtered variables is associated with a finite number of well-resolved scales so that any discrete model can induce significant alterations of the expected resolved flow dynamics. Despite the additional computational effort and loss of resolution, in order to filter numerical errors and unambiguously identify the shape, the use of explicit filtering technique (pre-filtering) was analysed in [5-7].

M.V. Salvetti et al. (eds.), *Quality and Reliability of Large-Eddy Simulations II*, ERCOFTAC Series 16, DOI 10.1007/978-94-007-0231-8_7,

This paper illustrates a part of a wider theoretical analysis on the use of the Finite Volume (FV) methods for performing LES. A FV-based approach is feasible for problems of engineering interest. Commercial codes supplied with modern SGS modelling (e.g., the dynamic Smagorisky modelling) are now available. These codes exploit the implicit FV-based filtering procedure. Following [8], for FV-based methods, one recognizes three different levels of approximation: interpolation, differentiation and integration. The FV approach leads to solve discrete equation models which allow any balanced quantity, to be *a-priori* conserved, no matter what the actual accuracy order of the method is in effect. On the contrary, it is well known that other methods such as Spectral Methods (SM) and Finite Difference (FD) methods do not automatically possesses such property.

When FV methods are based on the integral form of the equations (often the discretization of the divergence form is also denoted FV approximation), any flux reconstruction always drives us to approximate uniquely[1] the *top-hat* filtered variable associated to the local volume average over a small domain of linear measure Δ, e.g. see Refs. [9-11]. The computation of surface integrals of the fluxes must be considered acting as "volume filter". Furthermore, the use of the integral-based filtered equation requires a specific decomposition between resolved and SGS fluxes and it induces to re-expressing the Germano identity for the dynamic SGS modelling, e.g. see Refs. [11, 12, 19]. Actually, the use of a numerical method implies the introduction of the characteristic domain discretization length h, which further induces the implicit grid-filtering associated to the Nyquist cut-off frequency π/h. In LES practice, the only length h is defined by the user while the effective filter length Δ implicitly depends on the chosen numerical method. Thus, the scalar parameter $Q = \Delta/h$, also addressed in literature as *subfilter resolution* see Refs. [1, 13-15], is introduced. To perform the dynamic SGS modelling, a proper choice of the test-filtering length is consequent to the value Q resulting from the chosen FV method.

In order to analyze the filter shape induced by a discretization, in analogy with the *modified wavenumber* analysis, the 3D numerical transfer functions in effects for each non linear *numerical flux function* are deduced. Several FV formulas of second and higher order of accuracy are analysed, for both non-staggered and staggered centred collocation in the velocity components. The guideline of the analysis is the following: once deduced the exact FV-based transfer function its effective width is parameterized by the sub-filter parameter, that is $\Delta = Qh$. Then it is compared to the numerical ones, trying to find some value of the parameter Q that minimizes the error. This study on implicit filtering appears distinctive from those performed by other authors, e.g. see Refs. [14-16], that analysed combined filtering effects by considering the role of the local truncation error (LTE) in the *modified differential equation*. In fact, those studies are based on the discretization of the divergence formulation not of the integral one, confirming that such form induces a someway undeterminable filter shape.

[1] Deconvolution procedures that, while retaining the same range of frequencies, would be able to recover the spectral content of the resolved scale, are not considered.

A number of interesting issues, helpful for discerning the relevance of the built-in filter in a LES application, are addressed. They are the understanding of the practical scales separation and the consequent dynamic SGS modelling. The optimal choice of the test-filter length is influenced by the effective width of the primary filter that is induced by the FV discretization. Last, it is worthwhile remarking that using DNS data would be more realistic when the fields are post-filtered by means of a filter function that mimics at best the implicit filter in effect in the LES.

2 Integral versus differential-based filtered equations in continuous form

For the sake of simplicity, let us consider the multi-dimensional Burgers equation written in integral form

$$\frac{\partial}{\partial t}\int_{\Omega(\mathbf{x})} \mathbf{v}\, d\mathbf{x}' + \int_{\partial\Omega(\mathbf{x})} \mathbf{n}\cdot\mathbf{F}(\mathbf{v})\; dS(\mathbf{x}') = \mathbf{0} \quad , \tag{1}$$

being $\mathbf{v}(\mathbf{x},t)$ the point-wise velocity field, $\mathbf{F}(\mathbf{v}) = \mathbf{vv} - 2\nu\underline{\nabla}^s\mathbf{v}$ the total flux, Ω is an elementary volume, $\partial\Omega$ is its frontier, $\mathbf{n}$ is the normal unit vector outward oriented.

In LES approach, the low-pass filtering (no time-filtering is here considered) can be expressed as the convolution product between the unfiltered point-wise velocity and some suitable filter function G, e.g. see [1-4], that is $\overline{\mathbf{v}}(\mathbf{x}, t;\Delta) = \int_{R^3} G(\mathbf{x}-\mathbf{x}';\, \Delta)\; \mathbf{v}(\mathbf{x}', t)d\mathbf{x}' \equiv G * \mathbf{v}$, Δ being the characteristic filter width. The convolution product corresponds to have $\widehat{\overline{\mathbf{v}}}(\mathbf{k}_w,t;\Delta) = \hat{G}(\mathbf{k}_w;\Delta)\,\widehat{\mathbf{v}}(\mathbf{k}_w,t)$ in Fourier space, $\widehat{G}$ being the transfer function, $\mathbf{k}_w$ being the wavenumbers vector. Thus, if the elementary volume is homogeneous in space, the velocity field averaged over an elementary box-volume, having linear homogeneous measure Δ, is written as

$$\overline{\mathbf{v}}(\mathbf{x}, t;\Delta) = \frac{1}{\Delta^3}\int_{z-\Delta/2}^{z+\Delta/2} dz' \int_{y-\Delta/2}^{y+\Delta/2} dy' \int_{x-\Delta/2}^{x+\Delta/2} \mathbf{v}(\mathbf{x}', t)\, dx' \quad . \tag{2}$$

It is equivalent to the convolution with the top-hat filter kernel, its transfer function being a smooth function in the wavenumber space [1-4]. Note that, in the continuous form, the top-hat filtered field (2) remains formally composed by infinite Fourier components, the transfer function having an infinite number of zeros. Therefore, both filtered and unfiltered continuous velocity fields possess an infinite number of Fourier components. In such sense, they belong to the same functional space.

According to the historical Schumann's approach, the integral-based Eq. (1) has the mathematical counterpart expressed by the filtered differential equation

$$\frac{\partial\overline{\mathbf{v}}}{\partial t} + \overline{\underline{\nabla}\cdot\mathbf{F}(\mathbf{v})} = \mathbf{0} \quad . \tag{3}$$

Now, the key-point to bear in mind is that the LES equations are usually rewritten in differential form. This is obtained by commuting filtering and divergence operator in (3). Besides, when commutation property does not apply, one works using special commutative filters [1-4]. Thus, after the decomposition in resolved $\mathbf{F}_R(\overline{\mathbf{v}}) = \overline{\mathbf{v}}\,\overline{\mathbf{v}} - 2\nu\underline{\nabla}^s\overline{\mathbf{v}}$ and unresolved $\mathbf{F}_U(\overline{\mathbf{v}},\mathbf{v}) = (\overline{\mathbf{v}}\,\overline{\mathbf{v}} - \overline{\mathbf{v}\mathbf{v}})$ fluxes, the classical divergence form of the differential filtered equation is obtained

$$\frac{\partial \overline{\mathbf{v}}}{\partial t} + \underline{\nabla}\cdot\mathbf{F}_R(\overline{\mathbf{v}}) = \underline{\nabla}\cdot\mathbf{F}_U(\overline{\mathbf{v}},\mathbf{v}) \quad . \tag{4}$$

The closure procedure involves modelling of the non-linear SGS flux by introducing an approximation $\mathbf{F}_U(\overline{\mathbf{v}},\mathbf{v}) \cong \mathbf{F}_{SGS}(\overline{\mathbf{v}})$. If used, an explicit filtering procedure is generally applied for such a formulation, e.g. [5-7].

Conversely, let us remark that in FV-based approximation of (1), one never requires to discretize the divergence operator but only integrals of the flux. Furthermore, one does not need to commuting filtering and differentiation. Indeed, after a proper decomposition of the total flux $\mathbf{F}$ in the resolvable $\mathbf{F}_R(\overline{\mathbf{v}})$ and irresolvable fluxes $\mathbf{T}(\overline{\mathbf{v}},\mathbf{v}) = (\overline{\mathbf{v}}\,\overline{\mathbf{v}} - \mathbf{v}\mathbf{v}) - (2\nu\underline{\nabla}^s\overline{\mathbf{v}} - 2\nu\underline{\nabla}^s\mathbf{v})$, the integral-based equation (3) writes as[2]

$$\frac{\partial \overline{\mathbf{v}}}{\partial t} + \overline{\underline{\nabla}\cdot\mathbf{F}_R(\overline{\mathbf{v}})} = \overline{\underline{\nabla}\cdot\mathbf{T}(\overline{\mathbf{v}},\mathbf{v})} \quad . \tag{5}$$

One sees that $\mathbf{T}$ is very different from $\mathbf{F}_U$ both in the non-linear term and in the presence of a residual part coming from the diffusive flux. As for the differential form, SGS modelling means to approximate $\mathbf{T}(\overline{\mathbf{v}},\mathbf{v}) \cong \mathbf{T}_{SGS}(\overline{\mathbf{v}})$.

Note that in Eq. (4) the filtering operation is never explicitly applied, unless one decides to use an explicit filtering formulation, e.g. [1]. Provided that for Eq. (4) the top-hat filter function G is (formally) specified, it and Eq. (5) are mathematically equivalent, governing the same filtered velocity. Actually, if for (4) some other filter function is specified, such as the Gaussian or the spectral cut-off filter, that would let unchanged its expression but it is no longer equivalent to (5). Practically, the type of filtering in the differential Eq. (4) is driven only by the chosen SGS modelling. Conversely, the integral-based Eq. (5) univocally implies the application of the solely top-hat filter and it is characterised by the explicit presence of the filtered fluxes. As explained in the next section, only when the computational grid is introduced it implicitly acts as additional sharp cut-off at the Nyquist frequency π/h. This way, the expected continuous transfer function is actually modified by the numerical errors and truncated by the grid-filtering. The discrete filtered velocity vector will be considered equivalent to the exact solution of the continuous modified filtered equation. In this sense, the discrete set of velocity values associated to each FV is considered as the collocated-values of a modified filtered function.

In the framework of implicit filtering approaches, we can address some differences between the two (continuous) forms of equations: *a*) prescribing only the integral filtered equation (5) does allow us to univocally identify the type of filter, conversely this is not true for Eq. (4); *b*) the resolved flux is the same in both formulations but the unresolved flux is not; *c*) a consequent difference appears in developing the dynamic procedure for computing either the tensors $\mathbf{F}_{SGS}(\overline{\mathbf{v}})$ or $\mathbf{T}_{SGS}(\overline{\mathbf{v}})$.

---

[2] Use of integral or over bar symbolisms will depend on the necessity of exposition.

It is well known that, by modelling the former tensor, the classical Germano identity $\mathbf{M}-\widetilde{\mathbf{F}_U}=\mathbf{L}$ is used. Here, tilde symbol indicates the test-filtering over characteristic length $\widetilde{\Delta} > \Delta$, $\mathbf{M}$ being the subtest-scale tensor and $\mathbf{L} = \widetilde{\overline{\mathbf{v}}\,\overline{\mathbf{v}}} - \widetilde{\overline{\mathbf{v}\mathbf{v}}}$ is the known Leonard tensor [1-4]. Then, the identity for the eddy viscosity model rewrites as

$$-2C\widetilde{\Delta}^2 \left|\widetilde{\overline{\mathbf{S}}}\right| \underline{\nabla}^s\widetilde{\overline{\mathbf{v}}} + 2\Delta^2 \widetilde{\left(C\left|\overline{\mathbf{S}}\right|\underline{\nabla}^s\overline{\mathbf{v}}\right)} \cong \mathbf{L}_d, \tag{6}$$

$|\overline{\mathbf{S}}| = (2\underline{\nabla}^s\overline{\mathbf{v}} : \underline{\nabla}^s\overline{\mathbf{v}})^{1/2}$, $\mathbf{L}_d$ being the deviatoric part of $\mathbf{L}$. The modelling coefficient is then somehow arbitrarily extracted from test-filtering [1-4].

On the other hand, starting from the integral-based form (3), by applying the top-hat test-filter of width $\widetilde{\Delta} > \Delta$, one gets the new integral-based exact subtest-scale tensor [11, 12, 19] $\mathbf{M}' = -[2\nu(\underline{\nabla}^s\mathbf{v} - \underline{\nabla}^s\widetilde{\overline{\mathbf{v}}}) + (\widetilde{\overline{\mathbf{v}}}\,\widetilde{\overline{\mathbf{v}}} - \mathbf{v}\mathbf{v})]$. It is worthwhile noticing that $\mathbf{M}'$ drives us to get also a new resolved term $\mathbf{L}' = (\overline{\mathbf{v}}\,\overline{\mathbf{v}} - \widetilde{\overline{\mathbf{v}}}\,\widetilde{\overline{\mathbf{v}}}) - 2\nu(\underline{\nabla}^s\overline{\mathbf{v}} - \underline{\nabla}^s\widetilde{\overline{\mathbf{v}}})$, such that the Germano identity for the integral-based formulation writes as $\mathbf{M}'_d - \mathbf{T}_d = \mathbf{L}'_d$. The eddy-viscosity modelling coefficient can be determined by assuming the same coefficient and the same rate of dissipation at grid and test-level (scale-invariance hypothesis), that is $\nu_{LES} = C_\varepsilon \overline{\Delta}^{4/3}$ and $\nu'_{LES} = C_\varepsilon \widetilde{\Delta}^{4/3}$ being $C_\varepsilon \equiv C^{2/3}\varepsilon^{1/3}$ with ε the dissipation rate. Thus, the integral-based counterpart of (6) is expressed as

$$2\nu_{LES}\left(\underline{\nabla}^s\overline{\mathbf{v}} - \alpha^{4/3}\underline{\nabla}^s\widetilde{\overline{\mathbf{v}}}\right) = \mathbf{L}'_d \quad , \tag{7}$$

where $\alpha \equiv \widetilde{\Delta}/\Delta$, which is the only input parameter the dynamic procedure depends upon, is the ratio of the *test to FV-based implicit filter width* and $\nu_{LES} = \nu_{LES}(\mathbf{x}, t; \Delta; \widetilde{\Delta})$. Owing to the present integral formulation, the practical determination of ν_{LES} does not require now to arbitrarily extract the model coefficient out of the filtering operation. The problem with implicit filtering approaches is that we know exactly only the computational grid step-size h, not the effective filter length Δ thus a proper value α must be evaluated in some way. Besides, in order to practically evaluate the test-filtered velocity $\widetilde{\overline{\mathbf{v}}}$ on the computational grid, one needs to estimate the ratio $\widetilde{\Delta}/h = (\widetilde{\Delta}/\Delta)(\Delta/h) = \alpha Q$ thus an estimation of Q is required, the latter being implicitly defined by the type of numerical scheme. Thus, let us now introduce the discretization analysis.

3 Integral versus differential-based discrete filtered equations. Analysis of the implicit filtering

Let us briefly denote with $\overline{\underline{\nabla}\cdot(*)}^d$ a certain type of FV-based spatial discretization (from now on, superscript d will denote only spatial discretization), here considered according to [8] in terms of the numerical approximation of the integral-based

Eq. (5). The discrete integral-based LES equation that defines the evolution of the discrete filtered velocity, say $\overline{\mathbf{v}}^d$, is written as

$$\frac{\partial \overline{\mathbf{v}}^d}{\partial t} + \overline{\underline{\nabla} \cdot \mathbf{F}_R^d \left(\overline{\mathbf{v}}^d\right)}^d = \overline{\underline{\nabla} \cdot \mathbf{T}_{SGS}^d \left(\overline{\mathbf{v}}^d\right)}^d \quad , \tag{8}$$

having defined the discrete resolved flux $\mathbf{F}_R^d = \overline{\mathbf{v}}^d \overline{\mathbf{v}}^d - 2\nu \underline{\nabla}^d \overline{\mathbf{v}}^d$. The velocity $\overline{\mathbf{v}}^d$ is, therefore, representative of a finite-dimensional state-vector possessing only a finite number of length scales, bounded by the smallest computational size associated to the Nyquist frequency h/π. Actually, also the resolved Fourier components of $\overline{\mathbf{v}}^d$ are affected by the smooth shape of the built-in numerical filter, the relevance of the smoothing and its difference from the theoretical top-hat one, depending on the accuracy of the discretization of $\overline{\underline{\nabla} \cdot (*)}^d$.

It is worthwhile remarking that, in CFD literature, the discretization of the divergence form (4), that is

$$\frac{\partial \overline{\mathbf{v}}^d}{\partial t} + \underline{\nabla}^d \cdot \mathbf{F}_R^d = \underline{\nabla}^d \cdot \mathbf{F}_{SGS}^d \quad , \tag{9}$$

is also denoted as FV discretization, e.g. see [13, 14]. In "traditional" LES solution, Eq. (9) is used and it is accepted that both the computational grid and the discretization operators implicitly act as filtering to the characteristic turbulent wavenumbers range. For example, while using the SM approach, the filter is implicitly assumed to be the sharp cut-off associated to the truncated Fourier series that approximates the velocity $\overline{\mathbf{v}}^d$. Conversely, in FD approximations, the effective shape of the transfer function depends on the type of the discretization methods used for $\underline{\nabla}^d \cdot (*)$ and there is no theoretical reason to find an approximation to the exact top-hat transfer function[3], as it has been also reviewed in Refs. [1, 6, 9, 10, 14]. This ambiguity is reflected in the sub-filter stress that contains the full information on the filter. That is, defining an SGS model for the unresolved tensor is equivalent to define a filter function [13-16].

At a fixed time and in a discrete set of points, one can analyze the discrepancy between the exact (5) and the discrete (8) top-hat filtered acceleration fields according to

$$\frac{\partial \overline{\mathbf{v}}}{\partial t} - \frac{\partial \overline{\mathbf{v}}^d}{\partial t} = -\left[\left(\overline{\underline{\nabla} \cdot \mathbf{F}_R} - \overline{\underline{\nabla} \cdot \mathbf{F}_R}^d\right) + \overline{\underline{\nabla} \cdot \left(\mathbf{F}_R - \mathbf{F}_R^d\right)}^d\right] + \left[\overline{\underline{\nabla} \cdot \mathbf{T}} - \overline{\underline{\nabla} \cdot \mathbf{T}_{SGS}^d}^d\right] \quad . \tag{10}$$

The first term in the first square brackets, that is the difference between the continuous and the discrete filtered divergence operators, applied on the continuous resolved flux, is the classical LTE associated to the FV discretization, e.g. see [17].

---

[3] Rigorously speaking, one should distinguish the class of discrete filtered velocity $\overline{\mathbf{v}}^d$ in case of either FD or FV classes of discretizations. However, in order to not complicate the notation, no other symbols will be used.

The second term represents the difference of the discrete filtered divergence operator when applied on the continuous and the discrete resolved total flux due to a "sampling filtering" effect. In fact, the flux $\mathbf{F}_R^d$ is sampled on the computational grid, therefore possesses a finite numbers of length scales. The term in the second square bracket could be further expressible by means of two concurrent contributions. In fact, the so-called *computational turbulent stress tensor*, is the full closure problem at the level of the effective spatial discretization [13, 14].

For the sake of brevity, here only the LTE term in (10) is analysed. Six different FV-based schemes are considered. The 3D numerical transfer functions induced by each discretization are deduced and compared to the exact top-hat transfer function. Indeed, the continuous convective term in (5) can be expressed according to

$$\overline{\frac{\partial\left(\overline{u}_j\overline{u}_i\right)}{\partial x_j}} = I\sum_{l,m,n}^{\pm\infty}\sum_{p,q,r}^{\pm\infty} p'_j\widehat{G}_{\mathbf{x}}\left(p',q',r';\Delta\right)\,\widehat{\overline{u}}_j\left(p,q,r\right)\widehat{\overline{u}}_i\left(l,m,n\right)e^{I\sum\limits_{k=1}^{3}p'_k x_k}, \qquad (11)$$

having defined $\widehat{G}_{\mathbf{x}}(p',q',r';\Delta) \equiv \widehat{G}(p';\Delta)\widehat{G}(q';\Delta)\widehat{G}(r';\Delta)$, being$\widehat{G}(\chi;\Delta) \equiv \sin(\chi\Delta/2)/(\chi\Delta/2) = \sin(\chi Qh/2)/(\chi Qh/2)$ with $p' = (p+l)$, $q' = (q+m)$, $r' = (r+n)$ and $p' = p'_1$, $q' = p'_2$, $r' = p'_3$. The considered FV schemes are denotes as follows: NSF^{II}: a second order accurate scheme on non-staggered grid, obtained by choosing the *mean value formula* for the discrete integral and the *linear interpolation* for the non-linear flux reconstruction; NSV^{II}: as above, apart from the linear interpolation that is performed separately on each velocity components; SV^{II}: same as NSV^{II} but with the velocity components defined on staggered (MAC-like) grid; NSF^{II-IV}: a hybrid second-fourth order accurate scheme on non-staggered grid, obtained by choosing the *mean value formula* for the discrete integral operators and the *cubic lagrangian interpolation* for the non-linear flux function; NSR^{II}: the second order scheme on non-staggered collocation reported in Ref. [13], based on transversal reconstruction.; NSR^{IV}: same as NSR^{II} but at fourth order accuracy [13]. These latter two schemes are based on the discretization of the differential form (9).

The numerical transfer functions (for the sake of brevity the expressions are not reported here), obtained by means of a *modified wavenumber*-like analysis, are reported in Figs. 1. The plots are obtained by forcing the constraint $0 \le \xi = \eta = \zeta \le \pi$, where $\xi = p'h, \eta = q'h, \zeta = r'h$, and compared to the exact one $\widehat{G}_{\mathbf{x}} = [\sin(\xi Q/2)/(\xi Q/2)]^3$, plotted for Q=1 and 2. In the expression of the transfer functions induced by the NSV schemes there is the presence of the wavenumber l that remains outside the sum $p' = (p+l)$ and is therefore used as plotting parameter. The plots for NSF (1a) and NSV (1b) become coincident for $lh = 0$ ($\xi = p'h = ph$), as verified by inspection of expression of the respective transfer functions. Both curves stay close to the exact one for the Q= 1 value, this fact somehow contradicting the expected result Q= 2 one knows from the classical 1D analysis [1, 5, 6]. However, for non-vanishing values lh (that is $0 < \xi - ph \le \pi/2$), the case NSV produces spurious frequencies inherent to the appearance of aliasing terms (and triadic-interaction appearing from the non-linear terms, see [1]). A deep analysis of the aliasing errors in the case of either flux or velocity interpolation is addressed in Ref. [18]. The case

SV, plotted in Fig. 1c, has the same accuracy order of the *NSF* and *NSV* schemes, allowing us to discern the effect of the staggered collocation. Though, compared to *NSV*, there is the evidence of other wavenumbers that remain outside of the sums p', q', r'. Perhaps, on a side, when $l_j h = 0, p_i h = \xi$, the curve *SV* corresponds to the one *NSF* in Fig. 1a while on the other side, for lh=0, to *NSV* in Fig. 1b. Besides, when $l_j h = p_i h = \xi$, the curve *SV* tends to be close to the curve Q= 2 while becoming largely overestimated the curve Q= 1 for $l_j h = p_i h = 0$. Such effects are a manifestation of the supplementary transversal interpolations of the velocity product required by the staggered grid.

Considering now the NSF^{II-IV} and *NSR* schemes, they are characterized by transfer functions depending exclusively on the sums p', q', r'. Only the curve shown in Fig. 1e, for the NSR^{II} scheme, shows a good approximation of the exact top-hat filtering for Q=2. The accuracy order is the same of NSF^{II} but for NSR^{II} the trapezoidal rule, associated to the transverse second order accurate multidimensional interpolation, is used. This results in a more isotropic resolution of the transfer function for any wavenumber p', q', r', hence NSR^{II} appears a suitable multidimensional approximation of the top-hat filtering for Q=2. On the other hand, both schemes NSF^{II-IV} and NSR^{IV} are built in such a way that the accuracy is increased only for the reconstruction of the flux not for the discretization of the integrals. As a consequence, a better approximation of the sharp cut-off filter (at low wavenumbers) but not a better approximation of the exact top-hat filter is obtained. When compared to the exact transfer function at Q= 1, both plots in the Fig. 1d and 1f show an overestimation of the spectral content at low wavenumbers whereas Fig. 1e describes a slight underestimation of the exact case for Q= 2. Namely, despite the fact that the fourth order accurate formula NSR^{IV} is indicated in literature as a "finite-volume" one [13, 14], since there is no the congruent fourth order surface integration [8], it has to be considered as suitable discretization of the differential term in divergence form not of the integral based counterpart. In such case, the implicit filter can be considered by itself a higher order filter, closer to the identity operator not to the top-hat one, as addressed in Refs. [1-4, 9, 10, 14]. Accordingly, the figure 1f is indicative of the fact that the shape of the resulting transfer function appears, actually, like a "deconvolution-based" counterpart of the top-hat one.

The quantitative analysis of the effective filter width is a rather controversial issue and was object of several proposals, e.g., see Refs. [1, 6, 14], some exploiting one-dimensional expressions to characterize the filter kernel. A criterion is the 1/2 rule that consists in interpreting the wavenumber k_{eff}, for which the filter transfer function gets the value $\widehat{G}\left(k_{eff};\Delta_{eff}\right) = 0.5$, as the effective cut-off frequency. The filter width in physical space is then estimated simply as $\Delta_{eff} = \pi / k_{eff}$. Therefore, an estimation of the sub-resolution parameter Q can be obtained according to such criterion, by reconsidering the previous 1-D curves in Figs. 1 and evaluating $Q = \Delta_{eff}/h = \pi / \left(k_{eff} h\right) = \pi / \xi_{eff}$. Obviously, considering the constraint $\xi = \eta = \zeta$ is a further source of approximation. From the plots shown in Figs. $1a - f$ we can evaluate that the NSF^{II}, NSV^{II}, SV^{II} schemes (this latter two evaluated in absence of spurious wavenumbers) produce $Q \cong 1.67$. On the other hand, one gets $Q \cong 1.28$, $Q \cong 3.14$ and $Q \cong 1.99$ for NSF^{II-IV}, NSR^{II}, NSR^{IV}, respectively.

4 Further tasks and conclusions

The comprehensive analysis of the numerical filters induced by a FV method requires considering also the discretization of diffusive terms as well as the analysis of the SGS modelling. In order to discern the modelling errors from numerical errors, it is suitable to rewrite the SGS term according to

$$\left[\overline{\underline{\nabla}\cdot\mathbf{T}(\overline{\mathbf{v}},\mathbf{v})} - \overline{\underline{\nabla}\cdot\mathbf{T}^{d}_{SGS}}^{d}\right] = \left[\overline{\underline{\nabla}\cdot\mathbf{T}(\overline{\mathbf{v}},\mathbf{v})} - \underline{\nabla}\cdot\mathbf{T}_{SGS}(\overline{\mathbf{v}})\right] +$$

$$+\left[\underline{\nabla}\cdot(\cdot) - \overline{\underline{\nabla}\cdot(\cdot)}^{d}\right]\mathbf{T}_{SGS}(\overline{\mathbf{v}}) + \overline{\underline{\nabla}\cdot\left[\mathbf{T}_{SGS}(\overline{\mathbf{v}}) - \mathbf{T}^{d}_{SGS}\right]}^{d} \quad , \tag{12}$$

having denoted with $\mathbf{T}_{SGS}(\overline{\mathbf{v}})$ the modelled unresolved tensor expressed in continuous form. Therefore, the term in the first square brackets of (12) represents the modelling error (a continuous term), the second one is its LTE contribution and the third one is the equivalent of the discretization error due to grid sampling.

Both dynamic procedure and scale similarity SGS modelling require performing test-filtering over a certain length. Actually, the distinctive feature arises from the adoption of the FV-based integral formulation. Considering, for the sake of simplicity, only the modelling error of the scale similarity case, one obtains

$$e^{\text{mod}}_{ij}(\mathbf{x};\Delta,\Delta_{ex}) = \overline{\frac{\partial(\overline{u}_j\overline{u}_i - u_ju_i)}{\partial x_j}} - \overline{\frac{\partial\left(\overline{\overline{u}_j}^{\Delta_{ex}}\overline{\overline{u}_i}^{\Delta_{ex}} - \overline{u}_j\overline{u}_i\right)}{\partial x_j}}$$

$$= I\sum_{l,m,n}^{\pm\infty}\sum_{p,q,r}^{\pm\infty} p'_j\widehat{E}^{\text{mod}}_{\mathbf{x}}(p,q,r,l,m,n;\Delta,\Delta_{ex})\widehat{u}_j(p,q,r)\,\widehat{u}_i(l,m,n)\,e^{I\sum_{k=1}^{3}p'_kx_k} . \tag{13}$$

The modelling error function was expressed in the wavenumbers space according to

$$\begin{aligned}\widehat{E}^{\text{mod}}_{\mathbf{x}}(p,q,r,l,m,n;\Delta,\Delta_{ex}) &\equiv \widehat{G}_{\mathbf{x}}(p',q',r';\Delta)\Big\{\widehat{\Gamma}_{\mathbf{x}}(p,q,r,l,m,n;\Delta)\\ &-\left[\widehat{G}_{\mathbf{x}}(p,q,r;\Delta_{ex})\,\widehat{G}_{\mathbf{x}}(l,m,n;\Delta_{ex}) - 1\right]\left[\widehat{\Gamma}_{\mathbf{x}}(p,q,r,l,m,n;\Delta)+1\right]\Big\},\end{aligned} \tag{14}$$

being $\widehat{\Gamma}_{\mathbf{x}}(p,q,r,l,m,n;\Delta) \equiv \widehat{G}_{\mathbf{x}}(p,q,r;\Delta)\,\widehat{G}_{\mathbf{x}}(l,m,n;\Delta) - 1$. Here, the notation $\overline{f}^{\Delta_{ex}}$ takes in to account for the "explicit" test-filtering. In principle, the explicit filtering can be arbitrarily chosen to be the top-hat filter, having width $\Delta_{ex} = \alpha\Delta$. The optimal choice of the test-filter length is influenced by the effective width of the primary filter that is induced by the FV discretization. The minimizing of (14) depends therefore on the α ratio once an effective numerical filter length Δ has been estimated.

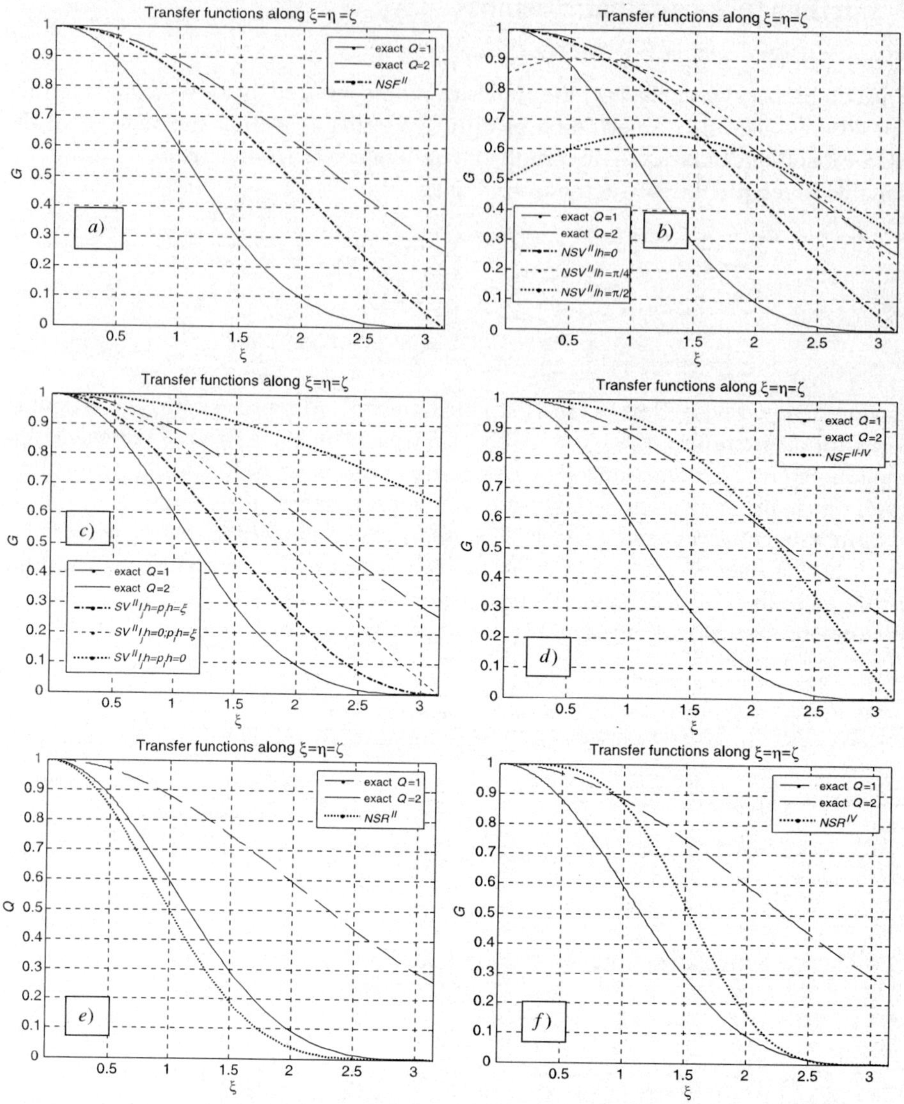

Fig. 1. Implicit transfer functions induced by the FV-based discretization of the convective term plotted versus the analytical one $\widehat{G}_x$ for $Q = 1$ (dot line) and 2 (continuous line), by defining $\xi = p'h$, $\eta = q'h$, $\zeta = r'h$ and forcing the constraint $0 \leq \xi = \eta = \zeta \leq \pi$.

Last but not least, comparisons with DNS data could be more realistic when the fields are post-filtered by means of a numerical filter function that mimics at best the effective implicit filter induced in effect by the FV scheme. Some preliminary FV-based LES solutions, obtained with the new dynamic procedure (7), are presented in Ref. [19].

References

1. P. Sagaut, Large Eddy Simulation for incompressible flows. An introduction. 3ed., Springer, 2006.
2. L. Berselli, T. Iliescu, W. Layton, Mathematics of Large Eddy Simulation of Turbulent Flows, Springer, 2005.
3. M. Lesieur, O. Métais, P. Comte, Large-Eddy Simulations of Turbulence, Cambridge, 2005.
4. B.J. Guerts, Elements of Direct and Large-Eddy Simulations, Edwards, 2004.
5. T.S. Lund, H.-J. Kaltembach, Experiments with explicit filters for LES using a finite-difference method. *Center for Turbulence Research, Ann. Research Briefs*, Stanford University, 91–105, 1995.
6. T.S. Lund, On the use of discrete filters for large eddy simulation. *Center for Turbulence Research, Ann. Research Briefs*, Stanford University, 83–95, 1997.
7. J. Gullbrand, F.K. Chow, The effect of numerical errors and turbulence models in large-eddy simulations of channel flow, with and without explicit filtering, *J. Fluid Mech.*, 495, 323–341, 2003.
8. J.H. Ferziger, M. Peric, Computational Methods for Fluid Dynamics, Springer, 2001.
9. G. De Stefano, F.M. Denaro, G. Riccardi, High order filtering for control volume flow simulations. *Int. J. Num. Methods in Fluids*, 37, 7, 2001.
10. P. Iannelli, F.M. Denaro, G. De Stefano, A Deconvolution-based Fourth Order Finite Volume Method for Incompressible Flows on Non-Uniform Grids. *Int. J. Num. Methods in Fluids*, 43, 4, 431–462, 2003.
11. A. Aprovitola, F.M. Denaro, On the control of the mass errors in Finite Volume-based approximate projection methods for large eddy simulations, Quality and Reliability of Large-Eddy Simulations, ERCOFTAC Series, Vol.12, Meyers, J.; Geurts, B.; Sagaut, P. (Eds.), XX, SPRINGER, 2008.
12. F.M. Denaro, G. De Stefano, D. Iudicone, V. Botte, A finite volume dynamic large-eddy simulation method for buoyancy driven turbulent geophysical flows. *Ocean Modelling*, 17, 3, 199–218, 2007.
13. J. Meyers, B.J. Guerts, P. Sagaut, A computational error-assessment of central finite-volume discretizations in large-eddy simulation using a Smagorinsky model. *J. Comp. Phys* 227, 2007.
14. B.J. Geurts, F. van der Bos, Numerically induced high-pass dynamics in large-eddy simulation, *Phys. Fluids*, 17, 125103, 2005.
15. J.C. Magnient, P. Sagaut, M. Deville, A study of built-in filter for some eddy viscosity models in large-eddy simulation, Phys. Fluids, 13, 5, 2001.
16. S. Hickel, N.A. Adams, Analysis of Truncation Errors and Design of Physically Optimezed Discretizations, Quality and Reliability of Large-Eddy Simulations, ERCOFTAC Series, Vol.12, Meyers, J.; Geurts, B.; Sagaut, P. (Eds.), XX, SPRINGER, 2008.
17. R.J. LeVeque, Finite volume methods for hyperbolic problems, Cambridge Press, 2002.

18. I. Fedioun, N. Lardjane, I. Gokalp, Revisiting Numerical Errors in Direct and Large Eddy Simulations of Turbulence: Physical and Spectral Spaces Analysis, *J. Comp. Physics*, 174, 816–851, 2001.
19. F.M. Denaro, G. De Stefano, A new development of the dynamic procedure in large-eddy simulation based on a Finite Volume integral approach. Application to stratified turbulence. *Theor. Comp. Fluid Dyn.*, submitted and revised, 2009.

Reduced interaction between numerical and model errors through anisotropic filtering

Marcello Meldi and Federico Perini

Dipartimento di Ingegneria Meccanica e Civile, Università di Modena, Italy.
marcello.meldi@unimore.it federico.perini@unimore.it

Summary. The work addresses to a better comprehension of the error assessment in LES due to the coupling between the model and the numerical discretisation.

The possibility to reduce the interactions between the error sources is investigated through the use of an algebraic function correlating the characteristics length of the subgrid model Δ_e to the subgrid scale Δ related to the grid discretization, incrementing the ratio between the two where the scales are poorly resolved.

The analysis, considering a range of grid resolutions as well as subfilter models, has been performed starting from database sets which have been reconstructed with ordinary kriging to estimate the sensitiveness of the strategy with respect to the simulation parameters. The results indicate that a reduction in the error cost function can be achieved for most subfilter models and that the approach looks quite stable for a moderate range of the grid resolution.

Key words: Large-eddy Simulation, anisotropic filtering, errors interaction

1 Introduction

The main concern regarding LES in industrial applications is the definition of criteria for the error assessment, as the increasing computational resources make LES an operable tool to the simulation of High-Reynolds complex flows.

In *traditional* LES the filter is associated to the grid discretisation and the numerical method: the scales smaller than the geometric width Δ are not captured by the simulation and are defined as *subgrid scales* [1].

While Δ is actually straightforward defined on structured isotropic grids, its definition may be arguable when dealing with anisotropic or unstructured grids, as in most industrial applications.

The problem has been approached by many authors ([2], [3]) but all the proposals till now presented about this topic focus on the geometry and do not take into account the flow behaviour.

A second characteristic length, referred as Δ_{SGS}, is introduced in the simulation by the *SGS model* to relate the modeled scales to the computed ones: a common approach is to set this width equal to Δ. As the SGS model is not $perfect$, i.e. it does

M.V. Salvetti et al. (eds.), *Quality and Reliability of Large-Eddy Simulations II*,
ERCOFTAC Series 16, DOI 10.1007/978-94-007-0231-8_8,

not reproduce all the physical characteristic of the *subgrid scales*, an error source is introduced in the system: the way it interacts with the numerical error is largely unpredictable due to strongly non-linear error dynamics. Furthermore Ghosal [5] showed that the numerical error is of the same order of the modelling error for most of the wave numbers and proposed as a possible solution the combination of pre-filtering and high order schemes. Since this approach increases noticeably the computational cost and the complexity of the problem, the main idea of our work is the adoption of a modified effective Δ_{SGS} , in the following Δ_e, computed considering a local increase with respect to the geometric width Δ and extending it only where needed. In particular, the Δ_e adopted in the SGS models is taken as the product of the geometric width by a function, f_{UAED}, which characterizes the local resolution in terms of presence of the smallest scales, where the interactions between numerical and model errors are particularly strong. This kind of approach should reduce the interaction between the two main sources of error and produce a lower discretisation dependence of the error itself.

This approach has been intensively tested to check its robustness when coupled with different subgrid-scales models, numerical discretisations or grids.

The paper is structured as follows: in section 2 LES and filtering approaches are presented as well as an exhaustive explanation of the *universal anisotropic* Δ_e (*UAED*) approach is furnished. The main results by the application of the anisotropic Δ_e to two different test cases are summed up in section 3, while in the last section 4 conclusions are drawn.

2 Theoretical LES

2.1 The filtering operation

We consider here Newtonian, incompressible, three-dimensional and time-dependent flows. The scales separation in LES is achieved by the application of a scale high-pass filter, which is mathematically formulated as a convolution product in the physical space. In *traditional LES* this separation is connected to the grid discretization and Δ is the correlated physical length that separates the resolved scales from the unresolved ones: in this context the scales not resolved due to a not sufficient grid resolution are supposed to be filtered and thus modelled adding the τ_{SGS} term to the momentum equation.

2.2 LES Models

In the following analysis four different models have been tested to check their sensitiveness to the application of the UAED filtering approach. The model chosen are the One-equation Turbulent Energy Model [6], the Smagorinsky Model [7], the Variational multi-scale approach [8] applied to the Smagorinsky one and the WALE model [9].

It is actually useful to remark that, as each model relate in a different way the ν_{SGS} with Δ_e, we can expect that the overall results will be sensitive of the different coupling between model and $UAED$. In the following sections, the models will be respectively refereed ans ONE, SM, VMS and $WALE$.

2.3 UAED Approach

This paper focuses on the possibility to mask the numerical error by means of the model one, as it generally looks like that $\Delta_e/\Delta > 1$ is beneficial for pursuing a global error reduction and the application of a Δ_e significantly higher than the Δ produces grid-independent numerical solutions [10].

A high ratio value is not anyway necessary on the whole physical domain, but just locally where the numerical error becomes large if compared to the model error, a situation most common when estimating the resolution of the smallest resolved scales. The strategy we are proposing relies indeed on the possibility to extend the effective width Δ_e only where the scales are poorly resolved through the definition of an algebraic function relating Δ_e and Δ:

$$f_{UAED} = \frac{\Delta_e}{\Delta} = 1 + C_{UAED}\, \Theta \, \tanh(\Psi)^{\frac{1}{\Theta}} \tag{1}$$

Θ and Ψ are respectively a parameter estimating the global and local resolution, while C_{UAED} is a constant. The shape of the function f_{UAED} for a value of $C_{UAED} = 5$ is presented in figure 1(a).

The parameter Ψ is an indicator of the wavelength of the resolved scales: as the parameter increases, smaller scales are locally present. Since vorticity is a quantity which clearly indicates the presence of small scales, we decided to define the parameter Ψ as the normalised magnitude of the vorticity vector

$$\Psi = \sqrt[4]{\frac{\omega}{\max(\omega)}} \tag{2}$$

the fourth root of Ψ being chosen as in that condition UAED results less case sensitive and a greater part of the volume is affected by the anisotropic filtering, fig. 1(b).

While the parameter Ψ is a field parameter, i.e. every computational cell has its own Ψ value, the parameter Θ is a scalar indicating the global resolution of the system, and his value determines the shape and the intensity of the function f_{UAED}. The form factor Θ, defined by the eq. 3, shapes the curve tending to flatten it out when vorticity increases (which is the case of well-resolved LES) while ensures a strong effect on most of the scales when the field is poorly resolved, as from picture 1(a).

$$\Theta = \frac{2 < \bar{U} >^2}{< \bar{U} >^2 + l^2 < \omega >^2} \tag{3}$$

l is taken as twice the characteristic length of the test case and $\bar{U}$ is the velocity magnitude. $<>$ denotes a spatial average.

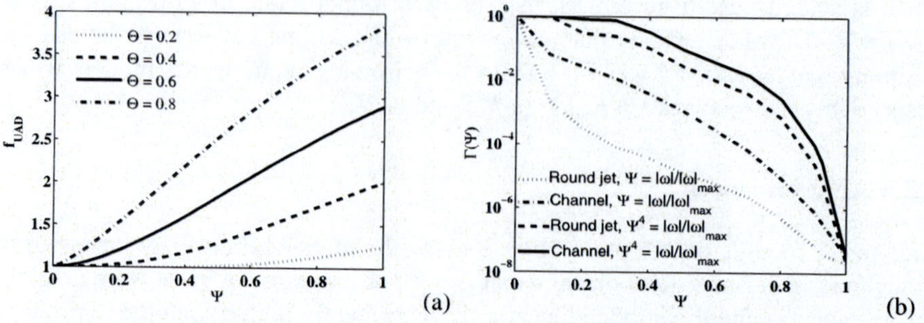

Fig. 1. Algebraic f_{UAED} function (a) and Volume pdf integrated from Ψ to 1 (b).

3 Applications and results

The approach described in section 2.3 has been tested on two different test cases, the plane channel flow and the round jet flow. Five different hexahedral structured grids, in the following named from 1 to 5, have been used to test the sensitiveness of the *UAED* approach to different resolutions.

As we followed the definition of geometric Δ proposed by Deardorff [2], we decided to build up the grids applying an isotropic coarsening between each one and the subsequent with a constant ratio $\sqrt{2}$, starting from a first grid able to well-resolve the flow field. Considering that grid 1 is the starting grid, its is quite simple to observe that grid 5 presents cells with the geometric Δ 4 times larger. Details regarding each test case grids are reported in the next sections. The time step for the simulations has been set constant and has been chosen to guarantee a Courant number lesser than 0.5; as the time step has a filtering effect over the solution, it has been scaled in the same way the physical dimensions have been coarsened.

C_{UAED}	0	2	4	8	16	32

Table 1. C_{UAED} constants used in the preliminary analysis.

The simulations have been performed on a finite volume solver in Gaussian formulation using the Crank-Nicholson second order scheme in time and a centred second order scheme in space.

A first set of preliminary simulations have been carried out, combining the 4 models with 6 different values of the constant C_{UAED}, as in table 1. A response surface to value the quality of the results has been generated through the interpolation of the starting matrix using ordinary kriging [11] with the goal to find an optimum value for each model.

A second set of simulations has then been performed, combining the 4 different models with the 5 computational grids. As in the preliminary case, a response surface has been generated interpolating the resulting matrix with ordinary kriging to describe the behaviour of the cost function F:

$$F = 0.5\varepsilon_U + 0.25\varepsilon_d + 0.25\varepsilon_s \tag{4}$$

$$\varepsilon_U = \int_V \frac{(U_{LES} - U_{DNS})^2}{U_{DNS}^2} dV \tag{5}$$

$$\varepsilon_d = \sqrt[3]{\varepsilon_{uu} \cdot \varepsilon_{vv} \cdot \varepsilon_{ww}} \tag{6}$$

$$\varepsilon_s = \sqrt{\varepsilon_{uv} \cdot \varepsilon_{uw}} \tag{7}$$

$$\varepsilon_{ij} = \int_V \frac{(R_{ij}^{LES} - R_{ij}^{DNS})^2}{(R_{ij}^{DNS})^2} dV \tag{8}$$

where U is the time averaged streamwise velocity and R_{ij} represents the generic component of the Reynolds stress tensor.

The main concern about kriging application was the definition of a consistent length scale regarding the constant, the models and the grids. First of all, in each analysis the length scale has been set equal for both axes, as to get *square matrices*, then we defined different partitions for each parameter. When dealing with the models, the length scale has been equally divided while when dealing with the constant it has been linearly divided with the constant value. The same approach has been conserved for the parameter set on the grid length scale, which has been divided linearly with the coarsening ratio.

Plane Channel Flow

The results of the simulations in a plane channel flow at $Re_\tau = 395$ are here illustrated.The computations were carried out on structured grids with constant spacing in the streamwise and spanwise direction and hyperbolic spacing in the wall-normal direction. The ratio between contiguous cells volumes is always lesser than 1.1 and it tends towards the unity approaching the wall. The main characteristics of the grids are presented in table 2. In the present analysis, the x axis is coincident with the streamwise direction, the y axis with the wall-normal direction, the z axis with the spanwise direction and the computational domain is taken equal to $2\pi h \times 2h \times \pi h$, being h the channel half-width.

As clear from the data in the table, not all the grids are able to capture the necessary scales to correctly simulate the flow at the wall: since our research activity aimed at checking how error behaves more than trying to reduce the error, we decided to strictly keep the concept of isotropically coarsening the grid even in the wall-normal direction.

The one-equation and the Smagorinsky model do not behave correctly approaching at the wall, as ν_{SGS} does not correctly scale: for this reason, the models have been corrected with the addition of a Van Driest damping term. The results are compared with the DNS data by Kim [12] and the statistical moments contributing to the definition of the cost function F have been normalised over the computed u_τ.

Two series of simulations have been carried out; in the first one both the convective and the diffusive term were discretised with centred second-order schemes,

Grid	nr of cells	Δx^+	Δy^+	Δz^+
1	80 X 100 X 100	30	$1.5 \rightarrow 15$	12
2	56 X 70 X 70	42.5	$2.1 \rightarrow 21$	17
3	40 X 50 X 50	60	$3 \rightarrow 30$	24
4	28 X 35 X 35	85	$4.2 \rightarrow 42$	34
5	20 X 25 X 25	120	$6 \rightarrow 60$	48

Table 2. Main characteristics of the grids used for the Plane Channel Flow case.

while in the second one they were discretised with a centred fourth-order scheme. Both sets of preliminary simulations where performed on grid 3 to limit the computational resources needed.

Figure 2 shows the cost function F isocontours normalised in each row by the value of F at $C_{UAED} = 0$. The application of $UAED$ in combination with the Van Driest damping deteriorates the quality of the results, and the phenomenon gets worse increasing the value of C_{UAED}: this is probably due to the fact that the beneficial effect that the Van Driest damping performs combined to the model, which imposes a decrease of Δ_e towards the wall, is counteracted by $UAED$ which increases the Δ_e dimensions in the boundary layer. This behaviour is connected to a sensible deterioration in u_τ predictions for the model corrected with the Van Driest Damping.A sensible reduction of the error is instead observed for both the discretisation schemes when $UAED$ is combined with variational multi-scale model, while WALE produces more accurate results only when second-order schemes are used. These results may be justified by the fact that in the variational multiscale-approach the model acts only over the small scales, while the WALE model acts on all the flow scales: a high value of the constant C_{UAED} combined with a high order scheme makes the model error to mask completely the smaller numerical error.

The optimum value for the constant C_{UAED} has been chosen through a minimum analysis of the cost function, represented by the dotted line in figure 2 (a) and (b): as the best value of C_{UAED} for the models corrected with the Van Driest damping is actually 0, we anyway set a symbolic value of 1 as, although we expect that the results will be worst than without the application of $UAED$, our aim is to test the stability of the error when Δ_e is modified.

Following the optimum C_{UAED} value definition for each model, a database of simulations has been generated crossing the model applied and the grid resolution. The same database has been replicated without the anisotropic filter and furthermore a set of coarse DNS has been simulated too, for a result of 45 simulations for each of the discretisation schemes used.

The u_τ response surface resulting from second order *traditional* LES database is presented in fig. 3 (a): the data, normalised on the u_τ value computed by Kim [12], look quite similar for all the models and deteriorate as the grid coarsens. Models used in combination with the Van Driest damping produce slightly more accurate results as scale resolution decreases if compared to the other models or coarse DNS. $UAED$ approach combined with the optimum constant did not perturb significantly u_τ predictions, while the fourth order scheme application produces more accurate results as shown in fig. 3 (b).

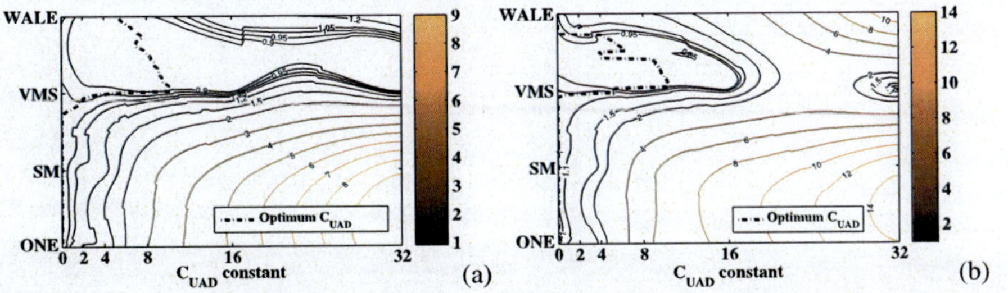

Fig. 2. *F* response surface on grid 3 using second (a) and fourth (b) order schemes.

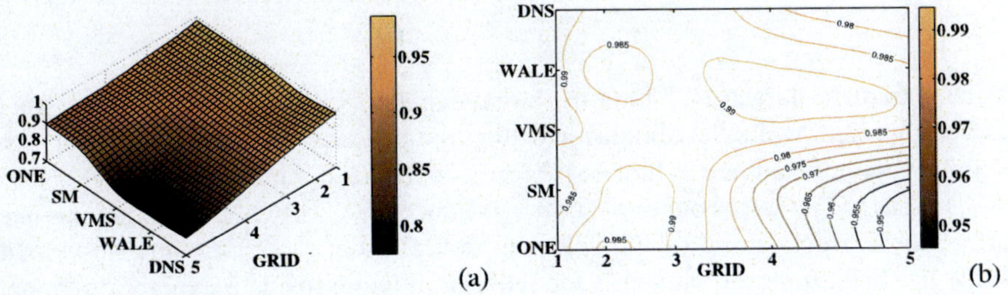

Fig. 3. Second order u_τ response surface (a) and u_τ ratio between second and fourth order prediction (b).

In picture 4 the cost function F normalised with the corresponding values computed through standard LES is presented. The most interesting result is a quite good consistency for the "*refined*" part of the response surface, meaning that the approach is actually able to adapt itself through the variation of the global shape parameter Θ to the correct level of smoothing. Conversely, the main problem emerging from this analysis is indeed that the behaviour becomes quite unpredictable when *UAED* is combined with coarse grids and it is arguable if this happens for a poor resolution of the boundary layer or if the approach does not work properly in presence of a large numerical error.

Round Jet Flow

In this section, the results of the analysis carried out on a round jet flow test case are presented. A maximum Reynolds number, referenced to the peak of the mean streamwise velocity profile and the initial jet diameter, up to $Re_D = 21000$, has been considered. The computations have been performed on five different structured grids, on the same regular domain defined by $[0, 15D] \times [-5.5D, 5.5D] \times [-5.5D, 5.5D]$, the most refined one being discretised as the finest grid of Sagaut and Lê [13]. The discretisation on each cross section (y, z planes in Table 3) along the streamwise direction has been kept constant in the jet zone, while gaussian-shaped coarsening

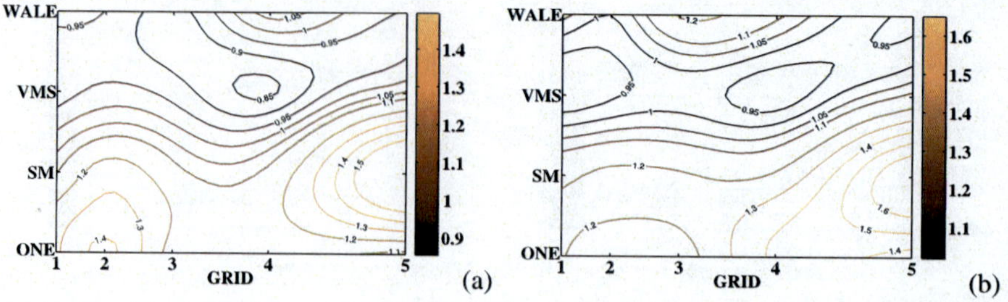

Fig. 4. Error response surface of *UAED* LES compared to traditional LES for second order (a) and fourth order (b) accuracy.

has been applied outwards. Along the streamwise direction x, a constant increment has been applied, while the dimension of the first cell layer has been chosen in order to comply with the advice in [14, 15]. More details are summarised in table 3.

As far as the jet inlet boundary surface is concerned, particular attention has been paid to the correct setting of the velocity profile: a flat, $1m/s$ streamwise velocity profile has been imposed outwards the jet region, while inside a hyperbolic-tangent function [14] has been set:

$$\frac{u(r)}{U_{ref}} = \frac{1}{2}\left(1 - \tanh\left[\frac{r_{\text{jet}}}{4\theta_{\text{jet}}}\left\{\frac{r}{r_{\text{jet}}} - \frac{r_{\text{jet}}}{r}\right\}\right]\right). \tag{9}$$

Moreover, a white-noise-spectrum turbulent fluctuation has been imposed on the whole inlet surface reference velocity field, adding up to, respectively, 4% along the x axis and 0.1% in the radial directions, in agreement with the tabulated experimental data [16].

Grid	nr of cells	Jet zone $\Delta y = \Delta z$	Outwards $\Delta y = \Delta z$	Jet inlet Δx	Outlet Δx
1	100 X 120 X 120	$D/36$	$1.00D$	$D/16$	$0.3D$
2	70 X 85 X 85	$D/25$	$1.15D$	$D/12$	$0.4D$
3	50 X 60 X 60	$D/18$	$1.35D$	$D/8$	$0.6D$
4	35 X 42 X 42	$D/13$	$1.60D$	$D/6$	$0.85D$
5	25 X 30 X 30	$D/9$	$1.90D$	$D/4$	$1.2D$

Table 3. Main characteristics of the grids used for the Round Jet case.

The results of the computations have been compared with detailed experimental data from the Ercoftac Database [16]. Since the test case does not present a periodic behaviour in the streamwise direction, the cost function F has been evaluated as the mean value of the cost functions defined on four yz planes at $x/D = 0.2; 2.0; 4.0; 8.0$ respectively.

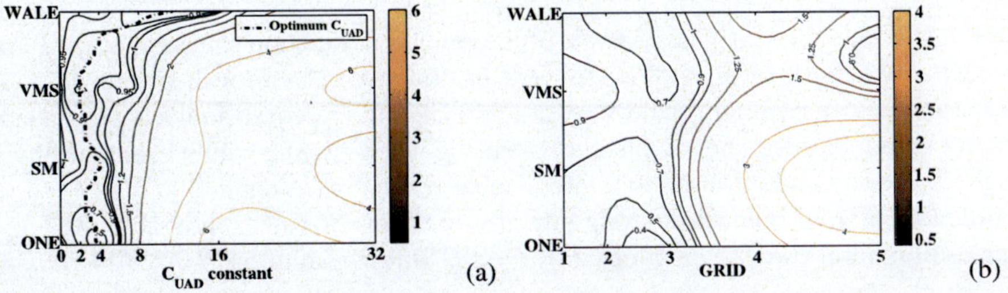

Fig. 5. *F* response surface on grid 3 (a) and on the five grids with optimum C_{UAED} values (b) for the Round Jet test case.

Figure 5 (a) shows the response surface of the cost function *F* on grid 3, crossing the four models and the six C_{UAED} values previously introduced, and normalised as in section 3. The thick dotted line identifies the optimum value of C_{UAED} on the response surface, depending on the model chosen. From the figure, the reduction in cost function *F* due to *UAED* is strong for each of the models, and approaches 30% and 50% circa, for the variational multi-scale and the one-equation models, respectively. Another interesting feature is that the optimum value of C_{UAED} seems to be quite independent on the model chosen, and anyway it is generally included in the $[2,4]$ range. The only slightly different behaviour is seen for the coupling of *UAED* with the WALE model, where the optimum is reached at $C_{UAED} = 12$ circa; nevertheless, in this case the behaviour of *UAED* is much more straight along the C_{UAED} range, and a reduction in *F* by approximately 10% can be achieved even for it being in the optimum range identified for the other models. The *UAED* performance deteriorates as C_{UAED} increases to values greater than eight.

The behaviour of the error cost function *F* on each grid, obtained setting the optimum C_{UAED} value for each of the models considered, has been reconstructed through kriging and plotted in figure 5 (b), normalised with respect to *traditional* LES. As it appears from the response surface, the application of UAED produces a stable global error reduction as the grid is refined; the total reduction in cost function adding up to more than 50% on the plain LES value. On the other side, a drawback of the UAED seems that, the approach unpredictably affects the results on coarse grids, as – even if in a framework of increasing error for each of the models – it appears much more stable when in presence of VMS or WALE, while instead the normalised cost function rapidly climbs up to values greater than 4 when adopting the one-equation and the Smagorinsky models.

4 Concluding Remarks

A local extension of the subgrid wavelength Δ_e in LES has been investigated.

Different grid resolutions, models and discretisation schemes have been considered on two well established test cases and the results have been compared to

DNS/experimental data. The analysis of the error cost function stresses how this approach is beneficial for shear flows and can be effective even for wall-bounded flows, if the model can correctly scale ν_{SGS} at the wall. The approach looks able to adapt itself to produce adequate wavelength extension when dealing with reasonable physical resolution, while it leads to unpredictable results when applied to really coarse grids: as a high Θ induces a strong effect even at low Ψ values, probably the large scales of motion come to be affected by *UAED* at low resolution.

References

1. G.S. Winckelmans et al., Physics of Fluids **13**, 1385-1403 (2001).
2. J.W. Deardorff, Journal of Fluid Mechanics **41**, 453-465 (1970).
3. J. Bardina, J.H. Ferziger, W.C. Reynolds, AIAA Paper 80-1357 (1980).
4. J.C. Magnient, P. Sagaut, M. Deville, Physics of Fluids **13**, 1440-1449 (2001).
5. S. Ghosal, Journal of Computational Physics **125**, 187-206 (1996).
6. A. Yoshizawa, Journal of the Physical Society of Japan, Vol. **54** (1985).
7. J. Smagorinsky, Monthly Weather Review **91**(3) 99-165 (1963).
8. T.J.R. Hughes, Physics of Fluids **13** (2), 505-512 (2001).
9. F. Nicoud, F. Ducros, Flow, Turbulence and Combustion **62**, 183-200 (1999).
10. P.J. Mason, N.S. Callen, Journal of Fluid Mechanics **162**, 439-462 (1986).
11. J.C. Jouhaud, P. Sagaut, Journal of Fluids Engineering **130** (2008).
12. J. Kim, P. Moin, R. Moser, Journal of Fluid Mechanics **177**, 133-166 (1987).
13. P.Sagaut, T.H. Lê, Direct and Large Eddy Simulation II, Kluwer, 81-92 (1997).
14. D.J. Bodony, PhD Thesis, Stanford University (2005)
15. P. Sagaut, Large-Eddy Simulation for Acoustics, Cambridge (2007).
16. F. Anselmet, L. Fulachier, ERCOFTAC database, Exp. C38 (1993).

Analysis of the subgrid models in the flow between rotating discs

Artur Tyliszczak and Andrzej Boguslawski

Institute of Thermal Machinery, Czestochowa University of Technology
Al. Armii Krajowej 21, 42-200 Czestochowa, Poland
atyl@imc.pcz.czest.pl, abogus@imc.pcz.czest.pl

Summary. The paper presents the results of the LES computations of the flow between rotating discs using high-order numerical code. The results of simulation for fully turbulent flow, obtained using various subgrid models (dynamic Smagorinsky [6], WALE [11] and HPF [18]) are compared with available DNS data. Besides the averaged quantities an additional attention was devoted to the temporal evolution of the flow field. The dynamic Smagorinsky and WALE models can be regarded as a parameter free while in the case of HPF model a high-pass filtering procedure requires some parameters which influence the model behavior. We found that in this type of flow this procedure should be directly related to the numerical scheme applied. Otherwise the results depend on the initial flow conditions and the predicted evolution of turbulent flow may be distorted.

Key words: subgrid modeling, rotating cavity, instability

1 Introduction

Flows in the cavities between rotating discs are the subject of a constant interest because they appear in many practical applications, among others in the turbomachines (turbines or compressors) but also in many applications in process engineering. Although, in general, both discs could rotate in a co-rotating or counter-rotating systems, the most common situation, at least in the turbomachinery, is a flow resulting from the rotor-stator interaction. In real applications the cavity between the rotor and stator most often must be classified as a complex geometry due to the fact, that the disc, shaft and shroud design is affected by the need for mechanical or thermal stress minimization. Numerical simulations of such a flow taking into account complex geometry and an unsteady flow behavior would require a prohibitively large computational time, especially when considering LES approach. Hence, the majority of numerical studies of this flow type was performed using a simplified geometry composed of two flat discs and two cylinders, the inner one called shaft and the outer one called the shroud. This type of configuration is one of the simplest where the the boundary layers are three-dimensional (3D) giving possibility to study the influence of the 3D mean flow structures on the transition process and turbulence

M.V. Salvetti et al. (eds.), *Quality and Reliability of Large-Eddy Simulations II*,
ERCOFTAC Series 16, DOI 10.1007/978-94-007-0231-8_9,

structure. A common feature of these flows is the coexistence of regions where the flow is laminar, transitional or turbulent with strong inhomogeneity effects - these are the reasons why the simulations with classical Reynolds Averaged Navier-Stokes approach had only limited success.

Despite the simple geometry the flow in the cavity between rotating and stationary discs is controlled by a variety of geometrical parameters like cylinders radii, cavity aspect ratio or type of the confinement understood as the cylinders attached to the rotor or stator. Although numerous theoretical and computational works on this flow type [2, 5, 8, 13, 14, 15, 21] have not led to full understanding of the transition process, it seems that they revealed fundamental flow features. In the limit of high rotation the flow between the infinite rotating and stationary infinite discs is of Batchelor type and consists of two layers separated by an inviscid rotating core. Close to the rotating disc the Ekman layer is established while on the stationary disc the Bödewadt type layer is formed. The transition in both layers is related to type I and type II generic linear instabilities studied by the stability theory [14] and identified experimentally [5, 8, 13]. The Ekman layer is more stable hence the first transition to unsteadiness appears in the boundary layer on stationary disc. Type I instability mechanism is manifested in the form of the system of travelling spiral vortices while type II instability leads to the system of axi-symmetric vortex rings traveling toward the inner disc. The particular flow structure, number of present vortices, critical Reynolds numbers, range of Reynolds numbers for which both flow structures could coexist depend on the confinement geometry but in many cases both unstable modes could be identified. The studies mentioned above considered mainly isothermal flows while in practical applications the non-isothermal conditions are also of primary importance. The weakly non-isothermal flow between the rotating discs was recently studied numerically using DNS/LES approach by Tuliszka-Sznitko and Zielinski [21].

The main goal of the present LES studies of the flow between the rotating and stationary discs was to evaluate sensitivity of the mean and transient flow characteristics depending on the subgrid scale model applied. Apart from the simple geometry the flow in rotating cavity is very demanding from the point of view of LES computations. This is because the subgrid models used in LES have to cope with laminar, transitional and turbulent flow conditions. Lygren and Andersson [1] performed comparative LES study in a sector of a rotor-stator cavity applying three eddy-viscosity type models, i.e. the dynamic Smagorinsky [6] and two mixed models of Zang [23] and Vreman [22]. Comparing to DNS results they found that the mixed model of Vreman gives the best overall results, however, they also suggested that in order to get better agreement the subgrid modeling should be improved. Severac et al. [17] and Severac and Serre [16] applied the spectral vanishing viscosity (SVV) technique [20] in their pseudo-spectral Chebyshev-Fourier code. As the SVV technique maintains spectral convergence of the pseudo-spectral code, their results obtained for moderate ($Re = 7 \times 10^4$ [17]) and high Reynolds numbers ($Re = \mathcal{O}(10^5, 10^6)$ [16, 17]) were in very good agreement with both DNS and experimental data. The pseudo-spectral methods together with SVV are however limited to relatively simple geometries, considerably different from those in real applications, where they cannot be applied without special treatment. Therefore, in most cases the finite volume or finite difference codes are used which do not imply special requirements on spectral behavior of the damping terms. Choosing from tens of existing subgrid models, in the present study we performed the LES computations using well known dynamic Smagorinsky model [6] and two relatively new but already well established models proposed by Nicoud and Ducros [11] (WALE - Wall Adapting Local Eddy) and Stolz et al. [18, 19] (HPF - High-Pass Filtered) Smagorinsky model. These models belong to the so-called eddy viscosity (subgrid viscosity) type [7, 12], where the subgrid stress tensor

$$\tau_{ij} = \overline{u}_i\overline{u}_j - \overline{u_i u_j} \tag{1}$$

is modeled as

$$\tau_{ij} - \tau_{kk}\delta_{ij} = 2\nu_t\overline{S}_{ij} \quad \text{with} \quad \overline{S}_{ij} = \frac{1}{2}\left(\frac{\partial\overline{u}_i}{\partial x_j} + \frac{\partial\overline{u}_j}{\partial x_i}\right) \tag{2}$$

The overbar in the above equations denotes LES filtering procedure and $\overline{S}_{ij}$ is the strain rate tensor. The differences in subgrid modeling rely on how the subgrid viscosity is evaluated, and so, in the case of dynamic Smagorinsky model the expression is $\nu_t = (C_d\Delta)^2\sqrt{2\overline{S}_{ij}\overline{S}_{ij}}$, where the model parameter C_d is adjusted dynamically depending on local flow parameters. In the case of WALE approach we would have [11]

$$\nu_t = (C_w\Delta)^2 \frac{\left(\mathfrak{S}^d_{ij}\mathfrak{S}^d_{ij}\right)^{3/2}}{\left(\overline{S}_{ij}\overline{S}_{ij}\right)^{5/2} + \left(\mathfrak{S}^d_{ij}\mathfrak{S}^d_{ij}\right)^{5/4}} \tag{3}$$

where

$$\mathfrak{S}^d_{ij} = \frac{1}{2}\left(\overline{g}^2_{ij} + \overline{g}^2_{ji}\right) - \frac{1}{3}\delta_{ij}\overline{g}^2_{kk} \quad \text{with} \quad \overline{g}_{ij} = \frac{\partial\overline{u}_i}{\partial x_j} \tag{4}$$

and the model constant C_w equals to 0.5. Finally, the HPF model reads as

$$\nu_t = (C_h\Delta)^2\sqrt{2\overline{S}^{HPF}_{ij}\overline{S}^{HPF}_{ij}} \tag{5}$$

where the strain rate tensor $\overline{S}^{HPF}_{ij}$ is calculated using high-pass filtered velocity field, which is obtained by subtracting low-pass explicitly filtered quantities from the unfiltered ones. Additionally, the deformation tensor in Eq.2 is also computed as $\overline{S}^{HPF}_{ij}$. The model constant C_h and the shape of the filter are the model parameters. The high-pass filtering procedure permits to select a range of scales which are important in calculation of the subgrid viscosity, however, as we will be shown in next sections, this freedom may also lead to unexpected model behavior.

The models chosen appear to be very accurate in channel and pipe flows but except for the dynamic Smagorinsky model the authors did not find literature data concerning their application in rotating geometries. In view of the LES computations of flow in rotating cavities this paper may be regarded as an extension of the work of Lygren and Andersson [1] .

2 Computational domain and numerical method

Computational domain is shown in Fig. 1, the radius of the inner (called shaft) and outer (called shroud) cylinders are denoted by R_0 and R_1. The radial extent of the annular cavity between the cylinder is $\Delta R = R_1 - R_0$ that related to the cavity height $2h$ defines the parameter called aspect ratio $L = \Delta R/2h$. The aspect ratio parameter together with the so-called curvature parameter, defined as $R_m = \frac{R_1+R_0}{\Delta R}$, fully describes the cavity geometry. The Reynolds number is defined as $Re = \Omega R_1^2/\nu$, where Ω is the angular velocity of the upper disc and ν is the kinematic viscosity. In general, both the inner and/or outer cylinder can rotate or can be stationary. The motion inside the cavity is described by the Navier-Stokes and continuity equations for incompressible fluid. The equations are written in the cylindrical coordinate system (r, θ, z) in the velocity (V_r, V_θ, V_z) and pressure formulation. The non-slip boundary conditions are assumed on the walls. The angular velocity of the upper disc and the velocity of the rotating cylinder is computed as $V_\theta = \Omega r$, where r is the local radius, the remaining velocity components are equal to zero. The numerical algorithm is based on the projection method [4], where the divergence free velocity field is obtained during the correction step after evaluation of the pressure field from the Poisson equation. The Poisson equation in the cylindrical coordinate system is first transformed into the periodic direction and then the diagonalization technique is applied leading to a set of one dimensional systems which are solved by a direct method. The Navier-Stokes equations are advanced in time with three step Runge-Kutta method, the spatial discretisation is performed using the Fourier pseudo-spectral method [3] in the periodic direction and 6-th order compact scheme [10] in the axial and radial directions. Without the subgrid model the algorithm becomes unstable for higher Reynolds number and therefore to prevent this instability we explicitly filter the solution every time step. The stabilizing low-pass filter was tuned in such a way that it acts on the smallest resolved scales only. The stabilizing filter was applied in the directions of the compact discretization (i.e. axial and radial) and it ensured stable high order solutions. Therefore, we could expect that the role of the subgrid modeling should lead to an improvement of the solutions accuracy but not to the stabilization of solution.

2.1 Parameters of HPF subgrid model

The high-pass filtering quantities u^{HPF} are obtained by subtracting low-pass explicitly filtered quantities $\widetilde{\overline{u}} = G * \overline{u}$ from the unfiltered ones $\overline{u}$, i.e. $u^{HPF} = H * \overline{u} = (I - G) * \overline{u}$, where G denotes low-pass filter. Detailed analysis of high-pass filtered Smagorinsky model together with the high and low-pass filtering procedures applied in subgrid modeling may be found in [9]. We remind that in our numerical code at the end of each time step we apply additional stabilizing low-pass filtering procedure ($u^{LPF} = L * \overline{u}$). This modifies a range of scales which are used in HPF model, because the quantities which are then high-pass filtered are in fact u^{LPF} and not $\overline{u}$. The transfer functions of the stabilizing low-pass filter ($\hat{L}(\overline{u})$), the second order low-pass filter $\hat{G}(\overline{u})$ and high-pass filters ($\hat{H}(\overline{u})$) and $\hat{H}(\overline{u}^{LPF})$) are shown in Fig. 2. The

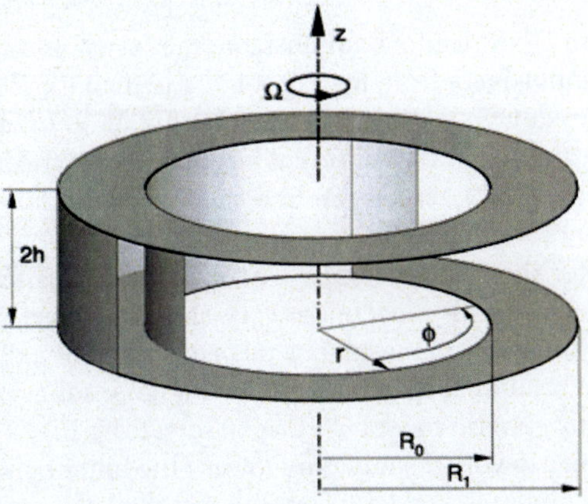

Fig. 1. Computational domain.

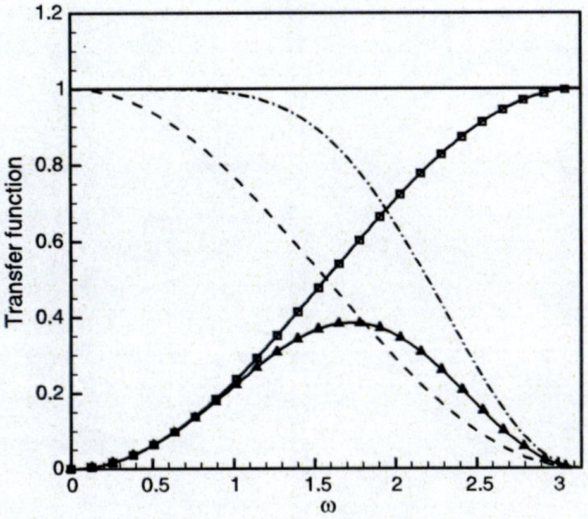

Fig. 2. Transfer function of the stabilizing low-pass filter ($\hat{L}(\overline{u})$) (dash-dot line), second order low-pass filter $\hat{G}(\overline{u})$ (dashed line) and high-pass filters: $\hat{H}(\overline{u})$ (line with squares) and $\hat{H}(\overline{u}^{LPF})$ (line with triangles).

combination of stabilizing low-pass procedure with HPF model was verified in periodic channel flow computations with $Re_\tau = 180$ and 395 leading to the accurate results.

3 Results and discussion

The test case studied corresponds to DNS results of Severac and Serre [16] for the cavity parameters $L = 5$ and $R_m = 5$ with $Re = 7 \times 10^4$ and both the inner and the outer cylinder are stationary. The Reynolds number chosen is above the critical value and according to the DNS results both unstable modes should coexist. The simulations were started impulsively from $Re = 2 \times 10^4$ and then the Reynolds number was increased every time step up to $Re = 7 \times 10^4$ after about 30 rotations. Computations were performed starting from two different initial velocity fields: 1) zero mean and zero fluctuating velocity; 2) zero mean and random fluctuating velocity with amplitude equal to 1% of the maximum angular velocity component. In this paper we present the results for the first case only, where substantial differences exist, in the second case the results for all models are very similar. Figure 3 shows complexity of the flow field, it illustrates the instantaneous isosurface of axial velocity after 150 rotations at constant rotational speed obtained with three subgrid models considered as well as without any subgrid model. As can be seen from Fig. 3 without the subgrid model the axi-symmetric vortices traveling toward the inner cylinder are reproduced while spiral vortices manifesting the instability of type I are not present. Similar re-

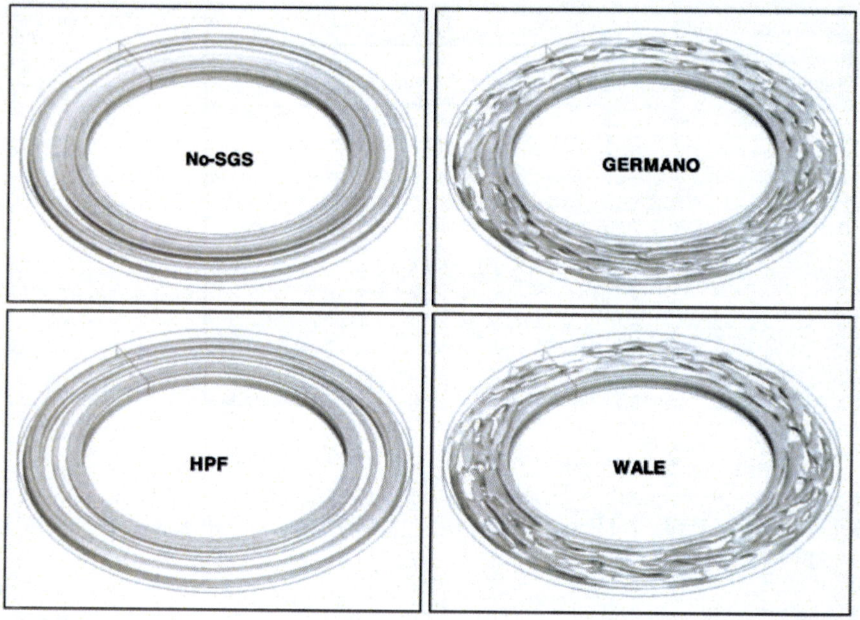

Fig. 3. Instantaneous isosurfaces of the axial velocity component for various subgrid models.

sults are obtained for HPF model, the spiral vortices are not seen in the figure and they can only be found after very careful selection of the isosurface, but even then they are revealed by a very weak wrinkles on the vortex rings located close to the outer disc. Very small influence of the HPF model is probably caused by the stabilizing filtering procedure, which to some extent removes the part of the spectrum which is then used in the HPF model. Indeed, when the the small scales are absent, then the subgrid viscosity is close to zero because in Smagorinsky model the strain rate tensor is calculated using velocity field with partially removed small scales. In the case of dynamic Smagorinsky and WALE models a significant influence of the model is visible. One can see that the regular spiral vortices corresponding to the instability of type I appeared, while the vortex rings are located only very close to the inner cylinder and reveal axi-symmetric shape. In this case one can easily distinguish between both unstable modes. The time series and the spectrum of the axial velocity component obtained for HPF and WALE models in the mid-radius of the cavity $r = R_0 + (R_1 - R_0)/2$ and approximately at the edge of the Bödewadt layer ($z = 0.2$) are shown in Fig. 4. A periodical character of the flow is established after the initial transient solution, it can be seen that in both cases there is a basic frequency equal to 0.44Ω. This frequency is related to the axi-symmetric vortex (type II instability) traveling toward the inner cylinder. In the case of the solution obtained with the HPF model this vortex is followed by smaller and also axi-symmetric vortices characterized by doubled frequencies. These smaller structures disappear in the results obtained with WALE and dynamic Smagorinsky models and therefore one may say that there are substantial qualitative differences between the HPF and WALE or dynamic Smagorinsky models. Additionally, we notice that the values of the radial velocity for the WALE model (the same concerns dynamic Smagorinsky model) are negative for the whole probing time while in the case of the HPF model the radial velocity oscillates close to zero. This is because the unstable stator layer is thicker

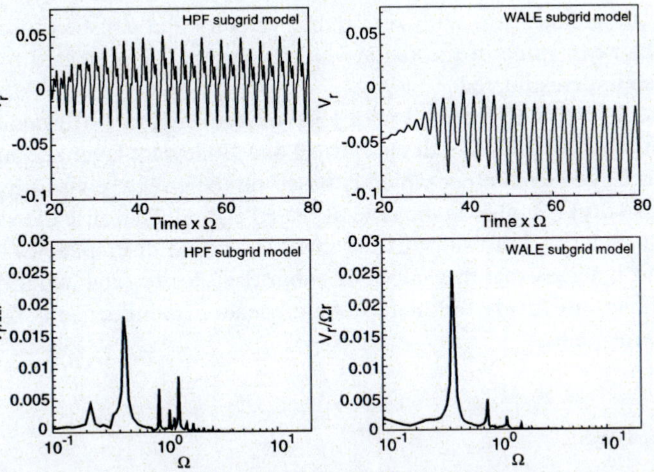

Fig. 4. Time series and spectrum of the radial velocity components at mid-radius of the cavity $r = R_0 + (R_1 - R_0)/2$ and $z = 0.2$ obtained for HPF and WALE models.

in the case of WALE model and this can bee seen in Fig. 5, where the instantaneous contours of the radial velocity component are presented for both subgrid models. The white regions correspond to the values lower than -0.05 and the arrow shows the point where the time signal shown in Fig. 4 was registered. Not surprisingly, the evident differences in instantaneous solutions have direct consequences in the mean velocity profiles, as can be seen from the comparison of the mean azimuthal and ra-

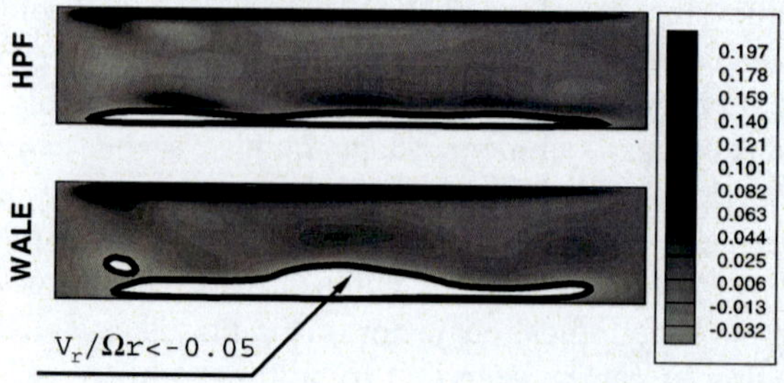

Fig. 5. Instantaneous contours of the radial velocity obtained for HPF and WALE models. White regions correspond to negative values lower than -0.05.

dial velocity profiles along the axial direction at mid-radius with DNS results, this comparison is presented in Fig. 6. One can notice that the HPF model gives results rather far from those obtained with the WALE and dynamic Smagorinsky models, which are in fact in good agreement with DNS data. The azimuthal velocity is best predicted by the WALE model, whereas the dynamic Smagorinsky model seems to be the most accurate in the case of predicting the radial velocity component. Noticeable discrepancies are only observed for radial velocity profile at $z = 0.2-0.5$, this corresponds to the transient region which encloses the Bödewadt layer and a part of the inner homogenous flow with zero radial velocity and constant mean azimuthal velocity. The peak value of the radial velocity at $z = 0.1$ predicted by the WALE model is also underestimated.

The velocity profiles obtained with HPF model seem to correspond to a higher Reynolds number for which both the Ekman and Bödewadt layer are thiner. Indeed, the effective Reynolds number which is based on the effective viscosity (molecular + subgrid visocity) is higher in the case of the HPF model because the values of subgrid viscosity are small. To some extent this can explain discrepancies in the results obtained and it seems that the values of subgrid viscosity (and subgrid tensor) are too small to act effectively on the flow field because its influence is suppressed by stabilizing procedure.

4 Conclusions

The LES computations of the flow inside the rotating cavity were performed showing the influence of various subgrid modeling approaches applied. In the case of dynamic Smagorinsky and WALE models the results obtained were similar and remained in good agreement with DNS data. On the other hand the results obtained with the high-pass filtered Smagorinsky model did not reproduce correctly the turbulence characteristics of the flow and were close to the results obtained without any subgrid model.This was however obtained for a particular case when the simulations were started from the zero initial velocity field, and a it was mentioned in the paper, in calculations starting from the disturbed flow field all the models produced to very similar results. This is because the turbulence "injected" at the initial stage is

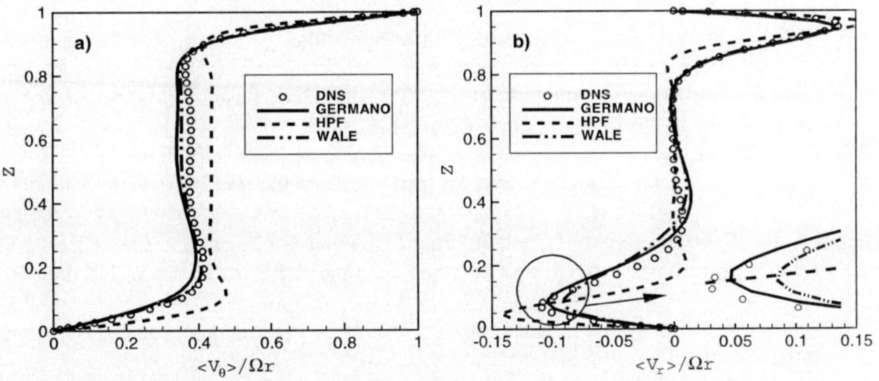

Fig. 6. Axial profiles of the azimuthal (a) and radial (b) components of the mean velocity at mid-radius.

strong enough to amplify both types of instabilities, i.e. the circular rings and the spiral structures. The latter structures were not present when computations were performed with HPF model or without any subgrid model. The simulations performed with HPF model could be probably improved by more careful choice of the high-pass filtering and stabilization procedure, which is planned as a future task.

The support for the research was provided within statutory funds BS-1-103-301/2004/P the research grant COST/78/2006 founded by Polish Ministry of Science. The authors are grateful to the Cyfronet Comp. Center and TASK Comp. Center for access to the computing resources.

References

1. Anderson H.I., Lygren M. Large eddy simulation of the turbulent flow between a rotating and stationay discs. *ZAMP*, 55:268–281, 2004.
2. Anderson H.I., Lygren M. LES of open rotor-stator flow. *I. J. Heat and Fluid Flow*, 27:551–557, 2006.
3. Canuto C., Hussaini M.Y., Quarteroni A., and Zang T.A. *Spectral methods in fluid dynamics*. Springer-Verlag, 1988.
4. Fletcher C.A.J. *Comp. Techniques for Fluid Dynamics*. Springer-Verlag, 1991.
5. Gauthier G., Gondret P., Moisy F. and Rabaud M. Instabilities in the flow between co- and counter-rotating disks. *J. Fluid Mech.*, 471:1–21, 2002.
6. Moin P., Germano M., Piomelli U. and Cabot W. A dynamic subgrid scale eddy viscosity model. *Phys. Fluids A*, 3:1760, 1991.
7. Geurts B.J. *Elements of Direct and Large-Eddy Simulation*. Edwards, 2003.
8. Itoh M., Yamada Y., Imao S., Gonda M. Experiments on turbulent flow due to an enclosed rotating disc. *Exp. Therm. Fluid Sci.*, 5:359, 1992.
9. Jeanmart H. and Winckelmans G. Investigation of eddy-viscosity models modified using discrete filters: A simplified "regularized variational multiscale model" and an "enhanced field model". *Phys. Fluids* 19, 055110, 2007.

10. Lele S.K. Compact finite difference with spectral-like resolution. *J. Comput. Phys.*, 103:16–42, 1992.
11. Nicoud F. and Ducros F. Subgrid-Scale Stress Modelling Based on the Square of the Velocity Gradient Tensor. *Flow, Turb. Comb.*, 62:183–200, 1999.
12. Sagaut P. *Large eddy simulation for incompressible flows*. Springer, 2001.
13. Schouveiler L., Le Gal P. and Chauve, M. Instabilities of the flow between a rotating and stationary disk. *J. Fluid Mech.*, 443:329–350, 2001.
14. Serre, E., Tuliszka-Sznitko, E. and Bontoux. Coupled numerical and theoretical study of the transition flow between a rotating and stationary disk. *Phys. Fluids*, 16:688–707, 2004.
15. Severac E., Poncet S., Serre E. and Chauve M-P. Large eddy simulation and measurements of turbulent enclosed rotor-stator flows. *Phys. Fluids*, 19?, 2007.
16. Severac E., Serre E. A spectral vanishing viscosity for the LES of turbulent flows within rotating cavities. *J. Comput. Phys.*, 226:1234–1255, 2007.
17. Severac E., Serre E., Bontoux P., Launder B. Large Eddy Simulations of Transitional Rotor-Stator Flows using Spectral Vanishing Viscosity Techniques. In *DLES VI*, editors: Lamballais E., Friedrich R., Geurts B.J. and Metais O., ERCOFTAC, 217–224. Springer, 2006.
18. Stolz S., Schalatter P. and Kleiser L. High-pass filtered eddy-viscosity models for large-eddy simulations of transitional and turbulent flows. *Phys. Fluids*, 17, 2005.
19. Stolz S., Schalatter P., Meyer D. and Kleiser L. High-pass filtered eddy viscosity models for LES. In *DLES V*, editors: Friedrich R., Geurts B.J., Metais O., KLUWER, 2004.
20. Tadmor E. Convergence of spectral methods for nonlinear conservation laws. *SIAM J. Numer. Anal.*, 26–30, 1989.
21. Tuliszka-Sznitko E., Zielinski A. DNS/LES of Transitional Flow in Rotating Cavity. *I. J. Transport Phenomena*, 10:223–234, 2008.
22. Vreman B., Geurts B., Kuerten H. On the formulation of the dynamic mixed subgrid-scale model. *Phys. Fluids*, 6:4057, 1994.
23. Zang Y., Street R., Koseff J.R. A dynamic mixed subgrid-scale model and its application to turbulent recirculating flows. *Phys. Fluids*, 5:3186, 1993.

A priori analysis of an Isothermal, Turbulent Two-Phase Flow

Alessandro Pecenko[1] and J. G. M. Kuerten[1]

Department of Mechanical Engineering, Eindhoven University of Technology
P.O. Box 513, 5600 MB Eindhoven, The Netherlands
a.pecenko@tue.nl   j.g.m.kuerten@tue.nl

Summary. A novel approach to the simulation of isothermal, turbulent two-phase (liquid/vapour) flow of one substance is presented. The two-phase flow is modeled with a "one-fluid" formulation by means of a *diffuse-interface* concept and of the frame-invariant *Korteweg tensor* of capillary stresses at the interface. A single system of compressible Navier-Stokes equations can be written for the whole flow domain, the fluid properties being described by continuous functions of the mass density. The Van der Waals equation of state accounts for the variation of pressure with density at a given value of temperature slightly below the critical isotherm. After description of the numerical method, results of the Direct Numerical Simulation of a classic benchmark problem are shown. Next, *two-phase subgrid terms* related to the nonlinear pressure and capillary terms are studied by means of an a priori analysis based on DNS results, and a subgrid model for these terms is proposed.

Key words: Two-phase, diffuse interface, Korteweg tensor, Van der Waals, subgrid terms, a priori analysis.

1 Introduction

The numerical simulation of multiphase flows generally involves physical and mathematical issues that are absent in single-phase flows since they arise from the presence of thin *interfaces*. Physically, they represent regions of the flow domain where, within a small but finite thickness, density, temperature, pressure, viscosity and other quantities vary rapidly but smoothly between their respective bulk-phase values.

The *diffuse-interface* approach, historically developed in a thermodynamically consistent theory by Van der Waals [1], seems, therefore, the most suitable for a two-phase flow of one substance in the vicinity of the critical point, which represents the upper limit of liquid-vapor equilibria before the liquid phase vanishes and the interface becomes infinitely spread. The only order parameter that describes the transition between the bulk phases is the mass density ρ. A unique set of governing equations for the whole flow domain can be written if the right-hand side of the momentum equation incorporates the divergence of a second-order, frame-invariant

M.V. Salvetti et al. (eds.), *Quality and Reliability of Large-Eddy Simulations II*,
ERCOFTAC Series 16, DOI 10.1007/978-94-007-0231-8_10,

tensor, called *Korteweg tensor*. This tensor represents capillary stresses at the interface, and depends on the density and its spatial derivatives [2]. It can be shown that such a tensor, even in the more general case of density-dependent fluid properties and of compressible flow, is non-dissipative. In isothermal conditions, the set of governing equations consists of the continuity and of the momentum equation, plus an equation of state, which is taken in the Van der Waals form.

When the two-phase flow is turbulent and a Large Eddy approach is adopted for the numerical simulation, the *filtering* operator yields as many *subgrid terms* as the nonlinear terms present in the equations. In a two-phase flow with Van der Waals equation of state and Korteweg tensor, the pressure and the components of the capillary tensor are nonlinear. Hence, they generate "non-classic" subgrid terms for which the literature does not provide as many well-assessed models as for the typical single-phase subgrid terms. Therefore, we have performed an a priori analysis for the new subgrid terms, based on a model that we have devised ad hoc and that we here call "*capillarity model*".

The contents of this paper are as follows. In Sect. 2, the capillary stress tensor and the governing equations are shown. In Sect. 3, the numerical method adopted for the Direct Numerical Simulation of an isothermal, two-phase, single-component flow with diffuse interface is described. Results for a classic problem are shown in Sect. 4. In Sect. 5, we propose a model for the subgrid terms related to the presence of the diffuse interface. The a priori analysis of the modeled subgrid terms is made on the basis of the test simulation previously discussed. Finally, in Sect. 6 conclusions are drawn and future steps are stated.

2 Governing equations

From a thermodynamic point of view, the notion of phase boundary as a diffuse interface can be related to the *interfacial Helmholtz free energy*, which can be expressed (see [1] [3]) by means of the local density gradient. Based on the variational concept that the extrema of density functionals like the free energy correspond to states of equilibrium, the two-phase equilibrium condition of a single-component fluid can be found by minimizing the total Helmholtz free energy of the system. In [4], it is recalled how such requirement, under the constraint of total mass conservation, is equivalent to the following conservation law:

$$\nabla \cdot \mathsf{T} = 0\,, \tag{1}$$

T being a second-order, frame-invariant tensor representing capillary stresses at the interface and commonly called *Korteweg tensor* [2]. In the most general case, it can be shown that this tensor is given by

$$\mathsf{T} = \{-p + \rho K \nabla^2 \rho + \frac{1}{2}(K + \rho K_\rho)|\nabla \rho|^2\}\mathsf{I} - K \nabla \rho \otimes \nabla \rho \tag{2}$$

where p is the thermodynamic pressure, I is the identity tensor, $K = K(\rho)$ is a capillarity coefficient commonly called *gradient energy coefficient*, and K_ρ denotes its derivative with respect to the density. Since our work is restricted to isothermal conditions, the dependence of K on temperature is ignored.

The whole fluid domain can be treated as a continuum, where mass density plays the role of phase-field function, provided that the fluid properties are reasonably well approximated in both phases by continuous functions of the density. Hence, a unique function $\mu(\rho)$ for the dynamic viscosity has to be assumed. Here we have taken the linear approximation

$$\mu(\rho) = c_1\rho\,,$$

where c_1 is a proportionality coefficient that can be obtained from experimental values of the dynamic viscosity at the given temperature and for the given substance. For $K(\rho)$ we take

$$K = \frac{c_2}{\rho}\,,$$

c_2 being a positive constant. Although there is no or little physical justification for this choice, it has two advantages. First, as can be seen from (2), this choice leads to the simplest form of the Korteweg tensor, as the term for the square gradient norm vanishes. Thus

$$\mathsf{T} = \{-p + c_2\nabla^2\rho\}\mathsf{I} - \frac{c_2}{\rho}\nabla\rho\otimes\nabla\rho\,. \tag{3}$$

Second, it leads to a linear highest order derivative term in T, which simplifies subgrid modeling. Moreover, the possible values of the macroscopic quantity related to capillarity, namely the surface tension, are not restricted by the choice made for $K(\rho)$.

The full system of governing equations reads in conservative form for an isothermal two-phase flow of a single-component fluid

$$\rho_t + \nabla\cdot(\rho\mathbf{u}) = 0 \tag{4}$$

$$(\rho\mathbf{u})_t + \nabla\cdot(\rho\mathbf{u}\mathbf{u}) = \nabla\cdot(\mathsf{d} + \mathsf{T}) \tag{5}$$

$$p = p(\rho)\,, \tag{6}$$

$\mathbf{u}$ denoting the velocity, d the usual viscous stress tensor for a Newtonian fluid, and $p(\rho)$ the Van der Waals equation of state.

3 Numerical method for DNS

The system of equations (4)-(6) is strongly nonlinear: besides the convective term, the capillary tensor components, even with the assumption (3), are nonlinear, as well the viscous stresses and the pressure term. The latter is of the form

$$p(\rho) = \frac{RT}{M - b\rho}\rho - \frac{a}{M^2}\rho^2 \tag{7}$$

where R is the gas constant, T is the prescribed value of temperature, M the molar mass, a and b two constant coefficients experimentally determined for the given substance. Although this equation represents qualitatively the behaviour of a fluid below the condensation point and above the saturation point, each isotherm in nearly critical conditions contains an unphysical region of negative compressibility $dp/d\rho < 0$. This is a possible source of intense, high-frequency oscillations, which can lead to severe stability problems in a numerical simulation. Hence, the choice of the time integration scheme has to be such that these oscillations are not amplified. To this purpose, we have adopted a three-stage, third-order accurate Runge-Kutta scheme based on the Total Variation Diminishing concept [5]. Such a scheme has proved to give stable results also in the unstable region of the solution domain.

For the space discretization we have used a central finite-volume scheme, globally accurate up to the second order, on a uniform Cartesian grid.

The time step Δt has been chosen according to the Courant-Friedrichs-Lewy condition

$$\Delta t \leq C \Delta x \left(\frac{dp}{d\rho} \right)^{-\frac{1}{2}} ,$$

where C is an empirical constant value smaller that unity, and $(dp/d\rho)^{-1/2}$ is the reciprocal of the speed of sound at the prescribed temperature. Other characteristic velocities are negligible in the test case that we have considered. Therefore, they do not affect the CFL condition significantly.

4 DNS results

The benchmark test simulation that we here present and that we have used as reference for the a priori analysis of the interfacial subgrid terms, is the so-called *single-drop retraction*. It is well known that in the absence of gravity and other external forces, a drop of liquid in a closed vessel filled with the saturated vapor of the same component must have, at stationary conditions, a spherical shape. In fact, the latter is the only shape that allows static mechanical equilibrium to hold in every point of the interface, as described by the classic Laplace-Young equation

$$p_l - p_v = \frac{2}{R} \sigma \tag{8}$$

where the left-hand side is strictly positive. The pressure of the liquid p_l and that of its saturated vapor p_v differ by a quantity $(2/R)\sigma$, where R is the radius of the drop and and σ represents the value of the surface tension at equilibrium.

In Figs. 1-2 we show the results of the simulation. The initial condition consists of a drop of ellipsoidal shape surrounded by vapor, with a diffuse interface separating the two phases. The velocity field is zero everywhere. The nonuniform curvature causes the onset of a reshaping of the drop, which eventually reaches the spherical form when local mechanical equilibrium is restored (Fig. 1). This state represents the steady solution since the total free energy has a minimum. A uniform pressure jump across the interface is established (Fig. 2), and surface tension can therefore be computed by means of (8) and compared with the analytical value given by [6]

$$\sigma = 2 \int_{\rho_v}^{\rho_l} \left[\frac{1}{2} K(\rho) \Delta f(\rho) \right]^{\frac{1}{2}} d\rho \,. \tag{9}$$

Here ρ_v, ρ_l are the values of the density in the vapor and in the liquid bulk phase respectively, and $\Delta f(\rho)$ represents the change in the Helmholtz free energy when a unit volume of a mixture of liquid and vapor at equilibrium of average density ρ is converted into a single-phase unit volume with the same density. The estimation of $\Delta f(\rho)$ can be easily done once the temperature is assigned and the equation of state known [6]. The numerical and the analytical value of the surface tension are found to differ by only 0.3% on a sufficiently fine grid.

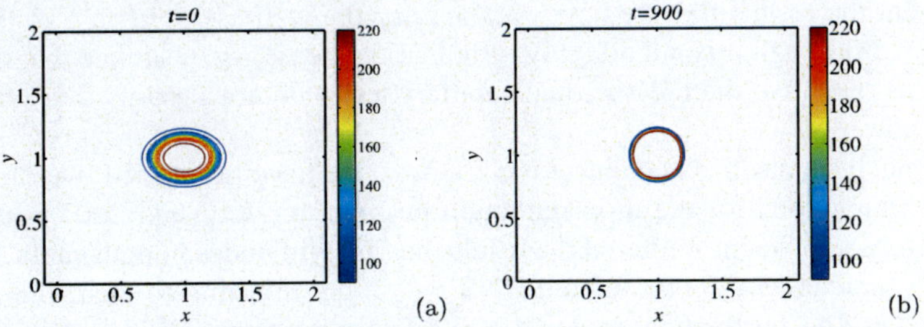

Fig. 1. Retraction of an ellipsoidal drop surrounded by quiescent saturated vapour, under the effect of capillary forces, represented on a symmetry plane. **a** initial state. **b** final equilibrium state. The interface is represented by means of density isolines. Time t is expressed in arbitrary units.

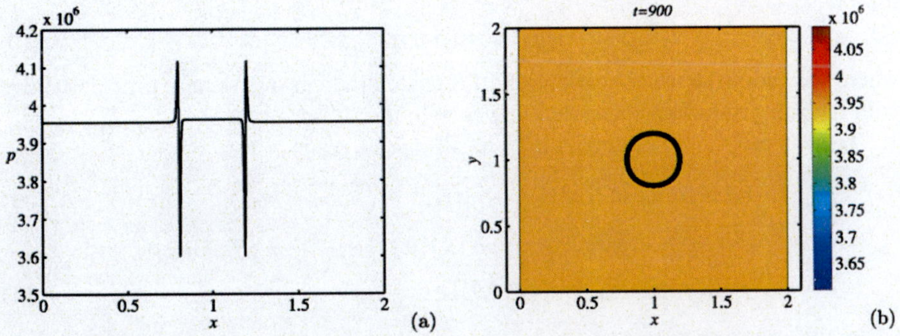

Fig. 2. Retraction of a drop in the absence of external forces. **a** steady solution for the pressure as a function of the Cartesian coordinate x on a line passing through the center of the drop. The spikes are physical and due to the local equilibrium stress tensor. **b** pressure field in the steady state in the same plane as in Fig. 1.

5 A priori analysis of the interfacial subgrid terms

In Sect. 4, we have briefly described a method of solution for the two-phase isothermal set of equations (4)-(6), and we have shown, as an example of numerical application, a classic test case.

If Large Eddy Simulation is applied to two-phase flow, new *subgrid terms* stem from the filtering operation of the nonlinear pressure and capillary terms. This implies that, apart from the well-known subgrid stress tensor for the case of a single-phase flow, a *subgrid pressure term* and several *subgrid capillary terms* appear. To devise a model for these unknown quantities, the first step is to carry out an a priori study of the subgrid terms based on results from Direct Numerical Simulation. Here we present the results that we have obtained for the well-assessed case of the drop retraction. Although it is not a highly turbulent flow case, we postulate that the subgrid terms related to the pressure and capillary terms do not strongly depend on the type of flow.

The analysis has been conducted as follows. We have performed two direct numerical simulations for the drop retraction problem on 400^3 and 200^3 grid nodes respectively, and we have filtered the results on 50^3 grid nodes in both cases. Hence, we can characterize the two filtered DNS by simply referring to their *filter width*, namely $\Delta = 8\Delta x$ for the finer grid and $\Delta = 4\Delta x$ for the coarser grid.

The type of filter that we have used is the so-called "top-hat" filter, which we denote with a bar sign over the filtered quantities. Since the flow is compressible, we use the Favre filter for the velocity components u_i, yielding

$$u_i = \frac{\overline{\rho u_i}}{\bar{\rho}} \,,$$

so that no subgrid terms appear in the filtered continuity equation. The latter, therefore, reads

$$\bar{\rho}_t = -(\bar{\rho}\, u_j)_{x_j} \,, \tag{10}$$

where we adopt the usual convention of repeated indices and denote time derivative with the subscript t and space derivatives with the subscript x_j.

The filtered momentum equation, recalling (3), reads

$$\begin{aligned}(\bar{\rho}\, u_i)_t &= -(\bar{\rho}\, u_i\, u_j)_{x_j} - (\bar{\rho}\tau_{ij})_{x_j} - [p(\bar{\rho})]_{x_i} - [\overline{p(\rho)} - p(\bar{\rho})]_{x_i} + [d_{ij}(\bar{\rho},\, u)]_{x_j} + \\ &+ [\overline{d_{ij}(\rho,u)} - d_{ij}(\bar{\rho},\, u)]_{x_j} - [K(\bar{\rho})(\bar{\rho})_{x_i}(\bar{\rho})_{x_j}]_{x_j} - [\overline{K(\rho)\rho_{x_i}\rho_{x_j}} + \\ &- K(\bar{\rho})(\bar{\rho})_{x_i}(\bar{\rho})_{x_j}]_{x_j} + (c_2\nabla^2\bar{\rho})_{x_i}\end{aligned} \tag{11}$$

where the unknown subgrid terms are the turbulent stress tensor τ_{ij}, which represents the subgrid convective term, the subgrid viscous tensor $(\overline{d_{ij}(\rho,u)} - d_{ij}(\bar{\rho},\, u))$, the subgrid pressure $p_{sg} = \overline{p(\rho)} - p(\bar{\rho})$, and the subgrid capillary terms

$$[K(\rho)(\nabla\rho\otimes\nabla\rho)]_{sg} = \overline{K(\rho)\rho_{x_i}\rho_{x_j}} - K(\bar{\rho})(\bar{\rho})_{x_i}(\bar{\rho})_{x_j} \,.$$

As a first step, the exact values of all subgrid terms have been evaluated, showing (Fig. 3) that the two-phase subgrid terms obtained by filtering the Van der Waals equation of state and the Korteweg tensor are the most important. Such a conclusion holds for both DNS on the finer and on the coarser grid, hence we will focus on the subgrid contributions arising from these two terms. Furthermore, the Figure shows that there is no substantial difference between finer and coarser grid for the pressure and the capillary subgrid terms. The viscous and the convective subgrid terms, instead, are affected by the grid size, since they depend also on the velocity field.

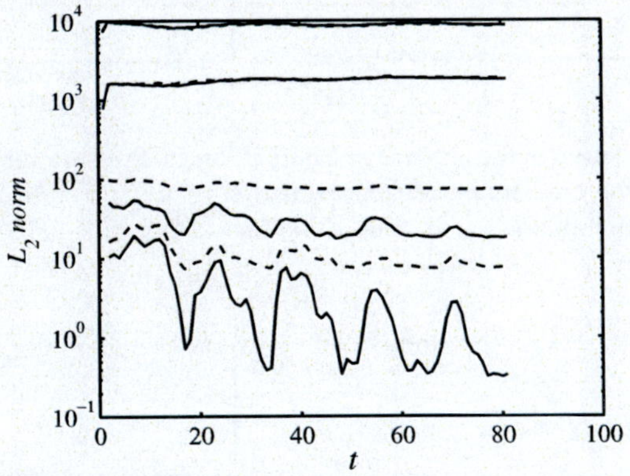

Fig. 3. Retraction of a drop in the absence of external forces: comparison of the L_2-norms of the exact subgrid terms for the DNS with 400^3 nodes (solid lines) and for the DNS with 200^3 nodes (dashed lines), versus time (in arbitrary units). From top to bottom: pressure, capillary, viscous, convective subgrid terms.

These subgrid terms have been compared with the values given by the *similarity model* [7] (denoted from now on with the superscript SM) and the *gradient model* [8] (superscript GM). The SM has provided the best approximation for the capillary terms, while the GM has given the best result for the pressure, as shown in Figs. 4-5. The figures show that the correlation coefficients hardly depend on the grid size used in the DNS.

However, we have noticed a significant difference in magnitude between the exact value and the approximation, both for the pressure and for the capillary subgrid terms. Therefore, an extra model that we have devised ad hoc and that we here call *capillarity model* (CM) has been added:

$$p_{sg} = p_{sg}^{GM} + C_p\,(Ca)^2\,K(\bar{\rho})|\nabla\bar{\rho}|^2 \tag{12}$$

$$[K(\rho)(\nabla\rho\otimes\nabla\rho)]_{sg} = [K(\rho)(\nabla\rho\otimes\nabla\rho)]_{sg}^{SM} + C_c(Ca)^2 K(\bar{\rho})(\nabla\bar{\rho}\otimes\nabla\bar{\rho})\,, \tag{13}$$

where $Ca \sim \Delta/L$ is a nondimensional parameter of the simulation called the *Cahn number*, representing the effect of the filter on the interface length scale, and C_p, C_c are the constants of the model, which we have evaluated by a least-square minimization of the following quantities (CM denoting the last terms on the r.h.s. of (12)(13))

$$(p_{sg} - [p_{sg}]^{GM+CM}) + [K(\rho)(\nabla\rho\otimes\nabla\rho)]_{sg}| - \{[K(\rho)(\nabla\rho\otimes\nabla\rho)]_{sg}\}^{SM+CM}|$$

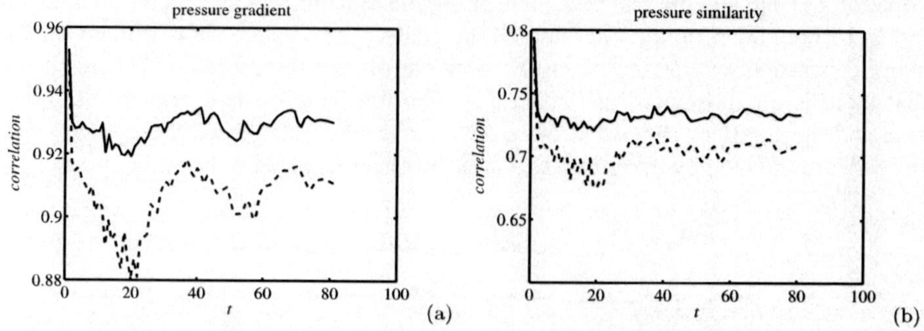

Fig. 4. Retraction of a drop in the absence of external forces. Pressure subgrid term. **a** correlation coefficients for the *gradient model* approximation vs. time. **b** correlation coefficients for the *similarity model* approximation vs. time. Solid lines: DNS with 400^3 nodes. Dashed lines: DNS with 200^3 nodes.

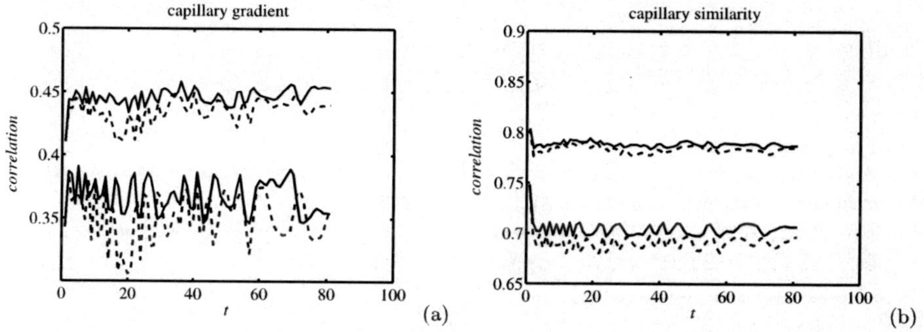

Fig. 5. Retraction of a drop in the absence of external forces. Capillary subgrid tensor. **a** correlation coefficients for the *gradient model* approximation vs. time. **b** correlation coefficients for the *similarity model* approximation vs. time. Solid lines: DNS with 400^3 nodes. Dashed lines: DNS with 200^3 nodes. Top lines: diagonal terms. Bottom lines: off-diagonal terms.

since they appear in each of the three scalar momentum equations (the superscripts being a short formal notation for the sum of the modeled subgrid contributions to pressure and capillary terms respectively). Note that the tensor of the subgrid capillary stresses also contains nondiagonal terms.

Since the test case is symmetric, we show results only for the scalar momentum equation in the x direction, where, for the sum of pressure and diagonal capillary subgrid terms, exact and modeled values are shown for the case $\Delta = 8\Delta x$ in Fig. 6 as isolines on the $(x,y)-$symmetry plane. The correlation coefficient between the total exact subgrid tensor and the total model value appears to be between 0.97 and 0.98.

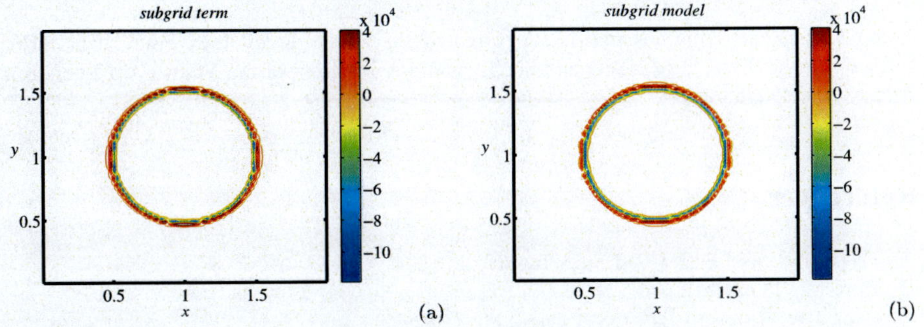

Fig. 6. Retraction of a drop in the absence of external forces. Sum of pressure and diagonal capillary subgrid terms on the $(x,y)-$symmetry plane of the drop. **a** exact distribution from filtered DNS with filter width $\Delta = 8\Delta x$. **b** modeled distribution on the basis of the approximations (12)(13).

6 Conclusions

We have explored the possibility of modeling the nonconventional two-phase subgrid terms stemming from a *diffuse-interface* approach, for use in Large Eddy Simulation of two-phase flows. To do so, we have first devised a stable numerical method that calculates the solution of the governing equations for an isothermal liquid-vapor flow of a pure component in the vicinity of the critical point, when the presence of the dividing surface is taken into account by means of a stress tensor. We have then made a few assumptions, concerning in particular the functional dependence on density of the capillarity coefficient, and we have presented results for Direct Numerical Simulation of a liquid drop retracting to its equilibrium spherical shape by capillarity. The choice of this test case is two-fold: it allows a clear validation of the numerical method by means of a comparison between the theoretical and the numerical value of the surface tension, and, after *filtering* the results, it emphasizes the importance of adequate modeling the pressure and capillary subgrid terms, since these terms appear to be the most important among the subgrid contributions, but also those for which well-assessed models are not, to our knowledge, yet available. The approximation that we have here proposed for the interfacial subgrid terms, is a combination of "classic" single-phase models, typically used to approximate the turbulent stress tensor, and of a "nonclassic" contribution that we have called *capillarity model*. An a priori analysis made for two simulations of the same test case with different mesh refinement has shown that such a combination of models provides a very high correlation, for both finer and coarser grid.

Currently, we are working on the a posteriori validation of our model for the same flow problem. Next, we will test the model on other turbulent, two-phase flow examples and, eventually, we will apply it to the LES of high Reynolds two-phase flows.

Acknowledgement. This research is supported by the Dutch Technology Foundation STW, applied science division of NWO and the Technology Program of the Ministry of Economic Affairs of the Netherlands.

References

1. Van der Waals J D (1979) J Stat Phys 20:200–244
2. Korteweg D J (1901) Arch Néerl Sci Exactes Nat Série II VI:1–24
3. Cahn J W, Hilliard J E (1958) J Chem Phys 28, 2:258–267
4. Anderson D M, McFadden G B, Wheeler A A (1998) Annu Rev Fluid Mech 30:139–165
5. Cockburn B, Gau H (1996) SIAM J Sci Comput 17, 5:1092–1121
6. Cahn J W (1959) J Chem Phys 30, 5:1121-1124
7. Bardina J, Ferziger J H, Reynolds W C (1984) Report No. TF-19, Stanford
8. Clark R A, Ferziger J H, Reynolds, W C (1979) J Fluid Mech 91:1–16

A Posteriori Analysis of Numerical Errors in Computing Scalar Variance

Colleen M. Kaul[1], Venkat Raman[1], Guillaume Balarac[2] and Heinz Pitsch[3]

[1] Dept. of Aerospace Engineering and Engineering Mechanics, The University of Texas at Austin, Austin, TX 78712, USA colleen.kaul@mail.utexas.edu, v.raman@mail.utexas.edu

[2] MOST/LEGI, BP53, 38041, Grenoble Cedex 09, France guillaume.balarac@grenoble-inp.fr

[3] Center for Turbulence Research, Stanford University, Stanford, CA 94305, USA h.pitsch@stanford.edu

Summary. The results of a novel procedure for *a posteriori* testing show that the predictions of a dynamic model for the subfilter variance are sensitive to accumulated errors in the evolution of the filtered scalar field as well as to numerical errors incurred in evaluating the variance model at a single time step. Tests were performed on homogeneous, isotropic filtered scalar fields simulated using exact and modified wavenumbers corresponding to second, fourth, and sixth order accurate finite difference schemes. The sixth order scheme showed close agreement with the spectral method in terms of the distribution of predicted subfilter variance values, which were assessed using nonparametric techniques. The second order scheme produced, on average, substantially higher values of subfilter variance. Valid methods for comparing models were sought that account for the variability of statistical estimates made from a spatially correlated sample, and some preliminary approaches were presented.

Key words: subfilter scalar variance, nonparametric statistics, subfilter model error analysis

1 Introduction

In conserved scalar methods for large eddy simulation of non-premixed combustion, the subfilter scalar variance indicates the degree of small-scale mixing. The subfilter variability of scalar values is approximated by a beta distribution, which is fully specified by the values of the filtered scalar and subfilter scalar variance. However, the subfilter variance, defined as $\overline{Z'^2} = \overline{Z^2} - \overline{Z}^2$, must itself be modeled. A common modeling approach uses an algebraic relation of the form $\overline{Z'^2} = \mathscr{C}\Delta^2\nabla\overline{Z}\cdot\nabla\overline{Z}$, where the coefficient $\mathscr{C}$ may be determined dynamically [9, 1]. Since combustion is highly sensitive to the mixing process, small errors in variance estimation can lead to large errors in species concentrations.

Because of the gradient-squared dependency of the dynamic model, it has been inferred that finite difference methods result in underestimation of the variance in

M.V. Salvetti et al. (eds.), *Quality and Reliability of Large-Eddy Simulations II*, ERCOFTAC Series 16, DOI 10.1007/978-94-007-0231-8_11,

actual simulations using grid-based filtering. However, DNS-based *a priori* tests [5] have indicated that this explanation is too simplistic. In fact, finite difference implementation of the dynamic procedure resulted in overestimation of the model coefficient, partially canceling the numerical underprediction of the gradient-squared quantity. While such static tests on DNS data are useful in isolating specific sources of error in subfilter variance estimation, they are not necessarily indicative of the errors encountered in LES simulations because they do not account for the interaction of the variance model with an LES-evolved filtered scalar field. Initial results suggest that the effect of errors in the filtered scalar evolution on subfilter variance prediction is significant relative to effects due solely to subfilter variance model implementation [6].

A second reason why *a priori* tests may lack extensibility to LES results is that a DNS field provides only one sample of the small scale turbulence that could correspond to a given filtered field. Generally, however, interest is focused on estimating one-point statistics rather than the statistics of the field per se, and spatial averages over homogeneous directions are used in place of ensemble averages. Most methods for estimating one-point statistics assume independent and identically distributed data. While the latter criterion is usually satisfied , the former cannot be. Both real and simulated turbulence fields, obviously, exhibit correlation in space and time. If methods based on independent samples are applied to correlated data, it should be with the awareness that the effective sample size is significantly reduced. With a smaller sample, it becomes more important to distinguish between sample estimates and the actual population value of a statistic and to provide some quantification of the uncertainty of the estimate. This is a point that has been largely neglected in *a priori* tests of subfilter models.

2 Details of LES Computations

To study the effect of numerical error in scalar evolution on subfilter variance prediction, we have formulated a novel *a posteriori* computational experiment. A pseudospectral code for simulation of homogeneous isotropic turbulence was modified to solve a governing equation for the filtered scalar field with exact and modified wavenumbers. A computational domain of 256^3 grid points was used for all simulations presented here. Second order central (CD-2), fourth order central (CD-4), and sixth order Padé (P-6) schemes were considered for both the convection and diffusion operators as well as computation of a common dynamic eddy diffusivity model [8]. The grid size Δx was set equal to the filter width Δ in the modified wavenumber expressions. Filter widths of 8η and 16η were considered, where η is the Kolmogorov length. Filtered velocities were supplied by explicit filtering of a DNS velocity field, which was forced at the large scales to maintain $Re_\lambda = 80$. The initial conditions for all filtered scalar evolutions were obtained by filtering the initial conditions of a DNS-resolution scalar with $Sc = 1$. All scalar fields were decaying, and simulations were carried out over a period of about one eddy turn-over time, τ, as calculated from the DNS velocity field.

Two analytically equivalent forms of the filtered scalar transport equation were considered, as the accuracy of the diffusion operator in finite difference implementations of the equation has been recently noted as an important factor in the prediction of subfilter variance [6]. The filtered scalar transport equation is frequently solved in the form

$$\frac{\partial \overline{Z}}{\partial t} + \overline{u}_i \frac{\partial \overline{Z}}{\partial x_i} = \frac{\partial}{\partial x_i}\left[(D + D_t)\frac{\partial \overline{Z}}{\partial x_i}\right] \tag{1}$$

which will be referred to as the conventional formulation. However, second derivative terms are approximated more accurately by a single application of a second derivative scheme than by two applications of a first derivative scheme. This suggests that Eq. 1 be written in a modified formulation as

$$\frac{\partial \overline{Z}}{\partial t} + \overline{u}_i \frac{\partial \overline{Z}}{\partial x_i} = \frac{\partial (D + D_t)}{\partial x_i}\frac{\partial \overline{Z}}{\partial x_i} + (D + D_t)\frac{\partial^2 \overline{Z}}{\partial x_i^2} \tag{2}$$

Modified wavenumbers for lower order finite difference schemes deviate significantly from the ideal value, causing the diffusion operator to be less effective at small scales [7]. Higher order schemes possess modified wavenumbers that are close to the ideal value for a larger portion of the spectrum, thereby increasing accuracy. The form of the diffusion term in Eq. 2 is motivated by the fact that the modified wavenumber-squared representation of a second derivative scheme is more accurate at high wavenumbers than the square of the modified wavenumber of a first derivative scheme of the same nominal order of accuracy.

A recently developed dynamic subfilter variance model [1] was implemented for each scalar with the same scheme used to solve its transport equation. This model, like its predecessor [9], calculates an averaged model coefficient over homogeneous directions according to $\mathscr{C} = \langle \mathscr{L}\mathscr{M}\rangle / \langle \mathscr{M}\mathscr{M}\rangle$ where $\langle\cdot\rangle$ indicates a volume average in the present context. The quantities $\mathscr{L}$ and $\mathscr{M}$ are calculated at the level of a test filter, $\widehat{\Delta}$. The Leonard term has the form $\mathscr{L} = \widehat{\overline{Z}^2} - \widehat{\overline{Z}}^2$, while the model's gradient-based term is given by $\mathscr{M} = \widehat{\Delta}\nabla\widehat{\overline{Z}} \cdot \nabla\widehat{\overline{Z}}$.

The predictions of the subfilter variance model in the *a posteriori* setting must be compared across scalar fields without recourse to the kind of point-wise measurements applicable in *a priori* tests because the different numerical schemes induce different subfilter scalar fluxes. We focus here on comparing distributions of subfilter variance values. Since there is no evidence for a parametric form of the distribution, non-parametric approaches, accounting for the correlation of the data, are necessary. Details of the methods are provided below.

3 Results

Fig. 1 compares spectra of filtered scalar fields at $t = 0.7\tau$ to assess the effects of errors in the diffusion operator. Comparison with the filtered DNS scalar spectrum indicates that the subfilter flux model is too diffusive when implemented exactly, as in the spectral LES solution. For reference, simulations were also performed using the exact convection operator and their spectra computed. Convective term errors most noticeably affect the middle range of wavenumbers, especially for the second order scheme, while errors in the diffusive term are manifested at the highest wavenumbers as an accumulation of energy. Lack of diffusion at these scales is apparent even when an exact convective operator is used, albeit less severe. The modified diffusion term in Eq. 2 clearly is more effective at removing small scale energy.

All results presented below are for scalar fields evolved using the same scheme for convection and diffusion operators. We next examine the extent to which the subfilter variance field is spatially correlated and how the degree of correlation is affected by the choice of filter width and numerical scheme.

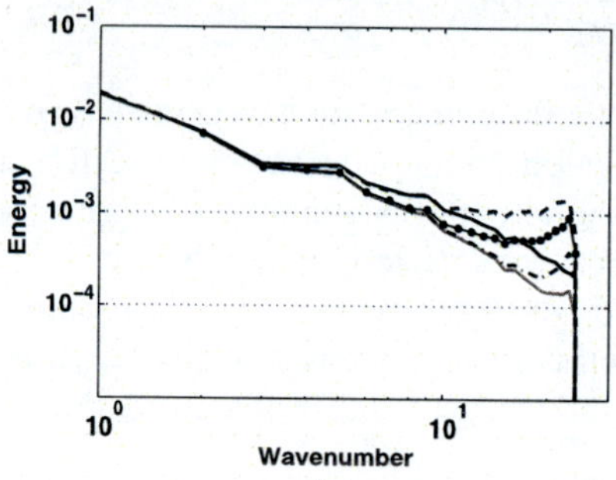

(a)

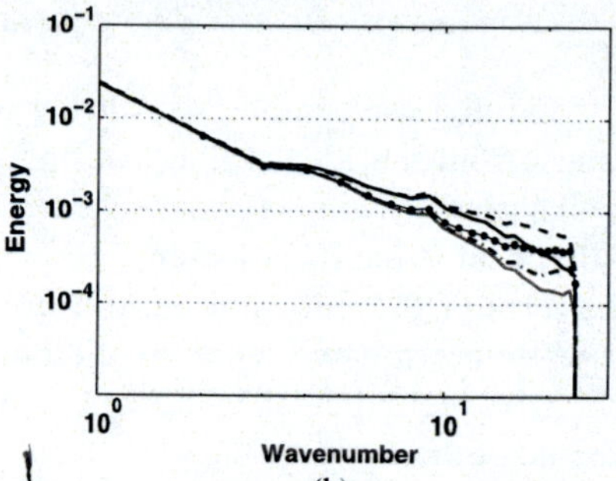

(b)

Fig. 1. Scalar spectra at t=0.7τ for filtered DNS (black) and LES evolved with spectral (gray), CD-2 (dashed), CD-4 (dotted), and P-6 (dash-dot) schemes using **(a)** Eq. 1 **(b)** Eq. 2

3.1 Patterns of spatial correlation

The semivariogram is widely used in geostatistical applications for describing the pattern of correlation in a spatial data set, such as some random scalar field $Y(\mathbf{x})$ which satisfies the conditions of intrinsic stationarity. That is, increments $Y(\mathbf{x}) - Y(\mathbf{x}')$ are second-order stationary, having constant (zero) mean and a covariance C dependent only on the lag vector [3, 10]. A homogeneous isotropic $Y(\mathbf{x})$ fulfills the requirements of second-order stationarity and so automatically satisfies the weaker conditions of intrinsic stationarity. The term semivariogram refers both to the quantity

$$\gamma(\mathbf{x}-\mathbf{x}') = \frac{1}{2}\mathrm{Var}\left[Y(\mathbf{x}) - Y(\mathbf{x}')\right] \tag{3}$$

and to the plot of γ versus the lag $\mathbf{h} = \mathbf{x} - \mathbf{x}'$; in the first sense it is identical to the structure function [3]. In the isotropic case, with $h = ||\mathbf{h}||$, $\gamma(h) = \mathrm{Var}[Y] - C(h)$, so the semivariogram provides the same information as the covariance function, and yields the variance of Y as its asymptotic value. Additionally, it is straightforward to calculate an unbiased estimate $\hat{\gamma}$. To prevent confusion with the $\widehat{(\cdot)}$ notation for test filtering, estimated quantities will hereafter use the same symbol as the true value. All results should be understood as estimated values.

In Fig. 2 a few points stand out. The lag, reported in terms of the grid spacing Δx, at which the asymptotic value is reached is greater for the $\Delta = 16\eta$ cases and the approach is uniformly more gradual, indicating greater smoothness of the field. This, however, does not mean that the range of subfilter variance values is narrower. On the contrary, the variances of the subfilter variance values are several times greater for the $\Delta = 16\eta$ data at all times analyzed. Between $t = 0.1\tau$ and $t = 0.3\tau$ (not shown), the variance increases for all calculation methods,. This reflects the transition of the scalar fields from their initial state to one mixed on intermediate scales. By $t = 0.7\tau$ the variance of $\overline{Z'^2}$ calculated from the second order central scheme increases, while it decreases for all other subfilter variance fields. The increase in $\mathrm{Var}[\overline{Z'^2}]$ is much greater for the case evolved using Eq. 1.

In the analysis of Markov fields, coding methods have been used to generate independent data. Essentially, the field is divided into disjoint sets of points containing no neighbors on the grid [2]. Although coding methods are obviously not strictly applicable here, they suggest winnowing the data to eliminate highly correlated pairs of points. In the current case, requiring zero correlation between points thins the data too severely. Instead, values are sampled at intervals of $16\Delta x$ apart so that only weak levels of correlation remain. The resulting data is approximated as independent.

It should be noted that failure to account for correlation of data points when analyzing isotropic data does not significantly modify the estimates of mean quantities. Because points are correlated to each other in a uniform manner, each carries the same statistical weight in calculating an average. More care, however, is due when calculating measures of dispersion about the mean [3, 10] (such as estimates of irreducible error in model analyses using optimal estimator concepts), although the specific implications for data characteristic of turbulence will require further study.

3.2 Evolution of subfilter variance distributions

Quantile-quantile (q-q) plots provide a means of comparing two distributions without assuming a parametric form while avoiding ad hoc bin assignments necessary for constructing histograms. The p quantile ζ_p indicates the value of a random variable ζ at which its cumulative distribution function, F, equals p, i.e. it satisfies $F(\zeta_p) = p$. In Fig. 3, quantiles of the modeled subfilter variances are plotted against quantiles of the exact DNS subfilter variance. A match between the distributions is indicated when the q-q plot forms a 45 degree line. Departures from that line can be used to diagnose the differences between the distribution. Provided the distributions are of the same shape, a linear plot is still formed. A difference in the location parameter (for example, the mean of the normal distribution) shifts the intercept of the line. Inequality of the scale parameter (e.g, the variance of the normal distribution) modifies the slope. More general differences in the shape of the distribution are manifested by

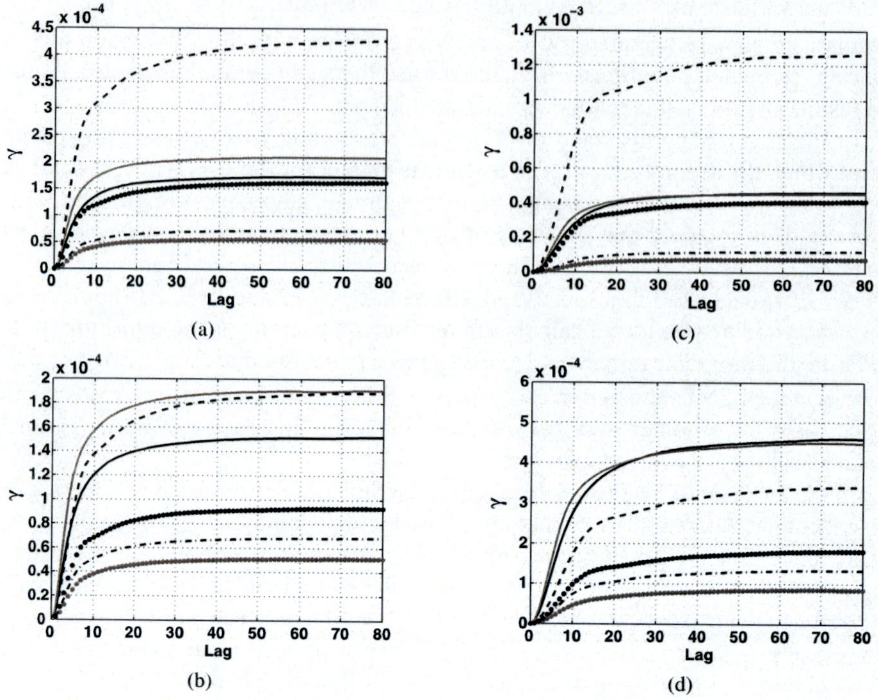

Fig. 2. Semivariograms of the subfilter variance field computed with exact DNS variance (black) and modeled variance from filtered DNS scalar (gray) and LES scalars evolved using spectral (gray dots), CD-2 (dashed), CD-4 (black dots), and P-6 (dash-dot) schemes at $t = 0.7\tau$. **(a)** Δ=8η, Eq. 1 **(b)** Δ=8η, Eq. 2 **(c)** Δ=16η, Eq. 1 **(d)** Δ=16η, Eq. 2

a nonlinear q-q plot [4]. It appears that the differences in the subfilter variance distributions can be largely explained by changes in the scale parameter, which accords with the results of Section 3.1.

In Figs. 3 and 4, symbols indicate the position of specific quantile estimates. These are the 0.05, 0.1, 0.2, 0.3, 0.4, 0.5, 0.6, 0.7, 0.8, 0.9, 0.925, 0.95, and 0.975 quantiles. The distribution of subfilter variance values is strongly left-skewed. Results are shown for $\Delta = 8\eta$ simulations only, as the $\Delta = 16\eta$ results are qualitatively very similar.

Initially, the differences between the modeled $\overline{Z'^2}$ quantiles are small for all but the highest quantiles examined and agree fairly well with the distribution of exact subfilter variance from the DNS. As the scalar fields evolve, the disparities between the distributions of modeled subfilter variance increase (Fig. 3).

Before pursuing the causes and implications of these findings, it should be recalled that the plots show estimated quantiles, based on a limited sample. Confidence intervals for the data can be obtained through bootstrap techniques [4, 3] if we continue to treat the data as independent. Fig. 4 plots differences of the estimated p

quantiles $\overline{Z'^2}_p^{\text{model}} - \overline{Z'^2}_p^{\text{exact}}$ and their 95 percent confidence intervals. These confidence intervals are obtained by forming 1000 random resamples of the original data and repeating the quantile difference calculations for each sample. The results are then sorted in ascending order, with the 25th and 975th values forming the lower and upper bounds. The behavior of the quantile differences is consistent with the hypothesis of a scale change between the modeled and exact differences, with some more complex changes perhaps occurring in the extreme right tails. The confidence intervals of the differences overlap at their edges in several areas, but do support the claim that the second order scheme results in a different subfilter variance distribution than the fourth or sixth order schemes. Fig. 5 provides another indication of the uncertainty of the quantile estimates. In addition to the data used in Fig. 3(a), alternative samples of the CD-2 and exact subfilter variance fields, formed by selecting points at the same interval but with a different starting point, are plotted as dashed lines. Intersample variability is clearly evident, but does not meaningfully alter our assessment of the scheme's performance. Also shown is the result when all points in the field are used for the quantile calculations. As expected, it does not differ greatly from the other estimates for this kind of statistic with this kind of data.

It is found that the model coefficient calculated by a dynamic procedure [1] increases with decreasing order of the numerical scheme. Qualitatively, the same relationship is found in *a priori* analysis [5]. However, the differences increase in time in the *a posteriori* case (Fig. 6), a feature that was not strongly apparent for *a priori* results at different times. Additionally, the coefficient predicted by second and fourth order accurate finite difference methods strongly depends on the form of the diffusion term in the filtered scalar equation.

In *a priori* tests, lower order schemes produce larger coefficient values by underestimating the gradient-based term $\mathscr{M}$ of the closure. In the *a posteriori* case, evolution using lower order schemes shifts the bulk of the distribution of the Leonard term $\mathscr{L}$ to higher values, as shown in Fig. 7(a). A higher value implies greater filtered scalar energy near the cutoff lengthscale, a feature that can be readily inferred from the scalar spectrum (Fig. 1). This result is a direct consequence of the increased errors in finite difference schemes at high wavenumbers, as quantified by the modified wavenumber of the discretization. Although higher values of $\mathscr{L}$ are associated with higher (exact) values of $\mathscr{M}$, finite difference errors cause underestimation of $\mathscr{M}$ that roughly balances the error of the scalar evolution (Fig. 7(b)).

4 Conclusions

A posteriori tests of a dynamic subfilter variance model show that errors in the numerical solution of the filtered scalar field can have significant effects on model predictions. Lower order schemes produce higher values of the model coefficient $\mathscr{C}$, as observed in *a priori* tests of finite difference model implementation. While in the DNS-based tests this effect resulted solely from underestimation of the gradient term $\mathscr{M}$, the most important factor in the current LES-based tests is the general increase in the Leonard term $\mathscr{L}$. The average rise in values of $\mathscr{L}$ can be attributed in part to

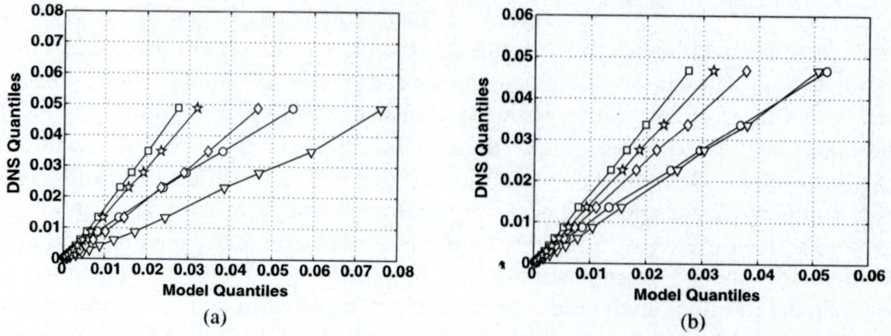

Fig. 3. Quantiles of modeled subfilter variance from Δ=8η filtered DNS (circle) and LES using spectral (square), CD-2 (triangle), CD-4 (diamond), and P-6 (star) schemes plotted against quantiles of exact DNS variance at t=0.7τ. (**a**) Filtered scalar evolved with Eq. 1 (**b**) Filtered scalar evolved with Eq. 2

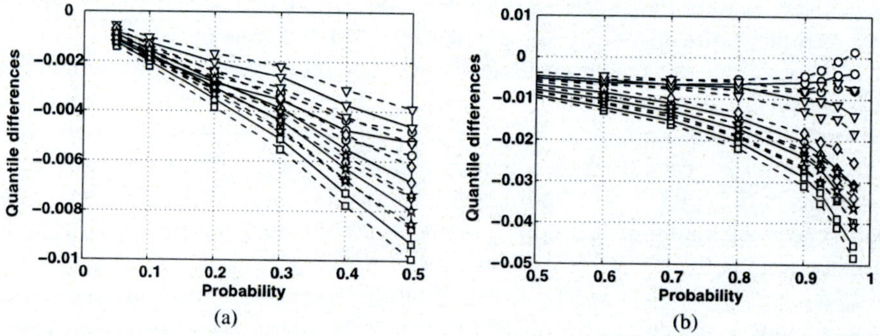

Fig. 4. Estimated quantile differences (solid lines) and 95% bootstrap confidence intervals (dashed lines) for subfilter variance of Δ=16η, Eq. 2 case at t=0.7τ. Exact DNS quantiles are subtracted from quantiles of modeled variance from filtered DNS (circle) and LES using spectral (square), CD-2 (triangle), CD-4 (diamond), and P-6 (star) methods. (**a**) 0.05 to 0.5 quantiles (**b**) 0.5 to 0.975 quantiles

poor representation of the diffusion operator at high wavenumbers by finite difference methods. This problem can be ameliorated by adopting the modified (although analytically equivalent) form of the filtered scalar transport equation given in Eq. 2.

Model comparisons were made on the basis of estimated one-point statistics. Values of subfilter variance were found to be highly correlated over a distance of several grid points, which can be problematic when using statistical methods that assume independent data. Although the statistical estimates considered here did not appear to be affected by the correlation of the data, this issue seems to warrant further attention for other quantities of interest in subfilter model tests. A method for estimating confidence intervals of the sample statistics was presented, which can guide expectations when generalizing the results of tests to other data.

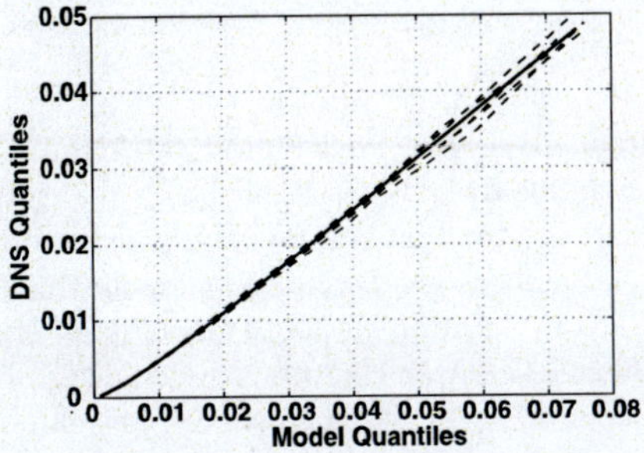

Fig. 5. Quantile plots of data from CD-2 and exact subfilter variance fields of equivalent size and spacing to the samples used in Fig. 3 (dashed) as well as to plot formed from full data set (solid)

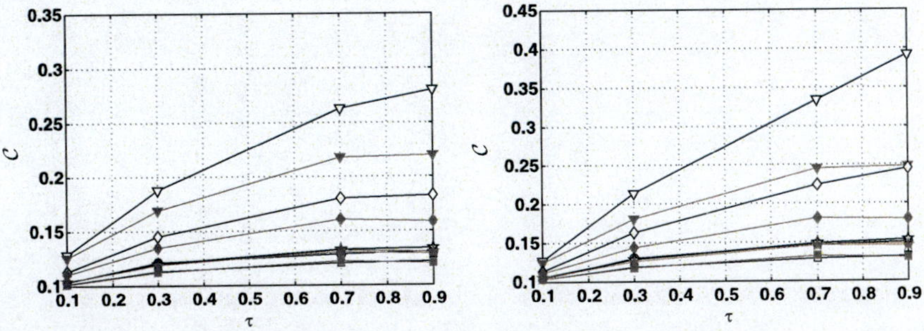

(a) (b)

Fig. 6. Model coefficient $\mathscr{C}$ plotted against normalized time τ. Results are for filtered DNS (circle) and LES using spectral (square), CD-2 (triangle), CD-4 (diamond), and P-6 (star) methods to solve Eq. 1 (open symbols) and Eq. 2 (filled symbols). (**a**) Δ=8η (**b**) Δ=16η

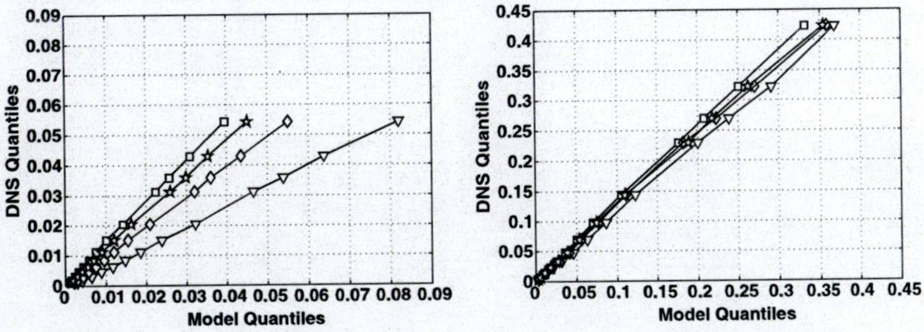

(a) (b)

Fig. 7. Quantiles of terms of the dynamic closure at t=0.7τ for Δ=8η filtered scalars evolved with Eq. 2 using spectral (square), CD-2 (triangle), CD-4 (diamond), and P-6 (star) schemes are plotted against quantiles of terms evaluated from filtered DNS. (**a**) $\mathscr{L}$ (**b**) $\mathscr{M}$

References

1. Balarac G, Pitsch H, Raman V (2008) Phys Fluids 20:035114
2. Besag J (1974) J Roy Stat Soc BMet 36:192-236
3. Cressie NAC (1993) Statistics for Spatial Data. Wiley, New York
4. Ferro CAT, Hannachi A, Stephenson DB (2005) J Climate 18:4344–4354
5. Kaul CM, Raman V, Balarac G, Pitsch H (2009) Phys Fluids 21:055102
6. Kaul CM, Raman V, Balarac G, Pitsch H (2009) Numerical errors in scalar variance models for large eddy simulation, Sixth International Symposium on Turbulence and Shear Flow Phenomena, Seoul, Korea
7. Kravchenko AG, Moin P (1997) J Comput Phys 131:310-322
8. Moin P, Squires K, Cabot W, Lee S (1991) Phys Fluids A 3:2746-2757
9. Pierce CD, Moin P (1998) Phys Fluids 10:3041-3044
10. Schabenberger O, Gotway CA (2005) Statistical Methods for Spatial Data Analysis. Chapman & Hall/CRC, Boca Raton

Quality of classical and variational multiscale LES simulations of the flow around a circular cylinder

Hilde Ouvrard[1], Maria Vittoria Salvetti[2], Bruno Koobus[1], and Alain Dervieux[3]

[1] Département de Mathématiques, Université de Montpellier 2, Case Courrier 051, Place Eugène Bataillon, 34095 Montpellier (France) houvrard@math.univ-montp2.fr, koobus@math.univ-montp2.fr

[2] Dipartimento di Ingegneria Aerospaziale, Università di Pisa, Via G. Caruso 8, 56122 Pisa (Italy) mv.salvetti@ing.unipi.it

[3] INRIA, 2004 Route des lucioles, BP 93, 06902 Sophia Antipolis (France) Alain.Dervieux@sophia.inria.fr

Summary. The role of subgrid scale (SGS) viscosity is investigated in LES and VMS-LES simulations of the flow around a circular cylinder at $Re = 20000$ on unstructured grids. The separation between the largest and the smallest resolved scales in the VMS formulation is obtained through a variational projection operator and finite-volume cell agglomeration. Two different non-dynamic eddy-viscosity SGS models are used both in classical LES and in VMS-LES. The sensitivity of the results to different parameters, such as the value of the SGS model input coefficient or the agglomeration level and to numerical viscosity in VMS-LES, is also addressed. Finally, a possible indicator of quality of the LES and VMS-LES predictions is examined.

Key words: Variational multiscale LES, circular cylinder, unstructured grids

1 Introduction

Thanks to the development of computational capabilities, nowadays large-eddy simulation (LES) is becoming an increasingly used tool also for engineering and industrial applications, at least for those flows for which the RANS approach encounters difficulties in giving accurate predictions. Paradigmatic examples of such flows are bluff-body wakes. The success of a large-eddy simulation depends on the combination and interaction of different factors, viz. the numerical discretization (which also provides filtering when no explicit one is applied), the grid refinement and quality, and the physical closure model. On the other hand, all these aspects can be seen as possible sources of error in LES, in particular in simulations oriented to industrial or engineering applications, which are often characterized by low-order non-conservative numerical schemes and coarse grid resolutions.

The present work is part of a research activity aimed at investigating the role of different sources of error, namely numerical viscosity, SGS modeling and unstruc-

M.V. Salvetti et al. (eds.), *Quality and Reliability of Large-Eddy Simulations II*,
ERCOFTAC Series 16, DOI 10.1007/978-94-007-0231-8_12,

tured grid resolution, both in *classical* LES and in Variational Multiscale (VMS) LES approaches, used together with an *industrial* numerical set-up. This industrial numerical set-up is based on a mixed finite-volume/finite-element discretization on unstructured grids, second-order accurate in space (see, e.g, [1]). The used VMS approach is the one proposed in [2], in which the projection operator in the largest resolved scale space is defined through finite-volume cell agglomeration. A key feature of VMS-LES is that the SGS model is only added to the smallest resolved scales. We adopt the so called small-small formulation, i.e. the SGS term is also computed as a function of the smallest resolved scales only. Two different eddy-viscosity SGS models are considered, both for classical LES and VMS-LES, viz. the classical Smagorinsky one [3] and the Wall-Adapting Local Eddy-Viscosity (WALE) model [4]. The same values of the model constant are used for both LES and VMS-LES.

In a previous work (see e.g. [5]), we applied the methodology briefly described above to the flow around a circular cylinder at a Reynolds number of 3900. This flow was chosen since it is a well documented benchmark and contains most of the difficulties encountered in the simulation of bluff-body flows also for more complex configurations at higher Reynolds numbers, at least for laminar natural boundary-layer separation. Moreover, the application of the VMS-LES approach to this class of flows is not documented in the literature, at least to our knowledge. In this previous study the used small-small VMS approach was found to significantly reduce the amount of SGS viscosity introduced by each of the considered SGS models with respect to that given by the same model in classical LES. This was observed for all the considered resolutions, and on the coarser grid this lead to a general improvement of the results. Conversely, on the finer grid, the VMS approach improved only the results obtained by the WALE model, which was observed to be generally more dissipative than the Smagorinsky one. The results of this previous analysis thus indicate that the SGS model, i.e. the amount of introduced SGS viscosity, has a significant impact on the results, also in the VMS-LES approach. Finally, the results of this previous study also gave an a posteriori support that the numerical viscosity introduced in our approach (which is proportional to high-order space derivatives and tuned by an ad-hoc parameter to the minimum amount needed to stabilize the simulations) does not give significant negative effects or interactions with the SGS model.

In the present contribution we present the results of the same analysis carried out for the flow around a circular cylinder, still in the laminar boundary-layer separation regime, but at $Re = 20000$, in order to investigate whether the previous findings hold when the Reynolds number is increased or, conversely, Reynolds number effects should be taken into account. Moreover, we examine a possible indicator of quality of the LES predictions [6] and we explore the possibility of using this type of indicator to devise criteria for adaptation of the SGS model constants both in LES and VMS-LES.

2 Methodology

The filtered Navier-Stokes equations for compressible flows and in conservative form are considered. In our simulations, filtering is implicit, i.e. the numerical discretization of the equations is considered as a filter operator (grid filter).

In modeling the SGS terms resulting from filtering the Navier-Stokes equations, we assumed that low compressibility effects are present in the SGS fluctuations and that heat transfer and temperature gradients are moderate. Thus, the only retained SGS term in the momentum equation is the classical SGS stress tensor. More details about the simplifying assumptions can be found in [7].

Two different eddy-viscosity SGS models are used in the present work to express the SGS stress tensor and to close the LES equations. The first one is the classical Smagorinsky model [3], in which the eddy viscosity is defined as follows:

$$\mu_{sgs} = \overline{\rho}\,(C_s\Delta)^2\left|\widetilde{S}\right| \quad , \tag{1}$$

where Δ is the filter width, C_s is a constant that must be *a priori* assigned and $\left|\widetilde{S}\right| = \sqrt{2\widetilde{S_{ij}}\widetilde{S_{ij}}}$ (repeated indexes imply summation and the tilde indicates the Favre filtering). The value typically used for shear flows of $C_s = 0.1$ is adopted herein. The Wall-Adapting Local Eddy-Viscosity (WALE) SGS model proposed by Nicoud and Ducros [4] is also considered. The eddy-viscosity μ_{sgs} of this model is defined by:

$$\mu_{sgs} = \overline{\rho}(C_W\Delta)^2(\widetilde{S_{ij}}^d\widetilde{S_{ij}}^d)^{\frac{3}{2}}[(\widetilde{S_{ij}}\widetilde{S_{ij}})^{\frac{5}{2}} + (\widetilde{S_{ij}}^d\widetilde{S_{ij}}^d)^{\frac{5}{4}}]^{-1} \tag{2}$$

with $\widetilde{S_{ij}}^d$ being the symmetric part of the tensor $g_{ij}{}^2 = g_{ik}g_{kj}$, where $g_{ij} = \partial\tilde{u}_i/\partial x_j$. As indicated in [4], the constant C_W is set to 0.5.

To complete the definition of the SGS viscosity, the grid filter width must be specified. Since filtering is implicitly applied by the numerical discretization, there is no unique rigorous definition of the filter width. The following expression is often used in LES on unstructured grids:

$$\Delta^{(j)} = Vol_j^{1/3} \tag{3}$$

where Vol_j is the volume of the jth grid element. Another option is to consider the length of the largest side of each grid element, as follows:

$$\Delta^{(j)} = max_{i=1,\dots,6}\left(\Delta_i^{(j)}\right) \tag{4}$$

$\Delta_i^{(j)}$ being the length of the ith side of the jth element.

The numerical solver (AERO) used for the present simulations is based on a mixed finite-element/finite-volume discretization of the flow equations on unstructured tetrahedral grids. The numerical scheme is vertex-centered, i.e. all the unknowns are computed at the nodes. Around each vertex i, a finite-volume cell C_i is built by using the rule of medians. The diffusive terms are discretized using P1

Galerkin finite elements on the tetrahedrons, whereas finite-volumes are used for the convective terms. The numerical approximation of the convective fluxes at the interface of neighboring cells is based on the Roe scheme with low-Mach preconditioning. To obtain second-order accuracy in space, the Monotone Upwind Scheme for Conservation Laws reconstruction method (MUSCL) is used, in which the Roe flux is expressed as a function of reconstructed values of W at each side of the interface between two cells. We refer to [1] for details on the definition of these reconstructed values. We just emphasize that particular attention has been paid to the dissipative properties of the resulting scheme, since this is a key point for its successful use in LES. The numerical dissipation provided by the scheme used in the present work is made of sixth-order space derivatives and thus is concentrated on a narrow-band of the highest frequencies. This is expected to limit the interactions between numerical and SGS dissipation, which is proportional to second-order space derivatives. Moreover, a parameter γ directly controls the amount of introduced numerical viscosity and can be explicitly tuned in order to reduce it to the minimal amount needed to stabilize the simulation.

Time advancing is carried out through an implicit linearized method, based on a second-order accurate backward difference scheme and on a first-order approximation of the Jacobian matrix.

More details on the numerical ingredients used in the present work can be found e.g. in [1] and [8].

In the VMS approach the flow variables are decomposed as follows: $W = \overline{W} + W' + W^{SGS}$, where $\overline{W}$ are the largest resolved scales (LRS), W' the smallest resolved scales (SRS) and W^{SGS} are unresolved scales. In VMS-LES the effect of W^{SGS} is provided by a closure model as in classical LES, but this model only acts on W', while the Navier-Stokes model is preserved for $\overline{W}$. In other words, given an approximation space V_h, the unmodified Navier-Stokes system for $\overline{W}$ is discretized on a coarser subspace of V_h by means of a Galerkin formulation. In the complementary space, a SGS model which applies only on the *SRS* variables is introduced. In the present study, we follow the VMS approach proposed by Koobus and Farhat [2] for the simulation of compressible turbulent flows through a finite volume/finite element discretization on unstructured tetrahedral grids. Let ψ_l be the finite-volume basis functions and ϕ_l the finite-element basis functions associated to the used grid. In order to obtain the VMS flow decomposition, these can be expressed as: $\psi_l = \overline{\psi}_l + \psi'_l$ and $\phi_l = \overline{\phi}_l + \phi'_l$, in which the overbar denotes the basis functions spanning the finite dimensional space of the LRS scales and the prime those spanning the SRS space. As in [2], the basis functions of the LRS space are defined through a projector operator in the LRS space, based on spatial average on macro cells [2]. The macro-cells are obtained by a process known as agglomeration [9].

For our VMS-LES simulations, the closure term is provided by the same SGS models as used in classical LES previously described. This term is computed as a function of the smallest resolved scales W', we thus adopt the so-called small-small approach [10]. The same values of the SGS model input parameters as for classical LES are used for VMS-LES. For the WALE model, we tested an additional value of C_w, in order to preliminarily investigate the sensitivity to this parameter. We emphasize again that in VMS-LES the closure term is added only to the smallest resolved scales; in our approach this is done through Galerkin projection in the SRS space above defined. We refer to [2, 11] for more details about this VMS-LES methodology.

3 Application and results

Classical and VMS large eddy simulations were carried out of the flow around a circular cylinder, at a Reynolds number, based on the cylinder diameter, D, and the free-stream velocity, equal to 20000.

The computational domain is such that $-10 \leq x/D \leq 25$, $-20 \leq y/D \leq 20$ and $-\pi/2 \leq z/D \leq \pi/2$, where x, y and z denote the streamwise, transverse and spanwise directions respectively, the cylinder center being located at $x = y = 0$. Characteristic based conditions are used at the inflow and outflow as well as on the lateral surfaces. In the spanwise direction periodic boundary conditions are applied and on the cylinder surface no-slip is imposed. The freestream Mach number is set equal to 0.1 in order to make a sensible comparison with incompressible simulations in the literature. As previously mentioned, preconditioning is used to deal with the low Mach number regime. The flow domain is discretized by an unstructured tetrahedral grid of approximately 1.8×10^6 nodes. The averaged distance of the nearest point to the cylinder boundary corresponds to $y^+ \approx 1$.

A first series of LES and VMS-LES simulations has been carried out with the two different SGS models. The effect of the filter width definition (see Sec. 2) has also been investigated in LES with the Smagorinsky model. Finally, a simulation with no SGS model, only relying on numerical viscosity, has been carried out. The parameter controlling the numerical viscosity, γ is set equal to 0.3 in all these simulations. The other main simulation parameters are summarized in Tab. 1, together with some bulk flow quantities, which are compared to the experimental data of [12] and [13] and to the results of previous LES in the literature [14], carried out with dynamic eddy-viscosity and mixed SGS models. Fig. 1a shows the mean pressure coefficient distribution at the cylinder obtained in the different simulations, compared to the experimental data of [12]. Mean quantities are obtained by averaging in the homogeneous spanwise direction and in time over 20 vortex-shedding cycles. From Fig. 1a and Tab.1 rather important differences can be observed in the predictions of the mean pressure and of the aerodynamic forces acting on the cylinder given by the present simulations. As analyzed in [14], the different predictions are related to different location, intensity and dimensions of the vortices detaching from the cylinders. As more these vortices form closer to the cylinder and are intense, as more their induction over the cylinder rear part is non-uniform, the $\overline{C_p}$ becomes non-uniform in this zone, the base pressure ($\overline{Cp_b}$) decreases, the drag ($\overline{C_d}$) increases and the amplitude of the time oscillations of lift (C_{Lrms}) increase. This type of behavior is obtained in the LES with the WALE model, in both the VMS-LES simulations and in the one with no SGS model. Conversely, the LES simulations with the Smagorinsky model, and in particular the one with the filter width defined by Eq. (3), are characterized by vortices forming more downstream and being less intense, and this gives a more uniform $\overline{C_p}$ distribution in the rear part of the cylinder, larger base pressure, lower drag and amplitude of the time oscillations of lift and longer mean recirculation bubble (l_r). These results are in partial disagreement to the observations made at Re=3900 ([5, 15]). Indeed, it was observed that, for an adequate grid resolution, the pressure distribution and the aerodynamic forces acting on the cylinder obtained in the different LES and VMS-LES simulations were quite similar, except than for the LES with the WALE model, which was found to be characterized by an excessive SGS dissipation also in the region near the cylinder. This low sensitivity at Re=3900 may be an effect of the very low Reynolds number: since at this Re the vortices detaching from the cylinder form rather downstream, differences observed in the location and intensity of these vortices obtained in the various simulations do not affect too much the pressure and the forces over the cylinder. Conversely, as previously discussed and as shown in [14], at Re=20000 differences in the location and intensity of these vortices, which form closer to the cylinder than at Re=3900, have a strong impact on pressure and aerodynamic loads.

Table 1. Simulation and bulk flow parameters obtained in classical and VMS-LES large-eddy simulations. $\overline{C_d}$ is the mean drag coefficient, C_{Lrms} is the r.m.s. of the time variation of the lift coefficient, l_r is the mean bubble recirculation length, $\overline{Cp_b}$ is the value of the mean pressure coefficient in the rear part of the cylinder.

Simulation	$\Delta^{(l)}$	$\overline{C_d}$	C_{Lrms}	l_r	-$\overline{Cp_b}$
LES Smagorinsky	Eq. (4)	1.22	0.388	0.97	1.16
LES Smagorinsky	Eq. (3)	1.03	0.097	1.60	0.92
VMS-LES Smagorinsky	Eq. (3)	1.29	0.523	0.81	1.26
LES WALE	Eq. (4)	1.34	0.606	0.74	1.26
VMS-LES WALE	Eq. (3)	1.36	0.690	0.67	1.36
With no model	–	1.38	0.648	0.68	1.40
Salvatici et al. [14]		[0.94,1.28]	[0.17,0.65]	[0.7,1.4]	[0.83,1.38]
Experiments [12, 13]		1.1	[0.4,0.6]	–	1.03

At Re=3900 a strong correlation between the SGS viscosity introduced in each simulation and the transition of the shear layers detaching from the cylinder was observed. As larger was the introduced SGS viscosity as closer to the cylinder the transition of the shear layers occured and the vortices formed, this leading to a smaller mean recirculation bubble length. Let us analyze now whether this correlation holds at Re=20000. Figs. 2 and 3 show instantaneous isocontours of μ_{SGS}/μ obtained in the different simulations in Tab. 1. The considered instants all correspond to peaks in the time variation of C_L. As observed at Re=3900, for the considered flow and the adopted values of the model constants, the WALE model is more dissipative than the Smagorinsky one, while the small-small VMS approach used herein significantly

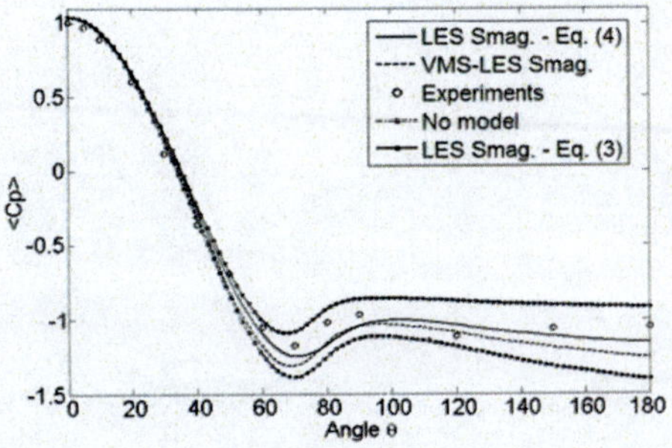

(a)

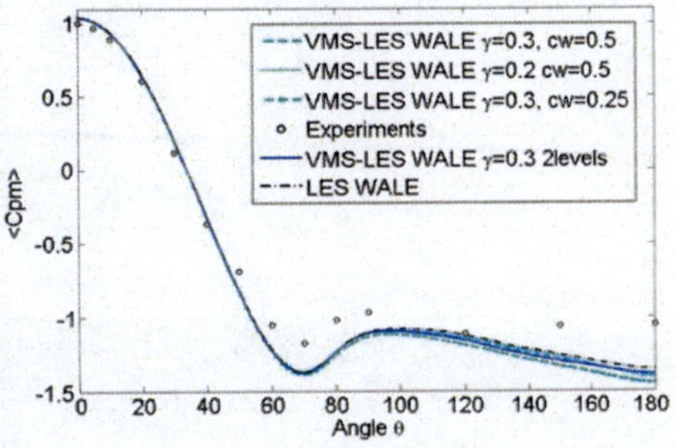

(b)

Fig. 1. Mean pressure coefficient distribution at the cylinder. (a) Simulations with the Smagorinsky model and with no SGS model. (b) Simulations with the WALE model.

reduces the introduced SGS viscosity for both the SGS models. We also recall that in VMS-LES the SGS viscosity acts only the smallest resolved scales. As for the sensitivity to the definition of Δ, it can be seen by comparing Fig. 2a and Fig. 2b that the definition (3) leads to a reduction of the introduced SGS viscosity with respect to (4), although less important than that observed in VMS formulation (Fig. 3a). The same qualitative observations can be made by looking at the mean value of μ_{SGS} (not shown here for sake of brevity). Keeping in mind this behavior of μ_{SGS}, from the analysis of Fig. 1a and Tab.1 it is clear that for the present flow there is no clear correlation between the introduced SGS viscosity and the shear-layer instability and consequently the formation of detaching vortices, the pressure distribution and the aerodynamic loads on the cylinder. For instance, LES and VMS-LES with the WALE model are characterized by a strongly different μ_{SGS} but they provide rather similar results, while significant differences are found in the predictions of LES and VMS-LES when the Smagorinsky model is used. Also, there is not a clear trend as far as the agreement with the experimental data is concerned.

Let us now examine a possible indicator of the quality of the LES predictions, based on the introduced SGS viscosity, as defined in [6]: $< QI_\mu >= \left[1+\alpha_\mu \left(< \mu_{sgs} > / \mu\right)^n\right]^{-1}$, with $\alpha_\mu = 0.05$ and $n = 0.53$. In [6] it was argued that a good LES simulation should be characterized by $\langle QI_\mu \rangle \geq 0.8$. In [6] also the numerical viscosity was accounted for in the definition of $\langle QI_\mu \rangle$; as a first approximation we neglect it herein. Moreover, for the considered type of flows, which are unsteady independently of turbulent fluctuations, it makes sense to also look at the instantaneous behavior of the SGS viscosity. We thus computed also the values of QI_μ for various instants, by using the instantaneous values of μ_{sgs}. Tab. 2 shows the minimum values in the computational domain of QI_μ (at the same instant as in Figs. 2-3) and of $\langle QI_\mu \rangle$ for the simulations in Tab. 1. First of all, note the differences between the values of the quality indicator computed by instantaneous and averaged μ_{SGS}; for instance, for the LES with the WALE model, if the instantaneous values of QI_μ are considered, there are zones in the field in which the *quality requirement* is not

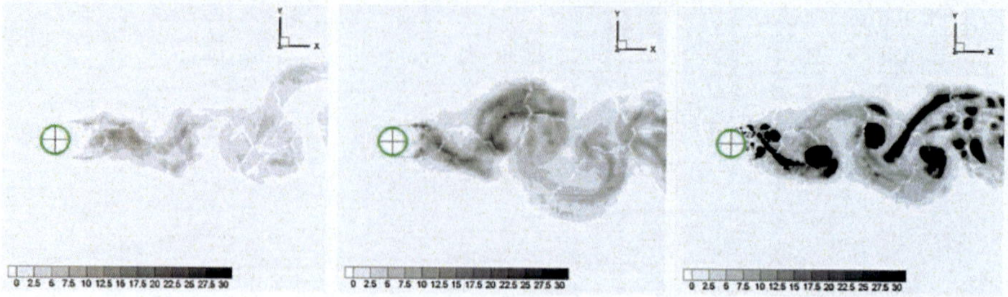

(a) (b) (c)

Fig. 2. Instantaneous isocontours of μ_{SGS}/μ in LES simulations. (a) Smagorinsky model, Δ defined by Eq. (3); (b) Smagorinsky model, Δ defined by Eq. (4); (c) WALE model. Isocontours range from 0 (white) to 30 (dark grey). The actual maxima in panel (c) are of the order of 220.

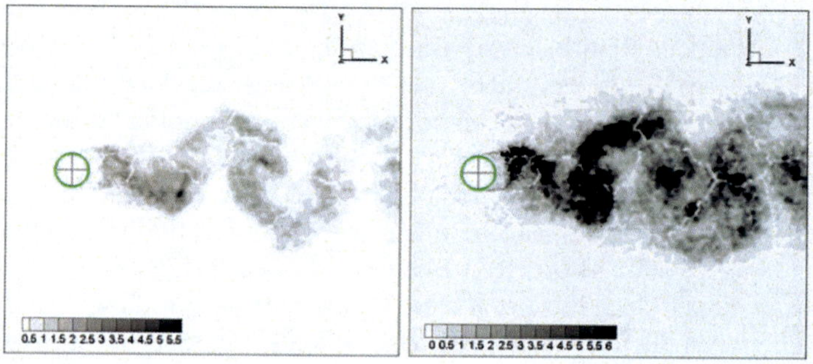

(a) (b)

Fig. 3. Instantaneous isocontours of μ_{SGS}/μ in VMS-LES simulations. (a) Smagorinsky model; (b) WALE model. The isocontours range from 0 (white) to 6 (dark grey). The actual maxima of μ_{SGS}/μ for the WALE model are however of the order of 30.

satisfied, in correspondence to the large peaks of μ_{SGS}/μ observed in Fig. 2c. Since, these peaks are smeared out by averaging, the *quality requirement* is conversely satisfied in the same simulation if we look at $< QI_\mu >$. In any case, this criterion does not seem to be of much help, since it is respected in most of the simulations.

Table 2. Minimum values in the computational domain of instantaneous and averaged QI_μ.

Simulation	$\min(QI_\mu)$	$\min(< QI_\mu >)$
LES Smag. + Eq. (3)	0.813	0.875
LES Smag. + Eq. (4)	0.776	0.826
LES WALE	0.533	0.833
VMS-LES Smag.	0.885	0.923
VMS-LES WALE	0.768	0.870

Finally, a preliminary analysis of the sensitivity to different simulation parameters was carried out focussing on the VMS-LES approach with the WALE model. The characteristics of the different carried-out simulations are reported in Tab. 3. The relative mean pressure distribution at the cylinder is shown in Fig. 1b. The sensitivity of the predictions to the considered parameters is rather low. As for γ, i.e. the parameter controlling the numerical viscosity, this is in agreement with the findings at Re=3900 and of our previous works [7, 1]. More surprising is the low sensitivity to the value of the SGS model constant and to the level of agglomeration, which controls the range of the scales at which the SGS model is added. This aspect deserves further investigation, possibly exploring a wider range of variation of the parameters.

Table 3. Bulk flow parameters obtained using VMS-LES WALE. See the caption of Tab. 1 for the meaning of symbols.

Simulation	γ	c_w	cell agglom.	$\overline{C_d}$	l_r	-$\overline{Cp_b}$
VMS-LES WALE 1	0.3	0.5	1 level	1.36	0.67	1.36
VMS-LES WALE 2	0.3	0.25	1 level	1.42	0.62	1.45
VMS-LES WALE 3	0.2	0.5	1 level	1.39	0.67	1.41
VMS-LES WALE 5	0.3	0.5	2 levels	1.38	0.68	1.39

The authors wish to thank Thomas Grinnaert for his help. CINECA (Italy) is gratefully acknowledged for having provided computational resources and support. Finally, HPC resources from GENCI- [CCRT/CINES/IDRIS] (Grant 2009-c2009025067 and 2009-x2009025044) are also gratefully acknowledged.

References

1. S. Camarri, M.V. Salvetti, B. Koobus, and A. Dervieux. A low diffusion MUSCL scheme for LES on unstructured grids. *Comp. Fluids*, 33:1101–1129, 2004.
2. B. Koobus and C. Farhat. A variational multiscale method for the large eddy simulation of compressible turbulent flows on unstructured meshes-application to vortex shedding. *Comput. Methods Appl. Mech. Eng.*, 193:1367–1383, 2004.
3. J. Smagorinsky. General circulation experiments with the primitive equations. *Month. Weath. Rev.*, 91(3):99–164, 1963.
4. F. Nicoud and F. Ducros. Subgrid-scale stress modelling based on the square of the velocity gradient tensor. *Flow Turb. Comb.*, 62(3):183–200, 1999.
5. H. Ouvrard, B. Koobus, A. Dervieux, S. Camarri, and M.V. Salvetti. The role of different errors in classical les and in variational multiscale les on unstructured grids. In *Direct and Large-Eddy Simulations 7*, ERCOFTAC. Springer, 2010. in press.
6. I.B. Celik, Z.N. Cehreli, and I. Yavuz. Index of resolution quality for large eddy simulations. *J. Fluids Eng.*, 127:949–958, 2005.

7. S. Camarri, M.V. Salvetti, B. Koobus, and A. Dervieux. Large-eddy simulation of a bluff-body flow on unstructured grids. *Int. J. Num. Meth. Fluids*, 40:1431–1460, 2002.
8. C. Farhat, B. Koobus, and H. Tran. Simulation of vortex shedding dominated flows past rigid and flexible structures. In *Computational Methods for Fluid-Structure Interaction*, pages 1–30. Tapir, 1999.
9. M.H. Lallemand, H. Steve, and A. Dervieux. Unstructured multigridding by volume agglomeration: current status. *Comput. Fluids*, 21:397–433, 1992.
10. T.J.R. Hughes, L. Mazzei, and K.E. Jansen. Large eddy simulation and the variational multiscale method. *Comput. Vis. Sci.*, 3:47–59, 2000.
11. C. Farhat, A. Kajasekharan, and B. Koobus. A dynamic variational multiscale method for large eddy simulations on unstructured meshes. *Computational Methods in Applied Mechanics and Engineering*, 195: 1667–1691, 2006.
12. S. Yokuda and R.R. Ramaprian. The dynamics of flow around a cylinder at subcritical reynolds number. *Phys. Fluids A*, 5:3186–3196, 1990.
13. R.I. Basu. Aerodynamic forces on structures of circular cross-section. part i: Model-scale data obtained under two-dimensional and low-turbulence streams. *J. Wind Eng. Ind. Aerodynamic*, 21:273–294, 1985.
14. E. Salvatici and M.V. Salvetti. Large-eddy simulations of the flow around a circular cylinder: effects of grid resolution and subgrid scale modeling. *Wind & Structures*, 6(6):419–436, 2003.
15. H. Ouvrard, B. Koobus, A. Dervieux, and M.V. Salvetti. Classical and variational multiscale LES of the flow around a circular cylinder on unstructured grids. *Comp. Fluids*, in press, 2010.

LES model intercomparisons for the stable atmospheric boundary layer

Arnold F. Moene[1], Peter Baas[1,2], Fred C. Bosveld[2], and Sukanta Basu[3]

[1] Meteorology and Air Quality Group, Wageningen University, PO Box 47, Wageningen, The Netherlands arnold.moene@wur.nl

[2] Royal Netherlands Meteorological Institute KNMI, De Bilt, The Netherlands

[3] Atmospheric Science Group, Department of Geosciences, Texas Tech University Lubbock, Texas, USA

Summary. Model intercomparisons are one possible method to gain confidence in Large-Eddy Simulation (LES) as a viable tool to study turbulence in the atmospheric boundary-layer. This paper discusses the setup and some results of two intercomparison cases focussing on the stably stratified nocturnal boundary-layer. Furthermore, options for the specification of new intercomparison cases based on data, but with less complex forcings are discussed.

Key words: Model intercomparison, Stable Boundary-Layer, Large-Eddy Simulation

1 Model intercomparison as a means of quality assurance

Model intercomparisons are one possible method to gain confidence in Large-Eddy Simulation (LES) as a viable tool to study turbulence in the atmospheric boundary-layer. By comparing the outcomes of different models run for a common set of initial and boundary conditions, as well as forcings, common strengths and weaknesses of models can be identified (regarding subgrid models, numerics and resolution). Since different atmospheric conditions emphasize the skill of different parts of the models, intercomparisons have been performed for a wide variety of situations: the convective boundary-layer [10], the neutral boundary-layer [1], different types of cloudy conditions (in particular stratocumulus and shallow cumulus cases, [9, 11]) and recently the stable boundary layer [5].

In the context of LES, one of the characteristics of stable boundary layers (SBL's) is the mere absence of large eddies. Furthermore, the levels of turbulent kinetic energy are low due to the suppression of vertical motion by the stable stratification. Due to these two aspects, the role of the subgrid model tends to be much larger in LES of SBL's than it is for convective or sheared boundary layers. Furthermore, the risk of losing any resolved turbulence exists if the subgrid model is overly dissipative.

M.V. Salvetti et al. (eds.), *Quality and Reliability of Large-Eddy Simulations II*, ERCOFTAC Series 16, DOI 10.1007/978-94-007-0231-8_13,

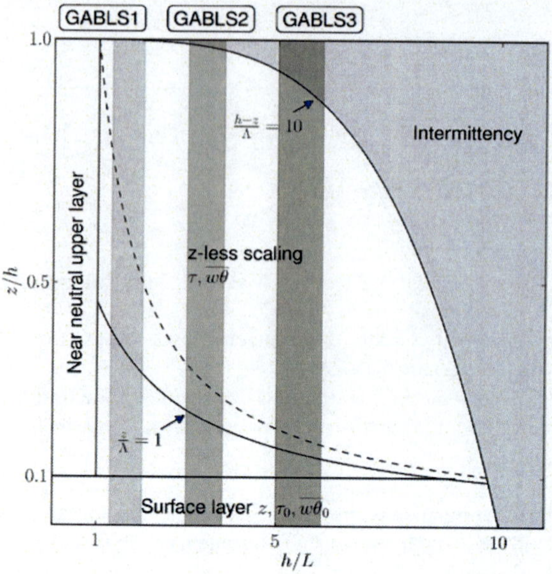

Fig. 1. Characteristic stability of the three GABLS intercomparison cases, depicted in the scaling diagram of [6].

2 GABLS intercomparisons

As a part of the GEWEX Atmospheric Boundary Layer Study (GABLS, [7]), a series of model intercomparisons has been organized for SBL cases. In all intercomparisons a single-column model intercomparison case was defined, whereas a LES case was defined in the first and third intercomparison (see Table 1 for some characteristics and Figure 1 for a comparison of the stability classes of the three cases). It is clear that the cases have increased in terms of the complexity of their setup.

2.1 Setup and results from GABLS-1

The GABLS-1 case [5] was inspired by the idealized setup of [8]: a moderately stable arctic SBL (with $z_i/L \approx 2$, where z_i is the height of the SBL and L is the Obukhov length). The forcing consisted of a constant geostrophic wind of 8 m/s and a constant cooling rate of 0.25 K per hour for the surface temperature.

A number of lessons were learned from the GABLS-1 intercomparison case. The subgrid model played a role down to resolutions of 2 meter. Furthermore, local similarity relationships were confirmed by the models. Finally, the fact that no observational data were available for direct validation proved to be dissatisfying. The importance of the GABLS-1 case as a benchmark is illustrated by the fact that [5] has been cited over 40 times in three years.

Table 1. Characteristics of the three GABLS model intercomparisons.

	GABLS-1	GABLS-2	GABLS-3
SCM/LES	both	SCM only	both
Forcing	$U_{geo} = 8ms^{-1}$, $\frac{\partial T_s}{\partial t} = -0.25Khr^{-1}$,	$U_{geo} = 3ms^{-1}$, $V_{geo} = -9ms^{-1}$, $T_s = f(t)$	$U_{geo} = f(t,z)$, $V_{geo} = f(t,z)$, $T_{0.25} = f(t)$, $q_{0.25} = f(t)$ (LES) surface scheme (SCM) Large-scale advection of u, v, T and q
Time	9 hours (night)	59 hours	24 hours (SCM) 9 hours (LES, night-day transition)
Character	idealized case	inspired by CASES data	as close to observations as possible

2.2 Setup and results of GABLS-3

The GABLS-3 case [4] is based on observations at Cabauw (The Netherlands) and not only covers the night-time period, but the night-to-day transition as well. Furthermore, the case is more stable than the GABLS-1, having a typical z_i/L of 4 to 5. In that sense, the case is more challenging for LES models.

In order to enable a validation of the results with observations, the initial conditions, boundary conditions and forcings have been chosen to resemble reality as closely as possible . The forcings (including time-dependent geostrophic wind as well as advection of heat, moisture and momentum) have been derived from a combination of surface pressure field observations, tower observations and 3D weather model simulations, but were idealized by omitting all short term variability (see also [3]).

The participating groups were required to run their model at least at a common resolution of 6.25 meter (equivalent to 128 points in each direction in a domain of 800 meter cubed). Some groups ran at higher resolutions, up to 1 meter, so that the dependence of the results on grid size (and relative importance of subgrid models) can be investigated. The present authors have also investigated the impact of the order of the advection scheme used.

In total 11 groups have submitted results by January 2009 (in the Figures the groups are indicated by a number, for the moment anonymously). Detailed analysis of the results from different models (focussing on subgrid models, numerics and resolution) is currently ongoing. Some preliminary results are shown in Figure 2 and 3. The profile of the components of the mean wind corresponds well between the different models, and match well with the observations below 200 m. A clear low-level jet is visible, but especially at the height of the velocity maxima the models show the largest discrepancy. Since the details of the inertial oscillation strongly depend on the initial conditions, this may in part be due to the way the models spin up from the initial profile with random variations superimposed. When we turn to some turbulence statistics, the profiles of the variance of horizontal and vertical wind speed show large differences between the models (no validation data included yet). Further analysis is required to clarify the causes of these discrepancies (and others) and to enable improvement of LES models.

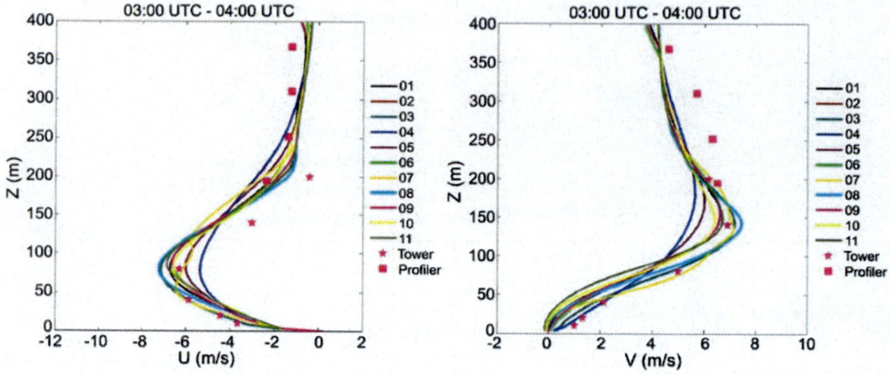

Fig. 2. Results for GABLS-3 LES intercomparison case: profile of mean horizontal wind speed east-west compeonent (left) and north-south component (average over 4th hour of simulation, around sunrise).

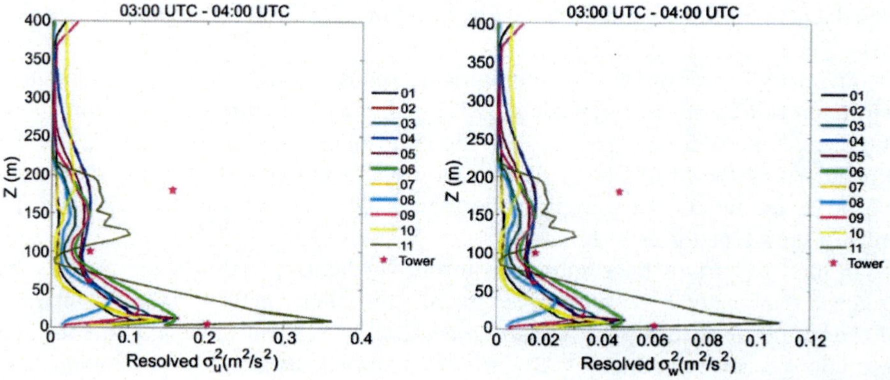

Fig. 3. Results for GABLS-3 LES intercomparison case: profile variance of horizontal wind speed (left) and vertical wind speed (right) (average over 4th hour of simulation, around sunrise). Note that only the resolved part of the variance is given.

3 How to deal with complex forcings

For the GABLS-3 case the forcings were derived from observations and 3-D model results. However, they were idealized for the definition of the final case setup. In this section we will explore other options regarding forcings (mainly momentum) for realistic cases.

3.1 More realistic, more complex forcings

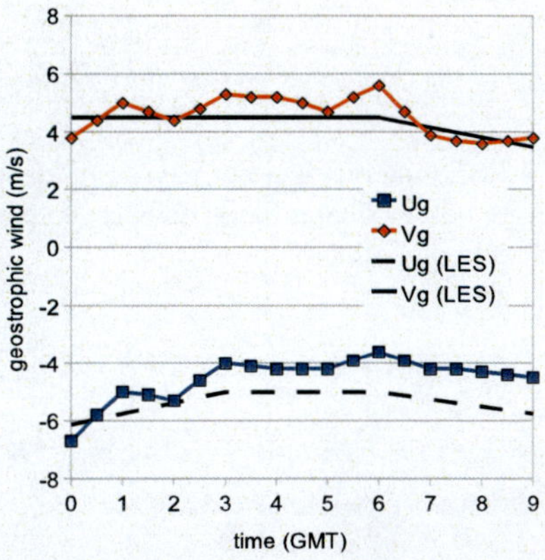

Fig. 4. Forcings for GABLS-3 LES intercomparison case: time-dependence of geostrophic forcing (lines with symbols are best estimates based on best available data; lines are idealized forcings as used in the LES intercomparison).

The momentum forcings as used in the GABLES-3 case were idealized versions of the diagnosed detailed forcings. The momentum forcings are shown in Figure 4 and 5. At certain moments there are large differences between the best estimates and the actual forcings used in the intercomparison case. These differences in forcings are reflected in the results of the LES simulations as well. This is shown in Figure 6 for the wind speed and direction at 200 meter height. The results for the original forcings show a nearly circular inertial oscillation (with an amplitude of 4 m/s), with a large discrepancy with the observations between 2 and 5 GMT. On the other hand, the modified forcings result in hodograph that more closely follows that observations. Hence, for a one-to-one validation of simulation results with data the details of the forcings matter. Still, a clear discrepancy remains. Thus there are variations in the wind field that are apparently not captured by the forcings.

3.2 Ensemble and composite cases

Although the time-dependence of the forcings for the GABLS-3 case seems to be complex, analysis of similar nights has revealed that some of the variation is system-

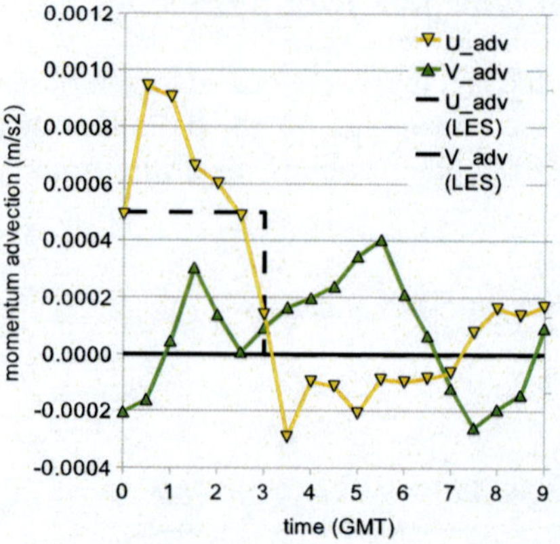

Fig. 5. Forcings for GABLS-3 LES intercomparison case: time-dependence of momentum advection (lines with symbols are best estimates based on best available data; lines are idealized forcings as used in the LES intercomparison).

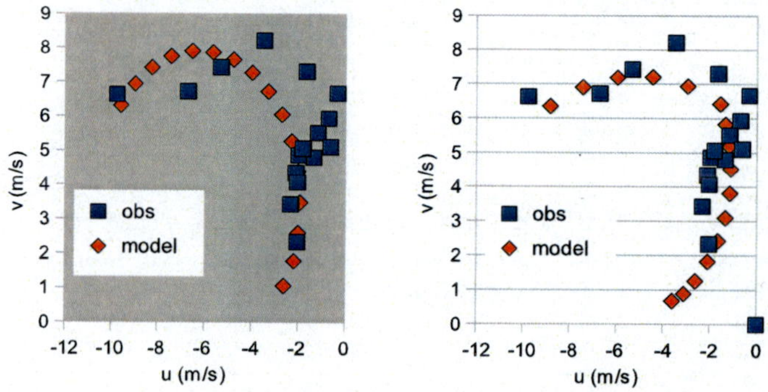

Fig. 6. Hodograph of windspeed at height of 203 meter (model) and 200 meter (data). Left: the result of the original forcings. Right, the same run, but with geostrophic forcings as well as momentum advection defined according to 4 and 5. Each symbol signifies a 5 minute mean and symbols are sepearated in time by 30 miniutes: first symbol at 0:30 GMT).

atic between nights (see [2]). In [2] two ways to deal with possibly incomplete forcings are dealt with (in the context of single-column modelling). The first method is to select a number of nights with similar characteristics. Although for the individual nights considerable discrepancies occur between model and observations, the mean of the model simulations appears to agree well with the mean of the observations.

This leads to the second method. The availability of a number of similar nights in the extensive data base of the Cabauw site enables the specification of a composite case description. In such a composite both forcings and validation data are averaged. This reduces the variability in the forcings and the observations, while retaining the realism of the case.

4 Conclusion

The main conclusions from the two LES intercomparison exercises and the developments with respect to the specification of composite cases are:

- Model intercomparisons for the stable boundary layer are useful exercises, but they should go beyond the validation of mean profiles, as these are strongly restrained by external forcings. In the analysis care is required to distinguish which part of the modelled dynamics is due to the exernal forcings and what part is produced by the model itself.
- To enable validation of model intercomparisons the open nature of the atmospheric system makes complex forcings necessary. However, this makes the case less useful as a simple benchmark.
- The use of an ensemble of similar cases will reduce random variations, while retaining a realistic case description.

5 Acknowledgement

This work was sponsored by the National Computing Facilities (NCF).

References

1. A. Andren, A.R. Brown, P.J. Mason, J. Graf, U. Schumann, C.H. Moeng, and F.T.M. Nieuwstadt. Large-eddy simulation of a neutrally stratified boundary layer: A comparison of four computer codes. *Quarterly Journal of the Royal Meteorological Society*, 120(520), 1994.
2. P. Baas, F.C. Bosveld, G. Lenderink, E. van Meijgaard, and A.A.M. Holtslag. How to design single-column model experiments for comparison with observed noctural lowlevel jets at cabauw? *submitted to Quarterly Journal of the Royal Meteorological Society*, 2009.
3. P. Baas, F.C. Bosveld, G.J. Steeneveld, and A.A.M. Holtslag. Towards a third intercomparison case for gabls using cabauw data. In *18th Symposium on Boundary Layers and Turbulence, Stockholm, Sweden, 9 - 13 June, 2008. - Boston : American Meteorological Society, 2008*, 2008.
4. S. Basu, G.J. Steeneveld, A.A.M. Holtslag, and F.C. Bosveld. Large-eddy simulation intercomparison case setup for GABLS3. In *18th Symposium on Boundary Layers and Turbulence, Stockholm, Sweden, 9 - 13 June, 2008. - Boston : American Meteorological Society, - p. - 8A.7.*, 2008.
5. R.J. Beare, M.K. Macvean, A.A.M. Holtslag, J. Cuxart, I. Esau, J.C. Golaz, M.A. Jimenez, M. Khairoutdinov, B. Kosovic, D. Lewellen, et al. An Intercomparison of Large-Eddy Simulations of the Stable Boundary Layer. *Boundary-Layer Meteorology*, 118(2):247–272, 2006.
6. A.A.M. Holtslag and F.T.M. Nieuwstadt. Scaling the atmospheric boundary layer. *Boundary-Layer Meteorol.*, 36:201–209, 1986.
7. B. Holtslag. Preface: GEWEX Atmospheric Boundary-layer Study (GABLS) on Stable Boundary Layers. *Boundary-Layer Meteorology*, 118(2):243–246, 2006.

8. B. Kosović and J.A. Curry. A large eddy simulation study of a quasi-steady, stably stratified atmospheric boundary layer. *Journal of the Atmospheric Sciences*, 57(8):1052–1068, 2000.
9. C-H. Moeng, W.R. Cotton, C. Bretherton, A. Chlond, M. Khairoutdinov, S. Krueger, W.S. Lewellen, M.K. MacVean, J.R.M. Pasquier, H.A. Rand, A.P. Siebesma, B. Stevens, and R.I. Sykes. Simulation of a stratocumulus-topped planetary boundary layer: Intercomparison among different numerical codes. *Bull. Amer. Meteor. Soc.*, 77:261–278, 1996.
10. F.T.M. Nieuwstadt, P.J. Mason, C-H. Moeng, and U. Schumann. Large-eddy simulation of the convective boundary layer: A comparison of four computer codes. In F. Durst, R. Friedrich, F.W. Schmidt, U. Schumann, and J.H. Whitelaw, editors, *Selected Papers from the 8th International Symposium on Turbulent Shear Flows, Munich, Germany*, number 8 in Turbulent Shear Flows, pages 343–367. Springer-Verlag, Berlin, sep 1993.
11. A.P. Siebesma, C.S. Bretherton, A. Brown, A. Chlond, J. Cuxart, P.G. Duynkerke, H.L. Jiang, M. Khairoutdinov, D. Lewellen, C.H. Moeng, E. Sanchez, B. Stevens, and D.E. Stevens. A large eddy simulation intercomparison study of shallow cumulus convection. *J. Atmos. Sci.*, 60(10):1201–1219, 2003.

Evaluating Subgrid-Scale Models for Large-Eddy Simulation of Turbulent Katabatic Flow

Bryan A. Burkholder, Evgeni Fedorovich, and Alan Shapiro

School of Meteorology, University of Oklahoma, 120 David L. Boren Blvd., Norman, OK, USA 73072 Bryan.A.Burkholder-1@ou.edu

Summary. The performance of commonly used subgrid-scale (SGS) models is evaluated for large-eddy simulation (LES) of turbulent katabatic flow. The very stable stratification and strong low-level shear in this flow provide a stringent test for SGS models. Using an *a posteriori* testing procedure, the SGS models' performance in reproducing turbulence statistics and spectra in katabatic flow is investigated.

Key words: Large-Eddy Simulation, Subgrid-Scale Models, Katabatic Flow, *A Posteriori* Testing

1 Introduction

Over the past few decades, large-eddy simulation (LES) has become an invaluable tool for investigating the structure and characteristics of atmospheric boundary layer flows [1]. While encouraging results have been obtained from LES of neutrally stratified and unstable (convective) boundary layers, questions still remain concerning the reliability of LES for reproducing stably-stratified turbulent boundary layers [2]. Under stably-stratified conditions, the characteristic length scale of the small-scale turbulent motions decrease, placing a larger burden on the subgrid-scale model employed.

A plethora of subgrid/subfilter scale (SGS/SFS) models exist for wide variety of applied problems in atmospheric dynamics. Reliable simulations have been performed in moderately stable boundary layers and have been successfully compared to observational data [3]. However, no single SGS model appears significantly more appropriate than another for a broad range of grid spacings. Typically, intercomparisons test LES performance against direct numerical simulation (DNS) or measurement data. However, LES is run at much coarser resolutions than DNS and outputs from the two simulation techniques may not be directly comparable. Similarly, LES may not incorporate all of the phenomena present in real flows when compared to observational data. To fairly evaluate the performance of the SGS models, we employ an *a posteriori* testing procedure [4][5]. To perform this type of procedure, the

M.V. Salvetti et al. (eds.), *Quality and Reliability of Large-Eddy Simulations II*, ERCOFTAC Series 16, DOI 10.1007/978-94-007-0231-8_14,

statistical behavior of the flow reproduced by LES is compared to filtered DNS data for an identically forced flow.

We employ the *a posteriori* testing procedure in simulations of a shallow jet-like flow developing along a cooled planar slope (katabatic flow). The earliest solutions to this flow type can be traced back to studies by Prandtl [6], where the Boussinesq equations of motion were solved analytically for a laminar slope flow in a stably-stratified environment. The Prandtl solution is characterized by a strong near-surface jet and weaker upslope flow aloft. It is no surprise that the turbulent counterpart of this flow is particularly difficult for LES, given the strong flow shear near the surface accompanied by strong stable stratification. To futher complicate matters, no complete similarity theory has been developed yet for flows along sloping terrain [7]. These factors only compound the problems LES is known to have near bounding surfaces, where the characteristic length scales of turbulent motions can be close to the LES filter width [8]. Katabatic and anabatic (heated slope) flows have actually been numerically investigated using LES with two different SGS closures [9][10], but the question regarding the optimal closure of LES for katabatic flows remains unresolved.

2 Governing Equations and Closures

Following [6][11], we simulate a katabatic flow over a doubly-inifinite sloping surface which is inclined at a constant angle α with respect to the horizontal. In DNS, we solve the Boussinesq equations of motion and thermodynamic energy in rotated coordinates analagous to the ones adopted in [12][13] (Fig. 1). In the rotated coordinate system, $x = x_1$ points in the downslope direction, $y = x_2$ lies in the cross-slope direction, and $z = x_3$ is oriented normal to the slope.

In LES, we use the same rotated coordinates and solve the filtered Boussinesq equations of motion and thermodynamic energy,

$$\frac{\partial \tilde{u}_i}{\partial t} + \tilde{u}_j \frac{\partial \tilde{u}_i}{\partial x_j} = \tilde{b}(-\delta_{i1}\sin\alpha + \delta_{i3}\cos\alpha) + \nu \frac{\partial \tilde{u}_i}{\partial x_j \partial x_j} - \frac{\partial \tau_{ij}}{\partial x_j} - \frac{1}{\rho}\frac{\partial \bar{\tilde{p}}}{\partial x_i}, \tag{1}$$

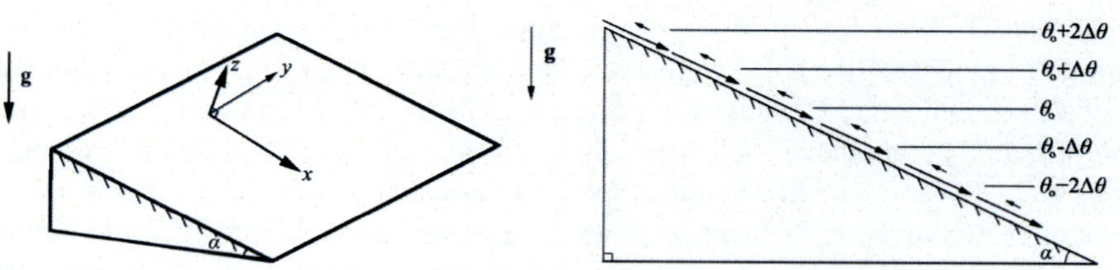

Fig. 1. Orientation of the coordinate system used (left) and a cross-sectional view of an idealized katabatic flow (right). The thin solid lines illustrate the stratification of the base-state environment (θ_∞), which is characterized by θ increasing linearly with height.

$$\frac{\partial \tilde{b}}{\partial t} + \tilde{u}_j \frac{\partial \tilde{b}}{\partial x_j} = -N^2(-\tilde{u}_1 \sin\alpha + \tilde{u}_3 \cos\alpha) + \kappa \frac{\partial^2 \tilde{b}}{\partial x_j \partial x_j} - \frac{\partial B_j}{\partial x_j}, \tag{2}$$

$$\frac{\partial \tilde{u}_i}{\partial x_i} = 0, \tag{3}$$

with

$$\tau_{ij} = \widetilde{u_i u_j} - \tilde{u}_i \tilde{u}_j, \tag{4}$$

$$B_j = \widetilde{b u_j} - \tilde{b}\tilde{u}_j, \tag{5}$$

$$\bar{\tilde{p}} = \tilde{p} + \frac{2}{3}E, \tag{6}$$

where $\widetilde{(...)}$ indicates a quantity filtered using a top-hat filter, N is the Brunt-Vaisala frequency (which is set to a constant value), τ_{ij} is the subgrid momentum flux (negative of the subgrid stress tensor), B_j is the subgrid buoyancy flux, $\tilde{p}$ is the filtered pressure, $\bar{\tilde{p}}$ is the modified pressure, and E is the subgrid/subfilter turbulence kinetic energy (TKE). The top-hat filter width, Δ, is taken to be equal to the LES grid spacing. Buoyancy is defined as

$$b = g\frac{\theta - \theta_\infty}{\theta_r}, \tag{7}$$

where g is gravitational acceleration, θ is potential temperature, θ_r is a constant reference potential temperature, and θ_∞ is the environmental potential temperature which is taken to vary linearly with height. To close the system of filtered equations, τ_{ij} and B_j must be parameterized in terms of the resolved flow fields.

One of the earliest SGS models used for meteorological applications was the Smagorinsky model [14], which employs the assumption of a balance between shear production and the dissipation of subgrid TKE. As a result, the SGS stress tensor is taken proportional to the resolved strain rate tensor:

$$\nu_T = [C_S\Delta]^2\tilde{S}, \tag{8}$$

$$\tau_{ij} = E\delta_{ij} - 2\nu_T\tilde{S}_{ij}, \tag{9}$$

where $\tilde{S} = \sqrt{2\tilde{S}_{ij}\tilde{S}_{ij}}$ and C_S is known as the Smagorinsky coefficient. Using a sharp cutoff filter in the inertial subrange and assuming the Kolmogorov scaling, C_S was found to be roughly 0.17 [15]. The value of C_S has also been found to vary significantly depending on the type of flow simulated [8]. A value around 0.2 is commonly used in atmospheric contexts [1]. While reasonable results can be attained using the Smagorinsky model, the underlying assumption that the SGS stress tensor is proportional to the strain rate is a critical one, and may not necessarily be valid in all applications. The model is also known to be overly dissipative, especially near the surface. Despite its inherent disadvantages, this SGS model is still implemented as a baseline model in LES of many atmospheric flows because of its simplicity.

Deardorff [16] considered another form of the subgrid TKE balance by including buoyancy effects and subgrid energy transport. The subgrid TKE is then calculated as a prognostic variable from a simplified version of its governing equation [17] and is used to parameterize the eddy viscosity locally through

$$\nu_T = C_D l \sqrt{E}, \tag{10}$$

$$\kappa_T = (1 + \frac{2l}{\Delta})\nu_T, \tag{11}$$

$$l = \begin{cases} \Delta & \frac{\partial \tilde{b}}{\partial z} \leq 0, \\ \min[\Delta, 0.5\sqrt{E/(\partial \tilde{b}/\partial z)}] & \frac{\partial \tilde{b}}{\partial z} > 0. \end{cases} \tag{12}$$

The adopted dependence of the mixing length on the stratification (Eq. 12) is used to decrease the turbulent length scale when stable stratification is encountered. The parameter C_D was set to 0.12 in [16]. Even though [16] used this l scaling factor within stratocumulus cloud layers, it is currently implemented in many SGS models that are used in general atmospheric applications.

Both considered formulations, however, do not account for the backscatter of energy from small scales to larger scales, which may be important in parameterizing the SGS motions [18][19][20]. To counteract the overly dissipative nature of the schemes proposed by [14] and [16], some form of non-linear or stochastic backscattering mechanism has been suggested for simulations of stable boundary layers in [1][21]. The original stochastic backscattering model of [19] was computationally exhaustive and did not match Monin-Obukhov similarity near the surface. Sullivan [22] argued that this backscatter behavior could be accounted for by using adequate grid resolution and less dissipative SGS closures. A two-part model was proposed, in which the near-surface region was controlled by the mean shear and the region away from the surface followed the closure [16]:

$$\tau_{ij} = -2\nu_T{}'\gamma\tilde{S}_{ij} - 2\bar{\nu}_T\langle\tilde{S}_{ij}\rangle + E\delta_{ij}, \tag{13}$$

where $\langle ... \rangle$ represents averaging in homogeneous directions (slope-parallel planes in the case of laterally homogeneous flow), $\nu_T{}'$ is the fluctuating field-eddy viscosity, $\bar{\nu}_T$ is the mean-field eddy viscosity, and γ is the isotropy factor:

$$\gamma = \frac{\tilde{S}'}{\langle\tilde{S}\rangle + \tilde{S}'}, \tag{14}$$

where $\tilde{S}'$ is the fluctuating strain rate. The calculation of the isotropy factor allows a smooth transition from an ensemble-average approach near the ground to the baseline subgrid TKE model of [16] in the main portion of the flow. The SGS buoyancy flux was found to be important in moderately stable conditions and was later included in the two-part formulation [23].

Another method to modify the classic Smagorinsky model is to dynamically calculate C_S. This approach employs a larger filter width, for example, $\widehat{\Delta} = 2\Delta$, and then uses the filtered fields to calculate the Smagorinsky coefficient using the Germano identity [24]. Rather than specifying the coefficient, the SGS model essentially calculates its optimal value from the resolved fields. To statistically minimize the error and find this optimal value, averaging over some homogeneous direction must be applied [25]. Preferably, this averaging should be performed over Lagrangian paths or over the nearest grid cell neighbors in homogeneous directions [26]. A third, even larger filter width, $\widetilde{\Delta} = 2\widehat{\Delta} = 4\Delta$, can be introduced to make the model scale-dependent using a power-law relation between filter widths [27][28]. Even though the dynamic versions of the original Smagorinsky type-closure can essentially adjust the characteristic length scale of the subgrid-scale motions, it cannot reproduce the backscattering and anisotropy necessary to correctly simulate the turbulence in the stable boundary layer [21].

3 Model setup and flow characteristics

The DNS code used in our study is described in [11]. It uses second-order in space finite difference approximations of spatial derivatives and a leapfrog scheme with a weak Asselin filter for time advancement. The modified pressure is computed diagnostically at every time step by a Poisson solver. The LES uses the same core of the computational code, but employs various SGS algorithms for solving the filtered Boussinesq equations. Table 1 lists the different subgrid closures that have been incorporated in the LES and tested in the present study. The naming convention adopted for the various SGS closures is as follows: Smagorinsky (S63) [14], the two-part model (S94) [22], and the scale-invariant dynamic model (PA00) [27][28]. In the case of the Deardorff [16] closure, two versions are employed; one using the l-scaling factor given in Eq. 12 (D80) and another that takes Δ to be the mixing length regardless of the stratification [D80 (no N)]. The LES with no SGS closure, commonly referred to as a quasi-DNS, is abbreviated as qDNS. For the filtered DNS output, the abbreviation is of fDNS is taken. The LES runs are carried out with grid sizes of $32 \times 32 \times 100$, while the DNS is run on the grid $128 \times 128 \times 400$. This gives the LES a degraded resolution by a factor of four. The filtering procedures used in the *a posteriori* test are described in Sect. 4.

To generate the katabatic flow in the simulations, a buoyancy flux of $-0.5\ \mathrm{m^2s^{-3}}$ is imposed at the surface. The boundary conditions for the DNS are taken to be no-slip and impermeable at the surface and zero-gradient for all variables at the top of the domain. The Prandtl number, $\mathrm{Pr} = \nu/\kappa$, is taken to be unity. In the LES, Monin-Obukhov similarity is applied to formulate surface boundary conditions in the runs with for D80, D80 (no N), and S94, while pure no-slip are used in the runs with S63, qDNS, and PA00. Small random perturbations are added to the surface forcing to excite instabilities and encourage the flow to become turbulent. Figure 2 shows the DNS output of the downslope velocity component, averaged over planes parallel to the surface, as a function of time and slope-normal height for a slope of 60°. The flow exhibits a characteristic oscillation in time with a period of $2\pi/(N\sin\alpha)$, as has been observed in previous katabatic flow studies [13].

Table 1. List of abbreviations for different DNS/LES runs. The asterisk indicates that the mixing length scaling factor in Eq. 12 was not used.

Abbreviation	Type (citation)
DNS	DNS
fDNS	DNS (filtered to LES grid)
qDNS	LES (no closure)
S63	LES [14]
D80	LES [16]
D80 (no N)	LES [16]*
S94	LES [22]
PA00	LES [27][28]

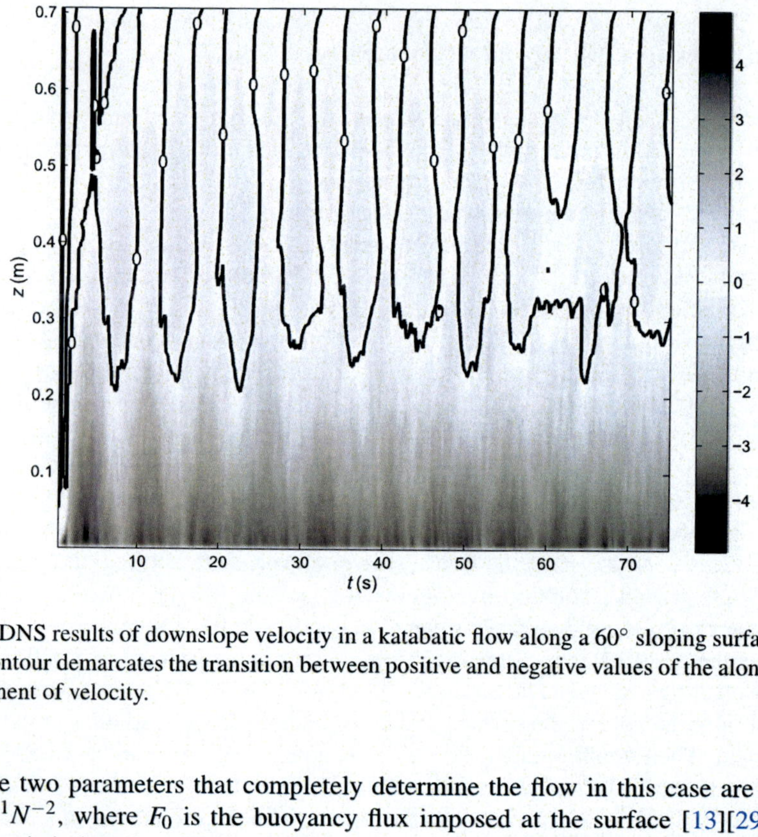

Fig. 2. DNS results of downslope velocity in a katabatic flow along a 60° sloping surface. The bold contour demarcates the transition between positive and negative values of the along-slope component of velocity.

The two parameters that completely determine the flow in this case are α and $|F_0|\nu^{-1}N^{-2}$, where F_0 is the buoyancy flux imposed at the surface [13][29]. The characteristic length, velocity, and buoyancy scales of the flow in the present study are given, respectively, by

$$l_s = \sqrt{|F_0|N^{-3}},\ u_s = \sqrt{|F_0|N^{-1}},\ b_s = \sqrt{|F_0|N}. \tag{15}$$

Taking the governing parameters for the flow to be $F_0 = -0.5\ \mathrm{m^2s^{-3}}$, $N = 1\ \mathrm{s^{-1}}$, and $\nu = 10^{-4}\ \mathrm{m^2s^{-1}}$, we obtain $|F_0|\nu^{-1}N^{-2} = 5000$ with characteristic scales $l_s = 0.71$ m, $u_s = 0.71\ \mathrm{ms^{-1}}$, and $b_s = 0.71\ \mathrm{ms^{-2}}$. Alternatively, we can choose other parameter values provided that $|F_0|/\nu N^2 = 5000$, which would result in exactly the same scaled flow solutions. For example, selecting values commonly found in atmospheric applications such as $N = 10^{-2}\ \mathrm{s^{-1}}$, $\nu = 0.1\ \mathrm{m^2s^{-1}}$, and $F_0 = -0.05\ \mathrm{m^2}s^{-3}$, our characteristic scales become $l_{s2} = 220$ m, $u_{s2} = 2.2\ \mathrm{ms^{-1}}$, and $b_{s2} = 0.022\ \mathrm{ms^{-2}}$. Throughout the paper, we use the first set of scales to present results in a dimensional form.

4 *A Posteriori* Testing

A posteriori testing is commonly used to compare LES flow statistics to the those provided by a DNS of an identically forced flow. Since DNS typically has a higher resolution than LES, the DNS fields must be filtered down to the LES grid. To do this, a top-hat filter that has the width of the LES filter is applied. The effect of filtering the DNS output to the LES grid is shown in the top two panels of Fig. 3. It is seen that some of the smaller-scale features are lost through the filtering operation, but the overall visual characteristics of the flow do not change significantly. Since the filtered DNS output would be the best field LES could ever reproduce at a degraded resolution, it is taken as a reference flow field.

Normally, differences found in comparing an LES output to the filtered DNS results could potentially originate from different SGS models, numerical discretizations in space, time advancement schemes, or resolutions [4]. In our case, since each LES run is completed using identical numerical schemes, virtually all of the differences are expected to originate from the SGS models. In Fig. 3, we can see that near the surface, all tested SGS closures tend to organize the near-surface flow into large-scale structures which are absent in the filtered DNS field. This could be due to the over-dissipative nature found in some of the schemes, but also may be an effect of the insufficient grid resolution near the surface. The level depicted in Fig. 3 is only the 8th point above the surface for LES, while it is the 32nd point above the ground for DNS. It is likely the problems that LES has near bounding surfaces in moderate to high Reynolds number flows are still observed at this level above the lower boundary. The dynamic closure and the qDNS share the most visual similarities with the filtered DNS output. This may be due to the the schemes being under-dissipative, which is briefly discussed below.

The one-dimensional velocity spectra shown in Fig. 4 demonstrate that most of the SGS closures produce the desired general spectral characteristics. The velocity spectra are calculated according to [30] and are averaged in time over at least six oscillations (see Fig. 2). This smooths the spectra making them easier to interpret. Near the surface, LES spectra obtained with S63 and S94 schemes are almost identical to the fDNS spectrum. The one spectrum that does not exhibit reasonable behavior is from the qDNS, which follows the -5/3 line at exceedingly high wavenumbers. In the plot with their dissipation spectra, $D_{11}(k) = 2\nu k^2 E_{11}(k)$, turbulence in the qDNS is

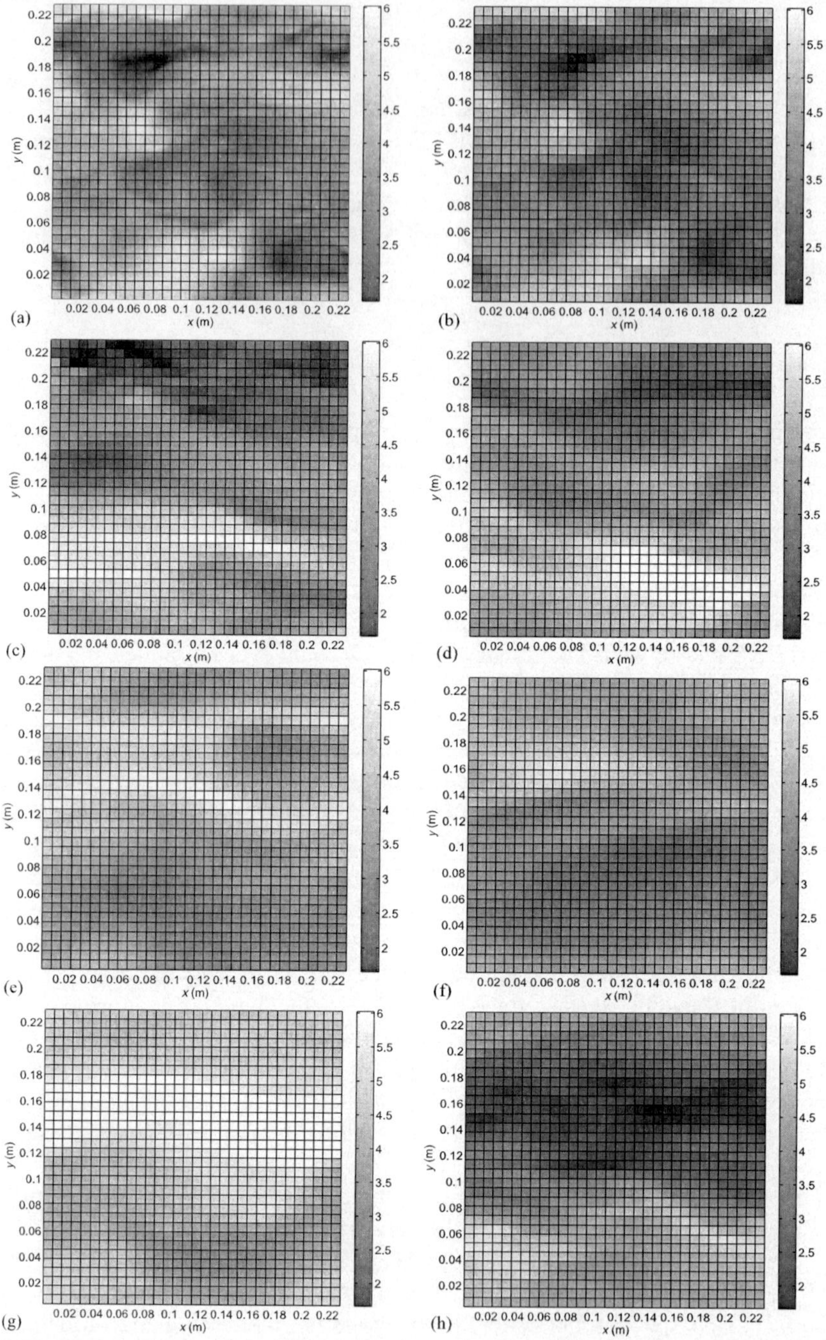

Fig. 3. Horizontal contour plot of downslope velocity on a plane parallel to the sloping surface. Each panel starting at the top left and going to the right and down, shows a) the output from DNS, b) the DNS output filtered to the LES grid, c) qDNS, d) S63, e) D80, f) D80 (no N), g) S94, and h) PA00. In each plot, the LES grid spacing is illustrated by the thin black lines.

found to be dissipating improperly. Most of the energy in this spectrum dissipates at wavenumbers very close to the Kolmogorov scale. All of the SGS closure schemes show reasonable spectral dissipation properties, as most of the dissipation occurs at wavenumbers much smaller than the Kolmogorov scale. The discrepancies between the different SGS closures become smaller with height, as the LES is able to resolve the motions at these levels. However, the D80 TKE-based schemes appear to be slightly over-dissipative, even at substantial distances from the slope. For katabatic flows, the S63 scheme performance appears to be reasonable, at least once turbulence has already developed. As has been previously reported, the dynamic model is under-dissipative close to the surface [27], but performs reasonably well in regions away from the surface.

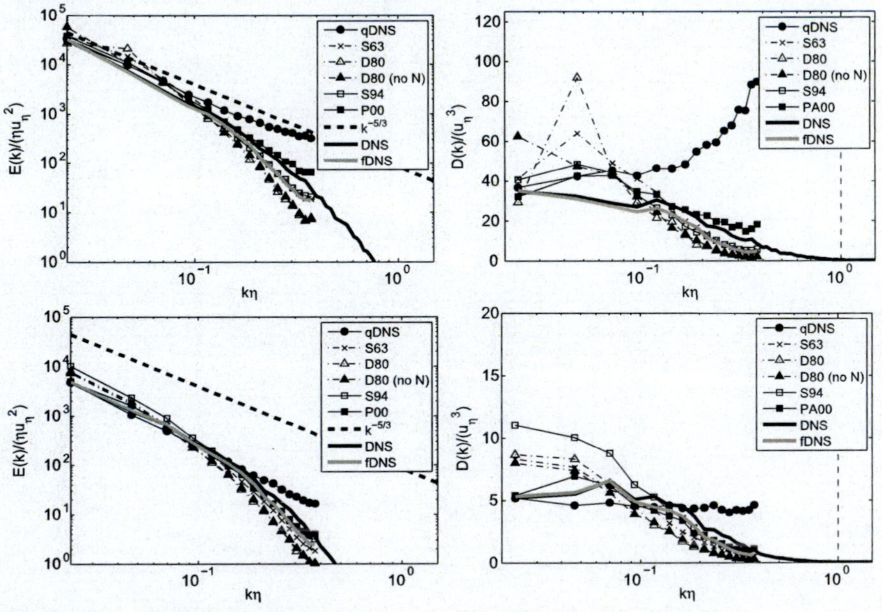

Fig. 4. Time-averaged normalized u_1 spectra at DNS level 32 (LES level 8) (top, left) and their respective dissipation spectra (top,right). The same is shown at level 160 in the bottom two panels.

From Fig. 5, we conclude that LES is roughly capable of reproducing the correct flow in terms of mean velocity and buoyancy which are obtained by averaging over planes parallel to the surface and over six oscillations in time. Not surprisingly, the LES performance is worse for higher order statistics, as is seen for the total turbulent fluxes and the variances, especially near the surface. When comparing the ratios of the total contribution of each velocity variance to the to total velocity component variance, we can see that the fDNS exhibits $\widetilde{\langle u_1' u_1' \rangle}$ that is about 3 times larger than $\widetilde{\langle u_3' u_3' \rangle}$, where the angle brackets here denote the combined spatial (over planes paral-

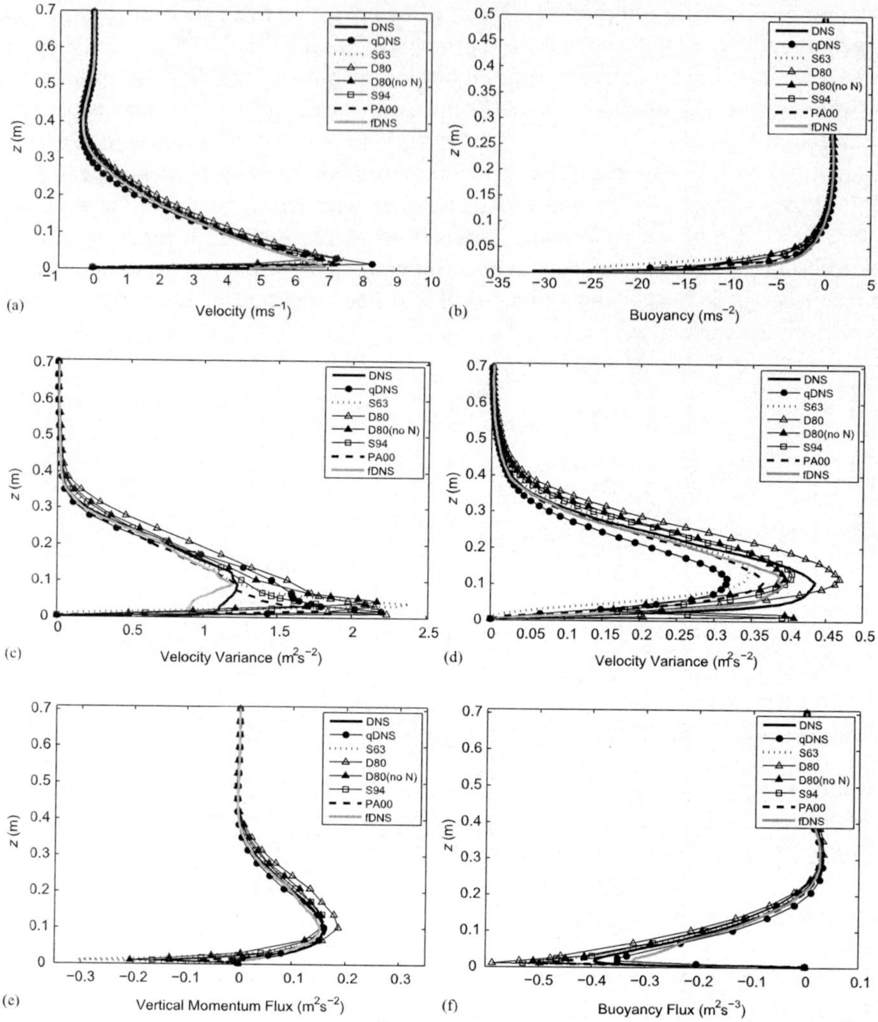

Fig. 5. Temporal and planar-averaged profiles ($\langle ... \rangle$) of a) $\langle \tilde{u}_1 \rangle$, b) $\langle \tilde{b} \rangle$, c) $\langle \widetilde{u_1' u_1'} \rangle$, d) $\langle \widetilde{u_3' u_3'} \rangle$, e) $\langle \widetilde{u_1' u_3'} \rangle$, and f) $\langle \widetilde{u_3' b'} \rangle$ as a function of z.

lel to the slope) and temporal (over several flow oscillations) averaging. Meanwhile, LES tends to have larger ratios of variance components, with $\langle \widetilde{u_1' u_1'} \rangle$ being about a factor of 4.5 - 10 times larger than $\langle \widetilde{u_3' u_3'} \rangle$. Near the surface, the TKE-based SGS closures generate a pronounced spike in vertical velocity variance. An analysis of the subgrid TKE budget (not shown) reveals that the near-surface shear production associated with the katabatic jet is responsible for the spike. Some of the SGS models tested (especially the S63 model) also tend to over-predict the slightly negative vertical momentum flux near the surface that is observed in the DNS output. Interestingly, for the buoyancy related fields, the LES with no SGS model (qDNS) outperformed some LES with SGS formulations. It should be noted that the correction of the Sullivan [22] scheme for scalars implemented in [23] was not tested.

5 Conclusions

Overall, the performance of the basic SGS closures employed by LES for atmospheric applications is acceptable for mean fields, but degrades noticeably when it comes to higher-order statistics (particularly buoyancy fluxes). Near the ground, where the SGS contributions compose the largest portion of the total fluxes and variances, the different closures produce highly variable results. The over-production of subgrid TKE by the shear close to the surface causes errors in the models based on [16] and especially affect the vertical velocity variance. Also, many of the tested schemes drastically overestimate the negative momentum flux near the ground. A majority of the tested schemes are able to capture the anisotropy of the flow in the along-slope direction, though this feature of the flow is also slightly overestimated by the LES.

With respect to velocity spectra, the [22] and [14] schemes perform the best near the ground, while the dynamic and two-part models perform better in turbulent regions of the flow away from the surface. Even though the solution produced by the qDNS remains stable, it exhibits a build-up of energy as well as unrealistic dissipation at high wavelengths. The schemes based on [16] are found to be over-dissipative, and flow statistics slightly improved when the mixing length stratification adjustment after [16] is not used.

The extension of testing toward other SGS models will constitute the scope of future studies. Updating the time advancement schemes in both DNS and LES is also planned. The influence of the varying slope angle on the performance of the SGS models in LES of katabatic flows should be studied as well.

References

1. Mason PJ (1994) Bound-Layer Meteor 120:1–26
2. Mahrt L (1998) Theoret Comput Fluid Dyn 11:263–279
3. Beare RJ, MacVean MK, Holtslag AAM, Cuxart J, Esau I, Golaz JC, Jiménez MA, Khairoutdinov M, Kosović B, Lewellen D, Lund TS, Lundquist JK, McCabe A, Moene AF, Noh Y, Raasch S, and Sullivan P (2006) Bound-Layer Meteor 118:247–272
4. Geurts B (2004) Elements of direct and large-eddy simulation. R. T. Edwards, Inc., Philadelphia, PA
5. Vreman B, Geurts B, and Kuerten H (1997) J Fluid Mech 339:357–390
6. Prandtl L (1942) Führer durch die Strömungslehre. Vieweg & Sohn, Braunschweig
7. Denby B (1999) Bound-Layer Meteor 92:67–100
8. Pope SB (2000) Turbulent flows. Cambridge University Press, Cambridge
9. Schumann U (1990) Quart J Roy Meteor Soc 116:637–670

10. Skyllingstad ED (2003) Bound-Layer Meteor 106:217–243
11. Shapiro A and Fedorovich E (2004) Int J Heat and Mass Transfer 47:4911–4927
12. Burkholder BA, Shapiro A, Fedorovich E (2009) Acta Geophysica (in press, DOI: 10.2478/s11600-009-0025-6)
13. Fedorovich E and Shapiro A (2009) Acta Geophysica (in press, DOI: 10.2478/s11600-009-0027-4)
14. Smagorinsky J (1963) Mon Wea Rev 91:99–164
15. Lilly DK (1967) Proc IBM Sci Computing Symp Environmental Sci 320-1951:195–210
16. Deardorff JW (1980) Bound-Layer Meteor 18:495–527
17. Piomelli U and Chasnov JR (1996) Large-eddy simulations: theory and applications. In: Henningson D, Hallbäck, Alfreddson H, and Johansson A (eds) Transition and turbulence modelling. Kluwer Academic Publishers, Dordrecht
18. Leslie DC and Quarini GL (1979) J Fluid Mech 91:65–91
19. Mason PJ and Thomson DJ (1992) J Fluid Mech 242:51–78
20. Brown AR, Derbyshire SH, Mason PJ (1994) Quart J Roy Meteorol Soc 120:1485–1512
21. Kosović B (1997) J Fluid Mech 336:151–182
22. Sullivan PP, McWilliams JC, Moeng C (1994) Bound-Layer Meteor 71:247–276
23. Saiki EM, Moeng C, and Sullivan PP (2000) Bound-Layer Meteor 95:1–30
24. Germano M, Piomelli U, Moin P, Cabot WH (1991) Phys Fluids A 7:1760–1771
25. Lilly DK (1992) Phys Fluids A 4:633–635
26. Stoll R and Porté-Agel F (2008) Bound-Layer Meteor 126:1–28
27. Porté-Agel F, Meneveau C, Parlange MB (2000) J Fluid Mech 415:261–284
28. Porté-Agel (2004) Bound-Layer Meteor 112:81–105
29. Fedorovich E and Shapiro A (2009) J Fluid Mech (in press)
30. Kaiser R and Fedorovich E (1998) J Atmo Sci 55:580–594

Large-eddy simulation of pyroclastic density currents

Tomaso Esposti Ongaro[1], Sara Barsotti[1], Augusto Neri[1], Maria Vittoria Salvetti[1,2]

[1] Istituto Nazionale di Geofisica e Vulcanologia, Sezione di Pisa, Via della Faggiola 32, 56126 Pisa, Italy
[2] Dipartimento di Ingegneria Aerospaziale, Università di Pisa, Via G. Caruso 8, 56122 Pisa, Italy

Summary. We investigate the dynamics of turbulent pyroclastic density currents (PDCs) by adopting a 3D, Eulerian-Eulerian multiphase flow model, in which solid particles are treated as a continuum and the grain-size distribution is simplified by assuming two particulate phases. The turbulent sub-grid stress of the gas phase is modelled within the framework of Large-Eddy Simulation (LES) by means of a eddy-viscosity model together with a wall closure. Despite the significant numerical diffusion associated to the upwind method adopted for the Finite-Volume discretization, numerical simulations demonstrate the need of adopting a Sub-Grid Scale (SGS) model, while revealing the complex interplay between the grid and the SGS filter sizes. We also analyse the relationship between the averaged flow dynamic pressure and the action exerted by the PDC on a cubic obstacle, to evaluate the impact of a PDC on a building. Numerical results suggest that the average flow dynamic pressure can be used as a proxy for the force per unit surface acting on the building envelope (Fig. 5), even for such steeply stratified flows. However, it is not possible to express such proportionality as a constant coefficient such as the drag coefficient in a steady-state current. The present results indeed indicate that the large epistemic and aleatory uncertainty on initial and boundary conditions has an impact on the numerical predictions which is comparable to that of grid resolution.

Key words: Large-Eddy Simulation, pyroclastic density currents, numerical simulation, multiphase flows

1 Introduction

Pyroclastic density currents (PDCs) are high-temperature, high-velocity particle-laden flows that propagate along the flanks of a volcano under the effect of their density contrast with respect to the atmosphere. They are made up of volcanic gases and fragments of magma and rocks, ranging in size from a few microns to several decimeters, with variable density and shape, which are the product of the *fragmentation* of the liquid magma during its decompression along the volcanic conduit. PDCs can be generated by the instability and collapse of a volcanic jet (pyroclastic flows and surges), by the collapse and crumbling of a lava dome (block-and-ash flows) or by the lateral explosion of a pressurized magma body (directed blasts). Solid particles

M.V. Salvetti et al. (eds.), *Quality and Reliability of Large-Eddy Simulations II*, ERCOFTAC Series 16, DOI 10.1007/978-94-007-0231-8_15,

within the current tend to segregate leading to a steep density stratification, with solid concentrations ranging from dense packing at the base (volume fraction $> 50\%$) to very dilute (volume fraction $<< 1\%$) on the top boundary [1, 9, 16]. PDCs dynamics are controlled by the competing effects of sedimentation and turbulent mixing. Particles are suspended by turbulence in the more diluted part of the current, whereas in the basal layer they are mainly supported by fluid pressure and particle collision, since the increasing solid concentration dampens turbulent fluctuations [2].

PDCs are among the most hazardous volcanic phenomena, due to their fast emplacement and destructive nature. One of the main objectives of volcanology is therefore to make a quantitative assessment of their dynamics, in order to mitigate their impact on the inhabited areas around active, explosive volcanoes. Unfortunately, PDCs are difficult to measure, even indirectly, and most of the information on their dynamics is related to the study of their deposits. On the other hand, analogue experiments are only partially useful, because of the difficulty of scaling. Theoretical and computational models thus represents a unique opportunity to deepen our knowledge of the fluid dynamics of these volcanic flows.

In the last years, thanks to the rapid development and availability of parallel supercomputers, 3D multiphase flow simulation of volcanic plumes and PDCs have become a viable tool for volcanological research [4, 10, 15]. Numerical results demonstrated the ability to catch the intrinsically 3D dynamics of the turbulent mixing, the instability of the gas-particle volcanic plume and the complex interaction of PDCs with 3D topographic features [5].

The need of simulating such non-steady-state processes over a wide range of spatial scales (from a few metres up to tens of km) and the difficulty of increasing the number of discretization elements to directly simulate all turbulent scales, make the Large Eddy Simulation (LES) approach promising [7, 12]. Nevertheless, the highly polydisperse nature of volcanic flows and the coexistence of several dynamic regimes (from dense to dilute, from high to low Mach number, from turbulent to granular flows), increases the complexity of the model and makes it difficult to achieve a high numerical accuracy. The estimate of the uncertainty associated to the numerical discretization and to the physical modeling becomes thus important to assess the quality of the results and the reliability of hazard estimates. The present work intends to give a contribution to this issue by investigating the role of grid resolution and of SGS modeling in the LES simulation of PDCs and of their impact on buildings.

2 Overview of the physical and numerical model

The dynamics of the eruptive mixture is modelled by adopting an Eulerian-Eulerian multiphase flow model. Accordingly, gas and particulate phases are treated as continua and balance equations for mass, momentum, and energy are solved accounting for advective transport, viscous dissipation, body forces and interphase momentum and energy transfers. An equation of state and a Newtonian stress tensor are prescribed for each phase in order to close the set of coupled partial differential equations (PDE). More details about the physical model can be found in [9].

A LES approach to turbulence is adopted where the Sub-Grid Scale (SGS) stresses for the gas phase are modeled through the Smagorinsky closure [13]. At the wall boundary, a roughness closure for the filter length is specified [7]. For solid particles, physical and rheological properties, as well as interactions between them, are described by using semi-empirical correlations validated in the laboratory, and no SGS model is imposed.

The transport equations are solved by a Finite-Volumes (FV) technique on a 3D, staggered grid in Cartesian coordinates. Convective fluxes are discretized through a second-order upwind method, based on MUSCL reconstruction of fluxes at the cell boundaries. Diffusive fluxes are computed explicitly by a second-order centered scheme. The non-linear system of discretized PDEs is solved by applying an iterative procedure based on the Implicit Multi Field (ICE-IMF) algorithm [3]. Mass and momentum equations and the interphase coupling are solved through a semi-implicit (predictor-corrector) algorithm, by adopting a point-relaxation (SOR) technique. Energy equations are solved explicitly by a first-order Euler scheme. Although it is well known that upwind FV schemes are affected by a considerable numerical diffusion [12], a cheap and robust numerical technique is required for the study of the dynamics of both subsonic and supersonic multiphase flows, with a low to high degree of phase coupling, such as those encountered in volcanic phenomena.

The numerical algorithm is parallelized by adopting a domain-decomposition strategy and the Message Passing Interface (MPI) [4].

3 3D simulation of a stratified PDC

The model above is applied to the numerical simulation of the propagation of a PDC in a rectangular box of size $L_x = 5$ km, $L_y = L_z = 1$, with steady-state inlet conditions on the left (x=0) boundary (Fig. 1a). Initial PDC thickness is equal to 100 m. In this application, the grain-size distribution is approximated with two particle phases of 30 and 500 μm, with densities of 2500 and 1000 kg/m^3, representative of volcanic ash and pumice, respectively. Initial conditions are comparable with those occuring in Plinian eruptions and derive from the large-scale simulations of the collapse of a volcanic jet [5], resulting in an estimated bulk Reynolds number of the current exceeding 10^7. The initial velocity of both gas and particulate phases is 25 m/s and temperature equals 573 K. The flow pressure at the inlet is equal to the atmospheric pressure, so that it must adjust to balance the mixture hydrostatic load immediately after the injection in the domain. The inlet volumetric fractions of particles of 30 and 500 μm are, respectively, 0.65×10^{-3} and 1.625×10^{-3}, corresponding to solid bulk densities of 1.625 kg/m^3 for both particulate and about 0.6 kg/m^3 for the gas. The resulting flow dynamics pressure $P_d = 0.5\rho_m v_m^2$ is about 1.25 kPa.

We analyze hereafter the influence of the computational grid size (dx, dy, dz), of the Smagorinsky coefficient (C_s) and of the filter width (Δ). The values of these parameters considered herein are summarized in Table 1.

The propagation of a PDC (see Fig.1) is characterized by the formation of a current head (the PDC *nose*), the development of a Kelvin-Helmoltz (KH) instability

Run name	dx=dy [m]	dz [m]	C_s	Δ [m]
A1	10	2-20	0.1	[5.8:12.6]
A2	10	2-20	0.0	[5.8:12.6]
A3	10	2-20	0.1	10
A4	20	2-20	0.1	[9.3:20.0]
B1	20	4-40	0.1	[11.7:25.2]
B2	20	4-40	0.0	[11.7:25.2]
B3	20	4-40	0.2	[11.7:25.2]
C1	10	10	0.1	10

Table 1. Grid and turbulence model parameters adopted in 3D simulations of pyroclastic density currents. Δ is the filter width, which is equal to $(dx \cdot dy \cdot dz)^{1/3}$ (minimum and maximum values are indicated) in all simulations except A3, where it is constant. The time step is 0.01 s, corresponding to a CFL of about 0.1 for the finest mesh. Roughness length is equal to 1m in all simulations.

(that generates transversal eddies at the upper interface between the current and the atmosphere) and the Lobe-and-Cleft (LC) instability (which is associated to the engulfment of air by the flow front that generates positive bouyancy at the current head [8]). The highest resolution that was affordable for 3D simulation (run A1 in Tab. 1) qualitatively reproduce the phenomenology of the PDC propagation, as shown in Fig.1.

Increasing the minimum vertical grid size to 4 m (B1 run, Fig. 2a) significantly reduces the intensity of the LC instability, although the overall PDC structure is captured. A dramatic change in the PDC large-scale behaviour is observed (Fig. 2b) when the vertical grid size is too coarse to describe the boundary layer and the PDC head structure (C1 run). In this case, LC instability is damped out, the flow is considerably faster and the number of KH rolls is largely reduced.

Concerning the effect of the SGS model parameters, model parameters, fixing the filter length scale to 10 m (A3 run, Fig. 2c) does not significantly influence the large-scale structure of the PDC, whereas removing the model by imposing the Smagorinsky constant $C_s = 0.0$ (A2 run, Fig. 2d) completely changes the PDC dynamics. In the latter case, the lower viscosity in the model produces a much thinner boundary layer profile, so that the horizontal momentum transferred to the basal layer by the effect of the sedimentation is not dissipated. As a result, the flow head develops a wedgelike shape that causes a suppression of the LC instability mechanism, since it inhibits the entrainment of atmospheric air from the bottom.

The vertical profile of a PDC results from the concurrent effect of the wall shear stress (that generates a boundary layer), sedimentation (that decreases the mixture density at the current top while concentrating particles at the base) and air entrainment. In Fig.3 we present the profiles of dynamic pressure of the mixture $P_d = 1/2\rho_m v_m^2$, averaged in time, at 1.5 km from the inlet and along the central axis (the uniform value at the inlet is also displayed for reference). Simulations A1 and B1

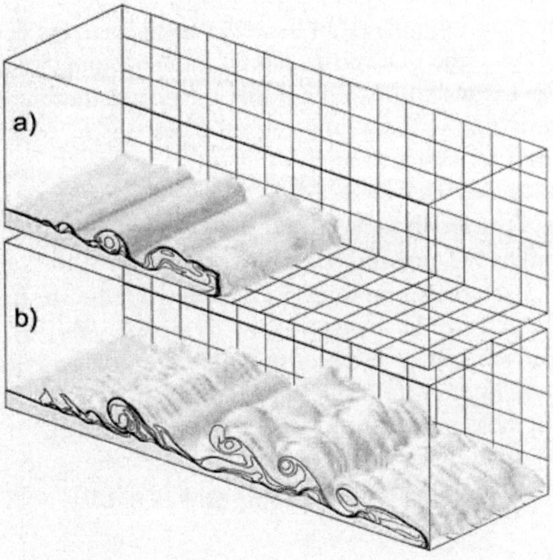

Fig. 1. Structure of the pyroclastic density current represented by the isosurface of the gas temperature (T=323 K) at 50 s (a) and 100 s (b) from the flow injection, for simulation A1. The isolines of the gas temperature, every 50 K, are also plotted on the front (y=0) plane. Gridding every 200 m.

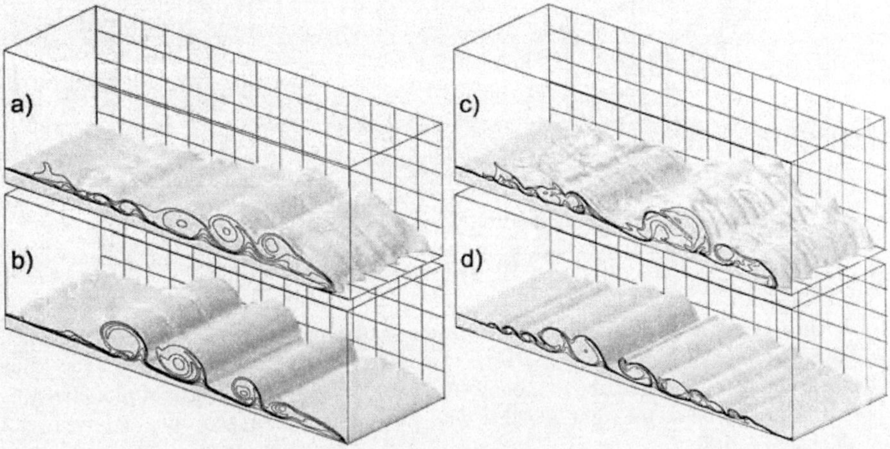

Fig. 2. Effect of the grid size and SGS filter size on the numerical results. Isosurface of the gas temperature for B1 at 100 s (a), C1 at 88 s (b), A3 at 100 s (c) and A2 at 75 s (d). See Table 1 and Figure 1 for parameters and comparisons.

give comparable results, whereas run C1 significantly underresolves the flow boundary layer and underestimates the concentration gradient near the wall. Interestingly, a similar net effect is observed when the SGS model is removed at higher resolution (A2 run). In this latter case (A2), the finer vertical grid size is responsible for the steeper concentration gradient and the reduced shear stress near the wall, which prevent the formation of the PDC nose, air entrainment from the head and the subsequent growth of the LC instability. As a result, the concentration profile is controlled by the sedimentation rate only and the current maintains a "constant settling zone" (with *top-hat* profile) for longer (see also Fig. 2d).

Simulations with a coarser mesh (B1-B2) seem less sensitive to the SGS model, probably because of the larger numerical diffusion associated to the grid.

The value of dynamic pressure in the first cell above the ground (reported in the legend of Fig. 3) increases on finer grids, reflecting the strong sensitivity of the concentration to the cell size (also observed in 2D simulations [9]). This value should then be considered carefully, also because the multiphase flow formulation at high concentration do not account for particle-particle friction.

Although direct measurements of PDC profiles are presently out of our technical possibilities, future studies should try to make the present results more quantitative, by comparing numerical to laboratory experiments (e.g. [6]).

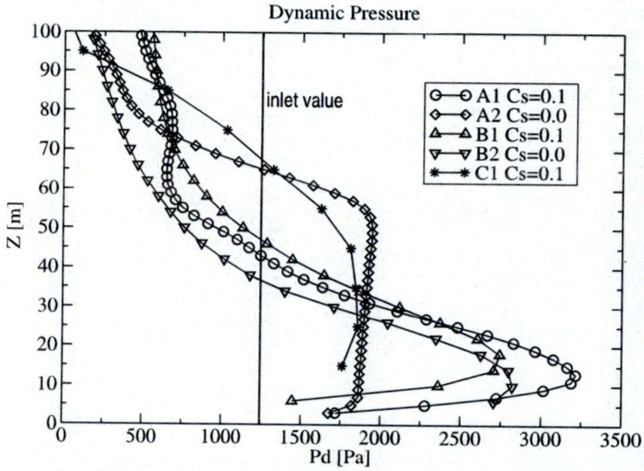

Run	P_d-av	P_d-base
A1	1570	5160
A2	1470	6550
B1	1390	2610
B2	1530	11500
C1	1450	2460

Fig. 3. Time-averaged vertical profile of the flow dynamic pressure at 1.5 km from the inlet. Time averaging is performed from the time of passage of the front up to 100 s. The value in the first computational cell above ground (P_d-base), omitted for the sake of plot clarity, is reported in the legend on the right, together with the vertically averaged value (P_d-av) over a flow thickness of 100 m.

4 Flow-building interaction

We finally present here the application of the 3D multiphase flow model to the analysis of the impact of a PDC on a building. Such study is mainly motivated by the need of estimating the action of the flow on a structure engulfed by a PDC and to design appropriate mitigation actions [19]. The damage on the infrastructures is also often utilized as an indirect measure of the maximum flow dynamic pressure [17]. However, the interaction between a PDC and a building is considerably complicated by 1) the presence of solid particles in a wide range of sizes and densities, 2) the stratified nature of PDCs and 3) the transient nature of the PDC emplacement. Therefore, the relationship between the (average) flow dynamic pressure and the action on the structure needs further investigation.

Numerical simulation have been performed by adding an obstacle of $20 \times 20 \times 20$ m^3 at 1.5 km from the inlet, in the same simulation conditions described above. Numerical results describe the flow separation on the building edge, the reattachment of the current downstream and the formation of a complex and unsteady eddy structure (Fig.4).

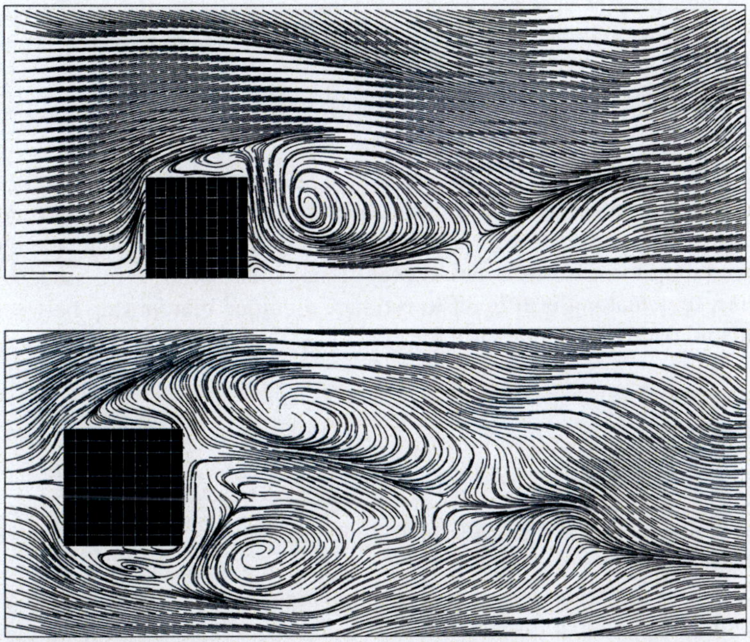

(a) (b)

Fig. 4. Vortex structure around a cubic obstacle engulfed by a PDC, within the A1 run. The streamlines represent gas velocity on the (a) longitudinal (xz) and (b) horizontal (xy) planes at half width and heigth, respectively and at 50 s. The grid size of 2 m is also represented within the obstacle.

The time-dependent action on the obstacle has been computed by integrating the pressure field along the building envelope. The PDC action on the building fluctuates around 2 kPa (consistent with the estimate of the average dynamic pressure, around 1.5 kPa) but it is significantly underestimated in the lowest resolution run C1.

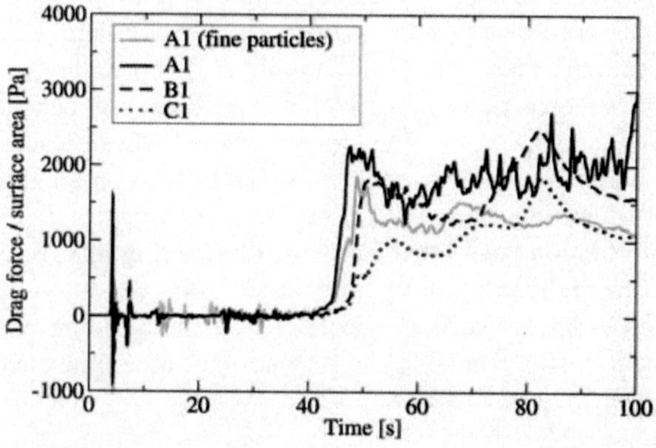

Fig. 5. Drag force per unit surface as a function of time. Black lines refer to A1 (solid), B1 (dashed) and C1 (dotted) runs. The grey, solid line refer to the A1 run with only one particle class of 30 μm and the same mixture density.

The effect of a change of the particle diameter has been also estimated for comparison and plotted (in grey) in Fig. 5, since the grain-size distribution represents one of the eruptive parameters most subject to uncertainty. The associated uncertainty in the computed drag force is of the same order of magnitude of the error associated to the grid size, thus making it difficult to estimate a unique relationship between the flow action and the dynamic pressure for PDCs.

5 Conclusions

Numerical results suggest that, despite the significant numerical diffusion associated to the upwind discretization, the LES subgrid model is needed to reproduce the qualitative behaviour of PDC (particularly the formation of a turbulent flow head with a *nose* structure and the development of KH and LC instabilities). In the adopted simulation conditions, the medium-resolution (4-40 m) mesh is able to resolve the flow boundary layer and to catch the qualitative behaviour of a PDC, giving a comparably good estimate of the flow action on a cubic obstacle. For the purpose of large-scale impact analysis (where the grid resolution cannot fully resolve the flow at an urban scale [5]) the averaged flow dynamic pressure results to be an acceptable proxy for the force per unit surface acting on the building envelope, although simulation C1 (the lowest resolution investigated with 10 m grid size) significantly underestimate it of a factor of 2-3. Present results also show that the effect of a change in the grain-size distribution may be comparable to that associated to the numerical grid and SGS filter size.

Physical and numerical models in volcanology are indeed subject to a variety of uncertainties. The multiphase formulation of the eruptive mixture dynamics is not univocal and intial and boundary conditions are subject to a large *epistemic* and aleatory uncertainty [18]. This implies that the absolute verification and validation of a model is inherently impossibile [14, 11]. Moreover, in the study of explosive eruptions it is difficult to test the congruence of numerical models to observational data, given the rarity of the events and their catastrophic nature.

However, numerical models can be used for sensitivity analyses, to elucidate the relative importance of model variables, and to compare single realizations in order to identify the most important eruptive parameters that define an eruptive scenario. Within this context, the assessment of quality and reliability of model results appears as an extraordinary challenge in which numerical benchmarking should be accompanied by an effort in combining modelling with uncertainty analysis, through statistical techniques leading to the construction of response surfaces relative to the variation of the different simulation parameters and possibly to their optimization. To this aim, however, the improvement of remote measurement techniques is also needed, to better characterize a natural phenomenon to which we have incomplete access.

Acknowledgements

This work was funded by Dip.to della Protezione Civile and Regione Campania, Italy, through the SPEED Project.

References

1. S. Dartevelle, "Numerical modeling of geophysical granular flows: 1. A comprehensive approach to granular rheologies and geophysical multiphase flows " Geochem., Geophys., Geosys. **5**, 8 (2004).
2. T. H. Druitt, "Pyroclastic density currents", in *The Physics of Explosive Volcanic Eruptions* (Eds J. Gilbert and R.S.J. Sparks), Geol. Soc. London. Spec. Publ., 145 (1998).
3. F. Harlow and A. A. Amsden, "Numerical calculation of multiphase fluid flow", J. Comput. Phys. **17** (1975).
4. T. Esposti Ongaro, C. Cavazzoni, G. Erbacci, A. Neri and M. V. Salvetti, "A parallel multiphase flow code for the 3D simulation of volcanic explosive eruptions", Parallel Computing **33** (2007).
5. T. Esposti Ongaro, A. Neri, G. Menconi, M. de' Michieli Vitturi, P. Marianelli, C. Cavazzoni, G. Erbacci and P. J. Baxter, "A transient 3D numerical simulation of column collapse and pyroclastic flow scenarios at Vesuvius", J. Volcanol. Geotherm. Res. **178** (2008).
6. S.A. Hosseini, A. Shamsai, B. Ataie-Ashtiani, "Synchronous measurements of the velocity and concentration in low density turbidity currents using an Acoustic Doppler Velocimeter", Flow Measurement and Instrumentation, **17** (2006).
7. P. J. Mason, "Large-eddy simulation: A critical review of the technique", Q. J. R. Meteorol. Soc. **120** (1994).
8. F. Necker, C. Hartel, L. Kleiser, E. Meiburg, "High-resolution simulations of particle-driven gravity currents", Int. J. Multiph. Flow **28** (2002),

9. A. Neri, T. Esposti Ongaro, G. Macedonio and D. Gidaspow, "Multiparticle simulation of collapsing volcanic columns and pyroclastic flows", J. Geophys. Res. **108**, B4 (2003).
10. J. M. Oberhuber, M. Herzog, H. F. Graf and K. Schwanke, "Volcanic plume simulation on large scales", J. Volcanol. Geotherm. Res. **87** (1998).
11. N. Oreskes, K. Shrader-Frechette, K. Belitz, Verification, validation, and confirmation of numerical models in the Earth Sciences Science **263** (1994).
12. M. V. Salvetti and F. Beux, "The effect of the numerical scheme on the subgrid scale term in large-eddy simulation", Phys. Fluid (1998).
13. J. Smagorinsky, "General circulation experiments with the primitive equations: I. The basic experiment", Mon. Weather Rev., **91** (1963).
14. R. S. J. Sparks and W. P. Aspinall, "Volcanic activity: frontiers and challenges in forecasting, prediction and risk assessment", In: The state of the planet: Frontiers and challanges in geophysics, Geophysical Monograph 150 IUGG, Volume 19 (2004).
15. Y. J. Suzuki, T. Koyaguchi, M. Ogawa and I. Hachisu "A numerical study of turbulent mixing in eruption clouds using a three-dimensional fluid dynamics model", J. Geophys. Res **110** (2005).
16. G. A Valentine, "Stratified flow in pyroclastic surges" Bull. Volcanol. **49** (1987).
17. G. A. Valentine, "Damage to structures by pyroclastic flows and surges, inferred from nuclear weapons effects", J. Volcanol. Geotherm. Res. **87** (1998).
18. G. Woo, *The Mathematics of Natural Catastrophes*, Imperial College Press, London (1999).
19. G. Zuccaro, F. Cacace, P. J. Baxter, R. Spence, Impact of explosive scenarios at Vesuvius, J. Volcanol. Geotherm. Res. **178** (2008).

Analysis of SGS effects on dispersed particles in LES of heated channel flow

Jacek Pozorski and Mirosław Łuniewski

Institute of Fluid-Flow Machinery, Polish Academy of Sciences,
Fiszera 14, 80952 Gdańsk, Poland; jp@imp.gda.pl

Summary. Large-eddy simulation of a turbulent, non-isothermal channel flow is performed. The Lagrangian approach is followed to compute the dispersed phase (heavy particles) under the assumption of one-way momentum and energy coupling with the carrier phase. A stochastic model for the residual fluid velocity along the particle trajectories is applied to account for subfilter flow effects on particles. It is shown that both the particle dynamics and temperature are affected by the model. Results for the carrier and dispersed phases are presented in terms of their velocity and thermal statistics. The need for further model improvement is discussed.

Key words: Two-phase dispersed flow, Heat transfer, Large-eddy simulation, Lagrangian-Eulerian approach, Subgrid-scale effects on particles

1 Introduction

Turbulent two-phase flows with the dispersed phase are quite common in various industrial situations, including chemical, process, and power engineering. Examples include liquid fuel and coal powder combustion, spray cooling, heterogeneous catalytic processes, spray dryers, etc. Since the physical and chemical phenomena occurring there are often controlled by the instantaneous flow structure, the statistical (RANS-based) approaches face difficulties. Nowadays, large eddy simulation (LES), although much more computationally demanding, becomes feasible for some of those applications [8].

In the Lagrangian-Eulerian studies of turbulent polydispersed flows, LES is used as the carrying phase solver to resolve more energetic (larger) flow scales. Naturally, questions arise as to the importance of the subgrid-scale (SGS), or subfilter, fluid flow and heat transfer on the behaviour of the dispersed phase. As discussed in a number of recent publications, the LES filtering and subsequent modelling of SGS terms may impact the statistics of particle locations (so-called preferential concentration) [17, 13, 6], turbulent energy and inter-particle collision rates [1], particle velocities and deposition rates in wall-bounded flows [21, 5, 14]. Consequently, the particle cooling, heating or evaporation rates may be affected as well [10].

M.V. Salvetti et al. (eds.), *Quality and Reliability of Large-Eddy Simulations II*, ERCOFTAC Series 16, DOI 10.1007/978-94-007-0231-8_16,

In the paper, both velocity and temperature of particles are of interest. Following our previous work on SGS particle dispersion models [14] and a recent DNS study on the particle thermal correlations [3], we consider a wall-bounded, geometrically-simple turbulent flow (plane channel). To the best of the authors' knowledge, the impact of SGS particle dispersion models on the particle thermal statistics has not been studied so far. The main aim of the present contribution is to discuss these issues and to present results obtained to date for the case of a heated/cooled channel flow (no net heating of the fluid) laden with particles. The SGS dispersion model is added to the particle equation of motion and its effects on particle velocity and temperature are illustrated for several values of particle inertia. Some comparisons with available DNS data are also provided.

2 Physical problem statement and closure relationships

We consider the particle motion one-way coupled with the wall-bounded, non-isothermal, turbulent fluid flow of constant density. The large-scale (filtered) fluid velocity and temperature fields, $\bar{\mathbf{U}}_f(\mathbf{x},t)$ and $\bar{T}_f(\mathbf{x},t)$, are computed in the LES approach [12]. The dynamic (Germano-Lilly) model is used to determine the sub-filter viscosity ν_t in the modelled SGS stress term of the filtered N-S equations, spatially-smoothed with a filter scale $\bar{\Delta}$. This is done through $\nu_t = C_G \bar{\Delta}^2 |\bar{S}|$ where $|\bar{S}| = (2\bar{S}_{ij}\bar{S}_{ij})^{1/2}$ is the scale of the resolved strain rate and the Germano parameter C_G is computed from double filtering and subsequent averaging over the homogeneity directions.

The temperature, treated as a passive scalar in the flow, is governed by

$$\frac{\partial \bar{T}_f}{\partial t} + \bar{U}_{f,i}\frac{\partial \bar{T}_f}{\partial x_i} = \frac{\partial}{\partial x_i}\left[\left(\frac{\nu_f}{\mathrm{Pr}} + \frac{\nu_t}{\mathrm{Pr}_t}\right)\frac{\partial \bar{T}_f}{\partial x_i}\right]. \tag{1}$$

In the filtered energy equation for fluid, Eq. (1), the subfilter thermal diffusivity ν_t/Pr_t is used to close the SGS heat flux term. It is determined with the assumed turbulent Prandtl number of $\mathrm{Pr}_t = 0.98$.

As far as the dispersed phase is concerned, the particle tracking approach is followed. We have used a simplified particle equation of motion where only the drag term is retained. This is generally accepted for the case of heavy particles, $\rho_p \gg \rho_f$. To improve the predictions of particle wall deposition, the use of the lift force term is recommended [21, 14]. For the present computation with no deposition, elastic rebound of particles with no temperature change upon the wall collision is assumed as the boundary condition. The drag force is based on the particle velocity $\mathbf{U}_p$ and the fluid velocity along particle trajectory, $\mathbf{U}_f^* = \mathbf{U}_f(\mathbf{x}_p,t)$. In practice, velocity $\mathbf{U}_f^*$ is interpolated from the fluid solution known at mesh points; trilinear interpolation is applied for the present work. The incurred error has not been assessed here; yet, lower-order interpolation may act as additional filtering/smoothing [5].

Analogously, the evolution of particle temperature T_p accounts for heating or cooling by the carrier phase and includes the fluid temperature at the particle location, $T_f^* = T_f(\mathbf{x}_p,t)$. So, for the present case the equations of particle evolution are:

$$\frac{d\mathbf{x}_p}{dt} = \mathbf{U}_p\,, \qquad \frac{d\mathbf{U}_p}{dt} = f_D \frac{\mathbf{U}_f^* - \mathbf{U}_p}{\tau_p}\,, \qquad \frac{dT_p}{dt} = f_\theta \frac{T_f^* - T_p}{\tau_\theta}\,. \tag{2}$$

The particle momentum relaxation time is defined as $\tau_p = (\rho_p/\rho_f)(d_p^2/18\nu_f)$; $f_D = 1 + 0.15\mathrm{Re}_p^{0.687}$ is the empirical drag correction factor; in this expression, $\mathrm{Re}_p = d_p|\mathbf{U}_f^* - \mathbf{U}_p|/\nu_f$ is the particle Reynolds number (based on the particle diameter d_p, the relative particle velocity, and the kinematic viscosity of the carrier fluid, ν_f). The particle Stokes number is conveniently defined in wall-bounded flows as $\mathrm{St} = \tau_p^+$ (viscous scaling). Then, the particle thermal relaxation time in Eq. (2) is expressed as $\tau_\theta = (\rho_p c_p/\rho_f c_f)(d_p^2/12\alpha_f)$ where the fluid thermal diffusivity is $\alpha_f = \nu_f/\mathrm{Pr}$. The correction factor f_θ in Eq. (2) is determined by the Ranz-Marshall correlation for the Nusselt number: $f_\theta = \mathrm{Nu}/2 = 1 + 0.3\,\mathrm{Re}_p^{1/2}\mathrm{Pr}^{1/3}$. So, apart from the particle Stokes number τ_p^+, another control parameter for a non-isothermal motion of the dispersed phase in the flow is the non-dimensional particle thermal relaxation time $\tau_\theta^+ = \mathrm{Pr}(3c_p)/(2c_f)\tau_p^+$.

The simplest treatment of the particulate phase ("regular LES") is to replace in Eqs. (2) the instantaneous fields "seen" by particle, $\mathbf{U}_f^*$ and T_f^*, by the filtered, or large-scale, fields $\bar{\mathbf{U}}_f^*$ and $\bar{T}_f^*$ with the residual scales neglected. Alternatively, with the account of SGS fluid motions, the velocity "seen" by the particles is taken as $\bar{\mathbf{U}}_f^* + \mathbf{u}_f^*$. In this case the residual velocity "seen", $\mathbf{u}_f^*$, is introduced to model the impact of the subfilter flow on the dispersed phase motion. The issue, here called the SGS particle dispersion, is a subject of ongoing debate and modelling efforts, eg., based on approximate deconvolution ideas [10, 5, 18] or stochastic approaches [13, 19].

In the present computations we used a stochastic model for the residual fluid velocity along particle trajectories. The model, originally proposed for homogeneous turbulence [13], is now briefly recalled. It is based on the Langevin-type equation and is constructed with the subgrid velocity and time scales of the fluid "seen" by the particle:

$$du_i^* = -\frac{u_i^*}{\tau_L^*}\,dt + \sqrt{\frac{2\sigma_{\mathrm{sg}}^2}{\tau_L^*}}\,dW_i \tag{3}$$

where dW_i is an increment of the Wiener process. To estimate the velocity scale of subfilter fluid motions, $\sigma_{\mathrm{sg}} = (\frac{2}{3}k_{\mathrm{sg}})^{1/2}$, the residual turbulent energy content of the carrier phase was approximated as $k_{sg} = C_I\bar{\Delta}^2|\bar{S}|^2$ and the dynamic procedure (with double filtering) was applied to determine the proportionality parameter C_I [16]. The subfilter time scale is estimated as $\tau_L^* = C_{sg}\bar{\Delta}/\sigma_{\mathrm{sg}}$. The model constant $C_{sg} = \mathcal{O}(1)$ accounts for the uncertainty concerning the time scale of the residual velocity autocorrelation. Various choices of C_{sg}, ranging from 0.002 to 1.0, were considered through its impact on the particle turbulent energy and deposition rate; cf. [14] for results.

Both treatments ("regular LES" and LES with SGS particle dispersion) were used in particle tracking reported in the present contribution. Note that the fluid temperature along particle trajectories, T_f^*, in Eq. (2c) is approximated by the filtered quantity $\bar{T}_f(\mathbf{x}_p,t)$. We argue that a part of the subfilter effects on particle temperature is already accounted for through the SGS dispersion model for velocity "seen". For a fully consistent approach, a correct choice would be $T_f^* = \bar{T}_f^* + \theta_f^*$ where the subfilter fluid temperature along particle trajectories, θ_f^*, should be modelled as well (like $\mathbf{u}_f^*$). Yet, stochastic temperature models of the diffusion type are more complex than those for velocity, because of the scalar boundedness constraint, cf. [15]. On the other hand, approximate deconvolution for temperature seems more straightforward, cf. [10]. In the present paper, no attempt to model θ_f^* has been made. We note, however, that the stochastic modelling of temperature in one-phase turbulent flows with heat transfer has been addressed both in the context of statistical approaches (one-point PDF method, [15]) and for the LES coupled with FDF scalar modelling [20]. In RANS of non-isothermal flows, stochastic models were proposed at the discrete time level, cf. [2] and references therein.

3 Numerical solution

A fully-developed turbulent channel flow is computed for $Re_\tau = 150$ (a benchmark test case of the COST Action LES-AID, [7]). The pressure-driven flow was assumed periodic in the streamwise and spanwise directions. The size of the flow domain in the streamwise (x), wall-normal (y) and spanwise (z) directions was $4\pi h \times 2h \times (4/3)\pi h$, discretised with $64 \times 84 \times 64$ FV meshes. Our earlier experience [16, 14] has shown that a coarser mesh deteriorates the quality of fluid statistics resulting from the finite volume code applied for the present LES computations. For the purpose, we used a finite volume, academic solver of second-order accuracy (`FASTEST3D` code – courtesy of Prof. M. Schäfer, TU Darmstadt, Germany). The mesh size Δy^+ (in wall units) varied from 0.17 at the wall up to 10 at the centerline. In the periodicity directions, the mesh was uniform with $\Delta x^+ = 29.5$ and $\Delta z^+ = 9.8$. At the walls, isotemperature b.c. were imposed with respective temperatures T_L and T_H, resulting in a heat flux from the colder to the hotter wall with no net heating of the fluid. The fluid Prandtl number was $\mathrm{Pr} = 0.7$ (air).

Computations were performed with the time step $\Delta t^+ = 0.11$; the total simulation time was $t_f^+ \sim 2.4 \cdot 10^4$ (about 240000 time steps) and the non-isothermal case was switched on from $t_T^+ \sim 0.5 \cdot 10^4$. A statistically-steady state has been achieved for the fluid velocity field. Results for thermal statistics suggest that a somewhat longer integration time may still be needed to collect quality time-averages. As the indicator of convergence, Fig. 1(a) shows the symmetrised profiles of the fluid temperature variance at some time instants.

Particle-related variables are determined by integrating Eqs. (2) in time until $t_p^+ \sim 2 \cdot 10^4$ since their injection in the flow. Altogether, 200000 particles were tracked for each St. Here again, we have first checked how the simulations attain a steady state. The fluctuating temperature variance at the channel centerline, $\langle \theta_p^2 \rangle_{CL}$, was computed for particles of various St; results are shown in Fig. 1(b) together with the corresponding quantity for fluid, $\langle \theta_f^2 \rangle$. Since $\langle \theta_p^2 \rangle_{CL}$ in a steady-state is expected to be a decreasing function of St, the statistics for smaller particles (St = 1 and 5)

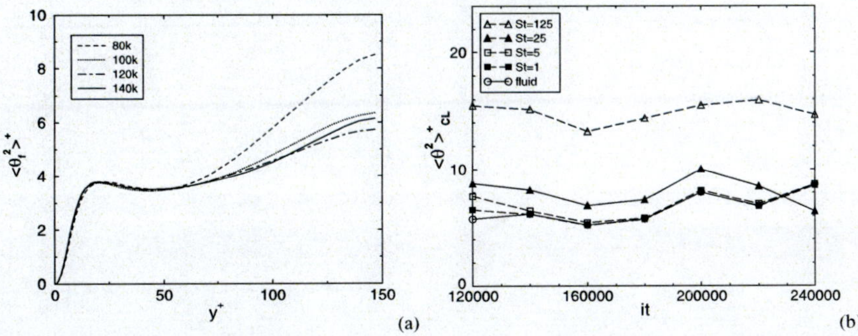

Fig. 1. LES of heated channel flow at $\mathrm{Re}_\tau = 150$: convergence of the computation. (a) fluid temperature fluctuation variance at selected time instants (from 80000 to 140000 time steps); (b) time records of the fluctuating temperature variance (fluid and particles) at the channel centerline.

seem close to converged, unlike those for $\mathrm{St} = 125$ where probably a much longer computation time is needed.

Then, statistics of the particulate phase: cross-stream profile of the particle number density, mean velocity and temperature profiles, intensity of velocity and temperature fluctuations, and fluctuating velocity-temperature correlations were computed. Results were taken as time averages over $t^+_{p,av} \sim 10^3$.

4 Computation results and discussion

The fluid and particle velocity statistics resulting from the LES have already been reported in [14]; the particle deposition mass flux has also been analysed there. Here, we recall and illustrate the issue of the SGS dispersion modelling effects on the r.m.s. particle fluctuating velocity in Fig. 2. Because of the isotropic character of the model, the increase of the wall-normal component is relatively larger. For the computations reported here, the model constant has been chosen to $C_{sg} = 1$ which is probably the upper bound of values, judging by the reconstruction of turbulent particle energy.

In the present contribution we focus on the thermal statistics. First, we report results for the fluid temperature, compared to the DNS reference data for the heated/cooled channel [9]. All statistical quantities are suitably normalised with u_τ and the friction temperature θ_τ (the non-dimensional wall surface heat flux q_w), defined as $\theta_\tau = q_w/(\rho_f c_f u_\tau) = (\alpha_f/u_\tau)(dT/dy)_w$ so that $T^+ = \mathrm{Pr} y^+$ in the viscous conductive sublayer.

The mean fluid temperature profile is shown in Fig. 3(a) and compared to DNS [11] for $\mathrm{Pr} = 0.7$. The r.m.s. of the fluctuating fluid temperature is shown in Fig. 3(b) and compared to available DNS data [9] at $Pr = 1$ (please note that the r.m.s. θ_f^+ values from LES are rescaled with the Prandtl number to confirm the correct profile in the viscosity-affected region). The correlation coefficients of fluctuating temperature and fluctuating velocity components are shown in Fig. 3(c,d). In general,

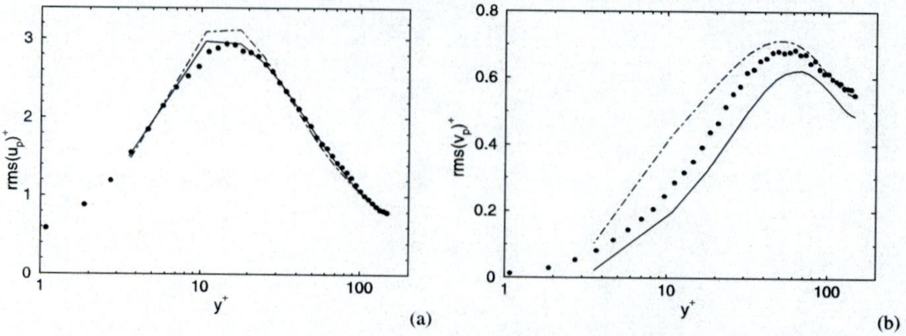

Fig. 2. Effect of the SGS particle dispersion model for the fluid "seen" on the r.m.s. fluctuating particle velocity for St $= 5$: a) streamwise component, b) wall-normal component. Regular LES: solid lines; LES with the SGS particle dispersion model (with $C_{sg} = 1.0$): dot-dashed lines; DNS data of Marchioli *et al.* [7]: (•).

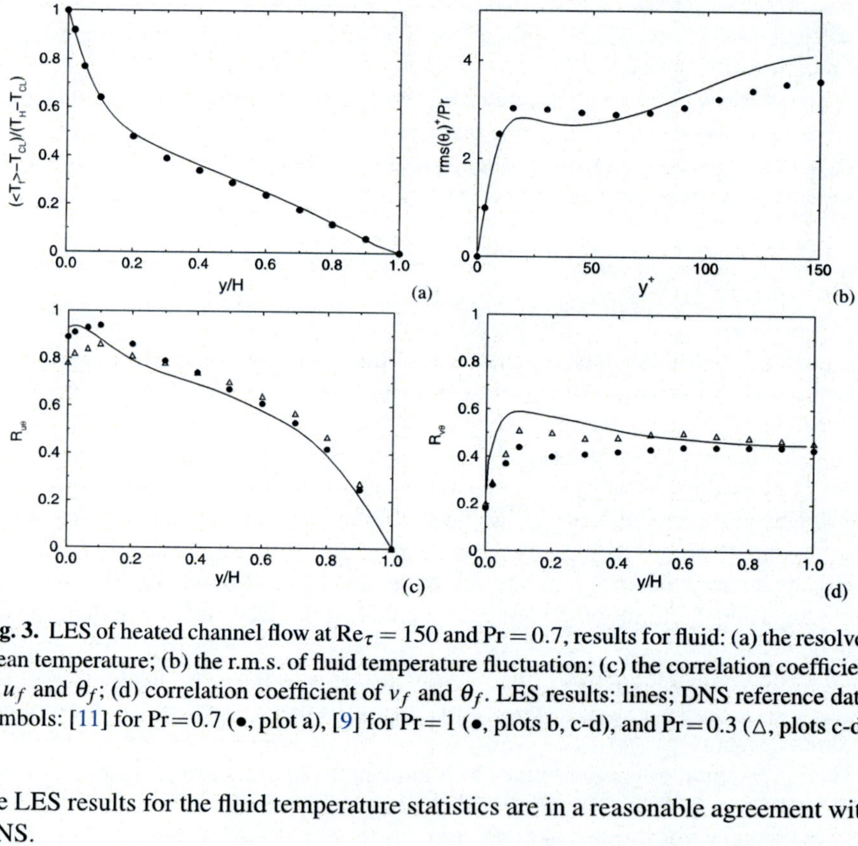

Fig. 3. LES of heated channel flow at $\mathrm{Re}_\tau = 150$ and $\mathrm{Pr} = 0.7$, results for fluid: (a) the resolved mean temperature; (b) the r.m.s. of fluid temperature fluctuation; (c) the correlation coefficient of u_f and θ_f; (d) correlation coefficient of v_f and θ_f. LES results: lines; DNS reference data, symbols: [11] for Pr=0.7 (•, plot a), [9] for Pr=1 (•, plots b, c-d), and Pr=0.3 (△, plots c-d).

the LES results for the fluid temperature statistics are in a reasonable agreement with DNS.

Results for regular LES of particle-laden flow are shown in Fig. 4. The mean temperature profiles for fluid and particles (plot a) are very close to each other and the SGS particle dispersion model does not affect them (results not shown). The profiles

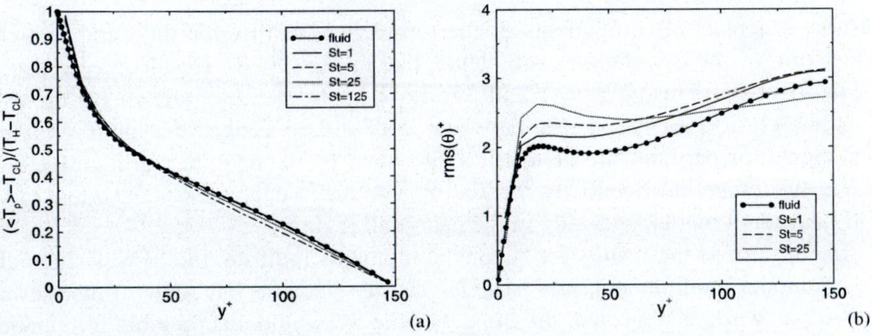

Fig. 4. Fluid and particle thermal statistics across the channel: a) the mean temperature profile, b) the r.m.s. of temperature fluctuations.

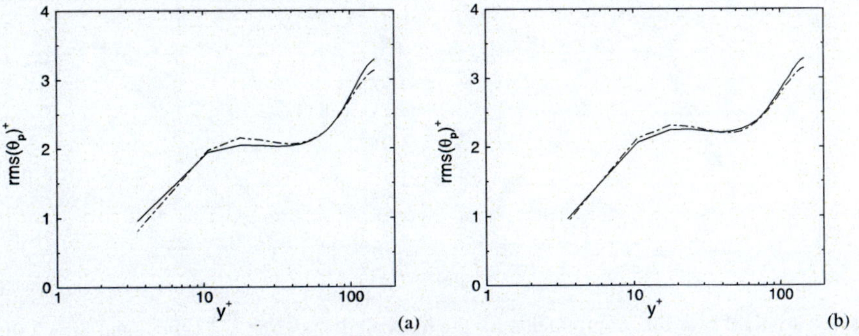

Fig. 5. Effect of the SGS particle dispersion model for the fluid "seen" on the r.m.s. fluctuating particle temperature, for: a) St = 1, b) St = 5. Regular LES: solid lines; LES with the SGS particle dispersion model (with $C_{sg} = 1.0$): dot-dashed lines.

of the fluctuating temperature r.m.s. (plot b) differ for fluid and heavy particles: with increase of St, the fluctuation intensity tends to increase in the wall region, but lies below that of fluid at the channel centerline (the results for $St = 25$, not yet fully converged, should be looked at with caution).

Next, we analyse the impact of the SGS dispersion model on the particle temperature. In particular, Fig. 5 shows the impact of the SGS dispersion model on thermal statistics from a regular LES study; the fluctuating temperature variance seems to become somewhat flatter throughout the channel (except at the walls, held isothermal). Because of the limited computer resources, we have taken the momentum and energy relaxation times equal for all particle classes: $\tau_\theta^+ = \tau_p^+ = \mathrm{St}$.

As far as the results for particle tracking with the SGS particle dispersion model are concerned, Fig. 6 contains the turbulent heat flux components of the dispersed phase. Here again (cf. Fig. 2), the impact of the model is more significant for the wall-normal component of turbulent heat flux. Moreover, particles of St = 5 seem to be more responding. This may be due to preferential concentration effects in the near-wall region [14], but further work is needed on this issue. Next, we computed the non-dimensional components of the turbulent heat flux for fluid and particles in the form of the correlation coefficients, like in Fig. 3(c,d). Results are presented in Fig. 7 for particles of St = 1 and 25; those for St = 5 (not shown for the sake of picture clarity) are situated in-between. It should be noticed that the correlation coefficients for particles are generally larger than those for fluid and increase with St. Yet, the temperature-velocity correlation is found to decreases as the effect of the SGS dispersion model, specially in the near-wall region.

To sum up: as the results for statistical moments indicate, the SGS particle dispersion model modifies not only particle velocity statistics but some of the thermal statistics as well; as expected, the impact of the SGS model is tangible for smaller-inertia particles (such as St = 5).

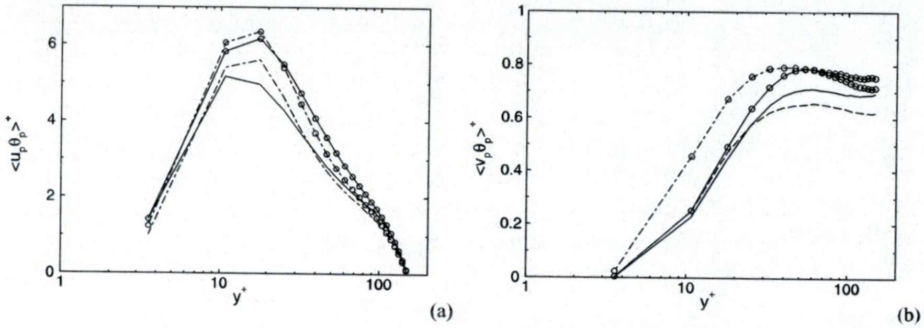

Fig. 6. Turbulent heat flux components for the dispersed phase, St = 1 (lines) and St = 5 (lines with open symbols): (a) streamwise, (b) wall-normal. Solid lines: regular LES, dash-dotted lines: LES with SGS particle dispersion.

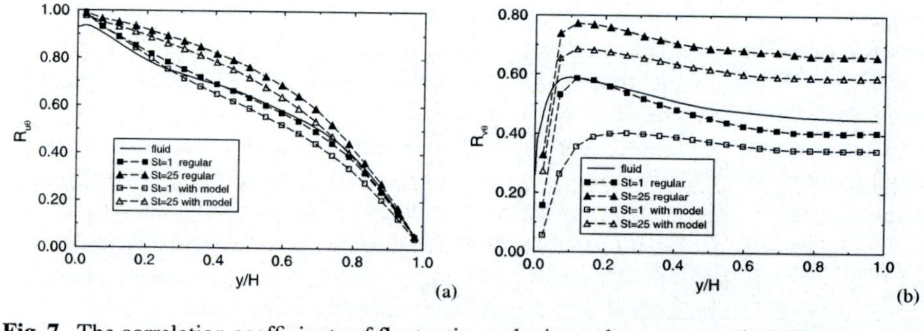

Fig. 7. The correlation coefficients of fluctuating velocity and temperature in LES of particle-laden flow, St = 1 and 25: (a) streamwise, (b) wall-normal. Fluid: lines. Particles: regular LES (lines with black symbols), LES with SGS dispersion model (lines with empty symbols).

5 Concluding remarks

In the paper, we considered a non-isothermal channel flow with the dispersed phase. In the LES computation with Lagrangian particle tracking, we applied the Langevin stochastic model for residual fluid velocity along the particle trajectories to study the impact of SGS dispersion effects on the particle thermal statistics. As already stated before [13], the model is able to restore some statistical quantities (like the turbulent kinetic energy of the particulate phase), yet it tends to be too diffusive as far as the particle preferential concentration is concerned.

We have presented LES results for the carrier and dispersed phases and compared them to available reference data from DNS of a heated/cooled channel. The effect of SGS particle dispersion modelling on the temperature r.m.s. and the fluctuating temperature-velocity correlations for the dispersed phase has clearly been shown. On the contrary, no SGS model for the subfilter temperature at particle locations has been proposed so far. The possible need for an additional SGS modelling of thermal effects via a specific closure, directly in the particle energy equation, remains an open issue.

A next-term objective will be a sound modelling of the SGS flow effects on mass transfer processes in the particulate phase (evaporation, chemical reactions). Also, the LES computation is underway for a channel flow heated at both sides. For that case, detailed DNS fluid data are available [4] and some DNS reference results exist for the dispersed phase for a selection of relaxation times τ_p and τ_θ. Comparison with [3] is intended to analyse the impact of the SGS dispersion model on particle thermal statistics.

Acknowledgments

The work has been supported by the Polish Ministry of Science and Higher Education through the research project SPB COST 163/2006.

References

1. Fede P, Simonin O (2006) Numerical study of the subgrid fluid turbulence effects on the statistics of heavy colliding particles. Phys Fluids 18:045103
2. Gao Z, Mashayek F (2004) Stochastic model for non-isothermal droplet-laden turbulent flows. AIAA J 42:255–260
3. Jaszczur M, Portela L (2008) Numerical data for reliability of LES for non-isothermal multiphase turbulent channel flow. In: Meyers J, Geurts B, Sagaut P (eds) Quality and Reliability of Large-Eddy Simulations (Springer) 342–354
4. Kawamura H, Abe H, Matsuo Y (1999) DNS of turbulent heat transfer in channel flow with respect to *Re* and *Pr* effects. Int J Heat Fluid Flow 20:196–207
5. Kuerten JGM (2006) Subgrid modeling in particle-laden channel flows. Phys Fluids 18:025108
6. Marchioli C, Salvetti MV, Soldati A (2008) Some issues concerning LES of inertial particle dispersion in turbulent bounded flows. Phys Fluids 20:040603

7. Marchioli C, Soldati A, Kuerten JGM, Arcen B, Tanière A, Goldensoph G, Squires KD, Cargnelutti MF, Portela L (2008) Statistics of particle dispersion in DNS of wall-bounded turbulence. Int J Multiphase Flow 34:879–893
8. Moin P, Apte SV (2006) LES of multiphase reacting flows in complex combustors. AIAA J 44:698–708
9. Na Y, Papavassiliou DV, Hanratty TJ (1999) Use of DNS to study the effect of Prandtl number on temperature fields. Int J Heat Fluid Flow 20:187–195
10. Okong'o N, Bellan J (2004) Consistent large-eddy simulation of a temporal mixing layer laden with evaporating drops. J Fluid Mech 499:1–47
11. Papavassiliou DV, Hanratty TJ (1997) Transport of a passive scalar in a turbulent channel flow. Int J Heat Mass Transfer 40:1303–1311
12. Pope SB (2000) Turbulent Flows. Cambridge University Press, Cambridge
13. Pozorski J, Apte SV (2009) Filtered particle tracking in isotropic turbulence and stochastic modeling of subgrid-scale dispersion. Int J Multiphase Flow 35:118-28
14. Pozorski J, Łuniewski M (2008) Analysis of SGS particle dispersion model in LES of channel flow. In: Meyers J, Geurts B, Sagaut P (eds) Quality and Reliability of Large-Eddy Simulations (Springer) 331–342
15. Pozorski J, Wacławczyk M, Minier JP (2004) Scalar and velocity-scalar PDF modelling of near-wall turbulent heat transfer. Int J Heat Fluid Flow 25:884–895
16. Pozorski J, Wacławczyk T, Łuniewski M (2007) LES of turbulent channel flow and heavy particle dispersion. J Theor Appl Mech (Warsaw) 45:643–657
17. Segura JC, Eaton JK, Oefelein JC (2004) Predictive capabilities of particle-laden LES. Rep. No. TSD–156, Dept. of Mech. Engng., Stanford University
18. Shotorban B, Mashayek F (2005) Modeling subgrid-scale effects on particles by approximate deconvolution. Phys Fluids 17:081701
19. Shotorban B, Mashayek F (2006) A stochastic model of particle motion in large-eddy simulation. J Turbulence 7:18
20. Wacławczyk M, Pozorski J, Minier JP (2008) New molecular transport model for FDF/LES of turbulence with passive scalar. Flow Turbul Combust 81:235–260
21. Wang Q, Squires KD (1996) Large eddy simulation of particle deposition in a vertical turbulent channel flow. Int J Multiphase Flow, 22:667–683

Relevance of approximate deconvolution for one-way coupled motion of inertial particles in LES of turbulent channel flow

Marek Jaszczur[1], Bernard J. Geurts[2,3], and J.G.M. Kuerten[4]

[1] AGH - University of Science and Technology, 30-059 Krakow, Al.Mickiewicza 30, Poland jaszczur@agh.edu.pl
[2] Multiscale Modeling and Simulation, Faculty EEMCS, University of Twente, P.O. Box 217, 7500 AE Enschede, The Netherlands b.j.geurts@utwente.nl
[3] Anisotropic Turbulence, Faculty Applied Physics, Eindhoven University of Technology, P.O. Box 513, 5600 MB Eindhoven, The Netherlands
[4] Department of Mechanical Engineering, J.M. Burgers Center, Eindhoven University of Technology, P.O. Box 513, 5600 MB Eindhoven, The Netherlands j.g.m.kuerten@tue.nl

Summary. The Euler-Lagrange approach, based on Direct Numerical Simulation (DNS) and Large-Eddy Simulation (LES) for the fluid, is applied to particle-laden turbulent flow in a channel. Explicit subgrid modeling of the turbulent stresses is adopted, while the particle motion includes small turbulent scales based on approximate deconvolution of the LES field. Results for turbulent flow in a channel at $Re_\tau = 150$ are discussed, focusing on one-way coupled point-particle statistics at three Stokes numbers. DNS provides a point of reference for assessing LES with different sub-filter eddy-viscosity models: Smagorinsky, Van Driest-Smagorinsky and the dynamic model are studied. Clustering and segregation of particles near the wall, due to turbophoresis, is strongly related to the quality of the LES velocity field and the approximate reconstruction of the smaller resolved scales. It is shown that deconvolution up to second order allows to better describe the particle statistics near a solid wall; deconvolution at higher order yields rather marginal additional improvements.

Key words: Large-Eddy Simulation, Direct Numerical Simulation, multiphase flow, approximate deconvolution

1 Introduction

Turbulent flows laden with a large number of small particles is fundamental to a multitude of environmental and industrial processes, e.g., the formation of clouds, combustion of coal, catalytic cracking of oil, etc. Accurate prediction of the behavior of particles in a turbulent flow can be obtained using Direct Numerical Simulation (DNS) [12]. However, such simulations are too costly for frequent application

M.V. Salvetti et al. (eds.), *Quality and Reliability of Large-Eddy Simulations II*, ERCOFTAC Series 16, DOI 10.1007/978-94-007-0231-8_17,

in problems of realistic complexity. This constitutes a major pacing item for the further development of Large-Eddy Simulation (LES) to also include the dynamics of large ensembles of discrete point particles [5, 18]. Such extensions of LES introduce challenges with respect to capturing the stresses that arise from small-scale turbulent motions, as well as challenges related to incorporating small-scale turbulence into the motion of the particles. Several important aspects of particle-turbulence interactions have been found by [13, 14, 15] and [16]. In this paper we focus on the relation between the quality of the LES field and the predicted point particle statistics. Special emphasis is given to assess improvements in the particle dynamics that can be obtained by adopting approximate deconvolution [2, 3] to represent small scale dynamics for the discrete particles.

In recent years, large-eddy simulation for turbulent flow has become focused on multiphase applications, including particle-laden flow [1]. This poses new challenges to the representation of small-scale turbulent motion, as the aim shifts toward capturing the dispersion of inertial particles that are embedded in the flow. On the one hand, the flow-coarsening in LES facilitates the simulation of the primary features of complex turbulent flow at high Reynolds numbers. On the other hand, this flow-coarsening takes out much of the small-scale turbulent contributions to the motion of particles. The problem that arises from the flow-coarsening is 'how to restore the *effects* of a range of turbulent scales on the particle motion' so that several important statistics of the embedded particle-swarm are adequately represented. We focus on the relevance of one particular strategy, that of approximate deconvolution of the LES velocity field [2, 3] and consider the effectiveness of this strategy for particles at low as well as high Stokes numbers, i.e, a very high and a rather small agility of the particles to respond to local changes in the turbulent flow.

The motion of particles that respond slowly to the flow because of their large inertia, i.e., large relaxation time, will not be influenced much by the small turbulent scales. At the other extreme, particles that closely follow details in the flow will require a highly resolved turbulent velocity field. The approximate deconvolution method can be used to reconstruct some of the smaller resolved scales in an LES. As shown in [4, 5] such reconstruction allows to better describe clustering of particles near a solid wall. This deconvolution approach should be adequate by itself for inertial particles with large relaxation time. However, with decreasing relaxation time, the dependence of the particle motion on small scales in the flow increases. We investigate to what extent an increase in the order of approximate deconvolution leads to an extension of the range of particle relaxation times that can be simulated accurately. We closely follow work in [6] and apply this strategy to turbulent flow in a channel while concentrating on statistics of the particle clustering and segregation near the wall. Deconvolution can only approximately reconstruct resolved scales on the LES grid. In cases where the motion of the particles also depends to a large extent on scales that are smaller than the LES resolution, or in case the LES field does not capture important flow structures near the wall, approximate deconvolution is not expected to yield improvements. In such situations additional sub-filter scale modeling for the particle motion may be required [7, 8].

The organization of this paper is as follows: in Section 2 we introduce the mathematical model, Section 3 is devoted to a discussion of the simulation results and concluding remarks are collected in Section 4.

2 Mathematical model

In this section we describe the Stokes dynamics of the discrete embedded particles and sketch the fluid-mechanical treatment of the continuous phase in DNS and LES context.

Detailed simulation of wall-bounded turbulent particle-laden flow can be based on the Euler-Lagrange point-particle approach [17]. Particles are dispersed in a pressure-driven fluid flow, assumed to be incompressible and Newtonian. We restrict to very small volume fractions and assume that the size of the particles is considerably smaller than the local Kolmogorov length-scale in the turbulent flow. In such situations the particles have a negligible feedback coupling on the turbulence and the one-way coupling formulation for the particle phase can be employed [12].

The motion of the particles in the turbulent flow is obtained by time-accurate tracking of their trajectories. The instantaneous transfer of momentum from the continuous fluid phase to the discrete particle phase is dominated by Stokes drag. This is determined by the velocity of the particle and that of the surrounding fluid at the location of the particle. For small heavy particles the equation of motion for the location $\mathbf{x}$ and velocity $\mathbf{v}$ can be written as [1]:

$$\frac{d\mathbf{x}(t)}{dt} = \mathbf{v}(t) \quad ; \quad \frac{d\mathbf{v}}{dt} = C_d \frac{Re_p}{24} \frac{1}{\tau_v} \left(\mathbf{u}(\mathbf{x}(t),t) - \mathbf{v}(t)\right) \tag{1}$$

where $\mathbf{u}(\mathbf{x}(t),t)$ is the velocity of the fluid at the location of the particle. The particle drag coefficient C_d and the hydrodynamic particle relaxation time τ_v, are defined as:

$$C_d = \frac{24}{Re_p} \quad , \quad \tau_v = \frac{\rho_p}{\rho_f} \frac{D_p^2}{18\nu} \tag{2}$$

in terms of the particle Reynolds number $Re_p = |\,(\mathbf{u}-\mathbf{v})\,|\,D_p/\nu$. In these expressions ρ_p and ρ_f are the particle and fluid mass densities, D_p is the diameter of the particles. The expression for C_d is valid only for small Re_p.

The continuous-phase, represented by conservation of mass and momentum, is solved using DNS and LES [1] for incompressible flow. We adopt a geometry of the problem as sketched in Fig. 1. The position, flow, and particle quantities are normalized by the channel half-width, δ, and the friction velocity, u_τ. The particle motion is obtained using a second-order Adams-Bashforth scheme for the time-advancement, and a tri-linear interpolation for the velocity. Periodic boundary conditions are imposed in streamwise and spanwise directions.

The filtered continuity and Navier-Stokes equations, that are the basis for LES of the continuous phase are:

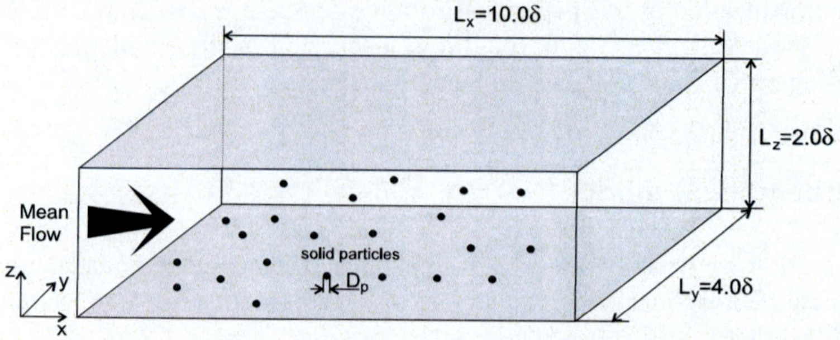

Fig. 1. Computational domain of particle-laden channel flow.

$$\nabla \cdot \overline{\mathbf{u}} = 0 \tag{3}$$

$$\frac{D\overline{\mathbf{u}}}{Dt} = -\frac{\nabla \overline{P}}{\rho_f} + \frac{1}{Re}\nabla^2 \overline{\mathbf{u}} - \nabla \cdot \tau \tag{4}$$

where the overbar denotes the spatially filtered flow variable, ρ_f is the fluid density and Re is the fluid Reynolds number, while P denotes the pressure. The influence of the subgrid motion on the resolved fluid-velocity is represented by the turbulent stress-tensor, τ. In components this tensor is given by

$$\tau_{ij} = \overline{u_i u_j} - \overline{u}_i \overline{u}_j \tag{5}$$

In LES the tensor τ is parameterized by a sub-filter model. Here we will consider eddy-viscosity models in which

$$\tau_{ij} - \frac{\delta_{ij}}{3}\tau_{kk} = -2\nu_t \overline{S}_{ij} \tag{6}$$

The trace τ_{kk} of the turbulent stress tensor is incorporated in the pressure term. In the Smagorinsky model [9] the turbulent eddy-viscosity ν_t is modeled by analogy to the mixing length hypothesis:

$$\nu_t = C_S \Delta^2 |\overline{S}| \tag{7}$$

where C_S is the Smagorinsky coefficient, Δ is the filter width, and $|\overline{S}|$ is the magnitude of the strain rate tensor defined as:

$$|\overline{S}| = \left(2\overline{S}_{ij}\overline{S}_{ij}\right)^{1/2} \quad \text{where} \quad \overline{S}_{ij} = \frac{1}{2}\left(\frac{\partial \overline{u}_i}{\partial x_j} + \frac{\partial \overline{u}_j}{\partial x_i}\right) \tag{8}$$

The Smagorinsky model contributes to the dissipative fluxes in the LES equations. Near the solid walls of the channel this model is known to exaggerate the dissipation, up to the point of preventing transition to turbulence, e.g., in a temporal mixing layer [19]. To counteract this tendency, the turbulent eddy-viscosity ν_t may be directly damped near the walls. A well-known example of this strategy is van Driest damping in which the constant C_S is replaced by $C_S(1-\exp(-y^+/A))$ where y^+ measures the wall-normal distance in terms of wall-coordinates and $A = 25$ [10].

A more elegant method to damp the overly dissipative Smagorinsky model in a turbulent boundary layer is obtained as a by-product of the dynamic procedure [11]. This approach is based on the well-known Germano identity which relates the turbulent sub-filter tensor τ and its 'test-filtered' analog to a resolved stress tensor. By requiring optimal compliance of an assumed model with this identity it becomes possible to automatically adapt the eddy-viscosity coefficient in response to the developing flow. So, instead of assuming a fixed valued C_S or a given damping of C_S near a wall, the dynamic procedure yields a self-adaptive dynamic eddy-viscosity coefficient. It is known that this eddy-viscosity reduces to zero in the viscous sublayer near a solid wall, thereby avoiding the excessive dissipation that arises with the Smagorinsky model.

After closure of the LES equations by assuming a particular sub-filter model, a smoothed representation of the turbulent flow will be obtained from actual simulations. This velocity field $\overline{\mathbf{u}}$ will contain a range of length-scales reflecting the turbulent flow. The smaller resolved scales will be considerably attenuated as a result of the sub-filter model. This lack of small-scale motion will have its impact on the dispersion of point-particles in the LES field. A strategy to recover some of the smaller scales involves approximate deconvolution or 'defiltering'. In this paper we quantify to what extent the recovered small scale dynamics improves the correspondence of LES and DNS. Denoting the filter by L, i.e., $\overline{\mathbf{u}} = L(\mathbf{u})$ we may express the approximate inverse in terms of a truncated geometric series expansion:

$$L^{-1} = \Big(I - (I - L)\Big)^{-1} \approx \sum_{k=0}^{N} (I - L)^k \equiv L_N^{-1} \tag{9}$$

where I denotes the identity operator. Here we introduced the N-th order approximation L_N^{-1} that can be implemented in terms of repeated filter application. The lowest order approximations are $L_0^{-1}(\overline{\mathbf{u}}) = \overline{\mathbf{u}}$, $L_1^{-1}(\overline{\mathbf{u}}) = 2\overline{\mathbf{u}} - \overline{\overline{\mathbf{u}}}$ and $L_2^{-1}(\overline{\mathbf{u}}) = 3\overline{\mathbf{u}} - 3\overline{\overline{\mathbf{u}}} + \overline{\overline{\overline{\mathbf{u}}}}$. The reconstruction of large scales, i.e., Fourier components for which $k\Delta \ll 1$ is very accurate. With increasing $k\Delta$ the reconstruction will be incomplete. In case higher order inversion is used one may expect this to be more accurate but also computationally more expensive. A proper balance should be struck between these conflicting requirements. We turn to this in the next section.

3 Turbulent statistics and deconvolution

In this section we first present turbulent statistics for the fluid flow as obtained with the Smagorinsky model, the Van Driest-Smagorinsky model and the dynamic model. Then we proceed with a discussion of the particle statistics, first without the use of the approximate deconvolution and subsequently with approximate deconvolution.

Simulations were performed in a computational domain as sketched in Fig. 1 with $64 \times 64 \times 48$ control volumes for LES and $128 \times 128 \times 64$ in case of DNS. For the streamwise and spanwise directions the grid spacing is uniform, and for the wall normal direction a hyperbolic-tangent stretching has been used. The shear Reynolds number of the flow was Re_τ=150 based on the shear velocity and half channel height. In order to obtain good statistics for particles in these simulations we used 1.5 10^6 particles for the DNS and one order less for LES.

The simulations were first advanced until a statistically steady turbulent flow was reached. In this flow the particles were initialized at random positions and with the local velocity as initial velocity. The particles need to adapt to the new velocity which usually takes a few times the particle response time. Subsequently the particle statistics is gathered, which requires significant computational effort, particularly since the particle concentration converges very slowly near the walls [1]. The statistics for the fluid and particles were averaged for $100\delta/u_\tau$.

In the following we compare results of DNS with various LES, using the following labeling: (a) LES with a value of C_S implies the use of Smagorinsky's model at that C_S, (b) with additional label 'V-D' in cases where Van Driest damping of the Smagorinsky eddy-viscosity is used and (c) dynamic if the dynamic eddy-viscosity model is adopted. In all simulations we use the LES velocity field in the Stokes drag law to advance the particles, unless otherwise stated. If use is made of approximate deconvolution in the particle dynamics then the Stokes drag law is based on $\mathbf{u}^* = L_N^{-1}(\overline{\mathbf{u}})$ where $\overline{\mathbf{u}}$ is obtained using the dynamic model. We incorporate $N = 0$, $N = 1$ or $N = 2$ for the approximate deconvolution reconstruction.

In Fig. 2 the streamwise fluid velocity component and its fluctuations are presented as a function of the wall normal coordinate for various sub-filter models. The Smagorinsky model yields considerable errors (Fig. 2(b)), which are mainly due to the overestimated near wall dissipation as can be seen by comparison with the results obtained with the dynamic and the Van Driest damped case (Fig. 2(a)). The streamwise velocity fluctuations are slightly under-predicted in case no sub-filter model is used ($C_S = 0$). Better results are obtained using the dynamic model and the Van Driest damped model at $C_S = 0.05$ and 0.1 (Fig. 2(c)). Too large values of C_S in combination with Van Driest yield too high fluctuation levels. The use of Smagorinsky's model is seen to affect mainly the location at which the turbulent fluctuation levels are highest (Fig. 2(d)).

In Fig. 3(a-d) the streamwise and wall-normal particle velocity fluctuations are presented for Stokes number $St = 1$. The streamwise velocity fluctuations are under-predicted without any model or in case the dynamic model is used. Van Driest damping can give under-prediction as well as over-prediction depending on the value of C_S - this allows to work with an optimized eddy-viscosity coefficient (Fig. 3(a)). The use of Smagorinsky's model is seen to yield a peak value further away from the wall as C_S is increased - the peak is also broader as C_S is higher (Fig. 3(b)). For the wall-normal particle velocity fluctuations all studied sub-filter models under-predict the fluctuation levels. In fact, the results without sub-filter model are best among the models studied (Fig. 3(c-d)). The particle wall-normal velocity is also closely related to the particle concentration profile near the channel walls. As can be seen in Fig. 3(e-f) all models under-predict the particle accumulation. The highest concentration was observed for the case when no sub-filter model was used; apparently the small-scale turbulent fluctuations near the wall are very important for proper prediction of turbophoresis.

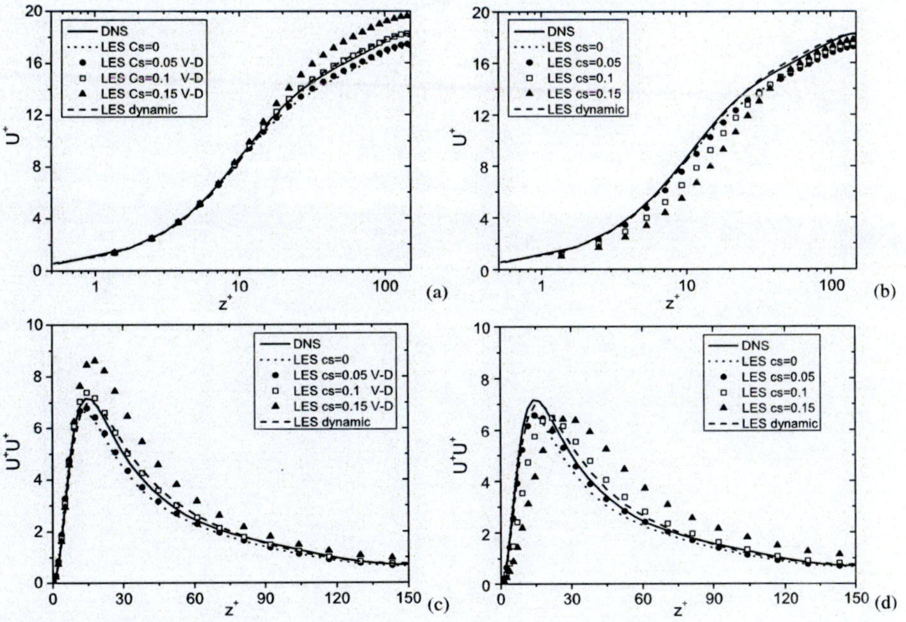

Fig. 2. Fluid velocity (a), (b) and fluctuations (c), (d) for the streamwise component comparing various sub-filter models with DNS results at $Re_\tau = 150$. For the labeling of the LES models see main text.

The introduction of approximate deconvolution can improve the particle statistics relative to the case where particle trajectories are based on the LES velocity field alone. This is illustrated for dynamic model in Fig. 4. All components of the velocity fluctuations show an improved correspondence with DNS when approximate deconvolution is used. Second order deconvolution yields the largest improvement. This is also reflected in a better representation of the particle concentration in the near wall region. The benefit of using $N = 1$ deconvolution relative to no deconvolution ($N = 0$) is larger than using one additional order of deconvolution. Using even higher order deconvolution seems not to be justified by the additional computational effort.

The benefit of the ADM for heavier particles at $St = 5$ and $St = 25$ was also studied. At $St = 5$ the deconvolution method yields improvements over using the dynamic model alone. At $St = 25$ the particles are so heavy that small turbulent scales are quite unimportant for their dynamics - using the dynamic model with or without deconvolution was then found to be close to using no model or DNS.

4 Conclusion

In this paper the Euler-Lagrange framework has been used to simulate the point-particle dynamics with fluid flow represented using LES and DNS. The relevance of the near-wall velocity fluctuations was studied in relation to the particle clustering.

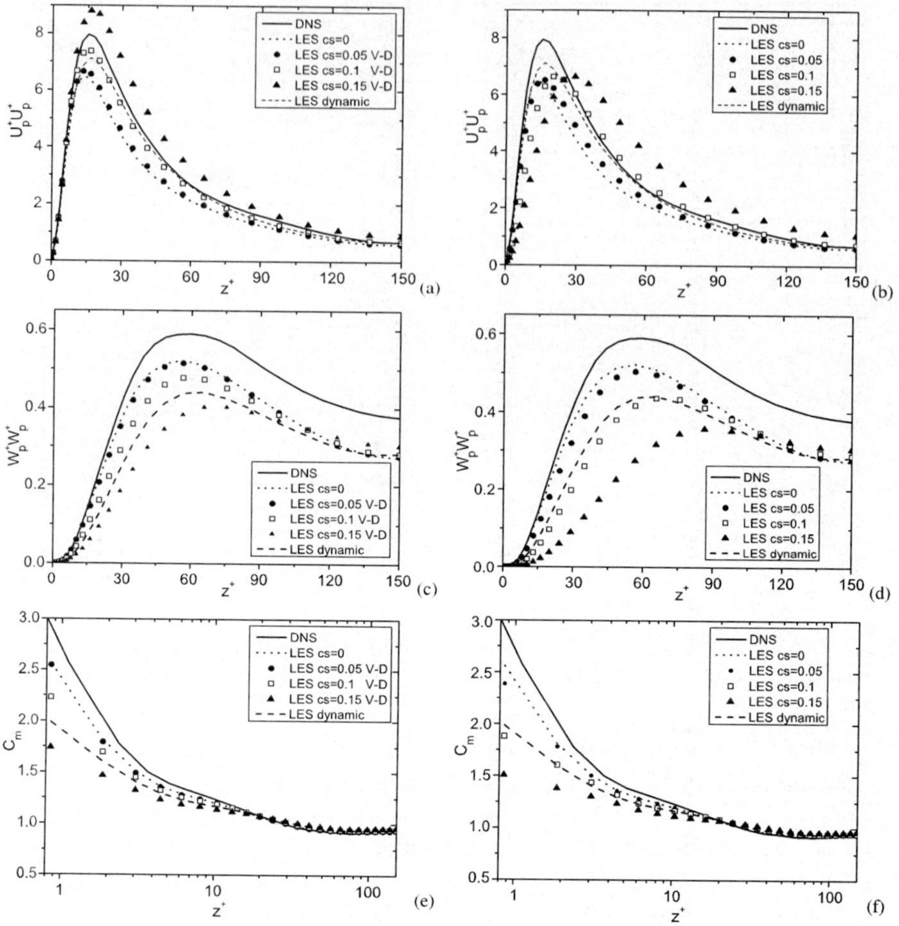

Fig. 3. Particle streamwise (a) and (b) and wall-normal (c) and (d) velocity fluctuations. Particle concentration (e) and (f). We used a particle Stokes number $St = 1$ and compare various LES models with DNS.

At low Stokes numbers the use of small-scale turbulent velocity deconvolution was found to be important for the particle statistics. The predicted particle dynamics using approximate deconvolution were found to agree better with DNS data compared to cases without deconvolution. However, the particle statistics based on the dynamic model including approximate deconvolution were found to be only just as good as those obtained in case no sub-filter model was used. Concerning the ADM approximation order it was found that N=2 give the best results for most of the cases studied and higher order doesn't improve results significant. At higher Stokes number the relevance of the smaller turbulent scales is less pronounced and LES results for turbophoresis were found to correspond closely to DNS data, quite independent of the sub-filter model that was used, as long as the near wall dissipation was not overesti-

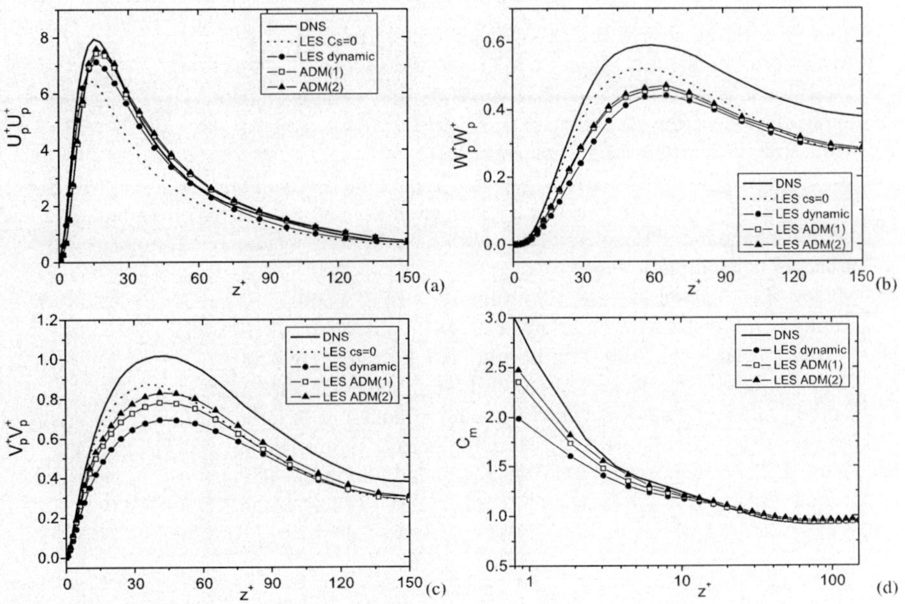

Fig. 4. Particle streamwise, wall-normal and spanwise velocity fluctuations (a, b, c respectively) together with particle concentration (d) at particle Stokes number $St = 1$ illustrating the use of the deconvolution method.

mated too much. This implies that Van Driest damping and the dynamic procedure proved to be quite reliable while the Smagorinsky model was found less accurate. In contrast to other work in present paper lower (second) order of interpolation for particle velocity has been used. But quality of the results compare to LES results of other Authors due to deconvolution remain the same as forth order interpolation[4]. Because the ADM can only models the effects of the resolved scales it is possible that with second order interpolation some additional benefit of defiltering was cancel and more scales can be retrieve with more accurate interpolation.

Acknowledgment

This research was supported by project COST P20 258/2006

References

1. Portela L.M., Oliemans R.V.A. (2003) Int J Numer Meth Fluids 43:1045–1065
2. Geurts B.J. (1997) Phys Fluids 9:3585–3587
3. Stolz S., Adams N.A. (1999) Phys Fluids 11:1699–1701
4. Kuerten J.G.M. (2006) Phys Fluids 18:025108-1–13
5. Kuerten J.G.M., Vreman A.W. (2005) Phys Fluids 17:011701-1–4

6. Shotorban B., Mashayek F. (2005) Phys Fluids 17:081701-1–4
7. Armenio V., Piomelli U., Fiorotto V. (1999) Phys Fluids 11:3030–3042
8. Marchioli C., Salvetti M.V., Soldati A. (2008) Phys Fluids 20:040603-1–11
9. Smagorinsky J. (1963) Mon Weather Rev 91:99–164
10. Moin P., Kim J. (1982) J Fluid Mech 118:341–377
11. Germano M., Piomelli U., Moin P., Cabot W.M. (1991) Phys Fluids A 3:1760–1765
12. Elghobashi S. (2004) An Updated Classification Map of Particle-Laden Turbulent Flows. In: Balachandar S., Prosperetti A. (eds) IUTAM Symposium on Computational Approaches to Multiphase Flow.
13. Maxey M.R., Riley J.K. (1983) Phys Fluids A 26:883–889
14. Kulick J.D., Fessler J.R., Eaton J.K. (1994) J Fluid Mech 277:109–121
15. Fessler J., Eaton J.K. (1994) Int J Multiphase Flows 20:169–209
16. Chung J., Koch D.L., Rani S.L. (2005) J Fluid Mech 536:219–227
17. Eggels J.G.M. (1994) Direct and large-eddy simulations of turbulent flow in a cylindrical pipe geometry. PhD thesis, Laboratory for Aero & Hydrodynamics, TU Delft
18. Geurts B.J., Vreman A.W. (2006) Int J Heat and Fluid Flow 27:945–954
19. Vreman A.W., Geurts B.J., Kuerten J.G.M. (1997) J Fluid Mech 339:357–390

Inertial particle segregation and deposition in large-eddy simulation of turbulent wall-bounded flows

Cristian Marchioli[1], Maria Vittoria Salvetti[2] and Alfredo Soldati[1]

[1] Dipartimento di Energetica e Macchine, Università degli Studi di Udine, Via delle Scienze 208, 33100 Udine, Italy marchioli@uniud.it, soldati@uniud.it
[2] Dipartimento di Ingegneria Aerospaziale, Università di Pisa, Via G. Caruso, 56122 Pisa, Italy m.v.salvetti@ing.unipi.it

Summary. Current capabilities of Large-Eddy Simulation (LES) in Eulerian-Lagrangian studies of dispersed flows are limited by the modeling of the Sub-Grid Scale (SGS) turbulence effects on particle dynamics. In this paper, the possibility is examined to account explicitly for SGS effects by incorporating *ad-hoc* closure models in the Lagrangian equations of particle motion. Specifically, a candidate model based on approximate deconvolution is considered and applied to particle-laden turbulent channel flow. Results show that, even if the fraction of SGS turbulent kinetic energy for the fluid velocity field (not resolved in LES) is recovered, quantitative prediction of local segregation and, in turn, of near-wall accumulation may still be inaccurate. This failure indicates that reconstructing the correct amount of fluid and particle velocity fluctuations is not enough to reproduce the effect of SGS turbulence on particles and that further information on the flow structure at the sub-grid scales must be incorporated.

Key words: Lagrangian Tracking of particles in LES, Approximate Deconvolution, Segregation, Deposition

1 Introduction

Turbulent dispersed flows in boundary layers are crucial in a number of industrial and environmental applications. In this context, a key information for practical applications is the rate at which the particles are transported to, deposited at, and re-entrained from the wall by turbulence. Direct numerical simulation studies (see [1] and references therein) have shown that all these phenomena are governed by the strong correlation existing between coherent flow structures, local particle segregation and subsequent deposition phenomena. Specifically, inertial particles are initially segregated and accumulated into specific flow regions close to the walls and only afterwards are driven to the walls, where deposition eventually occurs. Modeling these physical mechanisms in numerical methods coarser than in DNS is non trivial. Lagrangian tracking of particles in flow fields obtained from Large-Eddy Simulation (LES) represents a potentially useful tool for practical applications; however,

M.V. Salvetti et al. (eds.), *Quality and Reliability of Large-Eddy Simulations II*, ERCOFTAC Series 16, DOI 10.1007/978-94-007-0231-8_18,

the full capabilities of LES to accurately predict near wall accumulation and preferential segregation are still to be proven. Critical issues are: *(i)* How does the subgrid scale (SGS) turbulence affect particle dispersion? *(ii)* How should these SGS effects be modeled in order to obtain accurate prediction of the selective response of different-inertia particles?

In previous papers [2, 3], we investigated these issues focusing on a well-known archetypal instance of wall bounded flow, i.e particle-laden channel flow. Using a Eulerian-Lagrangian approach, we benchmarked a-priori and a-posteriori LES at shear Reynolds number[†] $Re_\tau = 150$ against reference pseudo-spectral DNS data performing tests with and without SGS closure models in the particle motion equations [2, 3]. In agreement with other recent studies (see e.g. [4]), it was found that LES tends to underestimate local particle segregation and, in turn, deposition fluxes and near-wall accumulation. The main conclusions drawn from [2, 3] were the following: *(i)* It is necessary to identify a way to model the effects of the filtered SGS flow velocity fluctuations on particle motion; and *(ii)* introduction of a SGS closure model for particles that is able to reintroduce the correct amount of fluid velocity fluctuations in the particle motion equations may not be sufficient to warrant accurate prediction of near-wall particle accumulation.

Based upon these conclusions, in this paper results obtained from a-posteriori tests performed at $Re_\tau = 300$ are discussed to investigate if previous findings also hold at higher Reynolds numbers. These tests may provide useful indications of the most important features which should be incorporated in SGS closure models for particles such as, for instance, the level of information about the flow field required at the sub-grid level and/or the necessity to include explicit Reynolds number/particle inertia dependencies in the model. Tests may also assist in highlighting the physical reasons for the observed lack of accuracy in predicting near-wall particle accumulation.

2 Problem Formulation and Numerical Methodology

The physical problem considered in this study is particle dispersion in fully-developed turbulent channel flow. The carrier fluid is air with density $\rho = 1.3\ kg\ m^{-3}$ and kinematic viscosity $\nu = 15.7\times10^{-6}\ m^2\ s^{-1}$. Both DNS and LES have been performed. The governing equations solved for are standard Continuity and Navier-Stokes equations written for incompressible Newtonian fluid. In LES, these equations were smoothed with a filter function of width Δ: The closure for such filtered equations is provided by the dynamic SGS model of Germano et al. [5]. A pseudo-spectral method [6], based on transforming the field variables into wavenumber space, using Fourier representations for the periodic streamwise and spanwise directions and

[†] The shear Reynolds number for the considered channel flow configuration is defined as $Re_\tau = u_\tau h/\nu$ where u_τ is the shear (or friction) velocity, h is the half channel height, and ν is fluid kinematic viscosity. The shear velocity is defined as $u_\tau = \sqrt{\tau_w/\rho}$, where τ_w is the mean shear stress at the wall and ρ is fluid density.

Chebychev representation for the wall-normal (non-homogeneous) direction, was used to integrate the governing equations. A two level, explicit Adams-Bashforth scheme for the non-linear terms, and an implicit Crank-Nicolson method for the viscous terms are employed for time advancement in both DNS and LES. All variables are taken in dimensionless form, and expressed in wall units by combining u_τ, ν and ρ. The reference geometry consists of two infinite flat parallel walls: the origin of the coordinate system is located at the center of the channel and the $x-$, $y-$ and $z-$ axes point in the streamwise, spanwise and wall-normal directions respectively. Periodic boundary conditions are imposed on the fluid velocity field in x and y, no-slip boundary conditions are imposed at the walls. The calculations were performed on a computational domain of size $4\pi h \times 2\pi h \times 2h$ in x, y and z respectively. In this paper, statistics obtained performing particle tracking in LES at a shear Reynolds number $Re_\tau = 300$ will be shown (for details on statistics - and simulation parameters - at $Re_\tau = 150$ the reader is referred to [2, 3]). The number of grid points used to discretize the computational domain is $256 \times 256 \times 257$ in DNS, $128 \times 128 \times 129$ in *fine* (e.g. well-resolved) LES, and $64 \times 64 \times 129$ in *coarse* LES, respectively.

Particles have density $\rho_p = 1000\ kg\ m^{-3}$ and are treated as pointwise rigid spheres injected at concentration low enough to consider dilute flow conditions: one-way coupling between the two phases is thus assumed and particle-particle collisions are neglected. Particle motion is described by a set of ordinary differential equations for particle velocity and position. For particles much heavier than the fluid ($\rho_p/\rho \gg 1$) the most significant forces are Stokes drag and buoyancy. However, our aim here is to minimize the number of degrees of freedom by keeping the simulation setting as simplified as possible; thus the effect of gravity has also been neglected. With the above assumptions, the following equations in vector form are obtained:

$$\frac{d\mathbf{x}}{dt} = \mathbf{v}\,, \qquad \frac{d\mathbf{v}}{dt} = \frac{\mathbf{v} - \mathbf{u}_{@p}}{\tau_p}(1 + 0.15 Re_p^{0.687})\,, \tag{1}$$

where $\mathbf{x}$ is particle position, $\mathbf{v}$ is particle velocity, $\mathbf{u}_{@p}$ is fluid velocity at particle position, $\tau_p = \rho_p d_p^2/18\mu$ is the particle response time, and $Re_p = d_p|\mathbf{v} - \mathbf{u}_{@p}|/\nu$ is the particle Reynolds number, ρ_p and d_p being particle density and particle diameter respectively. When no SGS model is used in the particle motion equations, $\mathbf{u}_{@p}$ is obtained by interpolation of the filtered fluid velocity $\bar{\mathbf{u}}$ issued by LES. When Approximate Deconvolution (AD) is applied to $\bar{\mathbf{u}}$, then $\mathbf{u}_{@p}$ is set equal to the deconvolved fluid velocity, $\mathbf{u}^*$, whose components are obtained as $u_i^* = \Sigma_{\alpha=0}^{N}(I-G)^\alpha * \bar{u}_i$, where G is the (invertible) filter kernel and N is the series truncation parameter (following [7] we fixed $N = 5$). Although the filter implicitly applied by the numerical discretization in the homogeneous directions is the Fourier cut-off filter, the deconvolution procedure requires a smooth filter to be effective (see e.g. [7]). Among the possible analytic and discrete smooth filters, a top-hat filter was applied in this work and deconvolution is carried out only in the homogeneous directions.

To calculate particle trajectories in the flow field, we have coupled a Lagrangian tracking routine with the DNS/LES flow solver. The routine solves for Eqs. (1) using 6th-order Lagrangian polynomials to interpolate fluid velocities at particle position; with this velocity the equations of particle motion are advanced in time using a 4th-

St	$\tau_p\ (s)$	d_p^+	$d_p\ (\mu m)$	$V_s^+ = g^+ \cdot St$	$Re_p^+ = V_s^+ \cdot d_p^+/\nu^+$
1	$0.283 \cdot 10^{-3}$	0.153	10.2	0.0118	0.00275
4	$1.132 \cdot 10^{-3}$	0.306	20.4	0.0472	0.01444
5	$1.415 \cdot 10^{-3}$	0.342	22.8	0.0590	0.02018
20	$5.660 \cdot 10^{-3}$	0.684	45.6	0.2358	0.16129
25	$7.075 \cdot 10^{-3}$	0.765	51.0	0.2948	0.22552
100	$28.30 \cdot 10^{-3}$	1.530	102.0	1.1792	1.80418

Table 1. Particle parameters for the simulations.

order Runge-Kutta scheme. The timestep size used for particle tracking was chosen to be equal to the timestep size used for the fluid, $\delta t^+ = 0.045$; the total tracking time was, for each particle set, $t^+ = 4050$. Particle initial distribution is homogeneous over the computational domain and their initial velocity is set equal to that of the fluid at the particle initial position. Periodic boundary conditions are imposed on particles moving outside the computational domain in the homogeneous directions, perfectly-elastic collisions at the smooth walls were assumed when the particle center was at a distance lower than one particle radius from the wall. For the simulations presented here, large samples of $\mathcal{O}(10^5)$ particles, characterized by different response times, were considered. Table 1 shows all relevant particle parameters, including the particle Stokes number $St = \tau_p^+ = \tau_p/\tau_f$, where $\tau_f = \nu/u_\tau^2$ is the viscous timescale of the flow.

3 Particle tracking in LES flow fields at $Re_\tau = 300$

In this section, we compare velocity statistics for the two phases to highlight the effect of using a closure model in the equation of particle motion. We thus show results obtained from LES with AD ($\mathbf{u}_{@p} = \mathbf{u}^*$) to those obtained from LES without AD ($\mathbf{u}_{@p} = \bar{\mathbf{u}}$), using DNS results ($\mathbf{u}_{@p} = \mathbf{u}$) as reference.

Fig. 1 compares the Root Mean Square (RMS) values of the fluid velocity streamwise and wall-normal components. The spanwise component is not shown for sake of brevity. For the *fine* LES, the LES velocity fields without AD (dotted line) are characterized by a fluctuation level in good agreement with the DNS one (solid line), except for the wall-normal component, for which LES noticeably underestimates the RMS values. As discussed in [4], this is already a source of error in predicting turbophoresis and may lead to significant underestimation of wall particle concentration. The use of AD (open circles) leads to an increase of the fluid velocity RMS with respect to standard LES. For the wall-normal component the correct RMS level is recovered. However, for the remaining components the deconvolved fluid velocity field is characterized by a slight overshoot of fluctuations. As discussed in [2, 3], in the *coarse* LES case the LES fluid velocity field is already characterized by quite large RMS values. AD increases further the fluid velocity fluctuations and, therefore, amplifies the disagreement with DNS.

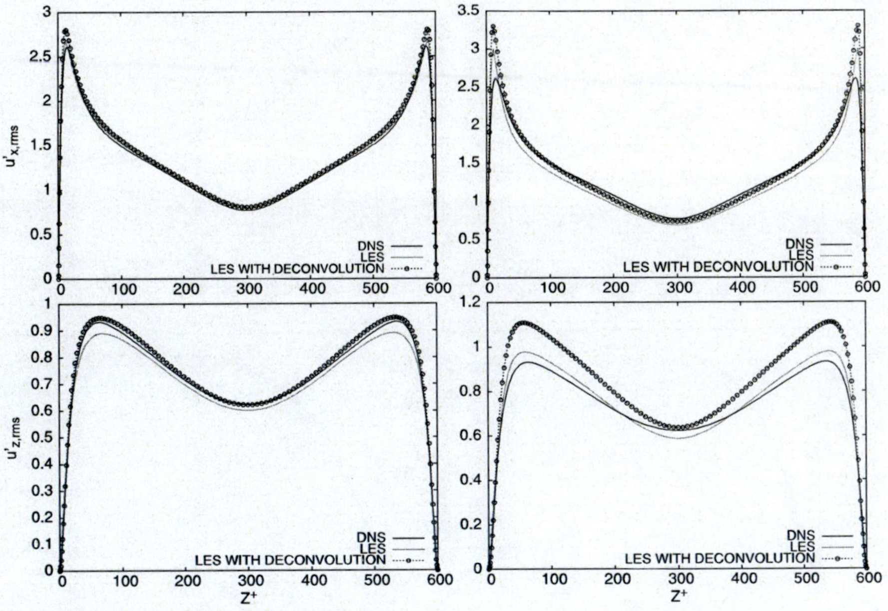

Fig. 1. RMS of fluid velocity fluctuations. Streamwise RMS at top; wall-normal RMS at bottom. Left-hand panels: *fine* LES; right-hand panels: *coarse* LES.

Similar considerations can be made for the RMS values of particle velocities. As an example, Figs. 2 and 3 show the RMS values of streamwise and wall-normal velocities for the $St = 1$ and the $St = 25$ particles. When LES is well-resolved (*fine* LES, Fig. 2), AD improves the agreement with DNS for particle wall-normal velocity, even though the streamwise component is still slightly overestimated. For *coarse* LES (Fig. 3), the RMS values are always overestimated, and the overshoot becomes larger when AD is applied to the fluid velocity fields. The effect produced by AD on velocity statistics in the wall-normal direction is also visible on the instantaneous particle concentration profiles, shown in Fig. 4. In all cases, AD improves the agreement with DNS [4, 8]. Prediction of the deposition velocity (not shown) also improves.

This is not surprising when LES is well-resolved, since AD proved its capability to recover the correct level of fluctuations for the wall-normal velocity component of both phases (see Figs. 1c and 2c-d). As previously mentioned and as discussed in [4], this is a key issue to have a good prediction of turbophoresis. In the case of *coarse* LES, this improvement was less expected and may be due to error compensation: the large overestimation of RMS values of both fluid and particle wall-normal velocity obtained with AD might compensate the inadequate description in LES of the physical mechanism leading to turbophoresis, due to inaccurate rendering of the interaction between wall turbulence structures and particles (see Sec. 4). Finally, the effect of AD on segregation is shown in Fig. 5 by means of the maximum value attained by the segregation parameter D in the near wall region. The segregation pa-

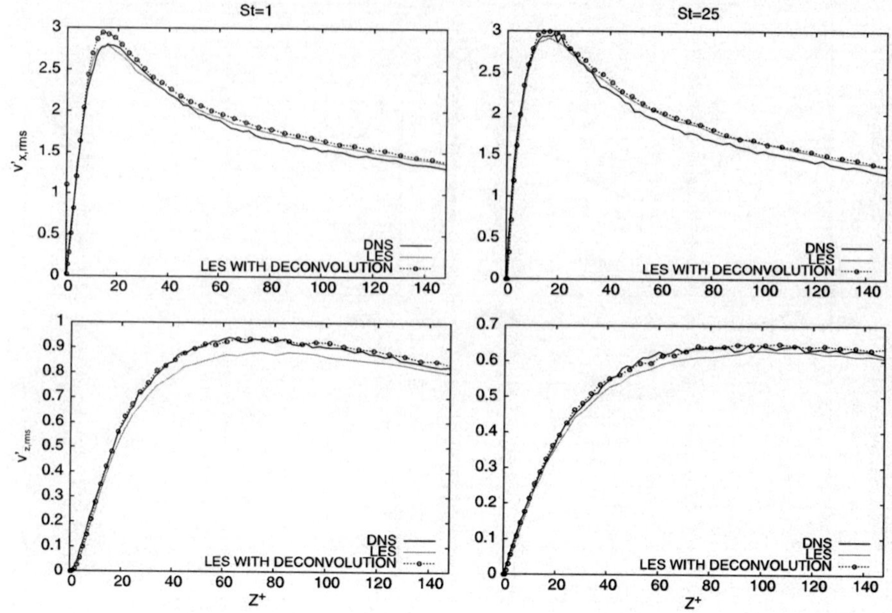

Fig. 2. RMS of particle velocity fluctuations: *fine* LES versus DNS. Top: streamwise RMS; bottom: wall-normal RMS. Left column: $St = 1$; right column: $St = 25$.

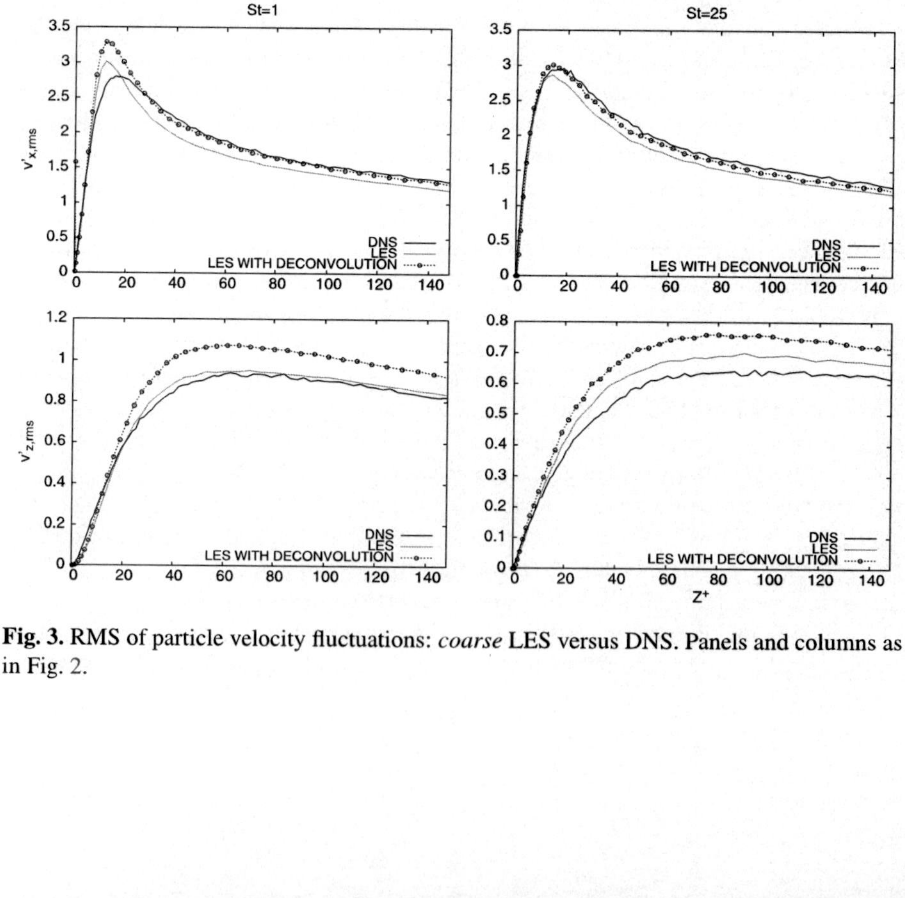

Fig. 3. RMS of particle velocity fluctuations: *coarse* LES versus DNS. Panels and columns as in Fig. 2.

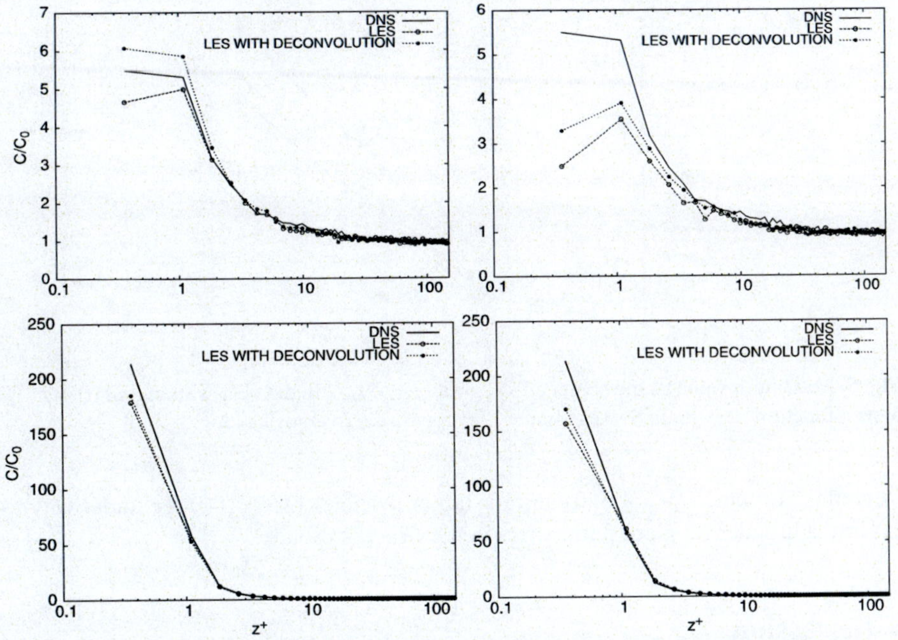

Fig. 4. Instantaneous particle concentration profiles. Top: $St = 1$; bottom: $St = 25$. Left column: *fine* LES; right column: *coarse* LES.

rameter is computed as $D = (\sigma - \sigma_{Poisson})/m$, where m is the mean particle number density and σ and $\sigma_{Poisson}$ represent the standard deviations for the measured particle number density distribution and for the Poisson distribution respectively. Technical details on the computation of its maximum value, D_{max}, can be found in [1]. Here, it suffices to remind that positive values of D_{max} are related to segregation of particles: specifically, larger values of D_{max} correspond to larger deviation from randomness in particle spatial distribution and thus indicate stronger segregation. Fig. 5 demonstrates that AD improves the agreement with DNS, yet segregation is still underestimated. This result, shown here for the near-wall region but observed also in the center of the channel, is of particular importance because it provides an explanation for the observed underestimation of particle near-wall accumulation (Fig. 4). As mentioned in the Introduction, particle dispersion in wall-bounded flows can be envisioned as a *hierarchical* multi-step process where segregation of particles by the near-wall turbulence structures assists in near-wall accumulation, which in turn assists in deposition [1]. According to this chain of events, if segregation is underpredicted, then all subsequent phenomena will be underpredicted.

Note how particles characterized by higher values of D_{max} show a stronger tendency toward near-wall accumulation. These results are consistent with previous findings at $Re_\tau = 150$ [2, 3], and indicate that Reynolds number effects are not important for small to moderate values of this parameter. Further simulations at higher

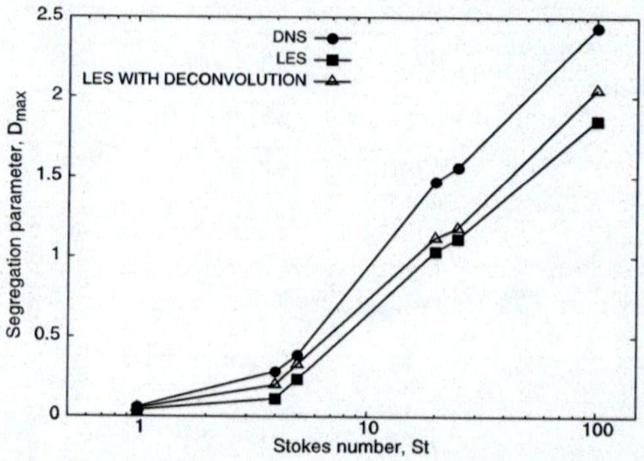

Fig. 5. Maximum value of the segregation parameter, D_{max}, in the near wall region ($0 \leq z^+ \leq 5$) as a function of particle Stokes number, St: *coarse LES* versus DNS.

Reynolds numbers ($Re_\tau^h \geq 600$) and for larger particles ($St \geq 100$) are under way to explore a broader range of parameters in the (Re_τ, St) space.

4 Discussion

In this Section we will try to provide possible physical explanations for the observed inability of Lagrangian particle tracking in LES fields to quantitatively reproduce segregation phenomena and to capture the wall-accumulation effect. The figures shown in the following have been obtained at $Re_\tau = 150$, but analogous considerations can be made for $Re_\tau = 300$. Fig. 6 shows a cross-sectional view of particles and vortical structures as captured by DNS (top panel) and by LES (bottom panel). Particles are drawn as circles, larger than the real scale for visualization purposes. Vortical structures are rendered using streamwise vorticity isosurfaces: the green isosurface identifies a clockwise rotating quasi-streamwise vortex, whereas the red isosurface identifies a counter-clockwise rotating quasi-streamwise vortex. Differences in the vortical structures between DNS and LES are quite evident, both near the wall and in the center of the channel: Subgrid scales are filtered in LES and, therefore, vortical structures are not properly captured if not fully ignored at the finest flow scales. As a consequence, modifications in particles interaction with these structures are expected: indeed, Fig. 6 shows that particles in LES exhibit a more persistent stability against non-homogeneous distribution and near-wall concentration, suggesting the occurrence of a less efficient interaction between the two phases.

Inaccurate rendering of the turbulence structures may be discussed also on a more quantitative basis. In terms of energy, for instance. In Fig. 7 the one-dimensional energy spectrum sampled from DNS and LES at two different locations of the channel ($z^+ = 150$ in the core flow region and $z^+ = 25$ in the near-wall region) are compared. Also shown (dot-dashed lines) are the estimated response frequencies which characterize each considered particle set, these frequencies being proportional to $1/\tau_p$. As obvious, LES resolves for a smaller range of frequencies, preventing particles from

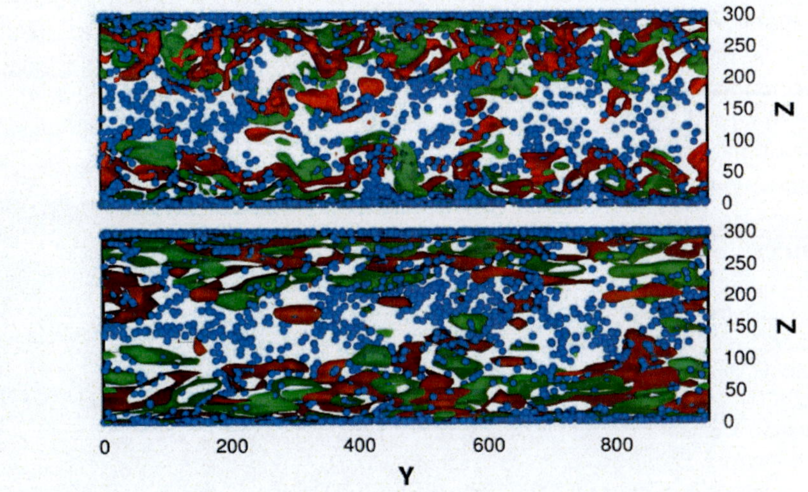

Fig. 6. Front view of particle instantaneous distribution and turbulent structures in the channel: DNS (top) versus LES (bottom: DNS (top) versus LES (bottom).

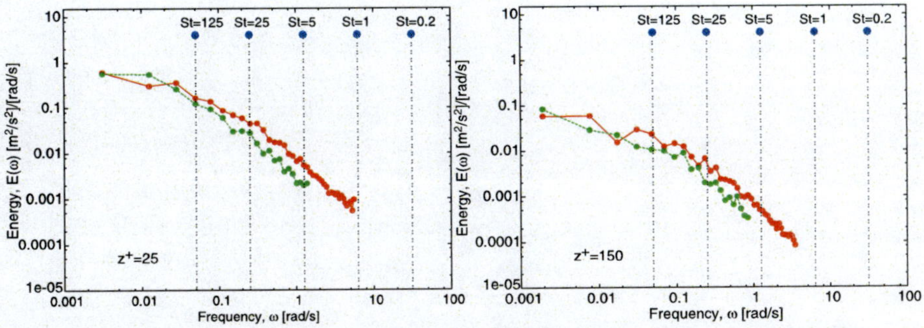

Fig. 7. One-dimensional (streamwise) frequency spectrum for turbulent channel flow computed at (a) $z^+ = 25$ and at (b) $z^+ = 150$. Red: DNS, green: LES.

being exposed to ever-increasing turbulent frequencies, i.e. to smaller and smaller flow scales. Clearly, this introduces a filtering error everywhere in the flow domain (not only near the wall). In the near-wall region, the frequencies removed in the LES flow field correspond to those with which smaller-inertia particles, characterized by low values of St, preferentially interact. Near the center of the channel, the LES cut-off frequency tends to decrease so that removed frequencies become close to those with which larger-inertia particles preferentially interact (see for instance the $St = 5$ particles).

Another source of error affecting particle behavior over a large range of St in the entire flow domain is due to the energy content at the resolved scales, which is lower in LES than in DNS. Even though there are cases in which AD can reintroduce the correct amount of fluid energy filtered out at the resolved scales, it is not expected (in principle) to retrieve the amount of energy filtered out at the subgrid scales. This means that use of AD as closure model for particle tracking in LES is not expected to provide the information on the flow field at the subgrid level which (in our opinion) is necessary to reproduce particle-fluid interaction and to ensure accurate quantitative prediction of the dispersion phenomena.

References

1. Soldati A, Marchioli C (2009) Int J Multiphase Flow, 35:827–839.
2. Marchioli C, Salvetti MV, Soldati A (2008) Phys Fluids 20:040603.
3. Marchioli C, Salvetti MV, Soldati A (2008) Acta Mech 201:277–296.
4. Kuerten JGM, Vreman AW (2005) Phys Fluids 17:011701.
5. Germano M, Piomelli U, Moin P, Cabot WH (1991) Phys. Fluids 3:1760–1765.
6. Pan Y, Banerjee S (1996) Phys. Fluids 17:2733–2755.
7. Stolz P, Adams NA, Kleiser L (2001) Phys. Fluids 13:997–1015.
8. Shotorban B, Zhang KKQ, Mashayek F (2007) Int. J. Heat Mass Tran. 50:3728–3739.

Scalar sub-grid energy in large-eddy simulation of turbulent flames: mesh quality criterion

Luc Vervisch[1], Pascale Domingo[1], Guido Lodato[1] and Denis Veynante[2]

[1] CORIA - CNRS and INSA de Rouen, Technopole du Madrillet, BP 8, 76801 Saint-Etienne-du-Rouvray, France vervisch@coria.fr
[2] EM2C - CNRS and Ecole Centrale Paris, Grande Voie des Vignes, 92295 Châtenay-Malabry, France

Summary. In Large-Eddy Simulation (LES), scalar fluctuations are decomposed into a resolved part and a complementary Sub-Grid Scale (SGS) part. Accordingly, it is usually assumed that the scalar energy contained in these two parts sum up, so that the time average of the scalar energy equals the time average of the resolved part of the scalar energy to which the time average of the SGS scalar variance is added. Conditions are discussed under which an additional residual term must be added to close this scalar energy budget. For this residual term to stay at a moderate level, the LES filter must be small enough compared to the integral length-scale of the scalar field, a condition that is verified from a canonical manufactured turbulent scalar solution. A mesh-quality criterion is derived from these observations and the minimum Reynolds number that a Direct Numerical Simulation (DNS) should feature for SGS scalar variance to be accurately studied from *a priori* filtering is obtained as a corollary.

Key words: Large-Eddy Simulation, Sub-Grid Scale scalar energy, Scalar variance

1 Introduction

In both chemically frozen flow mixing problems and reactive flows, as turbulent flames, scalars are used to trace the details of mass and energy distributions. Large-Eddy Simulation (LES) uses space filtered scalars; the instantaneous signal is decomposed into a space filtered quantity to which a Sub-Grid Scale (SGS) fluctuation is added. The precise modeling of this SGS fluctuation may be of great importance if scalars are associated to strongly non-linear phenomena, as chemical sources, whose space filtered values strongly deviate from the one calculated with LES resolved scalars only.

One option consists in introducing a measure of the SGS scalar activity based on the SGS scalar variance. The objective of this paper is to discuss spurious effects affecting SGS scalar variance when the LES mesh is too coarse. In particular, using a simple scaling analysis and a canonical turbulent scalar manufactured solution, it is discussed how the LES mesh can be optimized to ensure that the SGS scalar variance is fully representative of the SGS scalar energy.

M.V. Salvetti et al. (eds.), *Quality and Reliability of Large-Eddy Simulations II*, ERCOFTAC Series 16, DOI 10.1007/978-94-007-0231-8_19,

The scalar energy decomposition into its resolved and SGS parts is first introduced, then in a subsequent section a global scalar scaling is discussed on the basis of a turbulent resolution criterion. A manufactured scalar solution is proposed and the response of SGS scalar energy to filtering is examined to discuss a mesh optimization criterion. In a final part, some implications on Reynolds number of Direct Numerical Simulation (DNS) databases chosen to perform *a priori* studies of SGS variance are reported. Constant density is assumed in this study, Reynolds averaging is thus adopted for all the developments.

2 Scalar energy decomposition

Let us consider a scalar $\phi(\underline{x},t)$, which may be passive or subjected to chemical reaction, its balance equation reads:

$$\frac{\partial \rho\phi}{\partial t} + \nabla\cdot(\mathbf{u}\phi) = \nabla\cdot(\rho D_\phi \nabla\phi) + \dot{\omega}_\phi \tag{1}$$

where ρ denotes the density, $\mathbf{u}$ is the velocity vector and D_ϕ the diffusion coefficient used to express molecular diffusion with a Fick law. A spatial filter $\mathscr{G}_\Delta(\underline{x})$ of characteristic size Δ is applied to all the fields [1, 2, 3],

$$\overline{\phi}(\underline{x},t) = \int_{-\infty}^{+\infty} \phi(\underline{x}',t)\mathscr{G}_\Delta(\underline{x}-\underline{x}')d\underline{x}' \tag{2}$$

The scalar is then decomposed into an LES resolved part to which a SGS fluctuation is added:

$$\phi(\underline{x},t) = \overline{\phi}(\underline{x},t) + r_\phi \tag{3}$$

Squaring this relation and filtering it leads to an expression for the filtered unresolved energy of the scalar:

$$\overline{r_\phi^2} = \left(\overline{\phi^2} - \overline{\phi}^2\right) + \overline{\overline{\phi}^2} + \overline{\phi}^2 - 2\overline{\phi\overline{\phi}} \tag{4}$$

$$= \phi_v + r_v \tag{5}$$

where

$$\phi_v = \left(\overline{\phi^2} - \overline{\phi}^2\right) \tag{6}$$

is the SGS variance of the scalar ϕ and

$$r_v = \overline{\overline{\phi}^2} + \overline{\phi}^2 - 2\overline{\phi\overline{\phi}} \tag{7}$$

a residual contribution, which depends on the convolution between the exact topology of the scalar field and the filter $\mathscr{G}_\Delta$. In the particular case where

$$\overline{\overline{\phi}^2} = \overline{\phi\overline{\phi}} = \overline{\phi}^2 \tag{8}$$

the filtered unresolved scalar energy is equal to the SGS variance,

$$\overline{r_\phi^2} = \phi_v \tag{9}$$

a quantity that is widely used in turbulent combustion modeling to calibrate the impact of SGS unresolved fluctuations of species and temperature on chemical reactions and turbulent micro-mixing [4, 5, 6].

Time averaging of the scalar signal, is introduced and denoted $\langle\phi\rangle$; adding and subtracting terms to cancel their contribution, a straightforward relation is derived [7]:

$$\begin{aligned}\Phi_v = \langle\phi^2\rangle - \langle\phi\rangle^2 &= \left\langle\overline{\phi}^2\right\rangle - \langle\overline{\phi}\rangle^2 + \left\langle\overline{\phi^2}\right\rangle - \left\langle\overline{\phi}^2\right\rangle \\ &\quad + \langle\phi^2\rangle - \left\langle\overline{\phi^2}\right\rangle + \langle\overline{\phi}\rangle^2 - \langle\phi^2\rangle\end{aligned} \tag{10}$$

This may be cast in:

$$\Phi_v = \left\langle\overline{\phi}^2\right\rangle - \langle\overline{\phi}\rangle^2 + \langle\phi_v\rangle + R_v \tag{11}$$

with

$$R_v = \langle\phi^2\rangle - \left\langle\overline{\phi^2}\right\rangle + \langle\overline{\phi}\rangle^2 - \langle\phi^2\rangle \tag{12}$$

Or again using Eq. (5)

$$\Phi_v = \left\langle\overline{\phi}^2\right\rangle - \langle\overline{\phi}\rangle^2 + \left\langle\overline{r_\phi^2}\right\rangle - \langle r_v\rangle + R_v \tag{13}$$

In Eqs. (11) and (13), the time averaged scalar energy $\langle\phi^2\rangle - \langle\phi\rangle^2$ has been decomposed into a resolved part,

$$T_R = \left\langle\overline{\phi}^2\right\rangle - \langle\overline{\phi}\rangle^2 \tag{14}$$

plus a SGS part,

$$\langle\phi_v\rangle + R_v = \left\langle\overline{r_\phi^2}\right\rangle - \langle r_v\rangle + R_v \tag{15}$$

The usual decomposition into resolved and SGS parts,

$$\Phi_v = \langle\phi^2\rangle - \langle\phi\rangle^2 = T_R + \langle\phi_v\rangle \tag{16}$$

with the SGS part estimated from the SGS variance, is thus valid only if $R_v = \langle r_v\rangle = 0$, otherwise one should write:

$$\Phi_v = \langle\phi^2\rangle - \langle\phi\rangle^2 = T_R + \langle\phi_v\rangle + R_v \tag{17}$$

From the relation (10), this is achieved if [7]:

$$\langle\overline{\phi}\rangle = \overline{\langle\phi\rangle} = \langle\phi\rangle \tag{18}$$

and

$$\left\langle \overline{\phi^2} \right\rangle = \overline{\langle \phi^2 \rangle} = \langle \phi^2 \rangle \tag{19}$$

In Eqs. (18) and (19), the first equality is valid because space filtering and time averaging commute. The second equality holds only if the filtering operation has no impact on time average quantities, which is the case if:

$$\delta_T = \frac{\max(\langle \phi \rangle) - \min(\langle \phi \rangle)}{\max(|\nabla \langle \phi \rangle|)} >> \Delta \tag{20}$$

where δ_T is the characteristic length scale of the time average scalar signal (or scalar integral length scale) [7]. A critical value β_R of the ratio

$$\beta = \frac{\Delta}{\delta_T} \tag{21}$$

must therefore exists, so that for the condition:

$$\beta < \beta_R \tag{22}$$

the SGS scalar variance is a good approximation of the unresolved scalar energy, *i.e.* R_v and $\langle r_v \rangle$ vanish.

3 Scalar scaling

Simulation resolution criterions have been defined in the literature to measure the resolution departure between a given LES and its equivalent DNS [8, 9, 10]. The time average of the turbulent kinetic energy is denoted K and its SGS part k_{SGS}. They may be related as

$$k_{SGS} = \zeta K \tag{23}$$

to then assume that a sufficiently resolved LES is obtained for $\zeta < 0.2$ [9]. The LES filter is usually applied in the inertial range,

$$\frac{K^{3/2}}{\ell_T} \propto \frac{k_{SGS}^{3/2}}{\Delta} \tag{24}$$

Further assuming that the scalar field integral length scale is of the order of ℓ_T, these relation provides:

$$\beta = \zeta^{3/2} \tag{25}$$

The scalar criterion equivalent to the 80% of velocity fluctuations resolved in LES would then read:

$$\beta < \beta_R = 0.09 \; ; \; \Delta < 0.09\delta_T \tag{26}$$

It is anticipated that for $\beta < \beta_R$, the high resolution of the scalar field allows for neglecting the R_v term in Eq. (11). This is now verified from a manufactured scalar field solution.

4 Method of Manufactured Solution (MMS) for SGS scalar energy scaling

Canonical or synthetic problems have been used in various contexts to analyze the response of numerical methods to physical signals [11, 12]. To study SGS scalar energy, a scalar field may be manufactured so that it mimics the flickering of a scalar iso-surface within a turbulent flow. This is done by randomly moving over the characteristic length δ_T a one-dimensional zero-to-one scalar distribution that reads,

$$\phi(x,t) = \mathscr{F}_L(x - x_L(t)) \tag{27}$$

$$x_L(t) = x_o + \xi(t)\delta_T \tag{28}$$

with

$$\mathscr{F}_L(x) = \frac{1}{2}(1 - \tanh[2(x_o - x)/\delta_L]) \tag{29}$$

x_o is the reference position of the time average scalar signal, $x_L(t)$ is the position of the scalar signal within the manufactured turbulent scalar brush at time t, and, $\xi(t)$ is a random Gaussian distribution. δ_L is the characteristic thickness of the instantaneous scalar jump (Fig. 1).

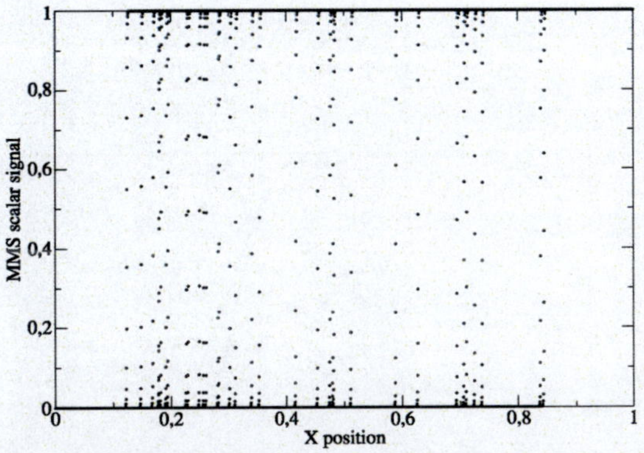

Fig. 1. Representative MMS scalar signal, $\delta_T/\delta_L = \alpha/\beta = 160$.

The corresponding time averaged signal is:

$$\langle \phi \rangle (x) = \int_{-\infty}^{x} P(x_L^*)dx_L^* \tag{30}$$

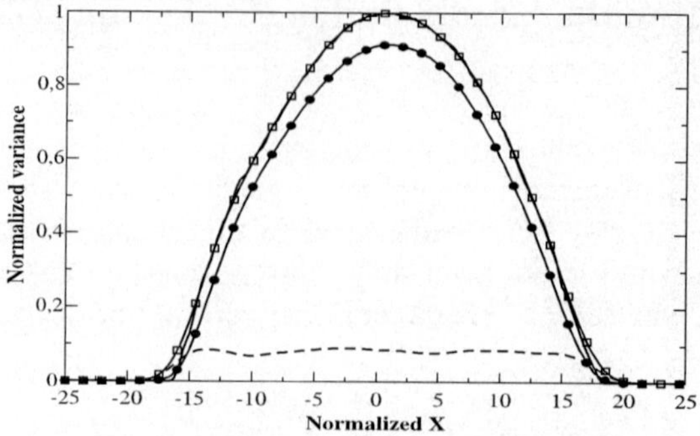

(a) $\alpha = \Delta/\delta_L = 5$, $\beta = \Delta/\delta_T = 0.15$

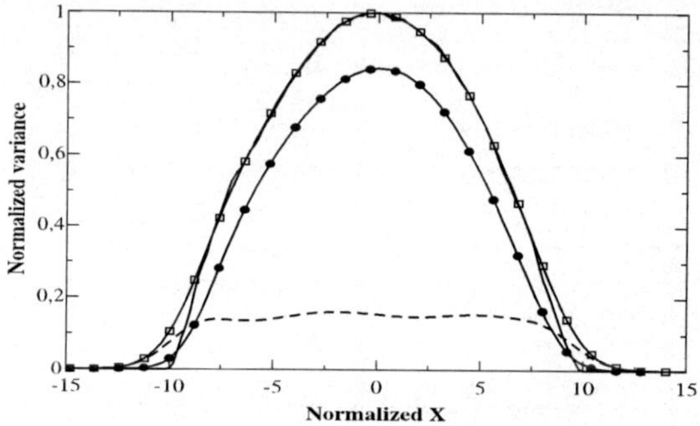

(b) $\alpha = \Delta/\delta_L = 3$, $\beta = \Delta/\delta_T = 0.25$

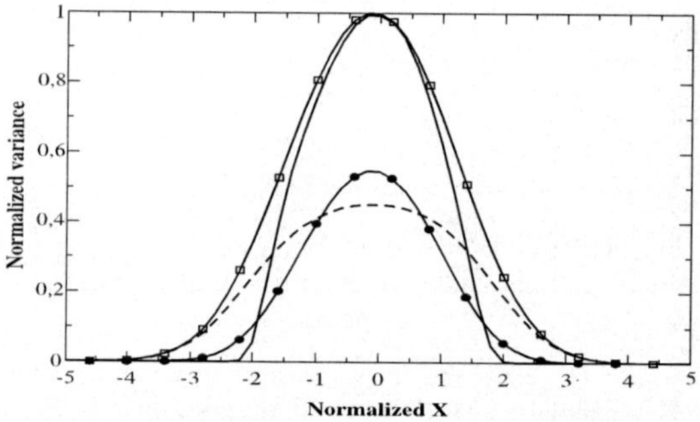

(c) $\alpha = \Delta/\delta_L = 3$, $\beta = \Delta/\delta_T = 0.75$

Fig. 2. Line: $\Phi_v/\max(\Phi_v)$. Line with circle: $T_R/\max(\Phi_v)$. Dash-line: $\langle\phi_v\rangle/\max(\Phi_v)$. Line with open square: $(T_R + \langle\phi_v\rangle)/\max(\Phi_v)$.

with $P(x_L^*)$ the probability density function (pdf) of the scalar signal positions. The manufactured solution is operated over a duration $t = T^c$, so that $P(x_L^*)$ is statistically converged.

The manufactured solution is convoluted with a Gaussian-type filter, the ratios $\alpha = \Delta/\delta_L$ and $\beta = \Delta/\delta_T$ are varied and the terms of relation (17) are collected.

5 Time averaged scalar variance signal and mesh optimization

Figure 2 shows profiles of Φ_v, T_R and $\langle\phi_v\rangle$ for three values of $\beta = \Delta/\delta_T$. As expected, the SGS variance grows with the filter size and for the smallest value $\beta = 0.15$, the relation $\Phi_v = T_R + \langle\phi_v\rangle$ is verified across the turbulent scalar brush, with however a small departure at both ends of the signal (*i.e.* for $\langle\phi\rangle \to 0$ and $\langle\phi\rangle \to 1$). Increasing β to 0.25, this departure spreads toward the center of the scalar brush, and for $\beta = 0.75$, there does not exist a single point where the SGS scalar energy is well captured by the time averaged SGS variance.

The residual term R_v of Eq. (17) is averaged across the turbulent scalar signal and normalized to get a single measure of its contribution for each α and β values,

$$[R_v] = \frac{\int_{-\infty}^{+\infty} R_v(x)dx}{\int_{-\infty}^{+\infty} \Phi_v(x)dx} \tag{31}$$

Figure 3 displays $[R_v]$ versus β for various α. Up to more than 20% of the scalar energy is contained in R_v for $\beta \approx 0.7$. Overall, when $\Delta \to \delta_T$, the difference between the SGS variance and the SGS energy grows. This trend does not depend much on the value of α. For $\beta < \beta_R = 0.1$, R_v represents less than 1% of Φ_v. Recovering the first estimation found in the above section from the resolution criterion.

A simple mesh optimization procedure emerges from these observations. After conducting a preliminary LES on a given mesh, a first set of statistics can be collected to compute the distribution $\delta_T(\underline{x})$. At locations where $\Delta > \beta_R\delta_T$, the mesh should certainly be refined to avoid approaching a too coarse LES regime, likely to be followed by spurious accumulation of residual contributions in the SGS energy budget. This is particularly true when the SGS scalar variance is used to presume the shape of a filtered probability density function, which is assumed to be representative of $\overline{r_\phi^2}$, the SGS fluctuations of the scalar field.

6 DNS minimum Reynolds number for R_v to vanish

Under the hypothesis $\delta_T \approx \ell_T$, the ratio α/β may be written:

$$\frac{\alpha}{\beta} = \frac{\Delta}{\delta_L} \times \frac{\delta_T}{\Delta} = \frac{\delta_T}{\eta_k} \times \frac{\eta_k}{\delta_L} \approx \frac{Re_{\ell_T}^{3/4}}{Ka^{1/2}} \tag{32}$$

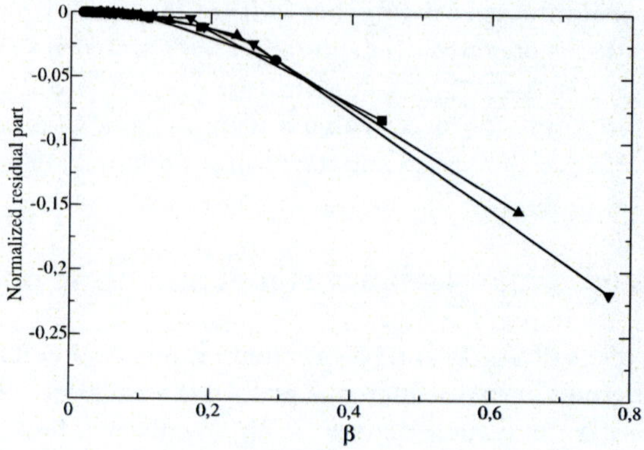

Fig. 3. Normalized residual term $[R_v]$ from MMS. Line with Circle: $\alpha = \Delta/\delta_L = 2$; Square: $\alpha = 3$; Triangle up: $\alpha = 4$; Triangle down: $\alpha = 5$.

where $Re_{\ell_T} = (u'\ell_T)/\nu$ is the turbulent Reynolds number based on ℓ_T, u' is the characteristic velocity fluctuation and ν is the kinematic viscosity. $Ka = (\delta_L/\eta_k)^2$ is a Karlovitz number defined as the ratio between the characteristic thickness of the scalar signal and η_k, the Kolmogorov scale. If the scalar is used to track a thin reaction zone, as for instance in the case of a premixed flame, δ_L is fixed by thermochemistry coupled with molecular diffusion and in the thin reaction zone turbulent combustion regime, $1 < Ka < 100$ [13]. It is therefore usual to observe values of $Ka > 1$ in real combustion systems.

Let us consider a free round jet whose bulk velocity is U_b and its injection diameter is D, the Reynolds number of this jet may be written:

$$Re_D = \frac{U_b D}{\nu} = \frac{U_b}{u'} \frac{D}{\ell_T} Re_{\ell_T} \tag{33}$$

With these relations, the condition $\beta < \beta_R$ brings the following constraint on the Reynolds number,

$$Re_D > Re_{D_R} = \frac{U_b}{u'} \frac{D}{\ell_T} \left(\frac{\alpha}{\beta_R} Ka^{1/2} \right)^{4/3} \tag{34}$$

For $U_b/u' = 10$, $D/\ell_T = 1$ and $\beta_R = 0.1$, this relation becomes:

$$Re_{D_R} = 215 \left(\alpha Ka^{1/2} \right)^{4/3} \tag{35}$$

Based on Eq. (35), Fig. (4) provides α values depending on Ka, that could be used to define a filter size $\Delta = \alpha\delta_L$ to perform *a priori* LES tests from a DNS of a round jet of bulk Reynolds number Re_{D_c}, ensuring that SGS scalar variance equals the SGS scalar energy. It is seen that for $Ka = 10$ and $\Delta = 10\delta_L$, the jet Reynolds number should be above 10,000, and even greater values of Re_D are required to filter the scalar signal above 10 times its characteristic thickness.

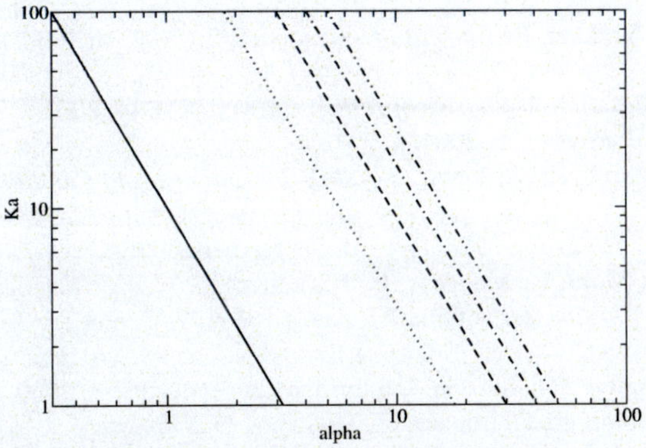

Fig. 4. α and Ka values imposed by Eq. (35) for $Re_{D_R} = 1000$ (line), $Re_{D_R} = 10{,}000$ (dotted), $Re_{D_R} = 20{,}000$ (dashed), $Re_{D_R} = 30{,}000$ (dotted-dashed) $Re_{D_R} = 40{,}000$ (double dotted-dashed).

7 Summary

Time averaging of SGS unresolved scalar fluctuations does not always sum up to zero and this has implication for scalar SGS energy. For instance, the SGS scalar variance is not systematically equal to the SGS scalar unresolved energy. However, from a simple scaling analysis and using a scalar manufactured solution, it is shown that applying a constraint on scalar resolution, the usual decomposition of scalar energy into resolved and SGS parts can be time averaged to recover the full time averaged scalar energy.

Typically, the LES filter size should stay ten time smaller than the characteristic length of the time averaged scalar signal. This result is useful to guide in scalar LES mesh optimization and to determine the minimum Reynolds number that DNS should feature to allow for accurate SGS energy *a priori* tests using SGS scalar variance. This analysis can be extended to cases where the flow mass density evolves with scalars [14].

References

1. Sagaut, P. *Large Eddy Simulation for Incompressible Flows: An Introduction*. Springer-Verlag, Berlin Heidelberg, 2nd edition, 2001.
2. Moin, P. *Int. J. Heat Fluid Flow*, 23(6):710–720, 2002.
3. Lesieur, M., Métais, O., Comte, P. *Large-Eddy Simulations of Turbulence*. Cambridge University Press, Cambridge UK, 2005.
4. Landenfeld, T., Sadiki, A., Janicka, J. *Flow, Turbulence and Combustion*, 68(2):111–135, 2002.
5. Domingo, P., Vervisch, L., Veynante, D. *Combust. Flame*, 152(3):415–432, 2008.
6. Vreman, A. W., Albrecht, B. A., van Oijen, J. A., Bastiaans, R. J. M. *Combust. Flame*, 153(3):394–416, 2008.

7. Veynante, D., Knikker, R. *Journal of Turbulence*, 7 (DOI: 10.1088/1468-5248/5/1/037), 2006.
8. Geurts, B. J., Fröhlich, J. *Physics of Fluids*, 14(6):L41–L44, 2002.
9. Pope, S. B. *N. J. Physics*, 6, 2004.
10. Celik, I. B., Cehreli, Z. N., Yavuz, I. *ASME Journal of Fluid Engineering*, 127:949–958, 2005.
11. Roache, P. J. *J. Fluids Eng*, 124(1):4–10, September 2002.
12. Domino, S. In Moin, P., Mansour, N. N., editors, *Studying Turbulence Using Numerical Simulation Databases*, volume XI, pages 163–177. Center for Turbulence Research, Stanford, 2006.
13. Peters, N. *Turbulent Combustion*. Cambridge University Press, 2000.
14. Vervisch, L., Domingo, P., Lodato, G., Veynante, D. *Submitted*.

Accuracy, Reliability and Performance of Spray Combustion Models in LES

Srikant Srinivasan, Andrew G. Smith, and Suresh Menon

School of Aerospace Engineering, Georgia Institute of Technology
270 Ferst Drive, Atlanta, GA, 30332-0150, USA
suresh.menon@ae.gatech.edu

Summary. The Large Eddy Simulation (LES) approach of tracking predetermined sets of droplets via individual computational parcels is assessed for its accuracy and applicability in the dilute regime. Non-reacting particle dispersion in isotropic turbulence and turbulent temporal mixing layers are initially investigated for a range of particles in parcels and Stokes numbers. It is found that the parcel approximation may be valid for a range of Stokes numbers depending upon the particle dispersion growth rate. Spray combustion in the wake of a bluff-body and swirl stabilized combustion in a gas turbine combustor are also studied and results suggest that a limited range of particles per parcels and cutoff radius can be used safely without incurring excessive errors in the predictions.

Key words: Large-eddy Simulation, Spray dispersion, Turbulent Combustion

1 Introduction

Simulation of spray combustion using LES poses modelling, computational and numerical challenges some of which are interrelated. Both Eulerian and Lagrangian methods are used to solve for the dispersed phase within an Eulerian gas phase. The Eulerian approach is computationally less intensive but poses problems in the application of boundary conditions, loss of droplet-identity and subgrid closure. The Lagrangian formulation is computationally more intensive since particles are explicitly tracked. However, applying boundary conditions is easier, polydisperse sprays can be simulated, and closures for particle-subgrid interaction can be included. Most importantly, comparison with droplet data is facilitated due to its discrete nature. In this paper, the Lagrangian method will be discussed.

When using the Lagrangian approach to simulate realistic fuel flow rate, particles of the order of 10^6 may have to be tracked [1], and this can be computationally intractable for most applications. To reduce the cost, statistically similar group of droplets are tracked by a "parcel" [2, 3, 4]. In this approach, the center of mass of parcel is tracked so that its constituent particles assume the properties of the parcel. This permits substantial computational savings while capturing the global features of droplet dispersion and combustion, but can also lead to erroneous predictions.

M.V. Salvetti et al. (eds.), *Quality and Reliability of Large-Eddy Simulations II*, ERCOFTAC Series 16, DOI 10.1007/978-94-007-0231-8_20,

Another cost saving assumption in reacting flows is to assume that particles or parcels below a cut-off radius, R_c undergo phase change abruptly. A larger R_c substantially reduces the number of small particles to be tracked but can also introduce errors if it is chosen too large. Although other simulation parameters such as grid resolution, test conditions, and geometrical complexity are all important, the sensitivity of the results to the chosen values for the particles-per-parcel ratio, ξ and the cut-off radius R_c is the primary focus of this paper.

2 Formulation

Using Favre-filtering, $\tilde{f} = \overline{\rho f}/\bar{\rho}$, the compressible LES equations for mass, momentum, energy, species and subgrid kinetic energy (k^{sgs}) are:

$$\frac{\partial}{\partial t}(\Theta\bar{\rho}) + \frac{\partial}{\partial x_j}(\Theta\bar{\rho}\widetilde{u}_j) = \tilde{\dot{\rho}}_s \tag{1}$$

$$\frac{\partial}{\partial t}(\Theta\bar{\rho}\widetilde{u}_i) + \frac{\partial}{\partial x_j}\left[\Theta\left(\bar{\rho}\widetilde{u}_i\widetilde{u}_j + \bar{p}\delta_{ij} - \bar{\tau}_{ij} + \tau_{ij}^{sgs}\right)\right] = \tilde{\dot{F}}_{s,i} \tag{2}$$

$$\frac{\partial}{\partial t}\left(\Theta\bar{\rho}\widetilde{E}\right) + \frac{\partial}{\partial x_i}\left\{\Theta\left[\left(\bar{\rho}\widetilde{E} + \bar{p}\right)\widetilde{u}_i + \bar{q}_i - \widetilde{u}_j\bar{\tau}_{ij} + H_i^{sgs} + \sigma_i^{sgs}\right]\right\} = \tilde{\dot{Q}}_s \tag{3}$$

$$\frac{\partial}{\partial t}\left(\Theta\bar{\rho}\widetilde{Y}_k\right) + \frac{\partial}{\partial x_j}\left[\Theta\left(\bar{\rho}\widetilde{Y}_k\widetilde{u}_i - \bar{\rho}\widetilde{Y}_k\widetilde{V}_{i,k} + \Phi_{i,k}^{sgs} + \theta_{i,k}^{sgs}\right)\right] = \tilde{\dot{S}}_{s,k} + \Theta\bar{\dot{\omega}}_k \tag{4}$$

$$\frac{\partial}{\partial t}(\Theta\bar{\rho}k^{sgs}) + \frac{\partial}{\partial x_i}\left[\Theta\left(\bar{\rho}\widetilde{u}k^{sgs}\right)\right] = \frac{\partial}{\partial x_i}\left(\bar{\rho}C_\mu\bar{\Delta}\sqrt{k^{sgs}}\Theta\frac{\partial k^{sgs}}{\partial x_i}\right) - \Theta\left(\tau_{ij}^{sgs}\frac{\partial\widetilde{u}_i}{\partial x_j} + C_\varepsilon\bar{\rho}\frac{(k^{sgs})^{3/2}}{\bar{\Delta}}\right) + F_{s,k} \tag{5}$$

Here, Θ is the fluid volume fraction, and ρ, u_i, E, p, Y_k and $\dot{\omega}_k$ are respectively, the density, the velocity components, the total energy per unit mass, the pressure, the kth species mass fraction and production rate. The closure of the subgrid quantities (denoted by superscript sgs) is obtained by a localized dynamic k^{sgs} model and given elsewhere [2, 5]. The dispersed phase source terms $\tilde{\dot{\rho}}_s$, $\tilde{\dot{F}}_{s,i}$, $\tilde{\dot{Q}}_s$, $\tilde{\dot{S}}_{i,k}$ and $F_{s,k}$ are discussed further below.

For reaction rate closure, a subgrid Eddy Breakup model (SEBU) [6] and a more comprehensive subgrid mixing- combustion closure based on the linear-eddy mixing model [2, 3, 5] have been used in the past and comparisons between the models as well as with experimental data have been reported. For the present study, the SEBU closure is employed due to its computational ease and since the current focus is only on the sensitivity of the modeling of the dispersed phase. This study considers a $n-$heptane-air mixture, with a global one-step mechanism for the combustion cases.

The equations for the dispersed phase are solved in the Lagrangian frame of reference. For injector applications, primary and secondary break-up models should be included [5] but since these models bring with them additional empirical parameters and uncertainties, we use the dilute approximation [7] here. It will be shown later that except for the immediate vicinity of the injectors in the reacting cases, this assumption is reasonable and valid.

The Lagrangian governing equations for the dispersed phase [8] are:

$$\dot{x}_{i,d} = u_{i,d} \tag{6}$$

$$\dot{u}_{i,d} = (3C_D\mu Re_d)/\left(16\rho_d r_d^2\right)(u_i - u_{i,d}) \tag{7}$$

$$\dot{m}_d = -\dot{m}_{Re_d=0}\{1 + [0.278\sqrt{Re_d}Sc^{1/3}]/[1 + 1.232/(Re_d Sc^{4/3})]^{1/2}\} \tag{8}$$

$$m_d\ C_d\dot{T}_d = h_d\pi d_d^2(\widetilde{T} - T_d) - \dot{m}_d L_v \tag{9}$$

Here, $\dot{()}$ indicates time derivative, and the subscripts, i and d correspond to coordinate index and the dispersed phase, respectively with $\dot{m}_{Red=0}$ in Eq. 8 representing surface vaporization rate under quiescent conditions as described in Patel and Menon [5]. The drag coefficient, C_D is obtained from empirical correlations and the Reynolds number is based on the slip-velocity between the gas and the droplets, droplet diameter (d_d) and the carrier viscosity [7]. These equations assume that the Kolmogorov length scale is larger than the size of the droplets.

The effect of turbulence on the dispersed phase, is included by the Stochastic Separated Flow (SSF) model [2, 7] in which, a velocity fluctuation is added to the resolved fluid velocities at the droplet locations according to $u_i = \widetilde{u}_i + X\sqrt{2k^{sgs}/3}$ with X, a random number generated from a uniform distribution with zero mean. It can be shown that this approach provides an important correction to the particle path due to the subgrid turbulent fluctuations. Finally, the source terms are obtained by a filtering process using a top hat filter:

$$\begin{pmatrix} \widetilde{\dot{\rho}_s} \\ \widetilde{\dot{F}_{s,i}} \\ \widetilde{\dot{Q}_s} \\ \widetilde{\dot{S}_{i,k}} \\ F_d \end{pmatrix} = \frac{1}{\Theta\hat{\Delta}^3} \begin{pmatrix} \sum n_d \dot{m}_d \\ \sum n_d \left[\dot{m}_d u_{i,d} - m_d \dot{u}_{i,d}\right] \\ \sum n_d \left[\dot{m}_d h_{v,s} - h_d \pi d_d^2 \left(\tilde{T} - T\right)\right] \\ \sum n_d \dot{m}_d \\ \sum n_d \left[\widetilde{\dot{F}_{s,i} u_i} - \widetilde{\dot{F}}_{s,i}\widetilde{u}_i\right] \end{pmatrix} \tag{10}$$

Here, n_d is the number of particles per parcel, $u_{i,d}$, ρ_d, r_d are the velocity, density and radius of the ith particle, respectively. The summation above is over all parcels within the computational volume defined by $\hat{\Delta}^3$.

3 Results and Discussions

The gas phase equations are discretized using a finite volume based scheme that is fourth order accurate in space and second order accurate in time. The particle equations are integrated using a fourth-order Runge-Kutta time-integration scheme. A massively parallel multi-block solver using a domain decomposition technique is employed for all studies. For two-phase LES, achieving true load-balance is challenging since large portions of the domain may have no droplets. Currently, there are two popular MPI communication methods to parallelize the Lagrangian phase: point-to-point and gather-scatter. In the former, only particles in the ghost cells are communicated to their adjacent neighbors. While the communications and the per-core memory requirement may scale linearly, book-keeping is time consuming. Furthermore, point-to-point communication involves three steps: (1) declaration of intent to communicate, (2) the size of the message to be communicated and (3) the actual message itself. These steps performed at the end of each time step induce additional latency. However, due to the smaller communication load relative to the method described below, the overall performance of this method is expected to be superior especially for a large number of parcels.

The gather-scatter communications (used here) is based on a master-slave paradigm that involves a designated processor taking charge of the majority of book-keeping operations. This requires entire droplet data to be gathered in that processor. To hold this information, a global buffer equal to the number of parcels in the system is allocated in all processors, and updated information on particle-location in the buffer are communicated back to all slaves at each time step. Doing so results in a core-invariant MPI communication time and book-keeping time. Although the gain in increasing the number of cores is only marginal (up to 20% speedup), the time for particle treatment nevertheless scales linearly with cores since this is a function of total parcel count of each core. However as mentioned above, the challenges in load-balancing the two phases simulataneously remain and are problem-dependent. Future studies will report on a more careful comparison of both these methods of communication in conjunction with optimal load-balancing strategies.

3.1 Isotropic Turbulence

LES of isotropic turbulence with a monodisperse distribution of non-evaporating inertial particles on a 32^3 grid is used to study the effect of Stokes number (defined as the ratio of droplet time to flow time, $St = \tau_p/\tau_f$) and ξ. Here, the particle response time is defined as $\tau_p = \left(d_p^2\rho_p\right)/(18\nu\rho)$ and the flow time is the Kolmogorov time. The flow is initialized by filtering a 256^3 DNS simulation at $R_\lambda = 140$, and a range of $St = 0.1 - 10$ are considered (although only a few are reported below). It is confirmed that neither the dilute assumption nor the particle size to Kolmogorov ratio limitation are violated. Four ξ cases: 1 (baseline), 2, 4 and 8 are considered in such a way that the total mass and volume of the dispersed phase remains constant.

For the cases studied, no appreciable differences are seen in the total kinetic energy decay of the gas phase. St effects are clearly seen and the trends match those of previous studies [9], e.g., small particles reduce the energy decay rate, larger particles increase the energy decay rate, and somewhere in-between there is no effect. The ξ also has very little effect on global particle velocity statistics but does effect the spatial distribution of particles within the domain. To quantify the differences the root mean square of particle number per cell [10]: $N_{rms} = \sqrt{1/N_c \sum_{i=1}^{N_c} N_i^2}$ is computed, where N_i is the number of particles inside a cell and N_c is the number of non-vacant cells. Figure 1 shows that based on dispersion characteristics, the parcel approximation is more valid for larger St. Though these differences in dispersion have little effect on the gas field for non-evaporating particles, they could be significant for reacting case since the location of evaporating fuel droplets impacts fuel-air mixing and the flame location. These issues are revisited below in the reacting studies.

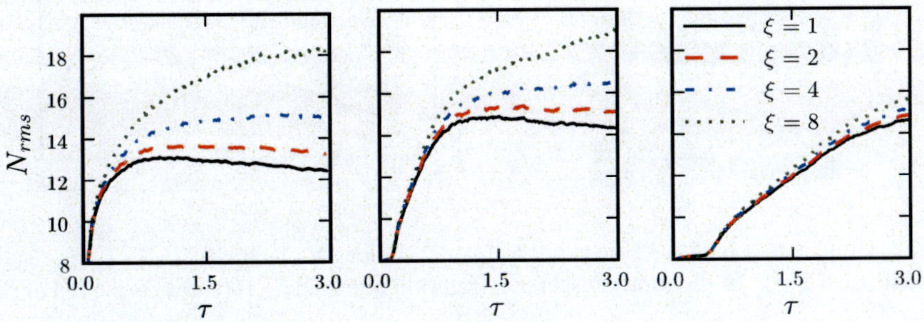

(a) $St = 0.1$ (b) $St = 1.0$ (c) $St = 10.0$

Fig. 1. N_{rms} for isotropic turbulence

3.2 Temporal Mixing Layer

The TML offers a simple test case to study the effect of large coherent structures on particle dispersion and, cases similar to [10] are simulated. A $4\pi^3$ box with a 64^3 mesh, and an initialization that combines the most unstable mode with 3D isotropic fluctuations [11] is used. The length scales and time scales are respectively chosen based on the initial vorticity thickness, δ_w and the free-stream velocity difference, ΔU. Three cases are chosen at $St_0 = 0.1$, 1 and 10 [$St_0 = \tau_p/(\delta_w/\Delta U)$] with an initialization of $64 \times 8 \times 64$ core of cells with monodisperse parcels. The behavior of parcel-dispersion as the initial shear-layer rolls up into coherent structures is St dependent; Parcels with small St follow the carrier phase more closely and therefore penetrate the region within the coherent structures, whereas parcels with larger St tend to accumulate in regions of high strain and low vorticity. The dispersion patterns shown in Fig. 2(a)] for different St are in good qualitative agreement with previous studies [10]. The distribution of droplets colored with their local St [$= \tau_p/(\delta_w/u_{rel})$] are shown for different St_0 in Fig. 2(a). The dispersion is quantified for different St_0 for the $\xi = 1$ case in Fig. 2(b), and calculated based on the average radii of separation of particle-groups originating from the same cell as a function of time. Eighty-five percent of all particle-groups have separation distances less than r_{85}, which is plotted on the y-axis against the normalized time. It is seen that the two lower St_0 cases disperse at a relatively rapid rate compared with the $St_0 = 10$ case. The slow growth rate of the $St_0 = 10$ case suggests that the parcel approximation for such heavy particles may be justified for a larger fraction of their overall residence time, whereas for the lighter particles the parcel approximation is valid for a relatively smaller fraction of their residence time. These results are in agreement with observations of Section 3.1. The general trends of Fig. 3 are comparable with the isotropic case (Fig. 1),

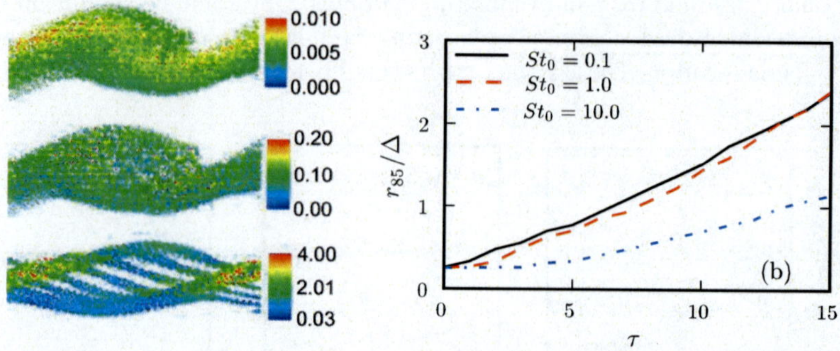

Fig. 2. Droplet distribution where every 4th parcel is plotted(left) for $St_0 = 0.1$ (top), $St_0 = 1$ (middle) and $St_0 = 10$ (bottom); Dispersion (right) in terms of r_{85} (see text) is plotted against time.

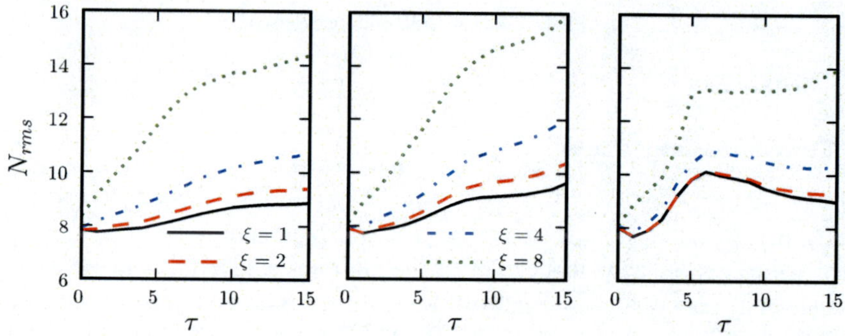

(a) $St = 0.1$ (b) $St = 1.0$ (c) $St = 10.0$

Fig. 3. N_{rms} for the temporal mixing layer

though relatively more error is seen for $\xi = 8$ for $St_0 = 10$. This can be explained by Fig. 2(a) showing $St \sim St_0/10$, differing from the isotropic case where $St \sim St_0$. Hence for $St_0 = 10$, there is large error for $\xi = 8$ *similar* to what is seen for $St_0 = 1.0$ in the isotropic case.

In the following two sections we address these issues by considering vaporization and combustion in relatively complex combustor configurations.

Bluff Body Stabilized Combustion

A blunt bluff-body of height D in a rectangular duct is considered with fuel injected through diametrically opposite injectors placed on the bluff-body shoulder with an injector radius of $0.014D$ located a distance, D upstream of the trailing edge. Characteristic boundary conditions for inflow and outflow, periodic conditions in spanwise, and adiabatic walls are used except for the bluff-body base where an isothermal condition of $T = 2100$K is imposed. The fuel mass flow rate is 0.576 g/s based on stoichiometric composition at 870K and 1 atm. The injection velocity and temperature are 10 m/s and 286K. A log-normal distribution with a Sauter Mean Diameter (SMD) of 30μm and σ of 0.5 is used for the spray. The computational domain consists of

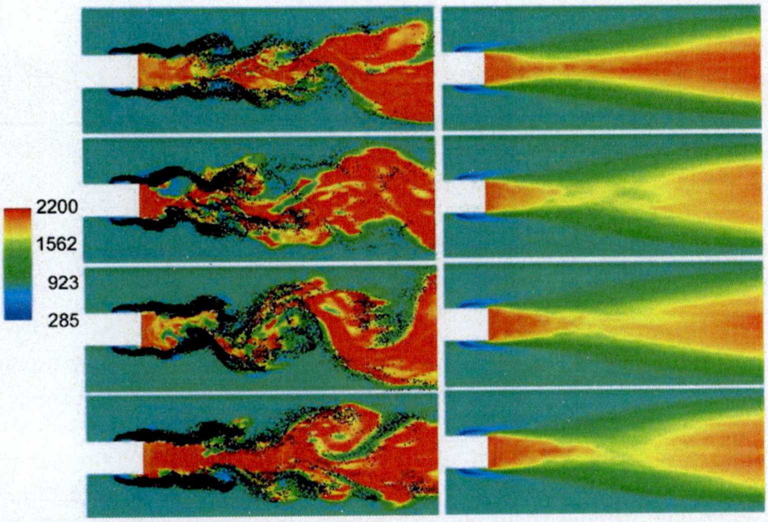

Fig. 4. Instantaneous (left) and time averaged (right) (over 2τ) temperature contours for Case 1 to 4 (top to bottom).

nearly 2.4M cells with grid clustering in regions of high shear. The dilute approximation is violated only in the immediate vicinity of the injectors in approximately 300 LES cells per injector. To allow for proper mass distribution, a filter width of $\hat{\Delta} = 3\Delta$ is employed near the injectors. Dispersion of droplets and the preheating effects result in satisfying the dilute regime requirements in the rest of the domain. Four cases are discussed: Case 1: $\xi = 8$, $R_c = 4$ μm, Case 2: $\xi = 8$ and $R_c = 1$ μm, Case 3: $\xi = 4$ and $R_c = 4$ μm, and Case 4: $\xi = 4$ and $R_c = 1$ μm. The instantaneous (overlaid with droplet distribution), and time averaged temperature distributions are shown in Figs. 4(a) and 4(b), respectively. It is seen that for all four cases, droplets tend to follow the peripheries of the large scale structures in the wake, and 95% of droplets are consumed within $4D$. Two distinct burning regions are identified in Fig. 4(b): a near-wake ($O(1-2D)$) just downstream of the recirculation and a far-wake region. Analysis of flame-brush plotted over droplet distribution (not shown) shows that particles near cutoff radii are concentrated in a v-shaped near-wake region within $1D$ whereas the large droplets are concentrated along the jet trajectory. However, little or no burning occurs in these regions hinting at an indirect effect of the cut-off radius through vaporization. The entrainment and mixing of the vaporized fuel in the region in-between leads to subsequent burning further downstream. The entrainment of droplets into the wake depends on the droplet St which in this case depends on the injection procedure (a small ξ leads to a better sampling of

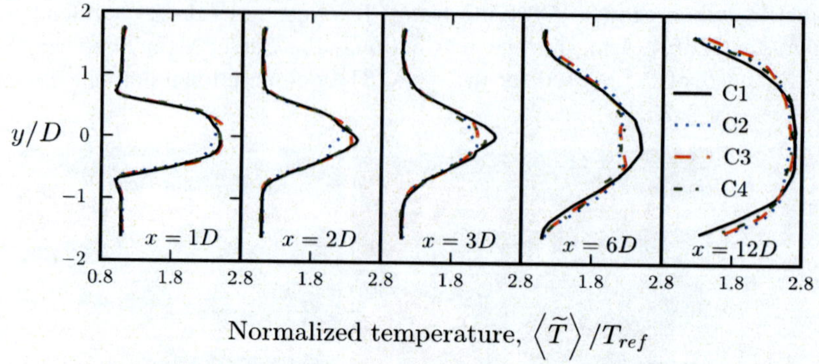

Fig. 5. Mean temperature profiles at different axial stations

the distribution) and the cut-off radius (as in the TML, small droplets follow flow-field closely). The combination of these two factors result in the observations seen in Figs. 4(a) and 4(b).

The Cases 1-4 involve tracking of 250,000, 310,000, 600,000 and 650,000 parcels and computational times of 1, 1.08, 1.9 and 2.16 units, respectively. The time-averaged temperature distribution for these cases are plotted in Fig. 5. It is seen that while some near-field differences are seen among the cases, at $x = 12D$ all cases except Case 1 collapse indicating reasonable exit plane profiles (typically a key design parameter). Thus, a relatively inexpensive engineering simulation with minimal error (when measured near the exit) can be performed with low R_c and a high ξ for this class of problems. The bluff-body configuration may qualify when considering the fact that dispersion of droplets is relatively limited in a cross-flow environment when compared to those found in swirling flows. Design considerations such as the distance of the entrainment region from the injectors and the droplet life-time need to be taken into account while fixing the cut-off radius.

3.3 Lean Direct Injection Combustor

The final case is LES of a swirl stabilized lean direction injection (LDI) combustor. This rig operates at atmospheric pressure with an air mass flow rate of 0.49 kg/min and a fuel mass flow rate of 0.0276 kg/min giving a global equivalence ratio of $\phi = 0.75$. Details of the combustor geometry can be found in [3]. Fuel droplets are injected based on a log normal distribution with SMD $= 25\,\mu$m. The R_c effect is minimal with the choice of $R_c = 0.1$ nm.

We compare results from three cases with $\xi = 1, 4$ and 8. The N_{rms} for $\xi = 1$ near the injector is slightly higher than one while downstream, it is closer to unity, indicating a more disperse flow. For $\xi = 8$, N_{rms} becomes 8 showing that the parcel representation is accurate, i.e., the parcels are spread out sufficiently enough to not compound the inherent errors associated with the parcel approximation. The average St based on slip velocity and swirl cup diameter for all ξ values were approximately

10 for all particles in the domain. Based on the problems analyzed earlier, this also suggests the parcel approximation is appropriate. At any given instant, over 90% of

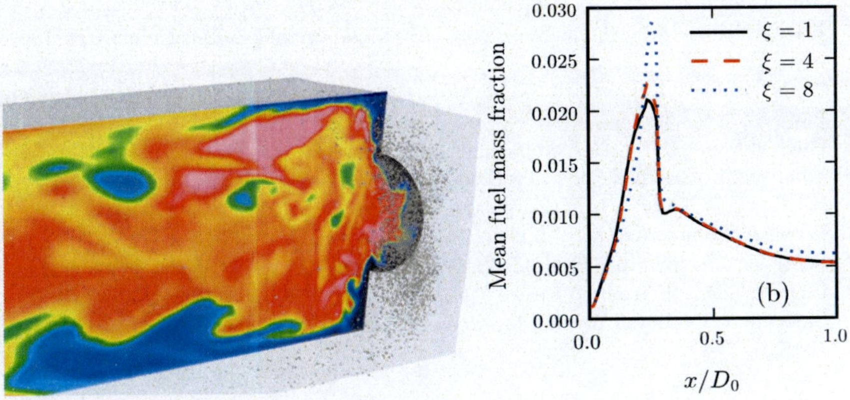

Fig. 6. Instantaneous temperature and particle locations; Mean fuel mass fraction

the particles vaporize and mix within one swirl cup diameter of the injector. Figure 6(a) shows a snapshot of the temperature and droplets in the combustor, while Fig. 6(b) shows the mean fuel mass fraction as a function of axial distance for all ξ. The average error for $\xi = 4$ is approximately 1.5% whereas the average error for $\xi = 8$ is nearly 12%. While the $\xi = 8$ simulation is within engineering accuracy limits, the $\xi = 4$ is more than twice as accurate for around twice the cost. These results and previous work with a kerosene fueled LDI using between $\xi = 1$ and $\xi = 5$ [3] show that swirl-stabilized combustors are ideal candidates for the parcel approach. The inherent compactness of the flame ensures relatively short drop lifetimes and high velocity swirling air at the injection point yield $St > 1$.

4 Conclusions

This paper discusses some of the issues for spray combustion LES. Both non-reacting and reacting cases are studied. It is shown in the non-reacting canonical cases that the dispersion within clusters of initially close single particle-parcels depend on the Stokes number. Simulations of spray combustion in a swirl combustor and bluff-body stabilized flame holder reveal the effects of the parcel/finite-drop size assumptions. For the bluff-body flame, with proper choice of R_c and ξ, the computational time can be brought down significantly at a relatively small expense of accuracy. The prolongation of droplet-life through a reduced R_c is especially helpful in simulations with limited dispersion characteristics thereby justifying the choice of a relatively high parcel density. In the LDI combustor, where the cutoff is near zero, increasing the ξ value is also shown to be acceptable in the 1-5 range.

References

1. Oefelein J, (2006) Progress in Aerospace Sciences 42:2–37
2. Menon S, Patel N (2006) AIAA Journal, 44(4):709–723
3. Patel N, Kirtaş M, Sankaran V, Menon S (2007) Proceedings of the Combustion Institute, 31:2327–2334
4. Apte S, Mahesh K, Gorokhovski M, Moin P (2009) Proceedings of the Combustion Institute 32:2257–2266
5. Patel N, Menon S (2008) Combustion and Flame, 153:228–257
6. Fureby C, Löfström C (1994) Proceedings of the Combustion Institute, 27:1257–1264
7. Faeth G (1987) Progress in Energy and Combustion Science 13(4):293–345
8. Crowe C, Sommerfeld M, Tsuji Y (1997) CRC Press, Boca Raton
9. Ferrante A, Elgobashi S (2003) Physics of Fluids, 15(2):315–329
10. Ling W, Chung J, Troutt T, Crowe C (1998) Journal of Fluid Mechanics, 358:61–85
11. Metcalfe R, Orszag S, Brachet M, Menon S, Riley J (1987) Journal of Fluid Mechanics, 184:207–243

LES of Triangular-stabilized Lean Premixed Turbulent Flames with an algebraic reaction closure: Quality and Error Assessment

B. Manickam[1], J. Franke[2], S. P. R. Muppala[3], and F. Dinkelacker[1]

[1] Institut für Technische Verbrennung, Leibniz Universität Hannover, Germany
manickam@itv.uni-hannover.de,dinkelacker@itv.uni-hannover.de
[2] Institut für Fluid- und Thermodynamik, Universität Siegen, Germany
franke@ift.mb.uni-siegen.de
[3] Faculty of Engineering, Kingston University, London, U.K
S.Muppala@kingston.ac.uk

Summary. In this LES study, an algebraic flame surface wrinkling model based on the progress variable gradient approach is validated for lean premixed turbulent propane/air flames measured on VOLVO test rig. These combustion results are analyzed for uncertainty in the solution using two quality assessment techniques.

1 Introduction

Lean premixed combustion technology offers reduction in NO_x and other pollutant emissions. Modelling of such combustion phenomena is a non-trivial task because of the interaction between turbulence and chemical reaction. In the last two decades, several reaction models namely eddy dissipation concept or turbulent flame speed closure [1] have been developed and tested for several flame configurations. In the present study, we use an algebraic flame surface wrinkling model [2] based on the reaction progress variable approach developed and validated against Bunsen-like flames and sudden expansion dump combustor is used here.

In general, the accuracy of numerical flame results depends substantially on the accuracy of the prediction of flow quantities. Therefore, in the present work, in addition to the investigation of flame characteristics using three turbulence closures and two combustion models, we also perform estimation of quality of these results using two different numerical grids.

The outline of the paper is as follows. In modelling part, the turbulence closures, reaction models and LES quality assessment methods are explained. In the next section, geometrical details, numerical mesh and boundary conditions are described. In Results and Discussion, simulated non-reacting and reacting flow results are compared with experimental data and analyzed for the quality and error of the solution. And, finally, of course, conclusions are drawn.

M.V. Salvetti et al. (eds.), *Quality and Reliability of Large-Eddy Simulations II*, ERCOFTAC Series 16, DOI 10.1007/978-94-007-0231-8_21,

2 Governing equations and modelling

In LES, the filtering process of Reynolds-based Navier-Stokes equations effectively filters out eddies whose scales are smaller than the filter width or grid spacing used in the computations. The resulting equations govern the dynamics of large eddies. The Favre filtering of the continuity and momentum equations results in [4][5][6][7]

$$\frac{\partial \overline{\rho}}{\partial x_i} + \frac{\partial}{\partial x_i}(\overline{\rho}\widetilde{u}_i) = 0 \tag{1}$$

$$\frac{\partial \overline{\rho}\widetilde{u}_i}{\partial t} + \frac{\partial}{\partial x_i}(\overline{\rho}\widetilde{u}_i\widetilde{u}_j) = -\frac{\partial \overline{p}}{\partial x_i} + \frac{\partial}{\partial x_i}\left[\tau_{ij}^{mol} - \tau_{ij}^{sub}\right] \tag{2}$$

$$\tau_{ij}^{sub} = \overline{\rho}\,(u_i u_j - \widetilde{u}_i\widetilde{u}_j) \tag{3}$$

Here the spatial commutation errors due to the varying, implicit filter width $\Delta = h = V^{1/3}$, where h is a characteristic grid width and V is the cell volume, have been neglected. The subgrid scale stresses are modelled with three different approaches, the standard Smagorinsky model with constant $C_s = 0.1$, the dynamic Smagorinsky model and a one equation model for the subgrid scale kinetic energy, K_{sgs}.

2.1 Premixed turbulent combustion model

A well-known approach to describe turbulent premixed flames is in terms of a combustion progress variable c, which is '0 ' in reactants and '1 ' in products. The transport equation for the Favre-filtered progress variable $\tilde{c}$, with gradient assumption for the scalar turbulent flux, is

$$\frac{\partial}{\partial t}(\overline{\rho c}) + \frac{\partial}{\partial x_k}(\overline{\rho u_k c}) = \frac{\partial}{\partial x_k}\left(\overline{\rho}\left(\frac{\nu_{sgs}}{Sc_{sgs}} + \nu\right)\frac{\partial \widetilde{c}}{\partial x_k}\right) + \overline{\dot{w}_c} \tag{4}$$

where ρ is the mean gas density, ν and ν_{sgs} are the molecular and subgrid kinematic viscosities, $Sc_{sgs}(= 0.7)$ is the subgrid Schmidt number and $\overline{\dot{w}_c}$ is the mean reaction source term, $\overline{\dot{w}_c} = \rho_u S_{L0} I_0 \Sigma$. Here S_{L0} is the unstretched laminar burning velocity, $I_0(= 1)$ the flame stretch factor and Σ the average flame surface density.

A RANS-based Algebraic Flame Surface Wrinkling (AFSW) model in which Σ is modelled with an algebraic relation for the flame-surface-wrinkling factor $A_T/\overline{A}$, with embedded pressure effects, is used

$$\overline{\dot{w}_c} = \rho_u S_L \frac{A_T}{\overline{A}} |\nabla \widetilde{c}| \tag{5}$$

the flame-wrinkling ratio transformed to subgrid quantities and the LES formulated reaction model is

$$\frac{A_T^{\Delta}}{\overline{A}} = \frac{S_T^{\Delta}}{S_{L0}} = 1 + \frac{0.46}{e^{(Le-1)}} Re^{\Delta 0.25} \left(\frac{u'_{sgs}}{S_{L0}}\right)^{0.3} \left(\frac{p}{p_0}\right)^{0.2} \tag{6}$$

Le - is the Lewis number of the fuel-air mixture
p/p_0 - normalized operating pressure
u'_{sgs} - subgrid turbulence fluctuation $u'_{sgs} = c_S\Delta\left|\widetilde{S}\right|$,
Re_t^Δ - subgrid Reynolds number $Re_t^\Delta = u'_{sgs}c_S\Delta/\nu$

In addition to the AFSW model, the Turbulent Flame speed Closure (TFC) of Zimont [1] is used in the LES context.

2.2 LES quality assessment

Two grid estimators

A few quality assessment methods are available for estimation of the uncertainty of LES results. Here, evaluation of data is performed using the Index of Quality for LES (LES_IQ), determined as a local quantity in the computational domain. In the following, quality assessment is based on the total kinetic energy relations. The LES_IQ is given as a function of total and resolved kinetic energies. More related details can be found in [9][10].

Combined model and grid variation

In the implicit filtering approach with $h = \Delta$, separation of individual contribution of modelling and numerics to the error is difficult as both strongly interact with each other. The approach assesses the quality of LES data by systematic grid and model variation. Again, more related details can be found in [11].

3 Geometry, grids and boundary conditions

The VOLVO test rig (Fig. 1) [3] has a channel of 1000 (25H) mm long, 240 mm (6H) wide and 120 (2H) mm high. The bluff body flame stabilizer is located approximately 318 mm from the inlet. It has a cross section of equilateral triangle with 40 mm (H) edge length and creates recirculation of hot burned gases which helps flame stabilization.

Initial cold flow conditions are inlet velocity $U_0 = 17\ m/s$, turbulence intensity 5% and temperature T at 288 K. For combustion flow, the inlet velocity and temperature of the premixed propane/air mixture with constant equivalence ratio of 0.6 at 1 bar are $U_0 = 35\ m/s$ and T = 600 K, at 1 atm. Inflow turbulence is generated with turbulence intensity of 5% and length scale 8.4 mm by using spectral code method. At the outlet a constant pressure is prescribed, with top and bottom as solid walls. In the lateral direction periodic conditions are imposed.

For the three-dimensional domain, two-block structured hexahedral meshes are created. Three numerical testing setups for computations use three types of grids having uniform maximum expansion ratio of 1.05 which are termed as fine, coarse

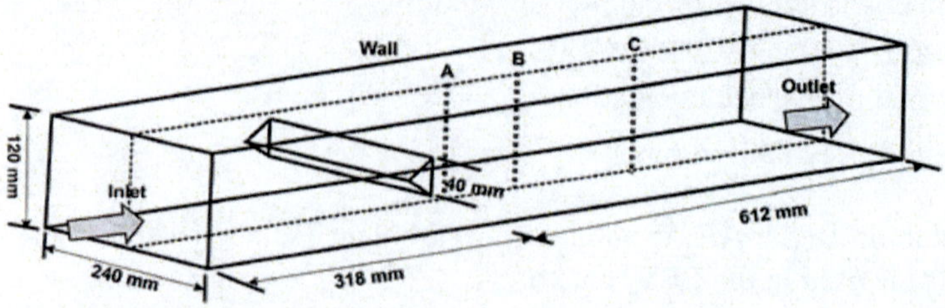

Fig. 1. Computational domain of VOLVO test rig, marked A, B and C data points are 38, 150 and 350 mm from the stabilizer

and very coarse with 2.4 million, 1.2 million and 300,000 grid points, respectively. With the first two grids contribute to the study of index of resolution quality and error assessment of LES results, the third grid type helps in the evaluation of effect of poor grid of same data. Simulations are carried out with FLUENT [12] and the two reaction models are implemented through user programming. For time and space discretization, implicit second order and bounded second order schemes are employed with convergence criteria and time steps of 1e-06s and 3e-05s, respectively, for CFL of 0.5. Each LES case is run for ten flow-through times of which the first half is used to get a quasi periodic flow and the second half for statistical averaging of output data.

4 Results and discussion

The LES results of non-reacting or cold flow are presented and discussed below, followed by combusting flow data. For cold flow, investigations are carried out in three stages: 1) by employing three popular turbulence subgrid Smagorinsky (SM), dynamic Smagorinsky (DS) and one-equation K_{sgs} (KSGS) closures with coarse grid, 2) with the Smagorinksy closure by grid variation between 1.2 and 2.4 million(2.4 *m*), and 3) with the Smagorinsky closure using very coarse grid. For sake of completeness, we present some interesting 3D URANS data to complement discussion of above results.

4.1 Non-reacting flow: subgrid scale closures

All the three turbulence subgrid scale closures whilst predict similar results for mean streamwise velocities at the axial positions A, B and C, the corresponding lateral mean velocities show deviation from each other especially at line B. At this position B, the KSGS model overpredicts significantly. These results are not presented here.

In Figure 2, the corresponding RMS velocities are shown. Both the DS and KSGS models overpredict the streamwise fluctuations close to the bluff body at line A. Here, the SM model shows better agreement with the experiments. At line B only

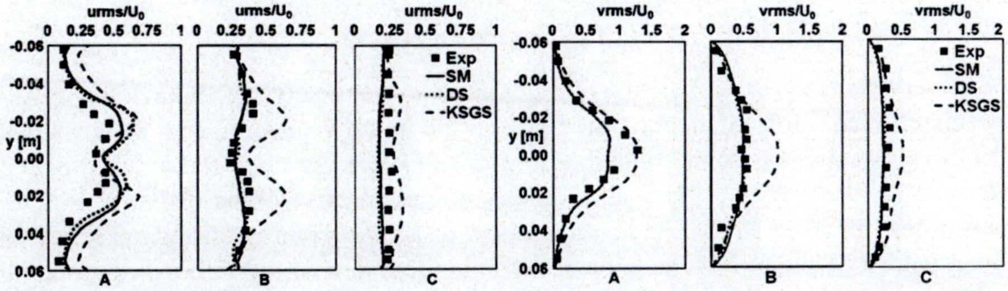

Fig. 2. Normalized RMS velocity $(urms/U_0)$ and $(vrms/U_0)$ at lines A, B and C for different subgrid scale models

the KSGS model over predicts $urms/U_0$, Figure 2. The lateral velocity fluctuations close to the bluff body are best reproduced by the KSGS model while the SM and DS model results are too small, Figure 2. Further away from the body these two models show better agreement with the experiments, while the KSGS model overpredicts $vrms/U_0$. URANS with standard k-ε model with 1.2 million grid points (1.2 *m*) predicts large differences in turbulence quantities in comparison with experiments [13].

4.2 Non-reacting flow: Influence of mesh resolution

The influence of the mesh resolution on the LES results is examined qualitatively for the standard Smagorinsky model. To that end, additional simulations were run on the fine mesh. These simulations results are compared with the ones obtained on the coarse mesh along the lines A, B and C. The local filter width changes from 2.1 *mm* to 1.65 *mm* at position A, from 2.7 *mm* to 2.15 *mm* and from 3.0 *mm* to 2.65 *mm*, at successive positions, respectively. Fig. 3 shows RMS $(urms/U_0, vrms/U_0)$ velocities. At position A noticeable difference in RMS velocities are observed between these two grids, while at positions B and C the grid influence diminishes. At position A, the $vrms/U_0$ peak could not be completely replicated in both mesh cases. However, further mesh refinement is likely to yield better prediction, as can be seen in Figure 3.

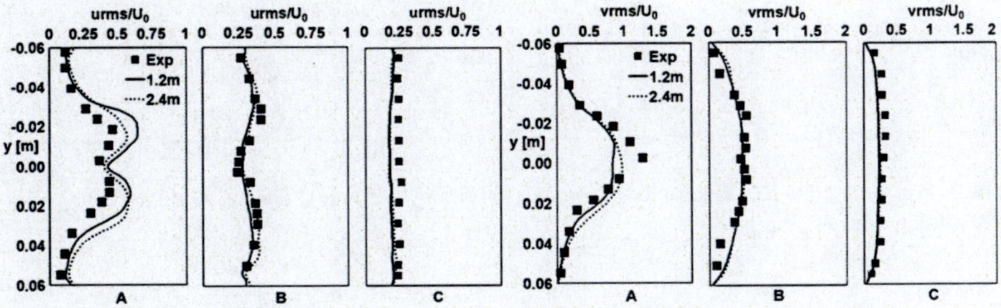

Fig. 3. Normalized RMS velocity $(urms/U_0)$ and $(vrms/U_0)$ at lines A, B and C for the standard Smagorinsky model for coarse and fine grids

4.3 Non-reacting flow: Effect of poor grid resolution

With very coarse grid, turbulent quantities are over predicted in comparison with experimental data. Instantaneous flow fields in the bottom image in Fig. 4 shows that the rotating eddies past the bluff body (of size H) are active in the axial direction of up to 8H downstream. Here, the eddies continue to exist longer distances as a result of low resolved strain rate leading to low dissipation rate, as discussed in [9]. In other words, owing to insufficient resolution in the spanwise direction, these rotating structures are not only incoherent but also disappear faster. Fig. 5, shows that normalized *urms* and *vrms* velocities are almost two times higher than the corresponding measured values in positions A and C. At position B, it is even higher by a factor of three.

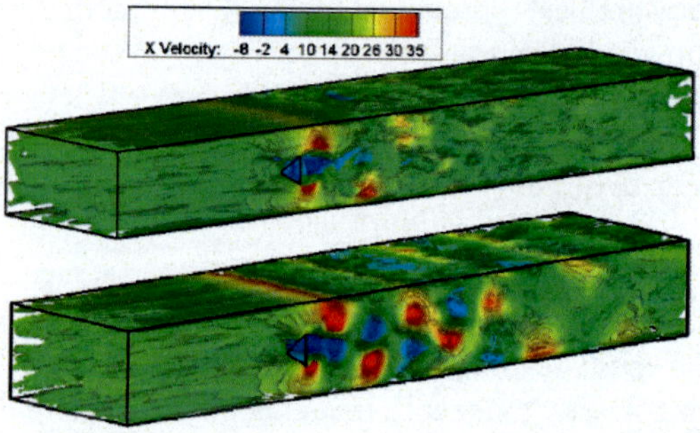

Fig. 4. Instantaneous isosurfaces of mean streamwise velocity with coarse grid of 1.2 m (top), and very coarse grid 300,000 (bottom) cells, Smagorinsky model

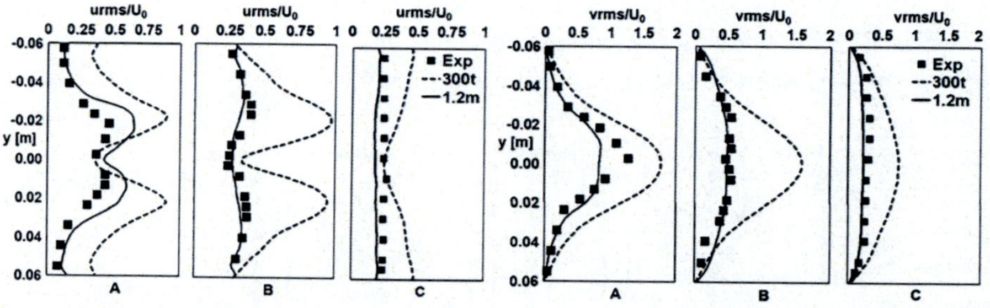

Fig. 5. Normalized u-RMS velocity $(urms/U_0)$ at lines A, B and C for the standard Smagorinsky model, very coarse grid

4.4 Reacting flows

Combusting flows in the bluff body stablized configuration are studied using the AFSW and the TFC reaction models with the Smagorinsky closure. It is interesting to compare the flow patterns using pressure fields past the stabilizer with that of cold case. Figure 6 depicts isosurfaces of pressure for both cases. For nonreacting flows, the negative pressure, with respect to reference value of 1 bar, is 150 *Pa* while it drops to as low as 900 *Pa* for reacting flows. In the latter case, a longer recirculation zone which is four times that in the former case, favours stabilization of triangular-stabilized flame.

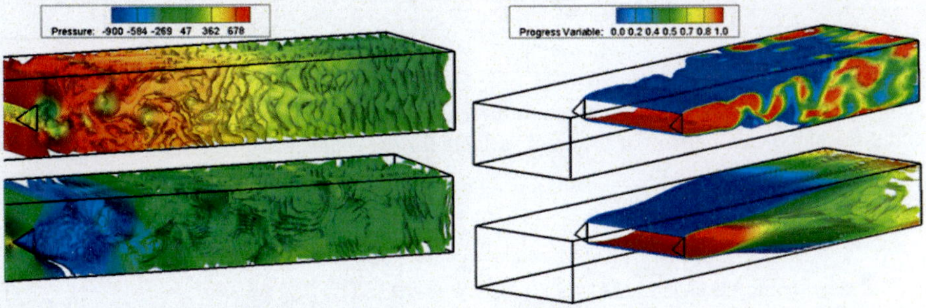

Fig. 6. Computed instantaneous isosurfaces of pressure for non-reacting flow (top-left), and instantaneous pressure (bottom-left), instantaneous and averaged isosurfaces of reaction progress variable (right) with coarse grid and AFSW model

Shown in Fig. 6 are the instantaneous as well as averaged reaction progress variable obtained with the AFSW reaction model. From these isosurfaces, the onset of vortex shedding behind the main combustion region is clearly visible. Close to the bluff body up to 4H, the flamelet is not disturbed by the length scales equal to bluff body dimension H. After distance of 4H along downstream, vortex shedding with three wrinkling periods expanding along width-wise and length-wise in the stream direction. Along downstream, within the length scale of H existence of small rotating eddies enhances fresh interaction between unburned and burned gases.

Figure 7 shows profiles of reaction progress variable in positions A, B and C from the AFSW and the TFC models with coarse grid. Both of them very well replicate experimental trends in first two positions A and B, but show larger differences for position C. The models show improved results with grid refinement.

For the very coarse grid situation, the recirculation region is 3H which is smaller than that observed with coarse grid. The poor resolution results pronounce that the attachment of the flame to the wall surface occurs relatively earlier. From Fig. 8, it is noticeable that due to relatively shorter region of combustion products, the phase of distribution and interaction of burned gas with fresh premixed fuel/air mixture diminishes, leading to under prediction of reaction rates at B and C.

4.5 LES quality assessment based on Celik et al. approach

Accuracy of the LES results is estimated using the quantity index of resolution quality LES_IQ using two popular techniques. First, the quality assessment technique by

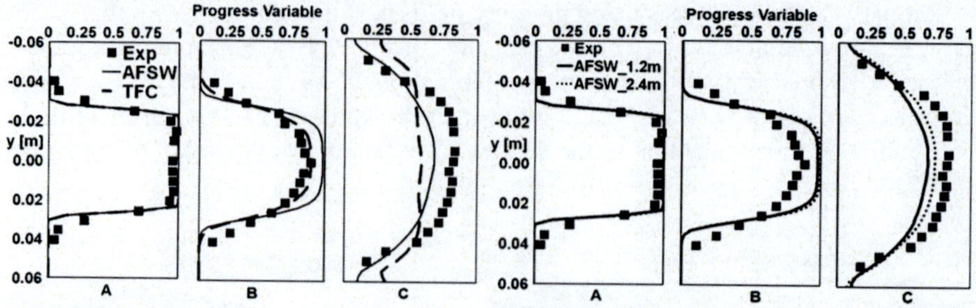

Fig. 7. Left-end three plots: Comparison of reaction progress variable using the AFSW and the TFC combustion models in combination the classic Smagorinsky turbulence closure. Right-end three plots: Comparison of progress variable for the coarse and fine grids with the AFSW and the Smagorinsky model

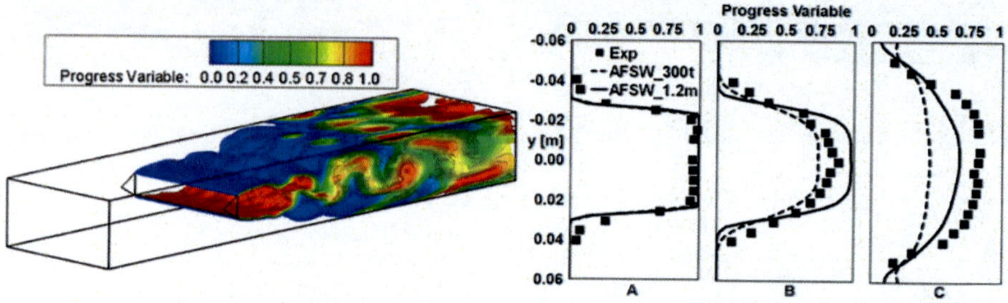

Fig. 8. Computed instantaneous isosurfaces of reaction progress variable with the AFSW reaction model for very coarse grid (left). The effect of grid on reaction progress variable, shown for coarse and very coarse grid points (right)

Celik et al. [9][10] is based on turbulent kinetic energy for combined data from two 1.2 *m* and 2.4 *m* grid cell cases. Figure 9 shows the quality of LES solution for the Smagorinsky model. The LES_IQ are calculated along y axis and averaged; for 2.4 *m* cells is 86%, 77% & 83% and for 1.2 *m* cells is 91%, 85% & 89% for positions A, B and C, respectively. This approach is conditioned for estimating the grid dependency error, where as it offers little choice for model errors.

Figure 9 gives some indication of necessary modifications that need to be carried out for mesh refinement around the flame zone where the quality drops down to 50-65%. For coarse grid at position B, where recirculation of hot gases terminates, LES_IQ falls to as low as 60%. For the farthest position C from the flame zone, LES_IQ again creeps up to 75-85%, perhaps because of decrease in turbulence level in this region.

4.6 LES error assessment based on the Klein approach

The three-grid based approach by Klein [11] gives more information on both numerical as well as model errors. The predicted errors are presented in Fig. 10 in absolute values, normalized with velocity. For non-reacting flow, both errors appear

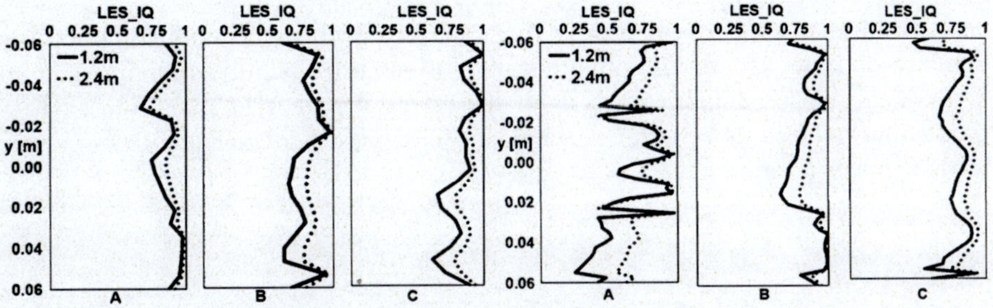

Fig. 9. Approach by Celik et al. prediction of LES_IQ for non reacting (left) and reacting flow (right) with 1.2m cells, Smagorinsky, AFSW models.

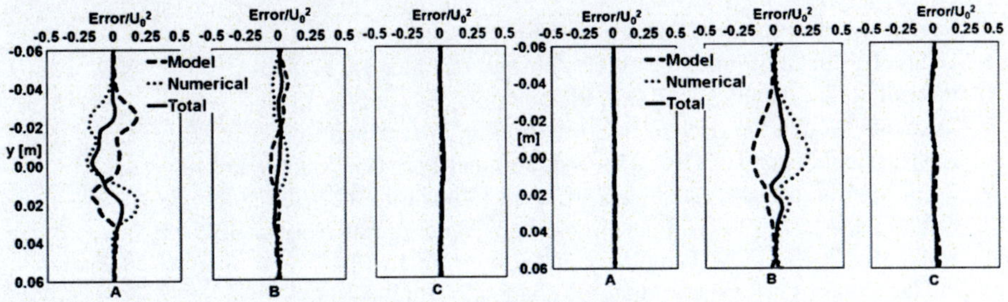

Fig. 10. Klein approach for predicting error with the Smagorinsky model for non reacting (left) and reacting flow (right) with AFSW model, 1.2 million cells.

to be considerably larger at position A close to bluff body. However, it appears to diminish along downstream sidetowards the outlet. These predicted error profiles are shown in Fig. 10. For reacting flows, the Klein approach shows that the total error increases from A to B due to shift in the recirculation zone towards the downstream side (Fig. 10).

5 Conclusion

The algebraic flame surface wrinkling model in the large-eddy simulation context was successfully compared and validated with the experimental data on VOLVO test rig triangular-stabilized flame configuration and with numerical results from the popular turbulent flame speed closure. The LES quality assessment of non-reacting and reacting flow data were performed using two techniques proposed by Celik et al. and by Klein. Whilst the non-combusting flows were well-resolved on the coarse mesh, it tends to show that even the fine mesh used in this study was found to be locally insufficient for the combusting flows. With the Celik approach, the LES index of resolution quality for the fine mesh decreased to approximately 50-60% in the flamelet region, whereas it was as high as 90% in cold flow scenario. The approach by Klein showed that depending on the recirculation region both numerical and modelling errors rise near the bluff-body region (38 mm from the stabilizer) in the non-reacting case, while they appear at relatively four times away from the stabilizing point in combustion flows as a result both preheating of unburned premixed mixture and reaction heat release. Finally, analysis of the effect of poor grid resolution on LES predictions showed that shorter attachment region past the flame stabilization zone leads to lower reaction rates.

References

1. Zimont V. L., and Lipatnikov A. N., (1995). Chem. Phys. Rep. 14: 993-1025.
2. Muppala S. P. R., Aluri N. K., Dinkelacker F., (2005). Com. Flam 140: 257-266.
3. Fureby C., (1995). AIAA, 33(7).
4. Sagaut P., Third edition 2006, Heidelberg: Springer.
5. Geurts B. J., and Fröhlich J., (2002). Physics of fluids 14: L41-L44.
6. Poinsot T., and Veynante, (2005). Philadelphia, Edwards.
7. Veynante D., and Vervisch L., (2002). Prog. in Energy and Combust. Sci. 28: 193-266.
8. Flohr P., and Pitsch H., CTR, Proceed. Summer Program 2000. Stanford.
9. Celik I., Cehreli Z. N., (2005) ASME, J. of Fluids Eng, 127(Sep): 949-958
10. Celik I., and Li J., (2005). Inter. J. for Num. Meth. In Fluids 49: 1015-1031.
11. Klein M., (2005). Flow Turbulence and Comb. 75: 131-147.
12. Fluent (2006). Fluent Incorporated. Lebanon, NH, USA.
13. Manickam B., Muppala S. P. R., Franke J., and Dinkelacker F., 4th ECM, 2009: Vienna, Austria.

Grid Effects on LES Thermo-Acoustic Limit-Cycle of a Full Annular Aeronautical Engine

Pierre Wolf[1], Laurent Y.M. Gicquel[1], Gabriel Staffelbach[1] and Thierry Poinsot[2]

[1] CERFACS, 42 Avenue G. Coriolis 31057 Toulouse Cedex, France
pierre.wolf@cerfacs.fr
[2] Institut de Mécanique des Fluides de Toulouse, Avenue C. Soulas 31400 Toulouse, France

Summary. Recent developments in large scale computer architectures allow Large Eddy Simulation (LES) to be considered for the prediction of turbulent reacting flows in geometries encountered in industry. To do so, various difficulties must be overcome and the first one is to ensure that proper meshes can be used for LES. Indeed, the quality of meshes is known to be a critical factor in LES of reacting flows. This issue becomes even more crucial when LES is used to compute large configurations such as full annular combustion chambers. Various analysis of mesh effects on LES results have been published before but all are limited to single-sector computational domains. However, real annular gas-turbine engines contain ten to twenty of such sectors and LES must also be used in such full chambers for the study of ignition or azimuthal thermo-acoustic interactions. Instabilities (mostly azimuthal modes involving the full annular geometry) remain a critical issue to aeronautical or power-generation industries and LES seems to be a promising path to properly apprehend such complex unsteady couplings. Based on these observations, mesh effects on LES in a full annular gas-turbine combustion chamber (including its casing) is studied here in the context of its azimuthal thermo-acoustic response. To do so, a fully compressible, multi-species reacting LES is used on two meshes yielding two fully unsteady turbulent reacting predictions of the same configuration. The two tetrahedra meshes contain respectively 38 and 93 millions cells. Limit-cycles as obtained by the two LES are gauged against each other for various flow quantities such as mean velocity profiles, flame position and temperature fields. The thermo-acoustic limit-cycles are observed to be relatively indepedent of the grid resolution which comforts the use of LES tools to provide insights and understanding of the mechanisms triggering the coupling between the system acoustic eigenmodes and combustion.

Key words: Large-Eddy Simulation, Turbulent Reacting Flows, Reliability, Complex Geometry

1 Introduction

The recent leap to petascale computer architectures (*cf.* TOP500[3]) allows to consider Large Eddy Simulation (LES) as a promising tool to predict turbulent reactive flows

[3] http://www.top500.org

M.V. Salvetti et al. (eds.), *Quality and Reliability of Large-Eddy Simulations II*,
ERCOFTAC Series 16, DOI 10.1007/978-94-007-0231-8_22,

in complex geometries. In the past decade, LES prooved to predict reacting flows [1, 2, 3] as well as the stability of flames [4] and flame-acoustic interactions [5, 6]. The development of massively parallel architectures permits to include more and more details in the simulations. In the particular field of annular gas turbines, which typically comprise ten to twenty sectors, computing the full chamber including all sectors stayed out of reach until very recently. Today, leadership class machines enables such computations by running LES codes on thousands of processors [7, 8, 9], providing key insights to phenomena such as ignition or azimuthal thermo-acoustic instabilities.

A fundamental but seldom addressed issue is the effects of mesh resolution on such LES results. As LES of such applications is based on an implicit filtering of the governing equations by the grid size, the question of mesh dependency is of utmost importance [10, 11]. Yet, most studies focusing on mesh dependency consider academic configurations [12, 13] or single sector configurations [14] and the right resolution for a complete burner remains an open question. Boudier *et al* [14] presented results obtained with the same numerical tool used here and draw the conclusion that independence upon the mesh resolution for the combustion quantities is difficult to obtain and that mesh resolution can have an impact on local combustion regimes.

In the present study, a mesh dependency study of a full annular helicopter chamber is performed with two grids respectively composed of 38 and 93 million elements. Evaluation of LES is made by comparing the results in terms of mean, fluctuating fields and Sub-Grid Scale (SGS) quantities. Finally and because the target configuration is subject to a self-excited azimuthal mode, grid effects on the resulting limit-cycle are gauged.

2 Numerical tools

In this paper, a fully compressible unstructured explicit code is used to solve the reactive multi-species Navier-Stokes equations [14]. A third order finite element scheme is used for both time and space advancement [15]. SGS tensor is modeled by a classical Smagorinsky approach [16]. Boundary conditions are implemented through the NSCBC formulation [17] and wall boundaries use a logarithmic wall-law approach.

Chemistry is computed by means of a reduced one-step mechanism for JP-10 / air flames, JP-10 being a surrogate for kerosen. This mechanism is fitted to match the full scheme behaviour for equivalence ratios between 0.4 and 1.5 [14]. Five species are explicitly transported and solved for: JP-10, O_2, CO_2, H_2O and N_2. To better capture flame/turbulence interactions, the Dynamic Thickened Flame (DTF) model is used [18]. The flame SGS wrinkling and interactions are supplied by an efficiency function [18].

3 Target configuration

The configuration (Fig. 1) considered throughout this study is a full annular reverse-flow helicopter gas turbine demonstrator designed by Turbomeca (Safran group). The combustion chamber is equipped with fifteen swirlers, each swirler consisting of two co-annular counter-rotating swirl stages. The whole chamber is computed with its casing, which helps avoiding uncertainties on the boundary conditions. Indeed, the calculated domain starts immediately after the compressor's outlet, and extends to a chocked nozzle corresponding to the throat of the high pressure distributor (Fig. 1). Fuel is supposed to be totally vaporized and only gaseous phase is computed. Air inflow feeds the combustion chamber through the swirlers, cooling films and dilution holes. Multi-perforated walls used to cool the liners are taken into account by a homogeneous boundary condition [19]. Note that this configuration, contrary to the one considered by Boudier *et al* [14], has strong swirled flow induced by each burner and is thus almost fully premixed.

Two grids with different resolutions are compared to assess the impact of mesh resolution on LES: a light one that comprises 37.7 million tetrahedral cells and a fine one, composed of 93.1 million tetrahedral cells. Refinement between the light and fine grid is homogeneous so as to retain a similar grid topology (Fig. 2).

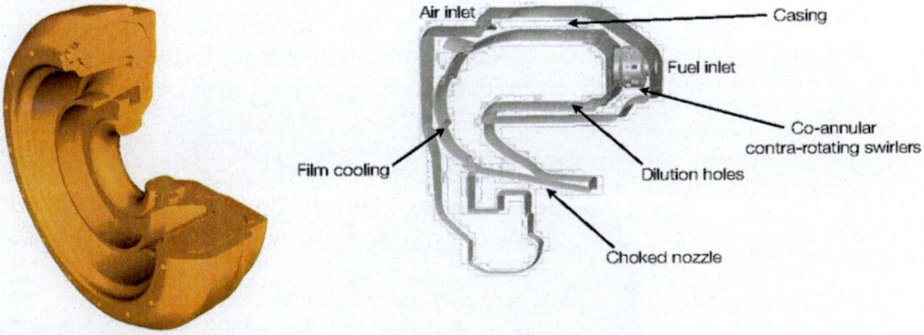

Fig. 1. 3/4 view of the fully annular combustion chamber and boundary conditions shown on a single sector.

4 Results and discussion

The lighter grid has been run on 2,048 cores on a SGI Altix Ice 3 equipped with Intel Xeon 3GHz CPUs[4]. The LES of 0.1 seconds physical time required around 400,000 CPU hours. The fine grid case has been run on 4,096 cores on a IBM Blue Gene/P[5] and almost 3,000,000 CPU hours were required to simulate 30 ms physical time. Figure 3 displays a resulting instantaneous prediction along with the geometry.

[4] http://www.genci.fr - http://www.cines.fr

[5] http://www.alcf.anl.gov

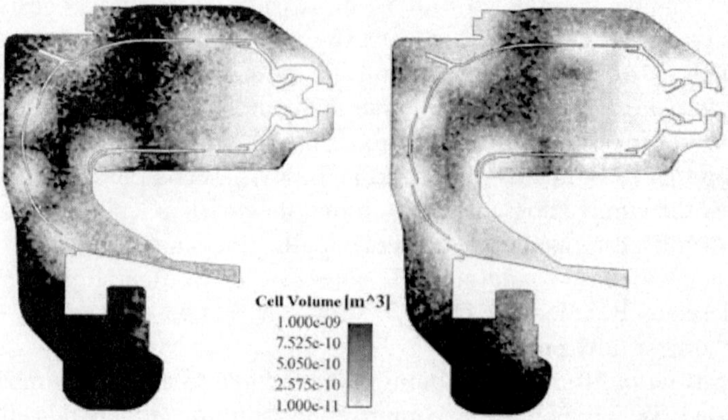

Fig. 2. Mesh resolution for the lighter (38 million cells - left) and the fine (93 million cells - right) grids, as obtained on a transversal cut passing through the swirlers' axis and showing only one sector.

Azimuthal pressure fluctuations can be observed on this figure, indicating the presence of an azimuthal mode. Convergence will first be checked in terms of mean flow quantities and behavior of the SGS models before conducting a brief analysis of the effects of mesh resolution on the azimuthal mode.

Fig. 3. Translucent part of the geometry along with pressure fluctuations on a cylinder containing all burners axis and a temperature isosurface representing the flame front, as obtained on the lighter grid.

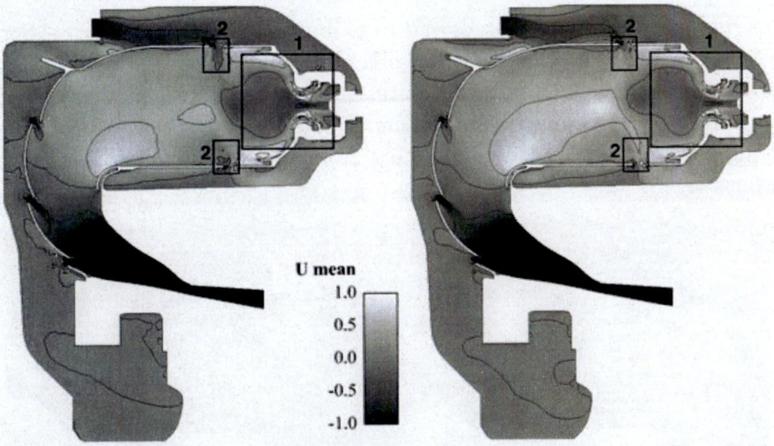

Fig. 4. Mean axial velocity fields for the lighter (38 million cells - left) and the fine (93 million cells - right) grids. Velocities are normalized by the inlet velocity.

4.1 Mesh dependency in terms of mean flow

Due to the intrinsic properties of LES, two instantaneous fields issued by two LES of the same problem may differ for various reasons [10, 20, 21]. However, one expects similar LES to converge towards the same statistics. Temporally averaged fields and standard deviation (RMS) fields are thus compared in this section to quantify the dependency of LES regarding mesh resolution.

The temporal averaging for both cases has been done over 13.8 ms, corresponding to 10 flow-through times, based on the mass flow rate and the primary zone volume, thus ensuring statistically converged values with regards to the main flow direction. However, a weak but non-negligible azimuthal motion exists in this annular configuration, the mean azimuthal velocity in the chamber casing being of the order of 1 m/s. An estimation of one flow-through time in the azimuthal direction is 0.5 s or 16 million iterations for the fine grid. Such a calculation would cost over 40 million CPU hours on a IBM BlueGene/P and remains out of reach. To improve the statistics, a spatial sector-to-sector averaging over the fifteen sectors is done.

Figure 4 shows the mean axial velocity scaled by the inlet velocity on a transversal cut passing through the swirlers' axis. The flow topology between the two grids is very similar with a strong recirculating zone that extends from the nose of the injector to the primary dilution jets (highlighted as zone 1 on Fig. 4). High velocity zones at the exit of the swirler are retrieved with both meshes, as well as the upper and lower high speed regions of the primary zone. The cooling films present similar shapes. The main discrepancies between the two fields are observed for the penetration of the primary dilution jets (highlighted as zones 2 on Fig. 4) and for the lower part of the high speed zone just downstream of the primary zone.

Figure 5 presents the mean temperature field scaled by the inlet temperature. Very good agreement is found between the two grids. The mean flame position is identical, indicating proper behavior of the combustion model. The thermal boundary layers created by the multiperforated walls and the cooling films are retrieved on both grids. Similarly to Fig. 4, dissimilarities between the two meshes are seen for the penetration of the primary dilution jets. Figure 6 confirms these conclusions by displaying profiles of the mean temperature at five different axial positions from the exit of the swirler.

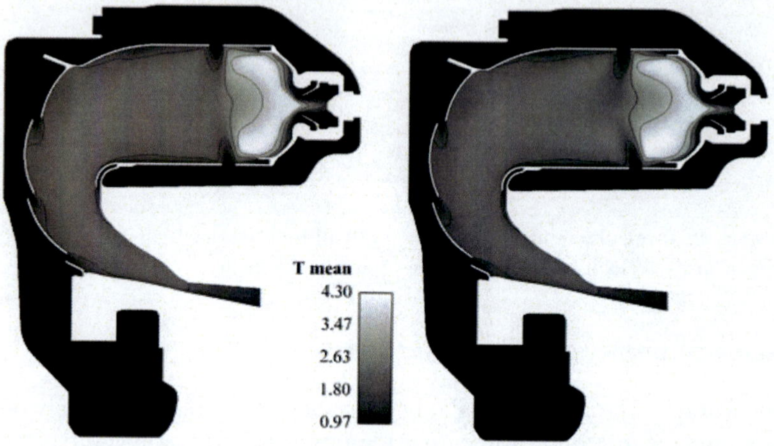

Fig. 5. Mean temperature fields for the lighter (38 million cells - left) and the fine (93 million cells - right) grids. Temperatures are normalized by the inlet temperature.

Figure 7 gives more insights on the way combustion occurs by showing the Probability Density Function (PDF) of the instantaneous resolved equivalence ratio in reacting zones, i.e. zones where the local heat release is higher than a threshold defined with regards to the maximum heat release [14]. PDFs obtained on both meshes are almost similar. Most of the combustion takes place in lean premixed zones, with a strong peak at an equivalence ratio of $\phi = 0.35$. This emphasizes the proper use of the fitted one-step chemical scheme as most flame elements burn under lean conditions. Another peak is located at $\phi = 1$ and stems from the presence of diffusion flamelets. A weaker peak appears around $\phi = 1.5$ and corresponds to rich premixed flames created close to the fuel injection. To quantify the effects of mesh resolution on the DTF model, Fig. 8 presents the percentage of total grid points as function of the thickening factor for both meshes. As expected, there is a shift towards lower thickening as the grid is refined while preserving the shape of both curves. On the lighter grid, 18.34% of the total grid points are affected by thickening (*i.e.* where the flame is localized), with a maximum thickening factor of 81.5 and a mean value of 18.77. For the fine grid, 14.93% of the total grid points involve reaction, with a maximum thickening factor of 62.5 and a mean value of 15.28. Even if the flame is obviously better resolved on the fine grid, the combustion regime appears to be the same for the two cases, which ensures independent flame response to acoustic perturbations. Note that resolved RMS velocity and temperature fields on the two

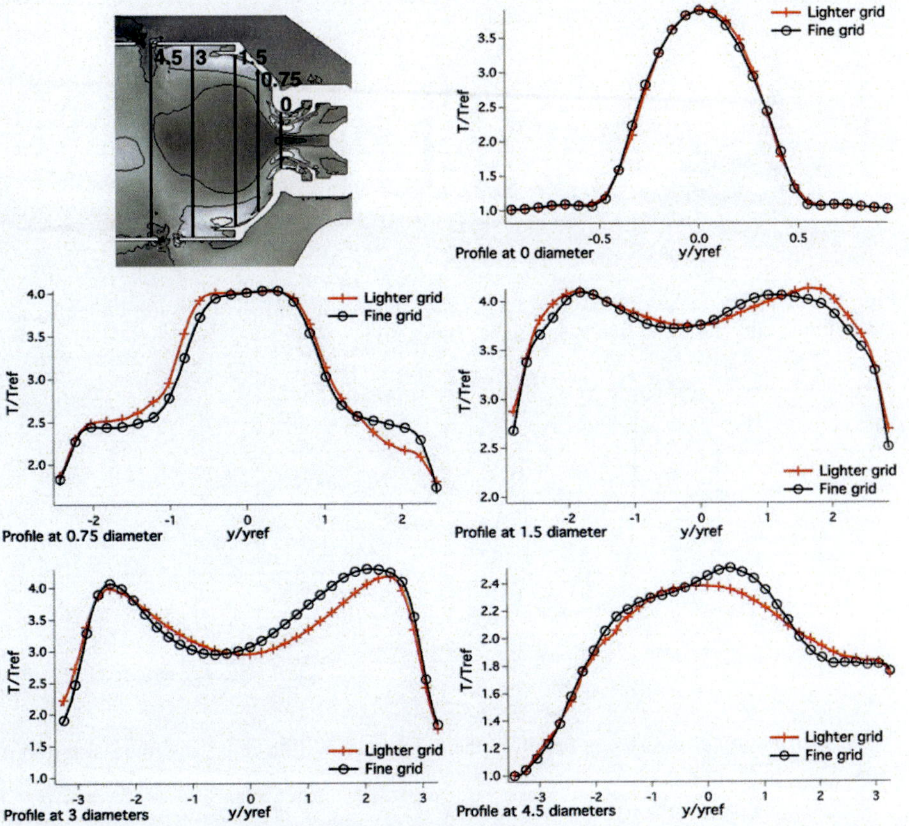

Fig. 6. Mean temperature profiles at 5 different axial positions from the exit of the swirler: 0, 0.75, 1.5, 3 and 4.5 swirler diameters. Temperatures are normalized by the inlet temperature and lengths are normalized by the swirler diameter.

considered grids (not shown here) exhibit very good agreement, confirming the previous findings. This conclusion is very different from the one drawn by Boudier *et al* [14] where a premixed inverted cane injection system is simulated. In this study, combustion shifts from mostly lean to a balance between diffusion and lean premixed flames as the grid is refined. This enforces the fact that proper behavior of the combustion model depends on the configuration.

4.2 Mesh dependency of the thermo-acoustic instabilities

The considered configuration is subject to self-established azimuthal thermo-acoustic instabilities. An observation of the pressure spectra (not shown here) reveals a mode around 750 Hz for both meshes. To get some insights on the impacts of mesh resolution on such a mode, $\frac{P_{RMS}}{P_{mean}}$ are plotted in Fig. 9 for both meshes along a line containing all swirlers' axis and located in front of all burners' exits. The same structure

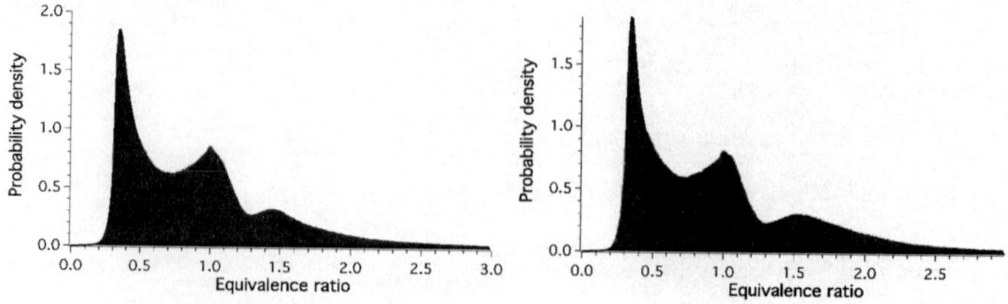

Fig. 7. Probability density function of local equivalence ratio in reacting zones on the lighter (38 million cells - left) and fine (93 million - right) grids.

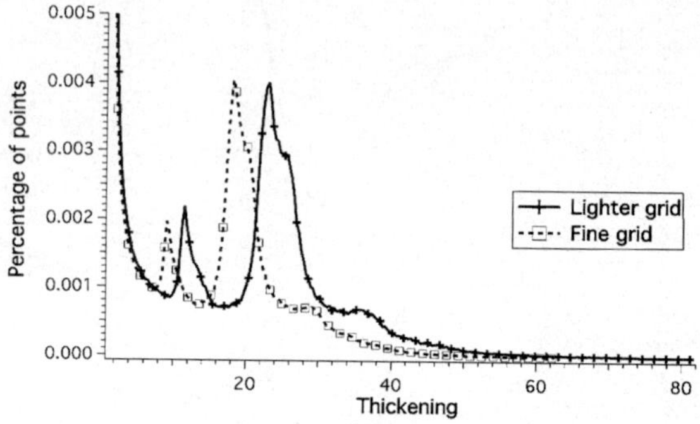

Fig. 8. Distribution of thickening factor on the lighter (38 million cells) and fine (93 million) grids.

is retrieved, with sinusoidally shaped pressure fluctuations, indicating sectors subject to almost no fluctuations and other sectors subject to strong pressure fluctuations (of the order of 2% of the chamber mean pressure). The amplitude of the pressure fluctuations is reduced for the fine grid when compared to the lighter one. Such observations ensure that the prediction obtained by LES for such an instability is quite robust and insensitive to grid resolution. Ongoing analyses are being conducted to identify the leading flow mechanisms triggering this instability.

5 Conclusion

A mesh dependency study of a fully annular helicopter gas turbine has been performed using massively parallel LES on two grids, respectively comprising 38 and 93 million tetrahedra. The predictions obtained are very similar in terms of mean flow. The DTF combustion model is shown to behave adequately on this type of almost fully premixed configuration. A self-established azimuthal mode naturally develops in both computations and the resulting thermo-acoustic limit-cycles are observed to be independent of the grid resolution, which partly qualifies LES for such critical industrial problems.

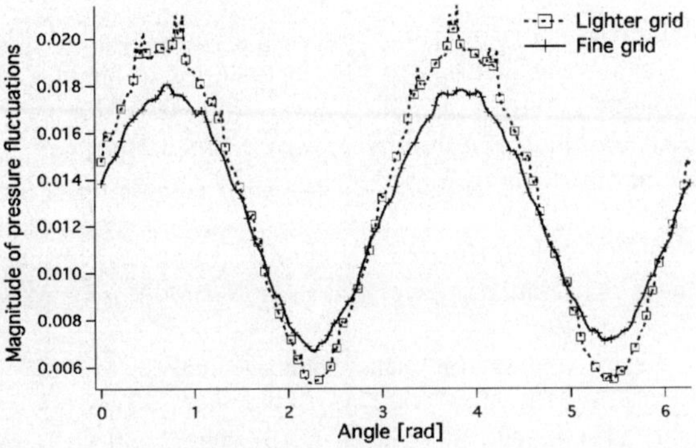

Fig. 9. $\frac{P_{RMS}}{P_{mean}}$ on a line passing by all burners' axis for the lighter (38 million cells) and the fine (93 million cells) grids.

Acknowledgments

The authors thank the Argonne Leadership Computing Facility at Argonne National Laboratory, which is supported by the Office of Science of the U.S. Department of Energy under contract DE-AC02-06CH11357, as well as GENCI (Grand Equipement National de Calcul Intensif) and CINES (Centre Informatique National de l'Enseignement Superieur) for providing the computing power necessary for these simulations.

References

1. T. Poinsot and D. Veynante. *Theoretical and Numerical Combustion*. R.T. Edwards, 2nd edition, 2005.
2. F. Di Mare, W. P. Jones, and K. Menzies. Large eddy simulation of a model gas turbine combustor. *Combustion and Flame*, 137:278–295, 2004.
3. H. Pitsch. Large eddy simulation of turbulent combustion. *Annual Review of Fluid Mechanics*, 38:453–482, 2006.
4. A. Sengissen, J. F. Van Kampen, R. Huls, G. Stoffels, J. B. W. Kok, and T. Poinsot. Les and experimental studies of cold and reacting flows in a swirled partially premixed burner with and without fuel modulation. *Combustion and Flame*, 150:40–53, 2007.
5. C. Martin, L. Benoit, Y. Sommerer, F. Nicoud, and T. Poinsot. Les and acoustic analysis of combustion instability in a staged turbulent swirled combustor. *AIAA Journal*, 44(4):741–750, 2006.
6. G. Boudier, N. Lamarque, G. Staffelbach, L.Y.M. Gicquel, and T. Poinsot. Thermo-acoustic stability of a helicopter gas turbine combustor using large-eddy simulations. *Int. Journal Aeroacoustics*, 8(1):69–94, 2009.
7. M. Boileau, G. Staffelbach, B. Cuenot, T. Poinsot, and C. Bérat. LES of an ignition sequence in a gas turbine engine. *Combustion and Flame*, 154(1-2):2–22, 2008.
8. G. Staffelbach, L.Y.M. Gicquel, G. Boudier, and T. Poinsot. Large eddy simulation of self-excited azimuthal modes in annular combustors. *Proceedings of the Combustion Institute*, 32:2909–2916, 2009.

9. P. Wolf, G. Staffelbach, A. Roux, L. Gicquel, T. Poinsot, and V. Moureau. Massively parallel LES of azimuthal thermo-acoustic instabilities in annular gas turbines. *C. R. Mecanique*, (in press), 2009.
10. P. Sagaut. *Large eddy simulation for incompressible flows*. Springer, 2002.
11. S. B. Pope. Ten questions concerning the large-eddy simulation of turbulent flows. *New Journal of Physics*, 6:35, 2004.
12. B. Vreman, B. Geurts, and H. Kuerten. Comparison of numerical schemes in large-eddy simulation of the temporal mixing layer. *International Journal for Numerical Methods in Fluids*, 22:297–311, 1996.
13. J. Meyers, B. J. Geurts, and M. Baelmans. Database analysis of errors in large-eddy simulation. *Physics of Fluids*, 15(9):2740–2755, September 2003.
14. G. Boudier, L. Y. M. Gicquel, T. Poinsot, D. Bissières, and C. Bérat. Effect of mesh resolution on large eddy simulation of reacting flows in complex geometry combustors. *Combustion and Flame*, 155(1-2):196–214, 2008.
15. O. Colin and M. Rudgyard. Development of high-order taylor-galerkin schemes for unsteady calculations. *Journal of Computational Physics*, 162(2):338–371, 2000.
16. J. Smagorinsky. General circulation experiments with the primitive equations: 1. the basic experiment. *Monthly Weather Review*, 91:99–164, 1963.
17. T. Poinsot and S. Lele. Boundary conditions for direct simulations of compressible viscous flows. *Journal of Computational Physics*, 101(1):104–129, 1992.
18. O. Colin, F. Ducros, D. Veynante, and T. Poinsot. A thickened flame model for large eddy simulations of turbulent premixed combustion. *Physics of Fluids*, 12(7):1843–1863, 2000.
19. S. Mendez and F. Nicoud. Adiabatic homogeneous model for flow around a multiperforated plate. *American Institute of Aeronautics and Astronautics Journal*, 46(10):2623–2633, 2008.
20. S. Ghosal and P. Moin. The basic equations for the large eddy simulation of turbulent flows in complex geometry. *Journal of Computational Physics*, 118:24–37, 1995.
21. J.-M. Senoner, M. García, S. Mendez, G. Staffelbach, O. Vermorel, and T. Poinsot. Growth of Rounding Errors and Repetitivity of Large-Eddy Simulations. *AIAA Journal*, 46(7):1773–1781, 2008.

Extension of LES approaches to conductive fluids and plasmas

P. Morel[1], D. Carati[1], F. Merz[2], T. Görler[2] and F. Jenko[2]

[1] Statistical and Plasma Physics Laboratory, Université Libre de Bruxelles, CP 231, Boulevard du Triomphe, 1050 Brussels, Belgium {pmorel,dcarati}@ulb.ac.be
[2] Max-Planck-Institut für Plasmaphysik, Boltzmannstr. 2, D-85748 Garching, Germany {flm,tbg,fsj}@ipp.mpg.de

Summary. The extension of LES methods to the MHD description of conductive fluids and plasmas as well as the kinetic description of small scales of turbulence in plasmas is presented. The results obtained during the last decade in the framework of MHD equations will highlight and motivate the current try to perform filtered gyrokinetic simulations.

Key words: Quality, Reliability, Large-Eddy Simulation, MHD, gyro-kinetics, Plasma.

1 Introduction

Large eddy simulations (LES) consists in an arbitrary scale separation between the large scales structures that are computed directly and the small scale phenomena that are not captured by the numerical grid and must be modeled. It is reasonably justified when turbulence is well developed and various modelling approaches have been proposed in order to take into account the effect of the small and unresolved fluid motions on the large scale dynamics. Although turbulent phenomena are observed in various type of systems, LES have been developed mainly for fluids described by the Navier-Stokes equations with considerable efforts to extend this technique to chemically reactive fluids and, in particular, to combustion.

In this work, we are interested in the extension of LES to electrically conductive fluids that may interact with electromagnetic fields. Electrically conductive fluids as well as plasmas can be described either as continuous media by fluid-type equations or as large sets of interacting particles using kinetic approaches. In both cases, the interactions between matter and electromagnetic fields have to be taken into account by coupling the equations of mechanics to the Maxwell equations and may lead to complex turbulent phenomena.

In the continuous media description, this coupling yields the magnetohydrodynamics (MHD). Extension of LES to MHD is technically easy and has been successfully proposed [1, 2]. Indeed, the evolution of the magnetic field is driven by the induction equation which is quite similar to the Navier-Stokes equation. However,

M.V. Salvetti et al. (eds.), *Quality and Reliability of Large-Eddy Simulations II*, ERCOFTAC Series 16, DOI 10.1007/978-94-007-0231-8_23,

because kinetic energy can be transformed into magnetic energy in MHD, the nonlinear transfers are physically more complex and modelling the small scales requires some reformulation that will be briefly reviewed in Section 2.

Despite kinetic and MHD equations describe the same system, they are significantly different as far as their structure and basic assumptions are concerned. In systems for which collisions are rare (low density plasmas or gases for instance), the continuous medium description is not necessarily valid. It is then often necessary to use kinetic approaches describing the evolution of the velocity probability distribution. This distribution corresponds to the probability to observe a particle at a given position with a given velocity, so that it is defined in a six-dimensional phase space. In presence of a strong magnetic field, taking advantage of the rapid helical motion of charged particles, one velocity variable can sometimes be eliminated establishing gyrokinetic model. The resulting five dimensional problem remains however extremely demanding in terms of computational power. Indeed, the resolution needed to capture all the physically relevant phenomena may be very high. In recent simulations [3], up to 768 x 384 x 16 grid points have been used for the spatial domain in order to describe both ion and electron typical scales and up to 64 x 16 points (or more) to discretize the velocity space. Also, differences in time scales characterizing various phenomena impose that converged simulations must use a very large number of time steps.

The objective of our study is to use a much lower resolution than the one imposed by the physics of the problem studied using the kinetic codes. This is achieved by developing a framework that is very similar to the one developed in LES for fluid turbulence. Adapting subgrid scale modelling approaches to kinetic equations however requires significant efforts that will be discussed into details in Section 3.

2 Extension of LES approaches to MHD

Many electrically conductive fluids – including plasmas – can be described within the framework of the magnetohydrodynamics (MHD). In this presentation, the flow will be assumed to be incompressible so that the balance equations read:

$$\partial_t \mathrm{v} = -\nabla \cdot (\mathrm{vv} - \mathrm{bb}) + \nu \Delta \mathrm{v} - \nabla p, \tag{1}$$

$$\partial_t \mathrm{b} = -\nabla \cdot (\mathrm{vb} - \mathrm{bv}) + \eta \Delta \mathrm{b}, \tag{2}$$

$$\nabla \cdot \mathrm{v} = \nabla \cdot \mathrm{b} = 0. \tag{3}$$

These equation are written here with the fluid velocity v and the magnetic field b both expressed in Alfvén speed units (the constant mass density is rescaled to unity). The total pressure p is obtained by imposing the incompressibility of v (3). The dissipation coefficients are the kinematic viscosity ν and the magnetic diffusivity η. The balance between the linear and the nonlinear term in the induction equation (2) defines the magnetic Reynolds number $Rm = v_0 \ell_0 / \eta$ (v_0 and ℓ_0 being a characteristic flow velocity and length scale, respectively). Simulation of MHD processes are thus not only limited by the grid requirement related to the velocity field but also by the necessity to properly capture the magnetic field. For instance, direct numerical simulations (DNS) of three-dimensional MHD turbulence on today's supercomputers [4, 5, 6] are limited to values of Rm that are still several orders of magnitude lower than the ones found in nature, e.g., in the solar convection zone. There is thus a double motivation for developing LES techniques for this type of problem.

As usual, the large scale fields that will be directly computed in LES for MHD are assumed to be obtained from the complete field by means of a filtering operator that will be noted $\overline{\cdots}$. The filtered equations for $\overline{\mathrm{v}}$ and $\overline{\mathrm{b}}$ thus read:

$$\partial_t \overline{\mathrm{v}} = -\nabla \cdot (\overline{\mathrm{v}}\,\overline{\mathrm{v}} - \overline{\mathrm{b}}\,\overline{\mathrm{b}}) + \nu \triangle \overline{\mathrm{v}} - \nabla \overline{p} - \nabla . \tau^{\nu} , \tag{4}$$

$$\partial_t \overline{\mathrm{b}} = -\nabla \cdot (\overline{\mathrm{v}}\,\overline{\mathrm{b}} - \overline{\mathrm{b}}\,\overline{\mathrm{v}}) + \eta \triangle \overline{\mathrm{b}} - \nabla . \tau^{b} , \tag{5}$$

$$\nabla \cdot \overline{\mathrm{v}} = \nabla \cdot \overline{\mathrm{b}} = 0 , \tag{6}$$

where two unknown terms, usually referred to as subgrid-scale or filtered-scale stress tensors, are introduced by the filtering operation: $\tau^{\nu} = (\overline{\mathrm{v}\mathrm{v}} - \overline{\mathrm{v}}\,\overline{\mathrm{v}}) - (\overline{\mathrm{b}\mathrm{b}} - \overline{\mathrm{b}}\,\overline{\mathrm{b}})$ and $\tau^{b} = (\overline{\mathrm{v}\mathrm{b}} - \overline{\mathrm{v}}\,\overline{\mathrm{b}}) - (\overline{\mathrm{b}\mathrm{v}} - \overline{\mathrm{b}}\,\overline{\mathrm{v}})$. These tensors account for the subgrid-scale effects that have to be modeled. The kinetic $E_v = \mathrm{v} \cdot \mathrm{v}/2$ and magnetic $E_b = \mathrm{b} \cdot \mathrm{b}/2$ energy balance can easily be derived and are expressed as follows:

$$\partial_t E_v = D_\nu - T_{vb} - D_{\mathrm{v-sgs}} , \tag{7}$$

$$\partial_t E_b = D_\eta - T_{bv} - D_{\mathrm{b-sgs}} , \tag{8}$$

where T_{vb} represents the transfer of energy from v to b in the large scales. Because the total energy is conserved by the nonlinearities, this terms is exactly the opposite of the transfer of energy from b to v : $T_{vb} = -T_{bv}$. The other terms represent the viscous dissipation D_ν, the Joule dissipation D_η and the subgrid scale dissipation of kinetic energy $D_{\mathrm{v-sgs}}$ and of magnetic energy $D_{\mathrm{b-sgs}}$. Like in Navier-Stokes turbulence, the main effect of the subgrid scales in MHD has been assumed to dissipate the correct amount of energy and the subgrid scale tensor have been modeled using eddy viscosity and eddy magnetic diffusivity terms:

$$\tau^{\nu}_{ij} = -\nu_t \left(\partial_j \overline{v}_i + \partial_i \overline{v}_j\right)$$

$$\tau^{b}_{ij} = -\eta_t \left(\partial_j \overline{b}_i - \partial_i \overline{b}_j\right)$$

Smagorinsky eddy viscosity models have been tested for ν_t and extended to the eddy magnetic resistivity. It should be noted however that, due to the structure of the induction equation, the subgrid scale tensor τ^b is anti-symmetric. It is thus assumed to be proportional to the anti-symmetric part of the magnetic field gradient tensor. Remarkably, the components of this tensor are simply the current j since $\mathrm{j} = \nabla \times \mathrm{b}/\mu_0$ where μ_0 is the magnetic permeability. This relation is directly derived from the Maxwell equations by neglecting the displacement current. Based on this analysis, the eddy magnetic diffusivity has been assumed to be proportional to the square of the grid size Δ and to the amplitude of the anti-symmetric part of the magnetic field gradient tensor which reduces to $\eta_t \propto \Delta^2 |\overline{\mathrm{j}}|$. The proportionality constant as well as the Smagorinsky constant that appears in $-\nu_t$ have been derived using the dynamic procedure [7]. This approach has been successfully tested for both forced and decaying MHD homogeneous turbulence [1, 2, 8].

3 Extension of LES approaches to kinetic equations

Kinetic Vlasov-type equations are more complex than the fluid balance equations. They represent the evolution of a velocity distribution function $g(\mathbf{r},\mathbf{v},t)$ that now depends explicitly on the velocity $\mathbf{v}$ as well as on the position $\mathbf{r}$. We consider only one species of particles for simplicity, but the following discussion trivially extends to multi-species systems. Kinetic equations can be written as follows:

$$\frac{\partial g(\mathbf{r},\mathbf{v},t)}{\partial t} = -\left(\mathbf{v}.\frac{\partial}{\partial \mathbf{r}} + \frac{1}{m}\mathbf{F}.\frac{\partial}{\partial \mathbf{v}}\right) g(\mathbf{r},\mathbf{v},t) + C, \tag{9}$$

where m is the mass of the particles and $\mathbf{F}$ is the force acting on these particles, which in plasma physics is related to the electromagnetic fields $\mathbf{E}$ and $\mathbf{B}$ through the Lorentz force. The force $\mathbf{F}$ may contain a part $\mathbf{F}_0$ that is independent of the distribution function when external electromagnetic fields are imposed. It also normally contains a part $\mathbf{F}_g$ that is proportional to g and that represents the average effect of the other particle. Finally, C represents the collision term. Without discussing the structure of the collision term (that might be represented by a linear or a quadratic function of g), a typical kinetic equation has thus the structure:

$$\partial_t g = L(g) + N(g,g) \tag{10}$$

where L is a linear operator than contains both the advection term $-\left(\mathbf{v}.\frac{\partial}{\partial \mathbf{r}}\right) g$ and the external force term $\mathbf{F}_0$. The nonlinear term is due to the averaged internal force term $\mathbf{F}_g$. Assuming, like in many gyrokinetic codes, that the distribution function g can be split into a local equilibrium distribution G^0 (independent of time but that can depend on the position) and a fluctuation function δg, the equation for δg reads:

$$\partial_t \delta g = \mathscr{S} + \mathscr{L}(\delta g) + \mathscr{N}(\delta g, \delta g) \tag{11}$$

The source term $\mathscr{S}$ is due to the gradient of the equilibrium distribution function. The linear term $\mathscr{L}$ contains contributions from the advection term, the external forcing and the coupling between the equilibrium distribution and δg in the nonlinear term N in (10). The nonlinear $\mathscr{N}$ term is basically the same as the original nonlinear term N but depends only on the fluctuating distribution function. Filtering this equation in space requires some care since the quantity $\mathscr{S}$ and the operator $\mathscr{L}$ may depend explicitly on the position because they depend on the local equilibrium distribution G^0. However, in most cases, it is reasonable to assume that G^0 varies very slowly with space when compared to the fluctuating distribution δg. Hence, it is reasonable to assume that $\overline{\mathscr{S}} = \mathscr{S}$ and $\overline{\mathscr{L}(\delta g)} = \mathscr{L}(\delta \overline{g})$ and applying a spatial filter to the equation (11) leads to:

$$\partial_t \delta \overline{g} = \mathscr{S} + \mathscr{L}(\delta \overline{g}) + \mathscr{N}(\delta \overline{g}, \delta \overline{g}) + \mathscr{T} \tag{12}$$

where the new term due to the subgrid-scale fluctuation of the distribution δg reads:

$$\mathscr{T} = \overline{\mathscr{N}(\delta g, \delta g)} - \mathscr{N}(\delta \overline{g}, \delta \overline{g}) \tag{13}$$

There is thus a clear formal analogy between equations (12) and (4, 5).

3.1 Conservation laws in Gyrokinetic equations

Gyrokinetic solvers deal with equations that have exactly the same structure as Eq. (11), we refer the reader for more explanations on foundations of gyrokinetic theory to the very exhaustive paper by Brizard and Hahm [9]. Applying a filter to the gyrokinetic equation will thus lead to equations similar to (12). The design of models for the subgrid-scale term $\mathscr{T}$ should, like in fluid equations, also be influenced by their expected role on ideal invariant balance. The difficulty is of course to identify which ideal invariant has to be considered. Fortunately, there has been a lot of efforts in the last years devoted to the derivation of conservation laws in gyrokinetic equations [9]. Independently of the present approach, the existence of conserved quantities by the gyrokinetic equations is very important in assessing the accuracy of the numerical solvers [10, 11].

In a recent paper [12], Schekochihin et al have discussed in details the energy balance in gyrokinetic simulations. They pointed out that, although conservation of energy is indeed important, the conservation of the free energy, mixing entropy and energy, presents much more similarities with kinetic energy conservation in fluid equations. In the kinetic description of a plasma, the power $\varepsilon = \int \mathbf{E}.\mathbf{J}_{\mathrm{ext}}$ injected in or removed from the system by the externally imposed electric current $\mathbf{J}_{\mathrm{ext}}$ modifies the sum of the kinetic energy,

$$K = \int d\mathbf{v}\, d\mathbf{r}\, g(\mathbf{r},\mathbf{v},t)\frac{m|\mathbf{v}|^2}{2} \tag{14}$$

and of the energy stored in the electromagnetic field

$$W = \int d\mathbf{r}\,\frac{|\mathbf{E}|^2 + |\mathbf{B}|^2}{8\pi\varepsilon_0}, \tag{15}$$

leading to the following simple equation:

$$\frac{d}{dt}\mathscr{E} = \frac{d}{dt}(W+K) \equiv \varepsilon \tag{16}$$

Using this energy balance for designing subgrid-scale models for $\mathscr{T}$ would lead to two difficulties. First, as pointed out by Schekochihin et al, the energy transfers between electromagnetic field and kinetic energy of the particles remain reversible. They do not lead to a mechanism in which an irreversible dissipation such as viscous effects in fluid mechanics ends a possible energy cascade mechanism. Second, it should be noted that the electric and magnetic fields are linearly proportional to $g(\mathbf{r},\mathbf{v},t)$ through the Maxwell equations. Hence, W is a quadratic function of $g(\mathbf{r},\mathbf{v},t)$ while K is linear in $g(\mathbf{r},\mathbf{v},t)$. The total conserved quantity $K+W$ mixes thus two terms with a quite different mathematical nature and this is very different from the kinetic energy in Navier-Stokes turbulence or the total (kinetic +magnetic) energy in magneto-hydrodynamics.

Let us now introduce the free energy $F = \mathscr{E} - TS$. Where the total entropy $S = -k_B \int d\mathbf{v}\, d\mathbf{r}\, g \ln g$ can be re-expressed in terms of the equilibrium and the fluctuating parts of the distribution function by keeping only terms up to order two in δg:

$$S = -k_B \int d\mathbf{v}\, d\mathbf{r} \left[g \ln G^0 + \delta g + \frac{1}{2} (\frac{\delta g}{G^0})^2 \right], \tag{17}$$

If we assume that the local equilibrium distribution G^0 is Maxwellian, then the first entropy term exactly cancels the kinetic energy K in the expression for F. The second entropy term simply represents the normalisation of the fluctuating distribution function and is assumed to be a constant that will disappear from the free energy balance. The evolution of F can then be re-written as:

$$\frac{d}{dt}(W + T\delta S) = \varepsilon - T\mathscr{K}_c \tag{18}$$

where

$$\delta S = -k_B \int d\mathbf{v}\, d\mathbf{r} \frac{1}{2} (\frac{\delta g}{G^0})^2, \tag{19}$$

This balance equation shares most of the properties of the kinetic energy balance in fluids described by the Navier-Stokes or the MHD equations (7, 8). The conserved quantity $W + T\delta S$ is not affected by the nonlinear term in the gyrokinetic equations. This conserved quantity is a sum of terms that are positive definite and quadratic in the distribution function δg. The evolution of $W + T\delta S$ is due to the effect of external mechanisms (that, in most of the interesting cases, will sustain the fluctuations) and an irreversible dissipation mechanism that tends to destroy the fluctuations through the collisional processes.

Due to the gyrokinetic ordering, the expression for the electromagnetic energy W takes a slightly different form in the gyrokinetic formalism. It is usually split into two parts W^ϕ and W^A, that are related respectively to the electric scalar potential ϕ and to the magnetic vector potential $\mathbf{A}$. Moreover, the entropic term $\mathscr{F} \equiv T\delta S$ is normally the sum over all species. Finally, in most gyrokinetic simulations, the source term is actually due to the equilibrium gradient and will be noted $\mathscr{G}$. However its role is the same and it injects free energy in the system. Finally, the temperature is usually lumped into the definition of the collision term $\mathscr{K} \equiv T\mathscr{K}_c$ so that the free energy balance in gyrokinetic equations is usually presented as:

$$\frac{d}{dt}(\mathscr{F} + W^\phi + W^A) = \mathscr{G} - \mathscr{K} \tag{20}$$

3.2 Filtered Gyrokinetic simulation

From the above analysis, it is easily understood that, in absence of collisions, the entropy balance will not reach a steady state. However, collisionless numerical experiences are routinely performed using gyrokinetic codes. In order to compensate for the increase of free energy due to the external constraints, artificial dissipative mechanisms are then added to the equations [11]. These artificial terms play in fact exactly the same role as the subgrid-scale term $\mathscr{T}$ and could be regarded as rather crude subgrid-scale models in filtered gyrokinetic simulation. The major difference is that the subgrid-scale models are intended to represent a larger fraction of the unresolved physics and not necessarily only the collisional scales. So, the equation that is effectively solved in gyrokinetic code reads:

$$\partial_t \delta g = \mathscr{S} + \mathscr{L}(\delta g) + \mathscr{N}(\delta g, \delta g) + \mathscr{D}(\delta g, \tilde{p}), \tag{21}$$

where the artificial dissipative term $\mathscr{D}(\delta g, \tilde{p})$ depends on the distribution function and on parameters $\tilde{p}$ that have to be adjusted to remove the correct amount of free energy. There is a striking analogy with the equation 12 if we interpret the artificial dissipative term as a subgrid-scale model. hence, the filtered gyrokinetic equation that we will analyse reads:

$$\partial_t \delta \overline{g} = \mathscr{S} + \mathscr{L}(\delta \overline{g}) + \mathscr{N}(\delta \overline{g}, \delta \overline{g}) + \mathscr{M}(\delta \overline{g}, \tilde{p}, \overline{\Delta}) \tag{22}$$

in which the unknown subgrid-scale term $\mathscr{T}$ is replaced by a model $\mathscr{M}(\delta \overline{g}, \tilde{p}, \overline{\Delta})$ that depends on the filtered distribution $\delta \overline{g}$ on some parameters $\tilde{p}$ and on the grid size needed for capturing the filtered distribution evolution.

In a preliminary study, the influence of a lack of resolution in gyrokinetic simulation has been analysed. These results have been obtained using the Cyclone Base Case set of parameters that is a reference commonly used in the Tokamak plasma community in order to compare the results obtained from various gyrokinetic solvers [13]. The CBC studies the ion temperature gradient turbulence (ITG) in magnetised plasmas typical of Tokamaks. ITG are controlled by the temperature (L_T) and density (L_n) gradient lengths along the minor radius of the Tokamak. The natural coordinates system in a Tokamak is defined by the radius r, the toroidal angle ϕ and the poloidal angle θ and the geometry is defined by the minor r_1 and the major r_2 radii. The magnetic field admits two components along poloidal and toroidal angles and one can then define the security factor $q(r) = \int_0^{2\pi} d\theta B^{\phi}/(2\pi B^{\theta})$ that measures the ratio between these components. The radial gradient of the security factor defines the magnetic shear parameter $\hat{s}(r) = (r/q)dq/dr$. The Cyclone Base Case key parameters are given in the following table:

parameter	R/L_T	R/L_n	q	$\hat{s}$
value	6.9	2.2	1.4	0.8

The simulations have been performed using the Gyrokinetic Electromagnetic Numerical Experiment (GENE) code, that is one of the most widely used Eulerian gyrokinetic codes. Eulerian method consists in computing the gyrokinetic equations on a fixed grid in phase space. A review of the numerical and theoretical key features of the GENE code can be found in reference [14][3]. The simulations presented here are made using a fluxtube approximation that consists in considering a simulation box very close to a magnetic field line [14, 15], allowing to neglect the variations of equilibrium quantities across the box. The anisotropy due to the presence of a strong magnetic field allows to distinguish between directions parallel (z for physical space, $v_{\parallel}$ for velocity) and perpendicular (x, y for physical space, μ for the magnetic momentum $\mu = mv_{\perp}^2/2B_0$ linked to the modulus of the perpendicular velocity) to the field lines. The standard grid for the CBC study with GENE is summarized in the following table:

[3] http://miami.uni-muenster.de/servlets/DerivateServlet/Derivate-4856/diss_merz.pdf

direction	x	y	z	$v_{\parallel}$	μ
number of grid points	64	16	16	32	8

Figure (1) represents the spectrum of the electrostatic potential $|\phi|$ along the minor radius of the torus. The fully resolved Cyclone Base Case is represented by the black curve with a grid of $n_x = 64$ points, $n_y = 16$ points for the spatial directions perpendicular to the magnetic field. We used three different factors to decrease resolution along both y and x: $3/4$ (dashed red), 2 (blue) and 4 (red). Without entering

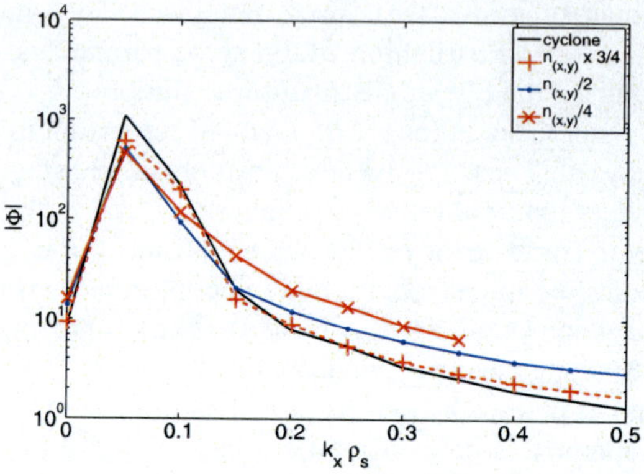

Fig. 1. Electric potential spectra for various grid resolution in the cyclone base case.

the details of the results, it is observed for instance that the spectrum of the electric scalar potential (very close to the energy term W_ϕ) is mostly affected in the smallest resolved scales while the large resolved scales are reasonably reproduced by an under-resolved simulation. This is reminiscent from under-resolved DNS of fluid turbulence in which the large scales are still reasonably reproduced while energy is piling up in scales close to the grid resolution.

4 Conclusions

The extension of LES ideas to describe turbulence in electrically conductive media has been discussed. When the system can be described using a fluid formalism, the extension of LES to MHD has already been discussed. However, when the fluid description is not appropriate (low density gases or plasmas), the LES framework needs to be extended to kinetic approaches. We have established a formalism for this purpose. It is based on a formal analogy between kinetic and fluid equations in terms of structure and of conserved quantities. This analogy is summarised in the following table.

description	variable(s)	conserved quantity
Navier Stokes	$\mathbf{v}$	kinetic energy: E^v
MHD	$\mathbf{v}, \mathbf{b}$	total energy: $E^v + E^b$
Gyrokinetics	δg	free energy: $T\delta S + W^{\phi} + W^{\mathbf{A}}$

Moreover, preliminary runs without model tend to show that the lack of resolution affects kinetic equations in a way that is quite similar to what is observed in fluid turbulence. In particular, a piling up of energy is observed close to the grid resolution. At this stage, no model has been tested yet. However, stabilising terms are often artificially introduced in kinetic codes when collisions are neglected. The structure of these terms should provide a reasonable guide for the expected structure of the subgrid scale terms while the approach developed here could provide some rational for the introduction of these artificial terms.

References

1. O. Agullo, W.-C. Müller, B. Knaepen and D. Carati, *Phys. Plasmas*, **8**, 3502 (2001).
2. B. Knaepen and P. Moin, *Phys. Fluids*, **16**, 1255 (2004).
3. T. Görler and F. Jenko, *Phys. Rev. Lett.*, **100**, 185002 (2008).
4. M.-M. Mac Low, R. S. Klessen, A. Burkert, and M. D. Smith, *Phys. Rev. Lett.*, **80**, 2754 (1998).
5. J. M. Stone, E. C. Ostriker, and C. F. Gammie, *Astrophys. J.*, **508**, L99 (1998).
6. D. Biskamp and W.-C. Müller, *Phys. Plasmas*, **7**, 4889 (2000).
7. M. Germano, *Journal of Fluid Mechanics*, **238** (1992) 325–336.
8. W.-C. Müller, D. Carati, *Phys. Plasmas*, **9**, 824 (2002).
9. A. J. Brizard, T. S. Hahm, *Rev. Mod. Phys.*, **79** No. 2 (2007), 421-468.
10. T.-H. Watanabe and H. Sugama, *Nucl. Fus.*, **46** (2006), 24-32.
11. J. Candy and R. E. Waltz, *Phys. Plasmas*, **13**, 032310 (2006).
12. A. A. Schekochihin, *et al.*, *Plasma Phys. Control. Fusion* **50** (2008), 124024.
13. A. M. Dimits, *et al.*, *Phys. Plasmas*, **7** 969 (2000).
14. Merz Florian (2008) Gyrokinetic simulation of mutlimode plasma turbulence, PhD Thesis, Westfälische Wilhelms-Universität, Münster, Germany.
15. F. Jenko, *et al.*, *Phys. Plasmas*, **7** 1904 (2000).

Part II

Assessment and reduction of computational errors

Grid Design and the Fate of Eddies in External Flows

Philippe R. Spalart[1], Mikhail Kh. Strelets[2], and Andrey Garbaruk[2]

[1]Boeing Commercial Airplanes, Seattle, WA 98124, USA
philippe.r.spalart@boeing.com

[2]New Technologies and Services, St.-Petersburg 197198, Russia
strelets@mail.rcom.ru

Summary. A few years ago we introduced a nomenclature for grid regions in Detached-Eddy Simulations (DES) of external flows: the RANS, Focus, Euler, and Departure Regions [1]. These may help organize grid generation and error estimation/reduction. We have four purposes here: to point out that most of these concepts apply to LES, if not to DNS, and are not truly limited to external flows; to discuss what really defines a DNS for external flows, and even for internal flows with inflow-outflow conditions; to ask whether the SGS model should be sensitized to increasing grid spacing along the flow, hoping to achieve stable simulations of "real-life flows" with energy-conserving differencing, which is definitely not the norm today; and finally to present an initial study of an explicit definition of the boundary between Focus Region and Departure Region. The long-term goal is to extend automatic grid adaptation to the turbulence-resolving approaches.

Key words: Grid Generation, Accuracy, Turbulence Modelling

1 Introduction

Large-Eddy Simulation and its hybrids with RANS are rapidly expanding towards complex applications, which is very gratifying but can also produce some chaos, and even a backlash when work that is not careful enough is publicized or introduced in designs which then prove faulty. This concern is central to the organisation of QLES (although complex flows are not the only ones considered), but curiously the list of topics either for 2007 or 2009 does not contain grid design, which is however our primary focus here. Fewer than ten out of thirty-three papers this week highlight grid issues, and only two have adaptation. In our opinion, the grid is just as important as the solver, and is arguably more important than the SGS model, except very near walls.

A factor in how research topics are chosen is that SGS research is more publishable and less dependent on heavy software than grid research. Grid-generation technology and practices have much control both over the accuracy of research simulations and over the practicality and reliability of LES in engineering practice. Once

M.V. Salvetti et al. (eds.), *Quality and Reliability of Large-Eddy Simulations II*, ERCOFTAC Series 16, DOI 10.1007/978-94-007-0231-8_24,

the geometry has any complexity, and even for simple shapes when they cause massive separation, there is much latitude on where to "spend" the grid points, and sharp differences in the true resolution requirements at different points. As a result, grids generated with too little regard for these facts are wasteful.

In addition, the long-term goal of the CFD community must be automatic generation of grids with little waste, grids which will necessarily depend on the flow and not just the geometry; in other words, be adaptive. For this to develop, the thinking needs to be organized as well as possible, and to attract some degree of consensus. Note that as of 2009, adaptation is not the norm even for steady RANS solutions, especially not in commercial solvers. The benefits in time savings *and* in accuracy will be extensive. It is an active research area, but an arduous one, with deep open questions including the balance between solver tolerance for very uneven grids and smoothness in grid generation, and the option between heuristic adaptation strategies and more systematic ones, often based on adjoint formulations. Of course, since many flows contain boundary layers, thin shear layers, and possibly shocks, isotropic adaptation cannot be sufficient.

A dominant interest of the authors is in hybrid RANS-LES methods and especially DES, but significant overlap with grid design in LES and DNS has become clear. Similarly, some of the thinking which is unavoidable in external flows because of the unbounded domain proved to have some logic in internal flows as well. This little-explored territory led to a paper containing more "vision/speculation" than "finished products," with three sections made up of remarks with strategic or "philosophical" aspects, and a fourth with a much more concrete nature and a proposal which may contribute in due time to grid adaptation.

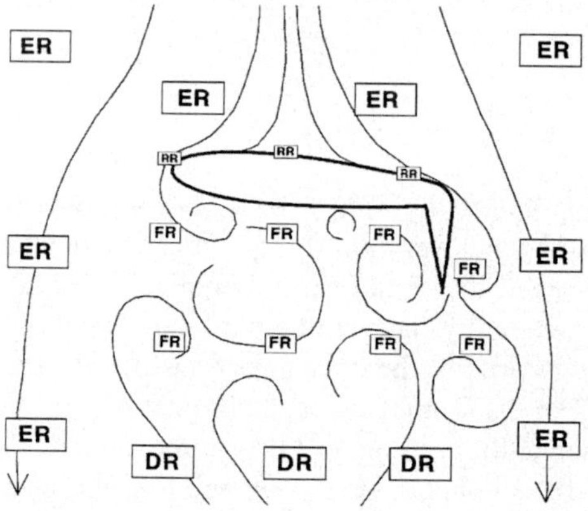

Fig. 1. Sketch of grid regions in the DES of an airfoil, with flow coming from above.

2 Grid Regions in DES, LES, and DNS

The "Young Person's Guide to DES Grids" is easily available as a NASA report [1], and the method contained in it will be called the "YPG approach." Figure 1 is a sketch from that report, and we use here the same nomenclature of RANS Region or RR, Focus Region or FR, Departure Region or DR, and Euler Region or ER, to which it will help to add VR for Vortical Region (this creates a conflict with "VR" for Viscous Region in the YPG, but the Viscous Region is not discussed here). At any instant, the VR is simply the complement of the ER, of course. However in the design of a stationary grid, the VR must be defined as the region which is visited by vorticity at any time during the simulation, and then the ER as its complement. This is because the resolution needs of VR are more demanding than those of the ER. This will be illustrated.

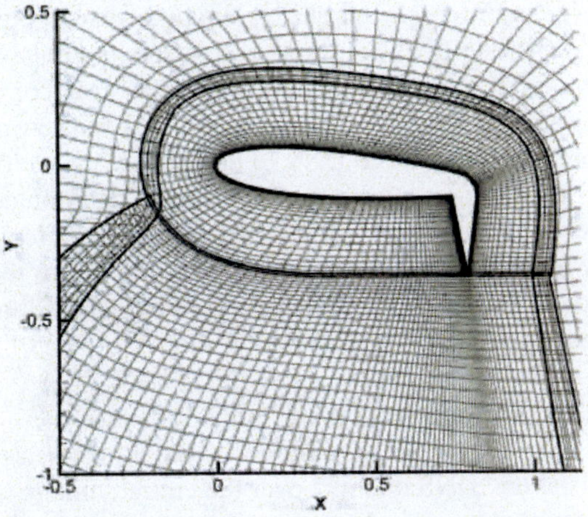

Fig. 2. Block-structured grid generated for airfoil [2].

Figure 2 shows an early example of grid design in DES, accomplished within the constraints of a block-structured solver with overset capability. Recall that the flow comes from above; this is the wing of a tilt-rotor aircraft during hover, with down-wash from the rotors. The Euler Region is most visible, in being coarser; the spacing in the z direction was also larger than in the other regions. The wasteful aspect that is not seen, of course, is that the entire simulation used a single time step, far shorter than the ER would need (however, the number of points in the ER is not dominant). The block in contact with the airfoil, with an unusual "snail" topology, covers the RANS region and part of the Focus Region. It extends into the ER above the airfoil, much farther than necessary for accuracy, but the advantage was to contain the number of blocks. The lower block covers part of the FR, and the Departure Region. Gradual coarsening is seen in the DR.

If an LES is considered, there is no pure RANS Region the way there is in a DES, where it includes the attached boundary layer on the upstream side of the airfoil. Still, if the LES has wall modelling, the very-near-wall layer will have a model with RANS character, and the grid will usually have RANS proportions, i.e., large cell aspect ratios. In a DNS, there is no such region at all; the cell aspect ratios are much larger than 1, but not unlimited nor Reynolds-number dependent. On the other hand, all simulations have an Euler Region, devoid of small eddies, and the most economical grid will have fairly coarse spacing in the ER.

The less obvious border is between the FR and DR. Recall that the DR is a region downstream of the bodies in which we accept poorer and poorer resolution, even though it contains turbulence. Having the grid spacing prevent the resolution of the smallest eddies is fundamental in LES, but not in DNS. Yet, all DNS of external flows has a DR, under another name or no name. The alternative to this would be to have a domain short enough to afford full DNS resolution everywhere, with an outflow boundary condition placed closely enough to keep the grid count under control. However, such conditions are crude and interfere with vortex shedding and similar phenomena, and the consensus seems clear that it is better to have a large domain with gradually expanding grid cells, and to push the outflow condition far away from the body. Most probably, in the DNS to date, the DR has effectively operated in Implicit LES (ILES) mode, that is, some feature of the numerics stabilized the simulation, as opposed to viscosity acting at the Kolmogorov length scale η which is not resolved any more. It could have been logical to declare that the DR was an LES region appended to a DNS, and activated an SGS model only in that region. Hoffman clearly recognized this concept, using the name DNS/LES, and counting on numerical stabilization in the effective DR [3].

Similar grid-design decisions face the scientist for some internal flows. Consider a backward-facing step flow, and the position of the downstream boundary of the simulation. If we insist on "DNS quality" over the entire domain, in simple terms on resolving η and the wall viscous scale ν/u_τ, much effort will be spent on the region, well past reattachment, where it is intuitively clear the small-scale turbulence cannot influence the step region. Again, one option is to end the domain close to the reattachment line with an outflow condition; the other option is to let the grid coarsen, arguing that only the larger eddies have a "dialogue" with the region upstream. The developing inaccuracy in skin friction, due to under-resolving the near-wall layer, will hardly be felt. Therefore, DNS resolution is not needed, and the best use of resources is to declare this region as a DR. This could also be viewed as a combination of DNS and LES. The resolution is testable in both regions: comparing results with different enough grid spacings provides meaningful indications of the residual errors (purely numerical in the DNS region, mixed numerical/SGS modelling in the LES region). Therefore, a thorough demonstration of accuracy is possible. This is although the order of convergence has no reasons to be the same in the two regions; in fact, to our knowledge, convergence according to a theoretical order of accuracy has never been demonstrated in a DNS or LES complex enough to be interesting. To begin with, it is usually impossible to reduce the scatter due to a finite time sample to levels such that grid convergence could be observed at all.

3 Requirements for a Simulation to be a DNS

DNS first developed primarily in homogeneous turbulence, with periodic conditions and uniform grids. This kept the criteria for "DNS Quality" very simple: the grid spacing (a single number) and time step had to be small enough in Kolmogorov viscous units. The assumption was not of an exact numerical solution, but that numerical errors were negligible. When DNS progressed to channels, the attitude was and still is much the same. There is agreement that *a DNS contains no empirical modelling, and numerical errors are negligible*, which is verified by running at least one case with a resolution finer than the standard, in all four directions. After that, usually when progressing to higher Reynolds numbers, guidelines are applied either in wall units or Kolmogorov units; in short, in viscous units. The simulation is highly accurate everywhere. Yet we have just argued that this attitude is not sustainable in external flows, and in some internal but inflow-outflow cases, and suggested that appending an LES region to a DNS is not illogical, at least downstream. It has not been formally applied to our knowledge; we rather presume that numerical dissipation has been at play. A proper study of that type would include grid-refinement studies in both regions, or for the DNS region the respect of established guidelines in viscous units, and also varying the DNS-LES interface position.

While planning a large DNS of pipe flow (for which they did not find the CPU time), the authors had reflections which are related to the ones just applied to external and inflow-outflow cases, even though the DNS was to be periodic in the streamwise direction. The overwhelming interest was in the Law of the Wall and Karman constant κ, so that the core region (say $r/R < 0.6$ with obvious notation) only needed to support the Reynolds shear stress that coupled with the wall layer. Therefore, resolving η in the core region of the pipe was not essential to the understood purpose of the study. Was it needed, then? Or could a genuine "DNS" aimed only at the wall layer be assembled with classical DNS in the wall region, and LES in the core region, thus reducing the cost? Boundary-layer DNS could lead to similar reflections: is the viscous superlayer resolved accurately, and does it need to be? This depends on the exact purpose of the study. Since some estimates make the viscous super-layer as thin as the viscous sub-layer, it certainly has not been resolved at any interesting Reynolds number.

Even in the law-of-the-wall region of the pipe, the dissipation range was not needed in itself; only the Reynolds shear stress, which is of course dominated by large eddies. If so, it made sense to let the grid spacing Δ grow proportionally to y, the distance from the wall, whereas η grows closely to $y^{1/4}$. At the very large y^+ values which were hoped for, this made some difference. Again, this does not deny the need for grid-refinement studies, but the difference between a systematic grid refinement with $\Delta \propto y$ and one with $\Delta \propto y^{1/4}$ is deep. This DNS may yet be conducted, although recent trends in DNS directed at the Karman constant are very discouraging; the Reynolds numbers needed are far higher than expected even a few years ago.

A similar and less controversial example is as follows. Some DNS studies have focused on the mean velocity and Reynolds-stress tensor. Others have instead focused on predictions of the dissipation tensor, to provide data for Reynolds-stress-transport models. It is clear that calculating these two tensors to the same accuracy does not require quite the same resolution; the dissipation tensor is more demanding. Thus, no DNS is exact, and the appropriate resolution depends on the purpose of the study, including the region of interest, the quantities of interest, and probably other features.

4 Sub-Grid-Scale Modelling in Coarsening Resolution

We have seen that resolved turbulence routinely needs to be convected into coarser grid regions, certainly for external flows and also for some internal ones. This demands the trouble-free destruction of the eddies too small for the grid spacing. The same applies to sound waves; it is not practical to track all such waves all the way to the outer boundaries, which gives great value to the Ffowcs-Williams-Hawkings equation, applied with permeable surfaces made rather tight around the turbulent region. The irrotational region outside the FWH surface can be viewed as a DR for waves, and jumps or even rapid increases in Δ pose a real danger of reflections. A study of the acoustic energy propagating into the outer layers of the grid revealed a steep decline, albeit for a wave with relatively short wavelength [4].

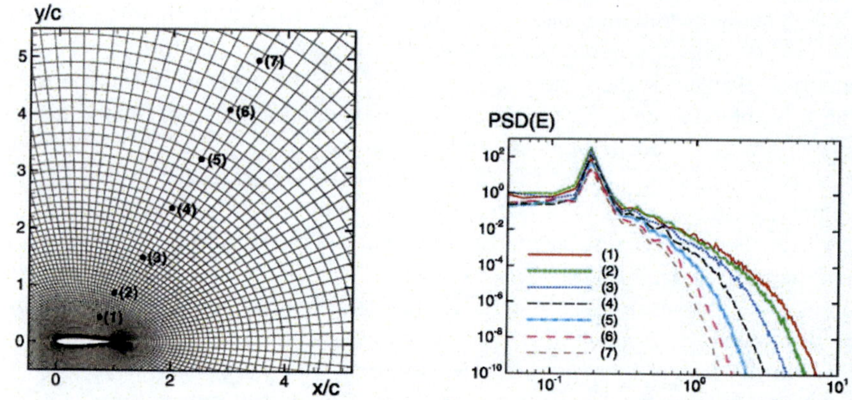

Fig. 3. Airfoil at 60° angle of attack. a) location of points in the wake, and b) corresponding spectra.

Vanella, Piomelli & Balaras carefully addressed this phenomenon, but for step changes in resolution, and that between regions each having a uniform grid [5]. They also considered refinement as well as coarsening; not surprisingly, refinement was trouble-free, as the energy cascade simply extended itself when the grid spacing and accordingly the eddy viscosity decreased. Their position was that, when coarsening, the SGS model ideally should be adjusted to make the energy removal uneventful. In some versions of the model, they used knowledge of the step in Δ a short distance upstream of that change, a similar idea to that of using the derivative of the grid spacing, as proposed here. At first sight, their use of centered differencing schemes left little room for error, but in the end they focused on accuracy and did not report any problems with numerical stability.

Figure 3a shows a line of points in the FR and DR behind a stalled airfoil, at which frequency spectra were calculated. For the present qualitative purposes, Taylor's hypothesis is sufficient to illustrate the behaviour of wave-number spectra, especially since the velocity does not vary widely. Still, flow visualisations show that these points are bathed in turbulence for most of the time, but not all: they spend brief periods in irrotational fluid. This intermittency means that the signals are not homogeneous in time; as a result, even results from very fine grids and high Reynolds numbers may not contain an inertial range. Here, the spectra for the first few points are actually rather consistent with a $-5/3$ slope.

The comparison in Figure 3b displays the gradual "shortening" of the spectrum as the grid coarsens farther from the airfoil, as well as the absence of energy pile-up. In this region the differencing scheme is very close to centered: a weighted average of 4th-order centered and at most 10% of the 5th-order upwind one. From the first to the last point, Δ increases by a ratio of about 7.5, and the frequency of the apparent steep spectral drop drops by a factor of about 5, which is not a perfect fit, but makes the connection plausible. The Nyquist frequency associated with the time step is $St \approx 17$, therefore well above the value where the spectrum drops; therefore, time-integration errors are not a factor in this region (the first spectrum reaches $St = 7$ in the figure, but the spectral density is down by 12 orders of magnitude).

This simulation was continued with a fully centered scheme, and rapidly deteriorated, with spurious short waves invading the wake and expanding into the irrotational region, as seen in figure 4b. This is in spite of implicit time integration. This confirms that the weighted scheme used is close to optimal [6].

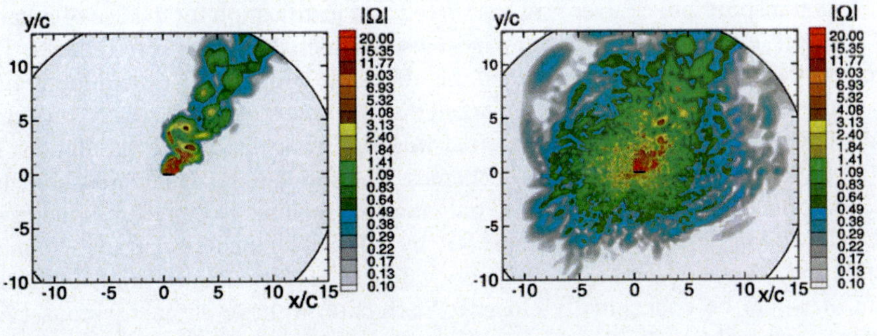

Fig. 4. Vorticity in: a) normal simulation, and b) continuation with centered differencing.

A revealing way to measure numerical dissipation may be as follows. Consider the simple convection of a wave in one dimension, and the effect of upwind differencing on its amplitude after it travels by one wave-length (in other words, after one period). This depends on the resolution, and we pick a resolution with 10 points per wave-length as a compromise. In the past, we have indicated that using 7 points provides useful accuracy with our schemes; on the other hand, some careful studies with other schemes have concluded to the need for 30 points, if not more. Some of these studies were based on very sensitive problems, such as the growth of near-neutral TS waves, and may therefore be unduly pessimistic.

For the first-order upwind scheme, the damping after one period is by 85% in amplitude. It is extreme. For the energy, that is 97.75%, but we will use amplitude. For the second-order upwind scheme, the damping is 31%, which is still intense. The third-order and fifth-order upwind-biased schemes give 12% and 1%, respectively. When a weighted average with any centered scheme is used, the damping can become almost negligible. Using a 10% weight for the fifth-order upwind-biased scheme, which has been common and stable enough for the focus region in our studies, leads to only 0.1% damping (and 0.5% with 7 points per wave). Thus the scheme is not rigorously energy-conserving, but comes extremely close to it for all but the shortest waves. For instance, a wave with only 3 points in it will have 38% damping; thus, the high order of accuracy makes the damping extremely selective.

Returning to the second-order upwind scheme, using 30 points gives a small damping of 1.5%, but comparing with 10, the number of grid points in 3D has increased by a huge ratio. The same second-order scheme with only a 10% weight gives only 3% damping with 10 points per wave. Thus, it is unfortunate that very few codes with the LES option offer the possibility of blending centered and upwind schemes. We recognize the extreme difficulty in obtaining better than second-order accuracy in unstructured codes, but blending two schemes does not appear as challenging. On the other hand, organizing the different blending weights in different regions of the domain is not trivial. Therefore, even to experiment with such an option requires access to the source code, and adds to the complexity of designing the LES study.

This transport into coarser grid has served as a justification for the use of numerical schemes, either monotone or upwind-biased, which do not conserve energy. This refers to the spatial discretisation, and the effects of temporal discretisation are left for future work, although we just remarked that the time step is effectively very short in the DR. It is very common for external-flow turbulence-resolving simulations not to be stable with energy-conserving schemes. The simulations which are stable then take on some of the nature of ILES, if only in some regions. Codes with switches between an upwind-biased scheme in the RR and ER and a much-less biased scheme in the FR and DR are in routine use and very valuable as just dicussed [6]. On the other hand, it would be conceptually closer to perfection to have an energy-conserving scheme, and task the SGS model with the energy removal (the amount removed would therefore be accessible to calculation, which is not the case with numerical dissipation). We will now attempt an estimate of the magnitude of this effect, in an inertial range.

In this concept, in simple terms of energy, the SGS model accounts both for the energy cascade, as usual, and the needed removal of short waves. Ideally, the first role represents interaction with non-resolved waves; it is fairly localized near the high end of the spectrum, but not in a simple way. Ideally, the second role would be very localized, acting only at the extreme end of the spectrum. Applied in Fourier space, it could be a "top-hat" filter, trimming the end of the spectrum. This implies that an eddy viscosity would not be very truthful, since it is far being local in Fourier space.

The SGS model in its first role consumes energy at the rate ε (we are assuming a large cell turbulent Reynolds number, so that the resolved dissipation is negligible). For the second role, let Δ be the grid spacing. If the grid cells are far from isotropic, let it be the larger dimension. The resolved turbulence is responding to a positive value of $D\Delta/Dt$, the Lagrangian derivative of the grid spacing (for waves rather than eddies, the rate of change of Δ at a speed $U \pm c$ would be relevant). We assume that the shape of the end of the resolved spectrum is invariant; in other words, the energy distribution in and beyond the inertial range is $C\varepsilon^{2/3}k^{-5/3}f(k/k_{\max})$ where C is the Kolmogorov constant, $k_{\max}$ is the cut-off wave-number and f a function over $[0,1]$ with $f(x) \to 1$ when $x \to 0$. Its shape is not trivial, because it can be controlled both by spatial and temporal discretisation, not to mention the Smagorinsky or similar constant. Figure 3b does not contradict the idea of an invariant shape, but a larger sample and a firm inertial range would be needed to be precise.

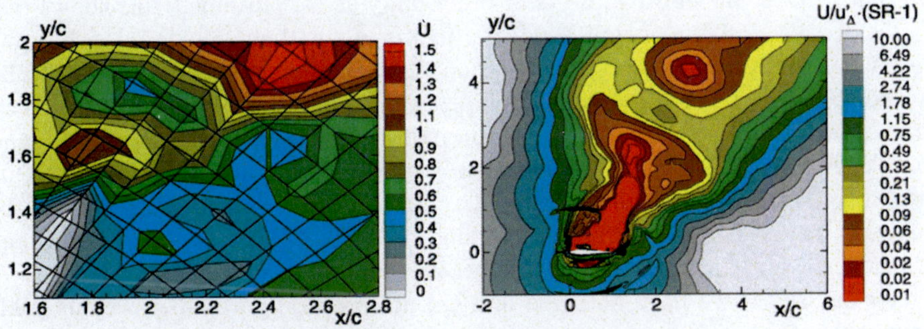

Fig. 5. Flow past an airfoil. a) Velocity contours, showing level of activity within grid cells; b) estimate of coarsening-related energy removal, normalized by true dissipation (equ. 3).

The grid-related energy-removal rate then results from integrating the spectrum up to $k_{\max}$, which is spatially-varying and inversely proportional to Δ so that $D(\log k_{\max})/Dt = -D(\log \Delta)/Dt$. We then have

$$\frac{Dk_{res}}{Dt} = -C' \, \varepsilon^{2/3} \, \Delta^{-1/3} \, \frac{D\Delta}{Dt} \qquad (1)$$

where C' is a constant of order 1, related to the Kolmogorov constant and to the LES spectrum shape f. This equation does not give the total derivative of k_{res}; only the rate related to grid spacing. The ratio of the two rates is

$$\frac{1}{\varepsilon} \frac{Dk_{res}}{Dt} = C' \, \varepsilon^{-1/3} \, \Delta^{-1/3} \, \frac{D\Delta}{Dt} \qquad (2)$$

and will be used to get a sense of the relative intensity of the two phenomena.

We first note that this ratio decreases if the grid is refined by a uniform ratio, because it is proportional to $D(\Delta^{2/3})/Dt$. This is reassuring: the issue vanishes during grid convergence. On the other hand, ε is not easy to estimate. We have $\nu_t S_{ij} S_{ij}$ where ν_{SGS} is the sub-grid viscosity and S_{ij} the strain tensor, but another measure may be more revealing and accessible. The velocity variations over one grid cell, which we will denote by u'_Δ, are proportional to $\varepsilon^{1/3}\Delta^{1/3}$, so that we have for the ratio in (2):

$$\frac{1}{\varepsilon}\frac{Dk_{res}}{Dt} = \frac{C''}{u'_\Delta}\frac{D\Delta}{Dt} = C''\frac{U}{u'_\Delta}\,(SR-1), \tag{3}$$

where C'' is also a constant of order 1, and SR is the stretching ratio from one cell to the next in the direction of the velocity U. A common value for SR in the wall layer, for the wall-normal spacing, is 1.25, but good grids in the FR often require values below 1.1, if not below 1.05 [7].

The values of u'_Δ, tentatively defined in each cell as $u_{\max} - u_{\min} + v_{\max} - v_{\min} + w_{\max} - w_{\min}$, as not as small as expected; see figure 5a, in which the quantity is often 1/2 of the freestream velocity. As a consequence, the values of $U/u'_\Delta\ (SR-1)$ are generally much below 1, as seen in figure 5b. If u'_Δ is defined as the rms within the cell instead of the max − min definition, the values are of course smaller, but still not $\ll U$. If the constant C'' is of order unity as we presumed, this shows that on the present grid, the energy-removal effect is of small magnitude relative to the dissipation, and therefore the need for stabilized differencing schemes may *not* be attributable to it. It should be easily compensated by a moderate turn-up of the end of the spectrum, which raises the dissipation of the resolved field (since both ν_t and $S_{ij}S_{ij}$ increase). This thought is consistent with the findings of Vanella *et al.* This issue may deserve more precise estimates, but the determination of f and C'' requires an inertial range, which is rarely found except in simulations of homogeneous turbulence, which in turn are done with centered differencing and uniform grids.

At this stage, we have not tried and are not proposing a model for this grid-coarsening effect, merely suggesting it for reflection. We consider that an ILES character is very acceptable in the DR, whereas ideally we would prefer to have no numerical dissipation at all in the FR.

5 Explicit Determination of Grid Regions: an Exercise

Automatic grid adaptation will, sooner or later, apply to LES and other turbulence-resolving approaches, like it has with great usefulness to the full-potential equations and is slowly developing for steady Navier-Stokes work. Adaptation shortens the user time and reduces the dependence of the results on the user's skills and patience. It can also improve accuracy for a given computing cost. However, as of today, we know of very few examples of automatic adaptation in DNS or LES [3]. We are only presenting a concept which may be one component of such a system. It relates to the YPG Approach.

The goal is to determine the Departure Region quantitatively. This information can then be used to allow gentle but sustained grid stretching in the DR, while the grid in the FR has a much more uniform target spacing. A plausible definition of the FR is that it contains turbulent eddies which may now or in the future interact with the region of interest. Here, this "region of interest" will be taken to be the immediate vicinity of the airfoil. For a backward-facing step, the analogous region could be the near-wall layer in the recirculation bubble, which is the sensitive region; it could also include a few lines of points across the separated region. The definition depends on the purpose of the study.

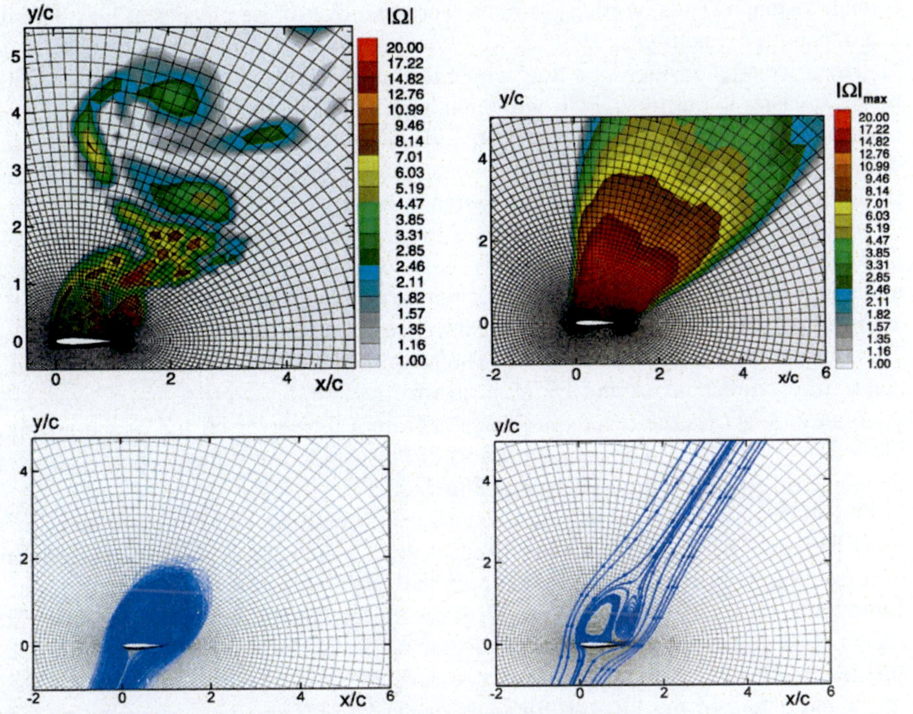

Fig. 6. Flow past an airfoil. a) Instantaneous vorticity; b) maximum vorticity during run; c) reverse trajectories; d) streamlines of the average flow field.

The idea, then, is to calculate reverse trajectories from that region of interest. In simple terms, together these trajectories will outline the region from which particles (eddies) may travel to impact the region of interest. We call this the Contact Region (CR). It is stationary; we are not envisioning time-varying grids yet. It is calculated *after* the simulation and is expensive in terms of storage, because a time-dependent 3D field is stored. As a result, it is practical only on a grid coarser than the one on which the primary simulation is conducted. This will be a precursor simulation. In the present one, the grid is 142*101*35 and the time step $0.03c/U_\infty$. Running such simulations is very good practice in any case, even if the grid generation is manual, and grid generation is rapidly reaching a capability such that little user time will be wasted in doing this (and of course little CPU time, the resolution being coarser in all four directions).

The CR is not a new concept added to the YPG list. It is a mathematical rather than a descriptive definition. Similarly, we are here seeking a mathematical description of the ER. The use of trajectories and of the concept of "contact" appears simplistic, knowing that pressure interactions occur at a distance, but recall that we are after the issue of suppressing small eddies, in other words of displacing vorticity by a small distance. Thus, vorticity is most representative of the physics at play, and it is a transported quantity.

Here, we are considering a flow which is homogeneous in the z direction, which makes it easier to outline the CR without a huge time sample. It is defined in the x-y plane and extruded. In this paper, it is only visual, but a workable definition in a code will be to tag every grid cell which has been visited by a trajectory, and then fill in the likely gaps to arrive at a contiguous area or volume.

Another step in the YPG Approach is to identify the Euler Region. Its elementary definition is simple: the region never visited by non-zero vorticity. For this, the maximum vorticity magnitude ω_{max} during the simulation is recorded, and the ER defined by a low threshold for ω_{max}. Again, here, the max is taken both over t and z. This is illustrated in figures 6a and 6b. The peak vorticity ω_{max} behaves as expected, and setting a threshold around $1 \times U_\infty/c$ is effective.

Figure 6c shows the reverse trajectories from a thin layer on top of the airfoil, located at $y = 0.007c$. The starting points are distributed uniformly spanwise, and the trajectories cover a time interval of $35c/U_\infty$.

The reverse trajectories naturally extend to ∞ upstream and therefore enter the Euler Region. Therefore, the Contact Region gives somewhat of an over-estimate of the extent of the Focus Region. In addition, they are seen to form a bundle, over one chord wide, upstream. The reason is engulfment of irrotational fluid by the turbulent region. This makes the CR based on trajectories overlap with the ER; the true FR is smaller than the CR. The indications it provides are still very concrete regarding the shape and extent of the FR: convincingly, it has an oblong shape truncated by the airfoil, and ends near $y/c = 2$.

When treating this flow with the YPG Approach, the RANS Region is easily outlined along the wall, and ω_{max} helps pick its thickness on the lower side of the airfoil. On the upper side, the RR blends into the FR, which is the intersection of the CR and the Vortical Region. The DR is then the intersection of the VR and the complement of the CR.

The exercise is convincing visually, which was helped by the homogeneity in z, but simple recipes to use the definitions in an automatic system were given. Naturally, such a system will need many decisions on the grid spacing itself, starting with the target FR spacing Δ_0, the first wall-normal spacing, stretching ratios particularly in the FR and DR, and so on. If using a flexible grid-generation package, the spacing in

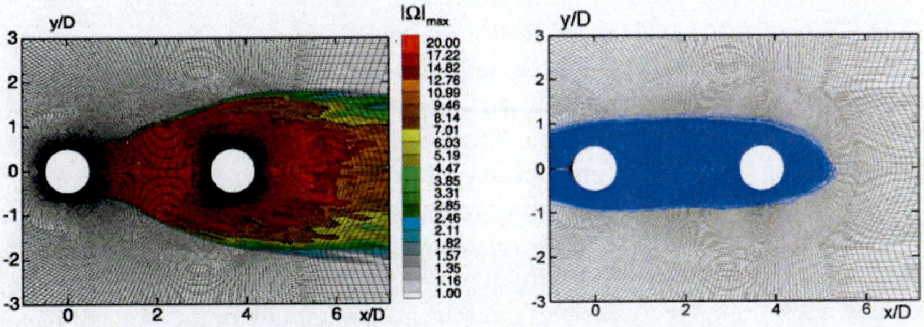

Fig. 7. Flow past tandem cylinders. a) Vortical and Euler regions; b) reverse trajectories.

the ER can be significantly larger than in the VR, including in z. Only the time step is fully shared.

The relationship between the CR and the more established concept of Recirculation Region is studied with figures 6c and 6d. The latter shows the streamlines of the Reynolds-averaged LES flow field. This mean flow being two-dimensional, a closed region or "bubble" dominated by two large circulation regions exists. Its extent is not too different from that of the CR, although it is smaller as could be expected, and the strictly-defined "bubble" (delimited by streamlines which make contact with the body) does not have the upstream "bundle." It also fails to include edge regions which are visited by vorticity and turbulence.

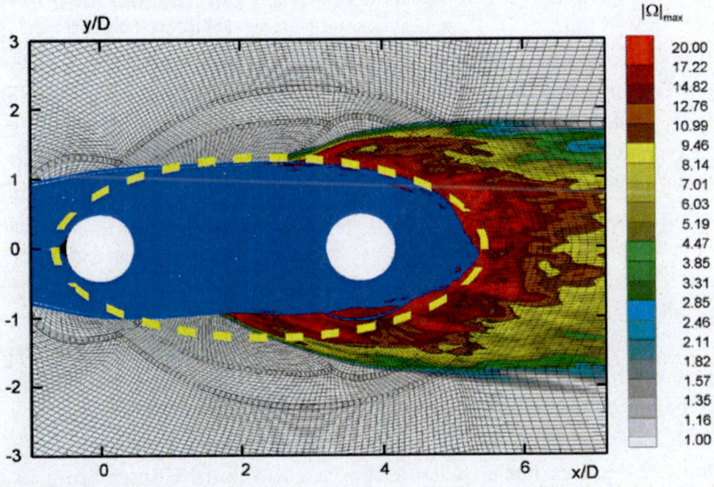

Fig. 8. Flow past tandem cylinders. Vortical region, reverse trajectories, and dashed line tentative Focus Region.

The CR concept also accounts for interference between bodies, as illustrated in figure 7 for tandem cylinders. Both cylinders were declared as Regions of Interest. The second cylinder then "emits" a fairly wide CR which bridges the two bodies. However, the CR is narrower than the VR, which is quite wide past the second cylinder; this will help save on the number of grid points. Note that the mean streamlines as in figure 6d would fail to bridge the two cylinders; thus the CR concept is demonstrating its potential, and plausibility relative to intuition.

Some day, simulations of an airplane will be aimed at whether the turbulence created by a wing spoiler causes buffet on the horizontal tail, or the wake of the nose landing gear disturbs an air inlet or ram-air turbine, and it is possible the present CR concept will be helpful. Other examples would be multiple road vehicles.

6 Outlook

We have provided material which has some preliminary and inconclusive aspects, but we hope is somewhat stimulating. We are fairly confident in this for the QLES community, but not as much about the wider turbulence/CFD community. It appears that only about 100 copies of the bound proceedings are sold outside of the QLES members, which is small. Naturally, this does not account for the "soft" copies of articles disseminated by internet, but presumably these reach only readers which are already connected to a QLES contributor. A valid question is how best to boost the quality of the turbulent CFD in the widest circles we can.

Journal articles derived from the work of QLES contributors must have a readership much larger than 100. Personal exchanges at QLES meeting are also often substantial and productive, thus contributing to research quality. Yet, informal surveys suggest that much of the "turbulence knowledge" CFD users outside academia apply actually travels with the CFD codes they use rather than literature they read. The users extensively consult the support staff of the CFD vendors or the government scientists who write and maintain the codes, for instance at DLR and NASA. The commercial companies have strong incentives to point at the ease of use of their products, which include the turbulence treatments. This creates a danger of overselling these treatments, which we all know to be shaky in many respects, turbulence being "the chief outstanding difficulty of our subject" according to Bradshaw's 1994 article, still fully applicable today [8]. One task of the turbulence research specialists is to decrease the incidence of poor "plug and play" practices, without discouraging the use of CFD or even the use of cutting-edge approaches.

Publishing accessible overview articles, with impartial and candid discussions of the state-of-the-art, is not so easy and is a good path. Contributing to the documentation which travels with the codes could be another. We find that most CFD companies have retained and grown some very competent turbulent-CFD experts, blending genuine scientific attitudes with commercial urges. The same is true for government code suppliers. QLES-type specialists should use every opportunity to collaborate with them, and possibly to directly contribute to the documentation they attach to their products. This appropriately goes beyond the intellectual pleasure and the prestige of publishing at high level, to educate "in real time" the ever-widening ranks of turbulent CFD practitioners.

References

1. Spalart P R (2001) "Young person's guide to Detached-Eddy Simulation grids." NASA CR-2001-211032. http://techreports.larc.nasa.gov/ltrs/PDF/2001/cr/NASA-2001-cr211032.pdf
2. Spalart P R, Hedges L S, Shur M, Travin A (2003) Simulation of active flow control on a stalled airfoil. Flow, Turb. and Comb., 361–373.
3. Hoffman J (2006) Adaptive simulation of the subcritical flow past a sphere. J. Fluid Mechanics Vol. 568, 77–88.
4. Spalart P R, Shur M L, Strelets M Kh (2009) Added sound sources in jets; theory and simulation. Int. J. Aeroacoustics, to appear.
5. Vanella M, Piomelli U, Balaras E (2008) Effect of local grid refinement on large-eddy simulation statistics and flow-fields. ETMM7.
6. Travin A, Shur M L, Strelets M Kh, Spalart P R Physical and Numerical Upgrades in the Detached-Eddy Simulation of Complex Turbulent Flows. In: Fluid Mechanics and Its Applications: Advances in LES of Complex Flows, pp.239–254. Proceedings of the Euromech Colloquium 412. R. Friedrich and W. Rodi, (eds.).
7. Shur M L, Spalart P R, Strelets M (2005) Noise prediction for increasingly complex jets, Part I: Methods and tests. International J. of Aeroacoustics, 4(3+4), 213–246.
8. Bradshaw P (1994) Turbulence: the chief outstanding difficulty of our subject. Expts. in Fluids, 16(3–4), 203–216.

How to estimate the resolution of an LES of recirculating flow

Lars Davidson

Division of Fluid Dynamics, Department of Applied Mechanics, Chalmers University of Technology, SE-412 96 Göteborg, Sweden lada@chalmers.se, www.tfd.chalmers.se/~lada

Summary. How should the resolution of an LES be estimated? There exist guidelines for boundary layers for how to create a grid in terms of the grid size expressed in viscous units. However, in other flow regions there are few – if any – guidelines for how to generate a grid that ensures accurate results. Worse, it is not even clear how to estimate the resolution after having carried out an LES. The present study evaluates the following quantities: energy spectra, dissipation energy spectra, two-point correlations, the ratio of SGS shear stress to resolved shear stress, the ratio of the SGS viscosity to the molecular and the ratio of the SGS dissipation due to the resolved fluctuating velocity gradients to that due to the mean velocity gradients. Two flows are analyzed, namely the flow in a plane asymmetric diffuser and decaying grid turbulence. The main conclusions are that two-point correlations are the best way to estimate the resolution and that energy spectra are not suitable. It is usually assumed that the SGS dissipation takes place at wavenumbers close to cut-off. The present work shows that this idealized picture is not true, but that the SGS dissipation takes place at rather low wavenumbers.

Key words: LES, resolution, two-point correlations, energy spectra, dissipation energy spectra, decaying grid turbulence

1 Introduction

After having carried out a Large Eddy Simulation, the question arises: how do we know that the resolution is sufficient? Or – at least – how do we estimate the resolution? For attached boundary layer flows, numerical experiments reported in the literature indicate that the streamwise and spanwise resolution in viscous units should be approximately 100 and 30, respectively; further, the center of the wall-adjacent cells should be located not more than one viscous unit away from the wall. However, in flow regions outside attached boundary layers, there are few guidelines for how to create a sufficiently fine mesh.

After having carried out an LES simulation, there are a number of ways to estimate the resolution. Energy spectra are frequently used to find out whether the

M.V. Salvetti et al. (eds.), *Quality and Reliability of Large-Eddy Simulations II*, ERCOFTAC Series 16, DOI 10.1007/978-94-007-0231-8_25,

resolved turbulence satisfies the $-5/3$ decay. The ratio of the SGS to resolved turbulence is another quantity; if is is small, it may indicate that the resolution is good. Two-point correlations are also useful for estimating the resolution. If they fall to zero over a separation distance of a few cells, it means that no turbulence is resolved at all; if, on the other hand, they fall to zero over five to ten cells, this is an indication that the largest scales are reasonably well resolved. Another option for evaluating the resolution is to evaluate the SGS dissipation. This dissipation takes place either through the mean flow or through the resolved fluctuations. In RANS, the entire dissipation takes place through the mean flow while in LES it takes places through both the mean flow and resolved fluctuations. Finally, dissipation energy spectra can be created that show the wavenumbers at which the SGS dissipation takes place.

The different ways of estimating resolution mentioned above were evaluated in [5] for fully developed channel flow. In the present study, we use these methods to evaluate the resolution of an LES of recirculating flow and of decaying grid turbulence.

2 Equations

2.1 The momentum equations

The incompressible momentum equation with an added SGS viscosity reads

$$\frac{\partial \bar{u}_i}{\partial t} + \frac{\partial}{\partial x_j}(\bar{u}_i\bar{u}_j) = -\frac{1}{\rho}\frac{\partial \bar{p}}{\partial x_i} + \frac{\partial}{\partial x_j}\left((\nu+\nu_{sgs})\frac{\partial \bar{u}_i}{\partial x_j}\right) \tag{1}$$

2.2 The turbulence model

The dynamic Smagorinsky is used, which reads

$$\nu_{sgs} = C\Delta^2|\bar{s}|, \quad \bar{s}_{ij} = \frac{1}{2}\left(\frac{\partial \bar{u}_i}{\partial x_j} + \frac{\partial \bar{u}_j}{\partial x_i}\right), \quad |\bar{s}| = (2\bar{s}_{ij}\bar{s}_{ij})^{1/2} \tag{2}$$

The dynamic coefficient is computed as

$$\begin{aligned} C &= -\frac{\langle \mathscr{L}_{ij}M_{ij}\rangle_z}{\langle 2M_{ij}M_{ij}\rangle_z} \\ M_{ij} &= \left(\widehat{\Delta}^2 |\widehat{\bar{s}}|\widehat{\bar{s}}_{ij} - \Delta^2\widehat{|\bar{s}|\bar{s}_{ij}}\right), \quad \mathscr{L}_{ij} \equiv \widehat{\bar{u}_i\bar{u}_j} - \widehat{\bar{u}}_i\widehat{\bar{u}}_j \end{aligned} \tag{3}$$

where $\widehat{\cdot}$ and $\langle . \rangle_z$ denote test filtering and spanwise averaging, respectively. The dynamic coefficient is limited to avoid negative total viscosity, i.e. $\nu + \nu_{sgs} \geq 0$.

2.3 The Numerical Method

An incompressible, finite volume code is used [6]. For space discretization, central differencing is used for all terms in the momentum equations. The Crank-Nicolson scheme (with $\alpha = 0.5$) is used. Convective boundary conditions are used at the outlet. The numerical procedure is based on an implicit, fractional step technique with a multigrid pressure Poisson solver [7] and a non-staggered grid arrangement.

2.4 Inlet boundary conditions

A fluctuating velocity field is generated each time step using a synthetic isotropic turbulence generator [4]. The velocity fields are independent of each other, however, and their time correlation will thus be zero. This is unphysical. To create correlation in time, new fluctuating velocity fields, $\mathscr{U}_i'$, are computed based on an asymmetric time filter.

$$\mathscr{U}_i'^m = a\mathscr{U}_i'^{m-1} + bu_i'^m. \tag{4}$$

Here, m denotes the time step number, and $a = \exp(-\Delta t/\mathscr{T})$; Δt and $\mathscr{T}$ denote the computational time step and a prescribed turbulent time scale, respectively. The asymmetric time filter resembles the spatial digital filter presented in [9]. The second coefficient is taken as $b = (1-a^2)^{0.5}$ which ensures that $\langle \mathscr{U}_i'^2 \rangle = \langle u_i'^2 \rangle$ ($\langle \cdot \rangle$ denotes averaging). The time correlation of $\mathscr{U}_i'$ will be equal to $\exp(-\Delta t/\mathscr{T})$, and thus Eq. 4 is a convenient way to prescribe the turbulent time scale of the fluctuations. The inlet boundary conditions are prescribed as

$$\bar{u}_i(0,y,z,t) = U_{i,in}(y) + \mathscr{U}_i'^m(y,z) \tag{5}$$

$U_{i,in}(y)$ denotes the mean inlet profile, which is taken from a DNS of fully developed channel flow at $Re_\tau = 500$. For greater detail, see [4].

The turbulent length scale and time scale are set to $L_t = 0.1H/2$ and $\mathscr{T} = 0.05(H/2)/u_{\tau,in}$ (see Fig. 1), respectively. The RMS amplitudes of the inlet fluctuations are scaled so that $u_{rms,in} = v_{rms,in} = w_{rms,in} = u_\tau$.

The synthetic fluctuations created with the method presented above yield homogeneous turbulence in the inlet plane, i.e. $u_{rms,in}$, $v_{rms,in}$, and $w_{rms,in}$ are constant (and equal) across the entire inlet plane. The fluctuations must be reduced near the wall so that they go smoothly to zero as the wall is approached; this is done for $n/H \lesssim 0.016$, where n is the distance to the closest wall.

3 Results

3.1 Diffuser

The configuration is an asymmetric plane diffuser with Reynolds number $Re = U_{b,in}H/\nu = 18000$, see Fig. 1. The opening angle is 10°. The inlet bulk velocity, the channel height at the inlet and the density are set to one, so that $\nu = 1/Re$. The dynamic Smagorinsky model is used, and predictions are compared with experiments [1]. The mesh in the $x-y$ plane has 256×64 ($N_x \times N_y$) or 512×64 cells (two cells in the x direction in the latter mesh correspond to one cell in the former mesh). In the z direction, $N_z = 32$, 64 or 128. In the inlet region, the friction Reynolds number based on half the channel height is approximately 500, which gives a spanwise grid size of $\Delta z^+ = (125, 62, 31)$ for $N_z = (32, 64, 128)$. The time step is $\Delta t = 0.023$ for $N_x = 512$ and $\Delta t = 0.039$ for $N_x = 256$; this results in a maximum instantaneous Courant number of 1.9. Time averaging is carried out over approximately 60 000 time steps. All simulations are made on a Linux PC using a single core. One or two global iterations are required each time step to reach convergence. The CPU time for the finest mesh ($512 \times 64 \times 128$) is 45s per time step (two iterations per time step).

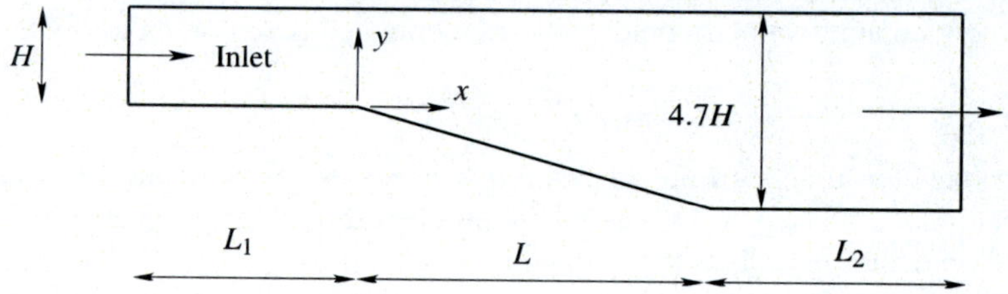

Fig. 1. Plane asymmetric diffuser (not to scale). $L_1 = 7.9H$, $L = 21H$, $L_2 = 28H$. The spanwise width is $z_{max} = 4H$. The origin of $x-y-z$ is at the lower wall at the entrance of the diffuser.

Figure 2 compares the predicted velocity profiles with experiments [1]. The agreement is not especially good for any resolution. LES simulations were also presented in [11]. Their meshes were similar; they employed the same turbulence model and their domain was shorter but had the same spanwise extent. They did use a finer mesh in the wall-normal direction and different inlet boundary conditions. However, the aim of the present work is not to achieve as good an agreement with experiments as possible but to evaluate the resolution for different grids.

The resolved shear stresses are presented in Fig. 3. The agreement is reasonable for $x \geq 13H$ but the stresses are overpredicted compared with experiments for $x \leq 6H$. All resolutions give very similar shear stresses.

Below we will make a detailed comparison of the flow at one streamwise position where the flow is attached ($x = -H$) and one where the flow exhibits incipient separation ($x = 20H$). At $x = -H$ the ratio of streamwise to the spanwise cell side is 0.3 for $N_z = 32$ and $N_x = 512$; at $x = 20H$ the corresponding ratio is one. Figures 4 and 5 show the shear stresses. As can be seen, the spanwise resolution is very important in the attached boundary layer, but the effect of the streamwise resolution is almost negligible. Also the spanwise resolution is unimportant in the incipient separation region, and all grids give virtually identical resolved shear stresses. Since it is seen that the streamwise resolution has no effect, we concentrate hereafter on comparing the spanwise grid resolutions for the coarse streamwise grid ($N_x = 256$).

Figure 6 presents the two-point correlation ($B_{ww}(\hat{z}) = \langle w'(z)w'(z-\hat{z})\rangle$). The correlations are presented at $x = -H$ (attached flow) and $x = 20H$ (incipient separation)

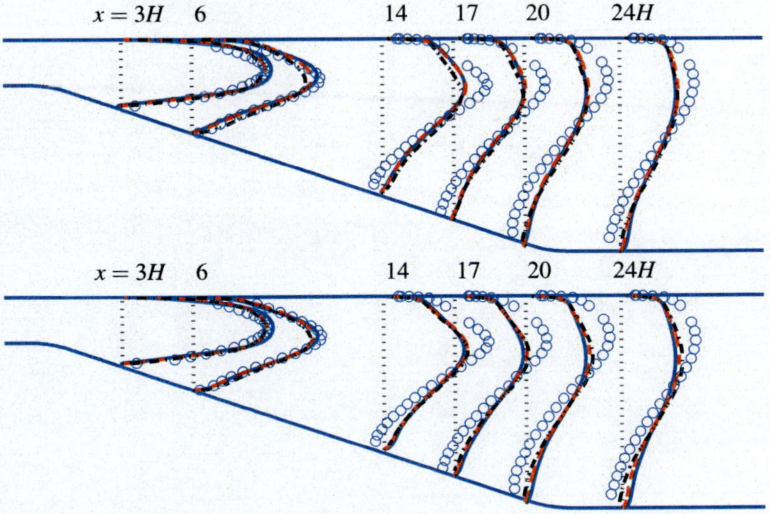

Fig. 2. $\langle \bar{u} \rangle / U_{b,in}$ profiles. Top: $N_x = 256$ cells; bottom: $N_x = 512$ cells. —— : $N_z = 32$; - - - : $N_z = 64$; - · - : $N_z = 128$; markers: experiments [1].

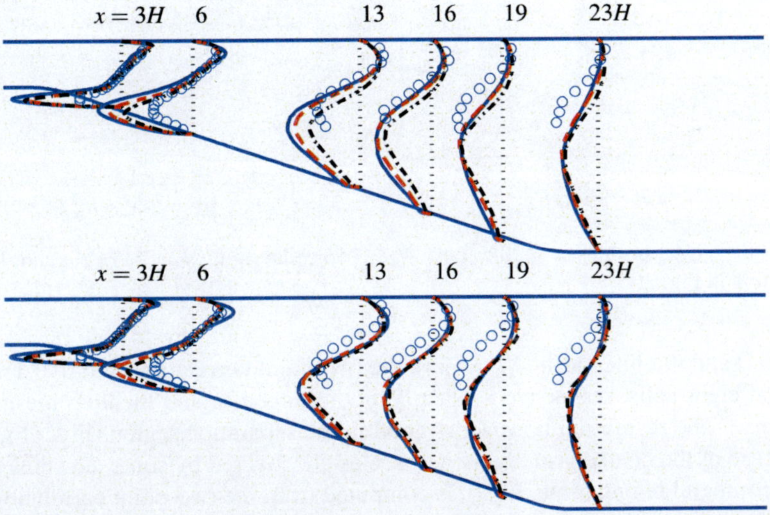

Fig. 3. $\langle u'v' \rangle / U^2_{b,in}$ profiles. Top: $N_x = 256$ cells; bottom: $N_x = 512$ cells. For legend, see caption in Fig. 2. Markers: experiments [1].

at y locations for which the magnitude of the resolved shear stress is large (see Figs. 4 and 5). The resolution for the coarse grid ($N_z = 32$) is indicated by markers in Fig. 6. As can be seen, the normalized two-point correlation at $x = -H$ (Fig. 6a) for $N_z = 32$ falls to 0.1 within a separation distance of two cells. This means that the largest scales are resolved by only two cells, i.e. they are not resolved at all. For the medium mesh

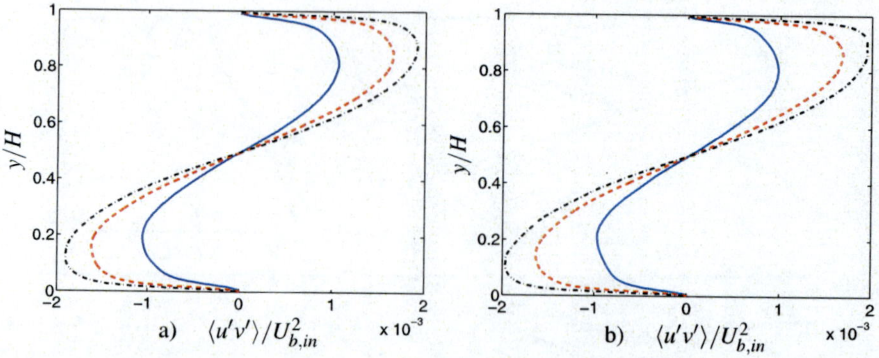

Fig. 4. $\langle u'v'\rangle/U_{b,in}^2$ profiles. $x = -H$. Left: $N_x = 256$ cells; right: $N_x = 512$ cells. For legend, see caption in Fig. 2.

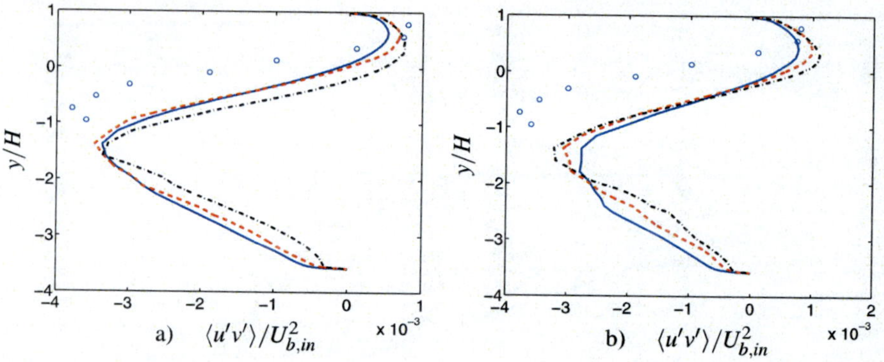

Fig. 5. $\langle u'v'\rangle/U_{b,in}^2$ profiles. $x = 20H$. Left: $N_x = 256$ cells; right: $N_x = 512$ cells. For legend, see caption in Fig. 2.

($N_z = 64$) and the fine mesh ($N_z = 128$), the two-point correlations fall to 0.1 within four and eight cells, respectively. Clearly, both the coarse and medium meshes are too coarse. The resolution however is good in the separation region (Fig. 6b). Even in the case of the coarse grid, the largest scales are covered by some ten cells.

The integral length scale, $L_{int,w}$, is computed from the two-point correlations as

$$L_{int,w}(x,y,\hat{z}) = \frac{1}{w(x,y)^2_{rms}} \int_0^{z_{limit}} \langle w'(x,y,z,t)w'(x,y,z,t)\rangle d\hat{z} \tag{6}$$

where the averaging $\langle . \rangle$ is done, as usual, in time (t) and the spanwise direction (z). The upper boundary of the integral, z_{limit}, is z_{max} or when the normalized two-point correlation falls below a small value ($= 10^{-4}$). As can be seen in Fig. 7, $L_{int,w}$ is reduced when the grid is refined. Figure 7 also confirms that the spanwise domain is sufficiently large, since $L_{int,w} < 0.5z_{max} = 2H$.

The energy spectra corresponding to the two-point correlations (Fig. 6) are shown in Fig. 8. The smallest wavenumber is $\kappa_{z,min} = 2\pi/z_{max} = 2\pi/4H = 1.57/H$. The

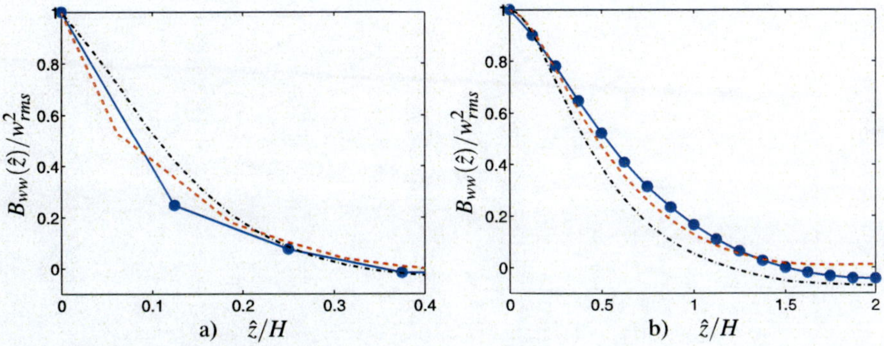

Fig. 6. Normalized two-point correlation $B_{ww}(\hat{z})/w^2_{rms}$. $N_x = 256$. Left: $x = -H$, $y = 0.15H$; right: $x = 20H$, $y = -2.9H$. Markers on the solid line show the coarse resolution. For legend, see caption in Fig. 2.

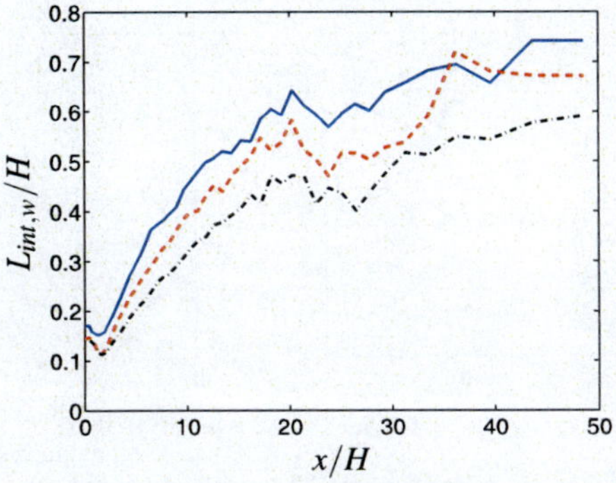

Fig. 7. Integral length scale, $L_{int,w}$ computed from B_{ww}, see Eq. 6. $(y - y_{low})/(y_{high} - y_{low}) = 0.15$. For legend, see caption in Fig. 2.

largest wavenumber must be resolved by more than two cells, which for $N_z = 32$ gives $\kappa_{z,max} < 2\pi/(2\Delta z) \simeq 25/H$. The decay of E_{ww} with wavenumber is small for the coarse mesh. However, the energy spectra for the medium and the fine mesh exhibit a decay versus wavenumber close to $-5/3$, indicating that the turbulence is well resolved. Still, as seen from the two-point correlation, the largest scales on the medium mesh are resolved by only four cells, which must be considered to be insufficient. Hence, it seems that energy spectra are not suitable for estimating the resolution. The energy spectra all exhibit a pile-up of energy at the largest wavenumber, and this indicates that the SGS dissipation is too small at these wavenumbers.

In the incipient separation region, Fig. 8b, all spectra show a decay slightly larger than $-5/3$, indicating a sufficient resolution. This indication agrees with the two-point correlations (Fig. 6b) for which it was seen that the largest scales are resolved by at least ten cells, even on the coarse mesh.

The energy spectra, $E_{ww}(\kappa_z)$, presented in Fig. 8 were obtained by Fourier transforming (FFT) the corresponding two-point correlation, $B_{ww}(\hat{z})$. This can be done

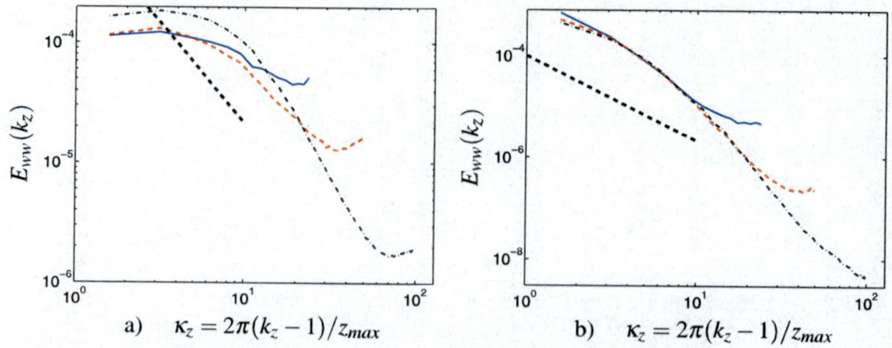

Fig. 8. Energy spectra $E_{ww}(k_z)$. The thick dashed line shows $-5/3$ slope. $N_x = 256$. Left: $x = -H$, $y = 0.15H$; right: $x = 20H$, $y = -2.9H$. For legend, see caption in Fig. 2.

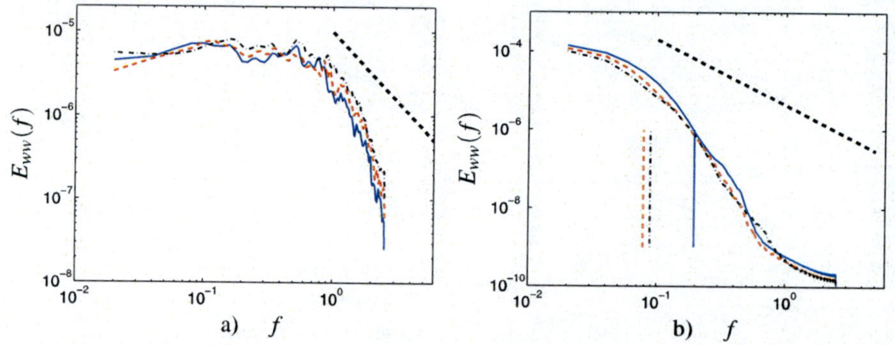

Fig. 9. Energy spectra $E_{ww}(f)$ vs. frequency. The thick dashed line shows $-5/3$ slope. $N_x = 256$. Left: $x = -H$, $y = 0.15H$; right: $x = 20H$, $y = -2.9H$. For legend, see caption in Fig. 2.

only if z is a homogeneous coordinate direction, i.e. if the predictions were carried out using periodic boundary conditions in the z direction. This is of course not the case in general. The only way to create energy spectra is generally to Fourier transform the time history of a variable (e.g. a velocity component) at a given point. The time signal is usually chopped up into small segments, making an FFT of each segment using an overlap of the segments and then averaging the spectra of all segments. Special treatment is given here to make the signal in each segment periodic. Here, the `pwelch` command in Matlab is used setting the length of each segment to $NT = 256$ (the time signal was created by sampling every fifth time step). The `pwelch` command reads

$$[\texttt{pw1 fw1}] = \texttt{pwelch}(\texttt{w1}, \texttt{NT}, [], [], 1/(5 * \texttt{dt}));$$

where $w1$, $fw1$, $pw1$ and dt denote the resolved spanwise velocity, w', the frequency, the square of the Fourier coefficients and the computational time step, respectively. The length of the `w1` vector is approximately 10000.

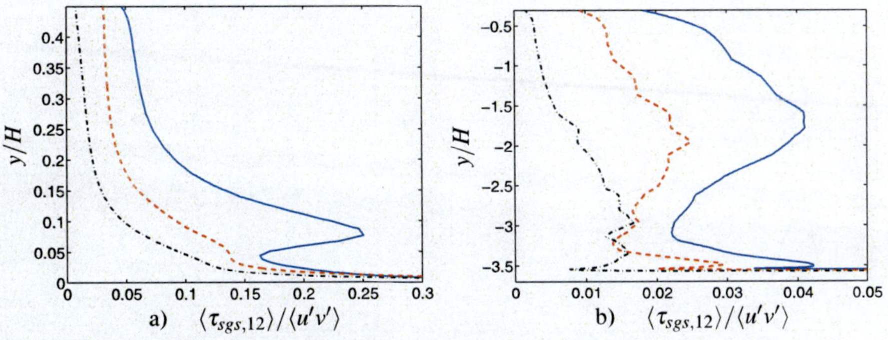

Fig. 10. Ratio of SGS to resolved Reynolds shear stress. $N_x = 256$. Left: $x = -H$; right: $x = 20H$. For legend, see caption in Fig. 2.

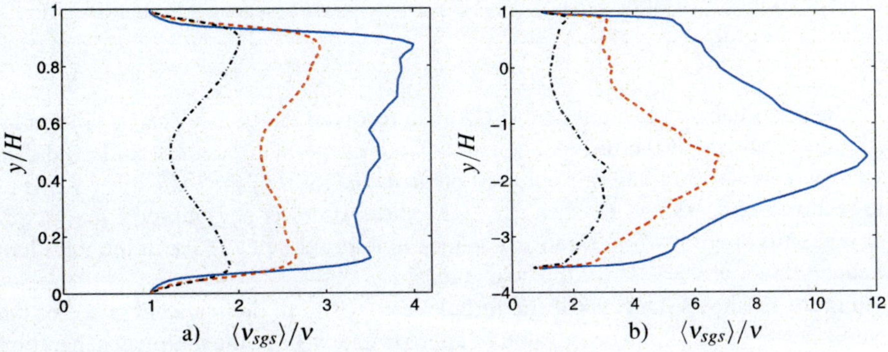

Fig. 11. Ratio of SGS to physical viscosity. $N_x = 256$. Left: $x = -H$; right: $x = 20H$. For legend, see caption in Fig. 2.

Figure 9 presents the spectra versus frequency. At $x = -H$, all spectra are flat for f smaller than $1Hz$; for larger frequencies, they exhibit a $-5/3$ decay or steeper. At $x = 20H$, the spectra for the three resolutions are very similar and exhibit a rather steep decay for $f > 0.1Hz$. The steep decay is related to the fact that the time signal has been sampled with a frequency higher than that corresponding to a local Courant number of one. The vertical lines in Fig. 9b show the frequency based on the time averaged velocity over the spatial grid size, $\langle \bar{u} \rangle / \Delta x$. Frequencies above this threshold correspond to "over-resolution" in time, i.e. the local Courant number is smaller than one. In Fig. 9a the Courant number based on the sampling frequency (every fifth time step), is close to three, i.e. the frequency, $\langle \bar{u} \rangle / \Delta x$, is equal to approximately 14. The conclusion drawn from the spectra presented in Figs. 8 and 9 is that they do not present a reliable picture of the resolution. Considering Fig. 9a it seems that all three simulations are well resolved, and Fig. 8a indicates that the prediction with the medium mesh is well resolved. However, the two-point correlation (Fig. 6a) shows that it is only the simulation on the fine mesh ($N_z = 128$) that is reasonably well resolved.

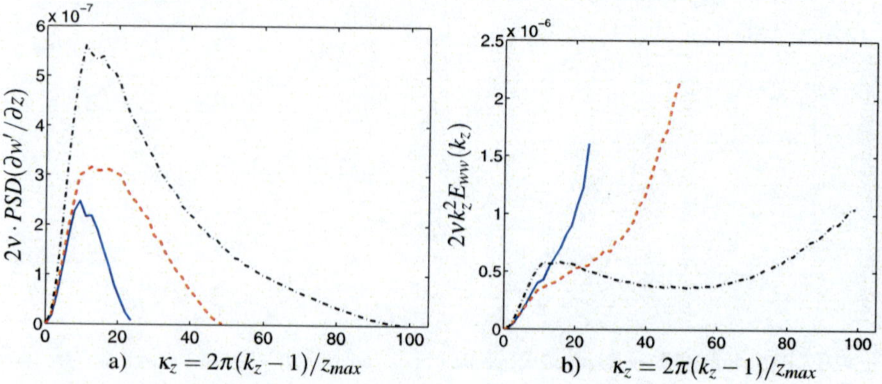

Fig. 12. Exact (left) and approximated (right) dissipation energy spectra of a spanwise component of viscous dissipation versus spanwise wavenumber. For legend, see caption in Fig. 2. $x = -H, y = 0.15H$

The relation between modelled (SGS) and resolved turbulence can be used as an estimate of how well the turbulence is resolved. Large SGS stresses could indicate a poorly resolved simulation. Figure 10 presents the ratio of the SGS shear stress to the resolved one. As can be seen, the SGS shear stress is negligible in the incipient separation region while it reaches values of approximately 25% in the turbulent boundary layer at $x = -H$ for the coarse mesh.

Figure 11 shows the ratio of the turbulent viscosity to the physical one. For the coarse mesh, the ratio reaches a value of approximately 12 in the incipient separation region while it is three times smaller (approximately 4) in the attached flow region. Both ratios $\langle \tau_{sgs,12} \rangle / \langle u'v' \rangle$ and $\langle \nu_{sgs} \rangle / \nu$ behave consistently: the better the resolution, the smaller they are. However, the ratio $\langle \nu_{sgs} \rangle / \nu$ is larger at $x = 20H$ than at $x = -H$ which should indicate that the turbulence is worse resolved at the former location than at the latter which is incorrect. Hence, this quantity is not reliable for estimating the resolution.

It is commonly assumed that the SGS dissipation is largest at wavenumbers close to the cut-off. To investigate at which wavenumber the SGS dissipation does takes place, we will investigate the dissipation energy spectra. Since the spanwise coordinate, z, is a homogeneous coordinate direction, it is suitable to investigate the energy spectrum of the spanwise component of the SGS dissipation including a spanwise derivative, for example $\partial w'/\partial z$. A discrete Fourier transform of $\partial w'/\partial z$ is formed as

$$\hat{D}_z(k_z) = \frac{1}{N_z} \sum_{n=1}^{N_z} \frac{\partial w'(n)}{\partial z} \left[\cos\left(\frac{2\pi(n-1)(k_z-1)}{N_z} \right) - \imath \sin\left(\frac{2\pi(n-1)(k_z-1)}{N_z} \right) \right] \tag{7}$$

where $\hat{D}_z$ are the complex Fourier coefficients of $\partial w'/\partial z$. Then the power spectral density (PSD) of $\partial w'/\partial z$, i.e. $\hat{D}_z * \hat{D}_z^*$, where superscript $*$ denotes a complex con-

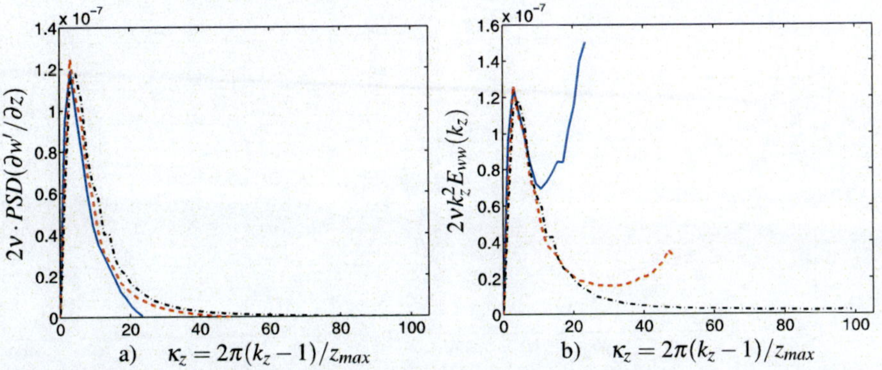

Fig. 13. Exact (left) and approximated (right) dissipation energy spectra of a spanwise component of viscous dissipation versus spanwise wavenumber. For legend, see caption in Fig. 2. $x = 20H$, $y = -2.9H$

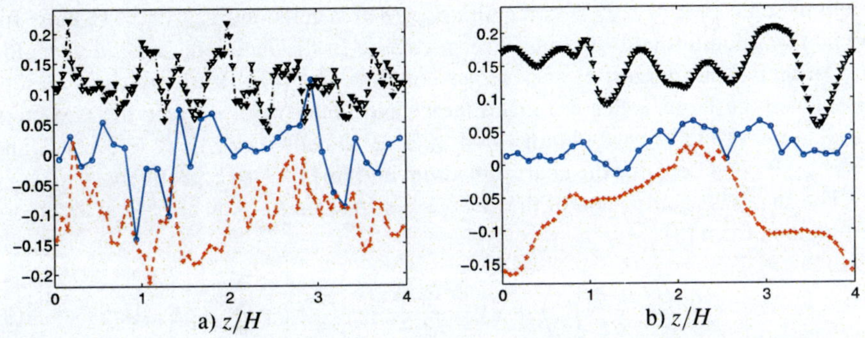

Fig. 14. Instantaneous velocity. Left: $x = -H$, $y = 0.15H$; right: $x = 20H$, $y = -2.9H$. ——, ○: $N_k = 32$, $\bar{w}$; - - -, +: $N_k = 64$, $\bar{w} - 0.1$; -·-, ▽: $N_k = 128$, $\bar{w} + 0.13$.

jugate, can be formed. The time-averaged value of $(\partial w'/\partial z)^2$ can be computed both in physical and wavenumber space, i.e.

$$\left\langle \left(\frac{\partial w'}{\partial z} \right)^2 \right\rangle = \sum_{k_z=1}^{N_z} \langle \hat{D}_z * \hat{D}_z^* \rangle = \sum_{k_z=1}^{N_z} PSD \left(\frac{\partial w'}{\partial z} \right) \tag{8}$$

The square of the Fourier coefficients, i.e. $\hat{D}_z * \hat{D}_z^*$, is computed, and time and spanwise averaged at run-time in the LES code.

The viscous dissipation corresponding to $\partial w'/\partial z$ for $x = -H$ is shown in Fig. 12a. As can be seen, it is largest at surprisingly small wavenumbers. In the inertial subrange with a $\kappa_z^{-5/3}$ behaviour, the viscous dissipation should – according to Eq. 9 – vary as $\kappa_z^{1/3}$.

For $x = -H$ the peaks of the spectra occur at $k_z = 7$, $k_z = 9$ and $k_z = 9$ for the coarse, medium and fine mesh, respectively (the first wavenumber represents the mean, which is equal to zero). This means that derivative $\partial w'/\partial z'$ is largest for

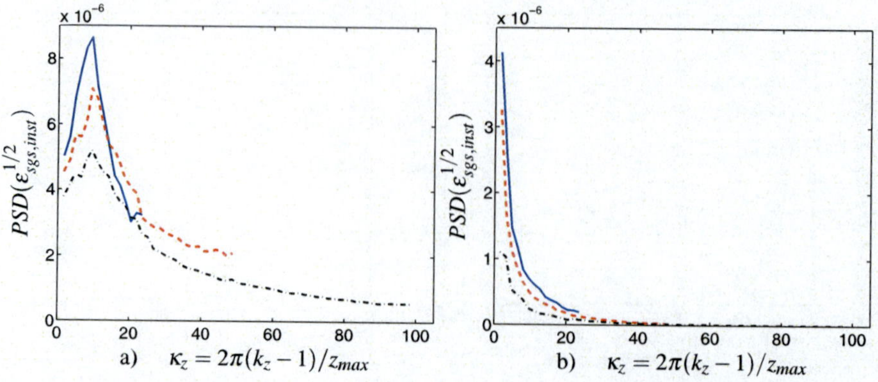

Fig. 15. Exact dissipation energy spectra of the SGS dissipation, versus spanwise wavenumber. Left: $x = -H$, $y = 0.15H$; right: $x = 20H$, $y = -2.9H$. For legend, see caption in Fig. 2.

a length scale of $\ell = z_{max}/6 \simeq 5$ cells, $z_{max}/8 = 8$ cells and $z_{max}/8 = 16$ cells for the coarse, medium and fine mesh, respectively. In the incipient separation region the derivatives are largest at even larger length scales (Fig. 13a), which is reasonable, since the large scales are much larger and better resolved (see the two-point correlations in Fig. 6); we obtain $\ell = z_{max}/3 \simeq 10$ cells, $\ell = z_{max}/2 = 32$ cells and $\ell = z_{max}/2 = 64$ cells for the coarse, medium and fine mesh, respectively.

The spanwise component of the viscous dissipation, ε_{wz}, can also — in theory — be obtained from [8]

$$\varepsilon_{wz} = 2\nu \left\langle \left(\frac{\partial w'}{\partial z} \right)^2 \right\rangle = 2\nu \frac{\partial^2 B_{ww}(\hat{z})}{\partial \hat{z}^2} \bigg|_{\hat{z}=0} = 2\nu \sum_{k_z=1}^{N_z} \kappa_z^2 E_{ww}(k_z) \tag{9}$$

where $\kappa_z = 2\pi(k_z - 1)/z_{max}$. However, this relation is not satisfied at the discrete level, because derivative $\partial w'/\partial z$ cannot be evaluated exactly in a finite-volume approach; the expression in Eq. 9 is based on an exact evaluation of the derivative. The viscous dissipation, ε_{wz}, for $x = -H$ is presented in Fig. 12b. In theory, the spectra presented in Figs. 12a and 12b should be equivalent. However, as can be seen, there is a large discrepancy between the spectra, especially at high wavenumbers. This discrepancy is a measure of the insufficient accuracy of the finite volume method for evaluating $\partial w'/\partial z$ at high wavenumbers (small scales).

Figure 13 presents the exact and approximated spectra in the incipient separation region. The picture is much the same: the maximum dissipation takes place at small wavenumbers, and there is a discrepancy between the exact and the approximated spectra, although the difference is smaller than for $x = -H$. The difference is actually very small for the fine mesh ($N_k = 128$), which indicates that the resolution is good (which is correct judging from the two-point correlation). Comparison of the exact and the approximated spectra could be an interesting approach estimating resolution. Unfortunately, it requires a homogeneous flow direction.

To get a more visual picture of how well the spanwise velocity component is resolved, Fig. 14 presents the w' velocity at one chosen instant. As can be seen, the coarse grid results are very poorly resolved and the profile exhibits odd-even oscillations. Each peak is resolved by at least two nodes at the medium mesh, but the velocity profile does not look physical. The finest mesh shows a tendency to be well resolved. The situation at $x = 20H$ is completely different. The velocity profiles on both the medium and the fine grid are very well resolved. The w' profile obtained with the coarse grid exhibits some odd-even oscillations but its behavior generally seems to be physical. It should be pointed out that the information from instantaneous pictures can be misleading and that it is safer to look at time-averaged quantities.

We present energy spectra of the spanwise derivative of w' in Figs. 12 and 13 and we assume that these spectra are representative for the spectra of the SGS dissipation. Let us check this assumption. To analyze the spectra of the total dissipation, a DFT is created by replacing $\partial w'/\partial z$ in Eq. 7 with the square-root of the instantaneous dissipation, i.e. $\varepsilon_{sgs,inst}^{1/2}$ (see Eq. 10). Figure 15 presents the energy spectra of the SGS dissipation. The first wavenumber, which corresponds to $\varepsilon_{sgs,mean}$ (see Eq. 11), is omitted. The peaks are located at approximately the same κ_z as those in Figs. 12 and 13, but the peaks in the former ones are more dominant $x = -H$. The amplitude in the SGS spectra also differ more for the different grids than those in Figs. 12 and 13; this is most likely related to the different SGS viscosities for the different grids, see Fig. 11.

Contrary to the case in RANS, where the main role of the turbulence viscosity is to act as a diffusion term in the momentum equation, the main objective of the SGS viscosity in LES is to dissipate resolved turbulent kinetic energy, $k_{res} = 0.5\langle u_i' u_i' \rangle$. The contribution of diffusion of the SGS viscosity in LES is usually negligible compared to that of the resolved turbulence. The SGS dissipation term in the k_{res} reads [3]

$$\begin{aligned} \varepsilon_{sgs}' &= \left\langle \left(\tau_{ij,sgs} - \langle \tau_{ij,sgs} \rangle\right) \frac{\partial u_i'}{\partial x_j} \right\rangle = \left\langle \left(\nu_{sgs} \frac{\partial u_i}{\partial x_j} \right)' \frac{\partial u_i'}{\partial x_j} \right\rangle \\ &= \left\langle \nu_{sgs} \frac{\partial \bar{u}_i}{\partial x_j} \frac{\partial \bar{u}_i}{\partial x_j} \right\rangle - \left\langle \nu_{sgs} \frac{\partial \bar{u}_i}{\partial x_j} \right\rangle \left\langle \frac{\partial \bar{u}_i}{\partial x_j} \right\rangle = \varepsilon_{sgs} - \left\langle \nu_{sgs} \frac{\partial \bar{u}_i}{\partial x_j} \right\rangle \left\langle \frac{\partial \bar{u}_i}{\partial x_j} \right\rangle \end{aligned} \tag{10}$$

where the right side on the first line is obtained because the cross-diffusion term has been omitted in the momentum equations (see Eq. 1). The SGS viscosity also appears in the dissipation term in the K equation, see Fig. 16, which reads [3]

$$\varepsilon_{sgs,mean} = \langle \tau_{ij,sgs} \rangle \frac{\partial \langle \bar{u}_i \rangle}{\partial x_j} = \left\langle \nu_{sgs} \frac{\partial \bar{u}_i}{\partial x_j} \right\rangle \frac{\partial \langle \bar{u}_i \rangle}{\partial x_j} \tag{11}$$

which sometimes may be approximated as

$$\varepsilon_{sgs,mean,approx} = \langle \nu_{sgs} \rangle \frac{\partial \langle \bar{u}_i \rangle}{\partial x_j} \frac{\partial \langle \bar{u}_i \rangle}{\partial x_j} \tag{12}$$

In the same manner, Eq. 10 may sometimes be estimated as

$$\varepsilon_{sgs,approx}' = \left\langle \nu_{sgs} \frac{\partial \bar{u}_i}{\partial x_j} \frac{\partial \bar{u}_i}{\partial x_j} \right\rangle - \langle \nu_{sgs} \rangle \left\langle \frac{\partial \bar{u}_i}{\partial x_j} \right\rangle \left\langle \frac{\partial \bar{u}_i}{\partial x_j} \right\rangle \tag{13}$$

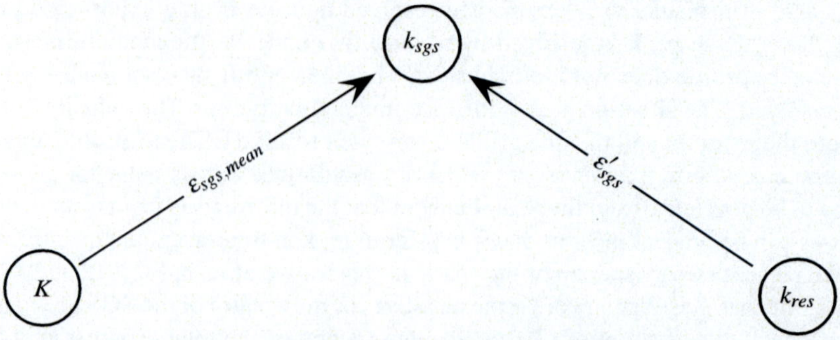

Fig. 16. Transfer of kinetic turbulent energy between time-averaged, resolved and SGS kinetic energy. $K = \frac{1}{2}\langle \bar{u}_i \rangle \langle \bar{u}_i \rangle$ and $k_{res} = \frac{1}{2}\langle u'_i u'_i \rangle$ denote time-averaged kinetic and resolved turbulent kinetic energy, respectively. k_{sgs} denotes time-averaged SGS kinetic energy. The dissipations, ε'_{sgs} and $\varepsilon_{sgs,mean}$, are defined in Eqs. 10 and 11.

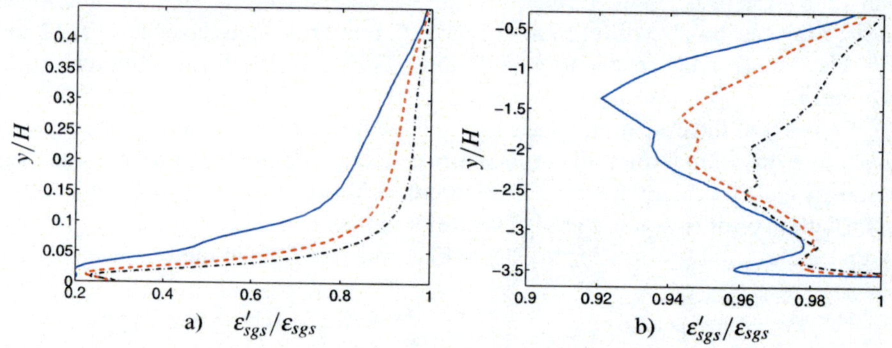

Fig. 17. Ratio of dissipation due to resolved fluctuations, ε'_{sgs}, to total SGS dissipation, $\varepsilon_{sgs} = \varepsilon_{sgs,mean} + \varepsilon'_{sgs}$. Left: $x = -H$; right: $x = 20H$. For legend, see caption in Fig. 2.

It is found that the approximations, Eqs. 12 and 13, agree within 10% with their corresponding exact expressions in Eqs. 11 and 10. However, when the turbulent viscosity is large – such as in the URANS region in hybrid LES-RANS [5] – the approximation is inaccurate.

The transfer of turbulent kinetic energy is illustrated in Fig. 16. The right part of the figure (i.e. ε'_{sgs}) vanishes in RANS whereas it dominates in a well-resolved LES. Hence, to estimate how well the turbulence is resolved, it may be interesting to compare ε'_{sgs} and $\varepsilon_{sgs,mean}$. This is done in Fig. 17. At $x = -H$, ratio $\varepsilon'_{sgs}/\varepsilon_{sgs}$ is larger than 60%, even for the coarse mesh, for $y > 0.1$. In the incipient separation region The ratio is larger than 94% in the incipient separation region for all three meshes. Ratio $\varepsilon'_{sgs}/\varepsilon_{sgs}$ – like that of shear stresses (Fig. 10) and viscosities (Fig. 11) – behaves consistently when the resolution is refined, but it is difficult to define a value above which the flow can be defined to be well resolved.

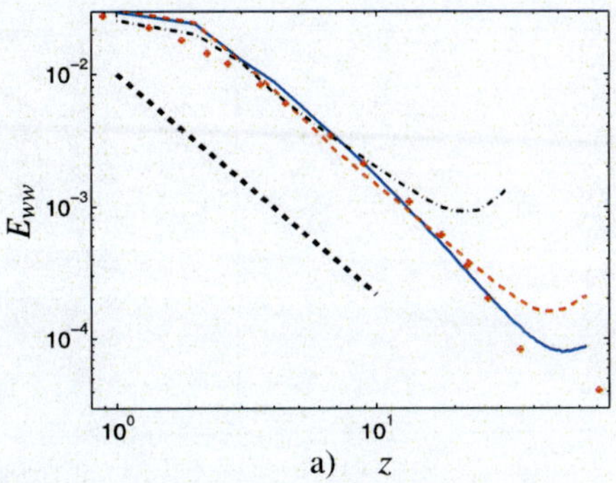

Fig. 18. Decaying grid turbulence. Energy spectra. $t = 2$. —— : fine LES; - - - : DNS; -·-·- : coarse LES; Marker: experiments [2]. The thick dashed line shows $-5/3$ slope.

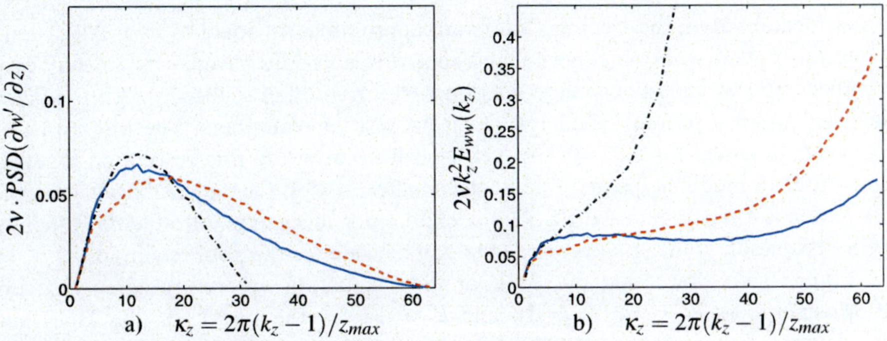

Fig. 19. Decaying grid turbulence. Dissipation energy spectra. $t = 2$. Exact (left) and approximated (right). —— : fine LES; - - - : DNS; -·-·- : coarse LES.

3.2 Decaying isotropic grid turbulence

Above it was found that the peak of the SGS dissipation for the diffuser flow is largest at relatively low wavenumbers. This section discussed LES and DNS of decaying isotropic grid turbulence. The objective is to investigate at which wavenumbers the dissipation takes place for DNS and well-resolved LES.

The initial velocity field at $t = 0$ is generated from the experimental energy spectrum [2] using the synthetic method in [4]. The domain is a cubic box of side 2π. Three computations were carried out:

1. fine LES using a Smagorinsky model with $C_S = 0.1$ on a 128^3 grid;
2. DNS on a 128^3 grid;
3. coarse LES using a Smagorinsky model with $C_S = 0.1$ on a 64^3 grid.

Figure 18 presents the one-dimensional energy spectra computed from the two-point correlations. The predicted amplitudes are scaled so that u_{rms}^2 agrees with the

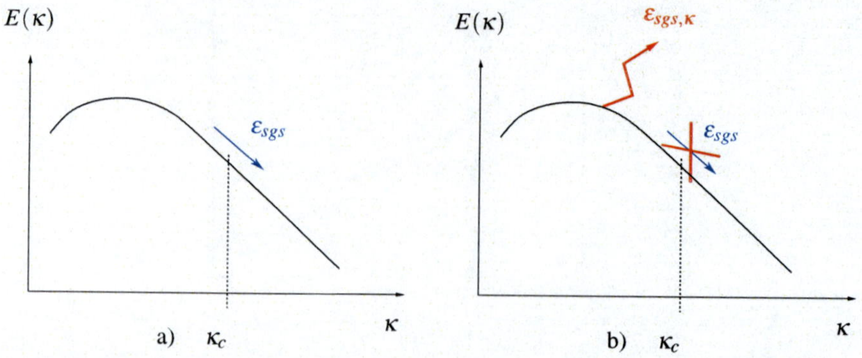

Fig. 20. Energy spectra. SGS dissipation according to the idealized picture (a) and SGS dissipation according to the present work (b).

experimental value; the scaling factors are approximately equal to two. All spectra agree fairly well with the experimental spectrum and they have a $-5/3$ decay over a rather large wavenumber range (the fine grid more than a decade). All predicted spectra exhibit a pile-up of energy at the largest wavenumbers. The pile-up is, as expected, smallest for LES on the 128^3 mesh because the fine resolution together with the SGS model helps to dissipate the smallest scales. On the coarse 64^3 grid, the resolution is too coarse and the SGS model does not succeed in introducing sufficient SGS dissipation at the smallest scales to compensate for the poor resolution.

Figure 19a shows that the peak of the fluctuating velocity gradients occurs at wavenumbers $k_z = 14$, $k_z = 14$ and $k_z = 12$ for the fine LES, the DNS and the coarse LES, respectively. This gives length scales of $\ell = z_{max}/13 \simeq 9$ cells, $z_{max}/13 \simeq 9$ cells and $\ell = z_{max}/11 \simeq 6$ cells, respectively. As expected, we find that the SGS model on both the the fine and the coarse mesh moves the location of the peak in the fluctuating velocity gradients towards slightly smaller wavenumbers compared to the DNS. Moreover, it can be seen that the LES on the fine and coarse grids gives the same location of the peak, indicating that – except for the smallest scales – the coarse resolution is sufficient.

The approximated spectra in Fig. 19b all exhibit a maximum at the highest wavenumbers. This is in accordance with the pile-up of energy in the energy spectra, confirming that the smallest scales are not well resolved on any grid.

In conclusion, it is found that also for DNS and well-resolved LES, the dissipation takes place at surprisingly small wavenumbers.

4 Concluding remarks

Various ways of estimating resolution in recirculating flow have been considered.

It is concluded that the most useful quantity for estimating resolution are two-point correlations. They show by how many cells the largest scales are resolved. It is then up to the CFD user to judge how many cells are required; at least eight to ten cells seems to be reasonable.

The energy spectra do not give reliable information about the resolution.

The ratio of the SGS to the molecular viscosity, $\langle \nu_{sgs} \rangle / \nu$, and the ratio of the SGS to the resolved shear stress behave consistently upon grid refinement, i.e. they decrease. However, the ratio $\langle \nu_{sgs} \rangle / \nu$ indicates incorrectly that the turbulence is better resolved in the attached flow region of the diffuser than in the incipient flow region. Hence, the ratio $\langle \nu_{sgs} \rangle / \nu$ is not a reliable quantity for estimating the resolution.

Energy spectra of resolved fluctuating gradients and SGS dissipation give information about the wavenumbers at which the SGS dissipation takes place. It is commonly assumed that the SGS dissipation occurs close to the cut-off wavenumber, see Fig. 20a. The present study (as was found also for channel flow in [5]), however, show that this idealized picture is not true. Instead the SGS dissipation takes place at rather low wavenumbers, see Fig. 20b. Note that this conclusion does not apply for *spectral* methods for which the relation Eq. 9 is valid; in these methods $\varepsilon \propto \kappa^{1/3}$ applies in the inertial region, see for example [10]. In the present study the length scales related to the peaks of the SGS dissipation correspond to approximately 10 cells (spanwise direction) in attached flow and much more in the incipient separation region. This was confirmed in DNS and well-resolved LES of decaying grid turbulence. The disadvantage of these quantities is that they can only be used in flows that possess a homogeneous direction, which is seldom the case in real flows.

Time-averaged velocity gradients and the resolved fluctuating velocity gradient both contribute to SGS dissipation, denoted by $\varepsilon_{sgs,mean}$ and ε'_{sgs}, respectively. The latter is zero in RANS, whereas it dominates in well-resolved LES. Hence, the ratio $\varepsilon'_{sgs}/(\varepsilon'_{sgs} + \varepsilon_{sgs,mean})$ may be useful in evaluating resolution. It is found that this ratio is very large in recirculating flow (more than 94%), revealing good resolution. However, as mentioned with respect to the ratio of the shear stresses and and the viscosities, it is difficult to give any recommendations for threshold values.

Acknowledgments

The financial support of SNIC (Swedish National Infrastructure for Computing) for computer time at C3SE (Chalmers Center for Computational Science and Engineering) is gratefully acknowledged.

References

1. Buice, C., Eaton, J.: Experimental investigation of flow through an asymmetric plane diffuser. Report No. TSD-107, Thermosciences Division, Department of Mechanical Engineering, Stanford University, Stanford, California 94305 (1997)
2. Comte-Bellot, G., Corrsin, S.: Simple Eularian time correlation of full- and narrow-band velocity signals in grid-generated "isotropic" turbulence. Journal of Fluid Mechanics **48**(2), 273–337 (1971)
3. Davidson, L.: Transport equations in incompressible URANS and LES. Report 2006/01, Div. of Fluid Dynamics, Dept. of Applied Mechanics, Chalmers University of Technology, Göteborg, Sweden (2006)

4. Davidson, L.: Using isotropic synthetic fluctuations as inlet boundary conditions for unsteady simulations. Advances and Applications in Fluid Mechanics **1**(1), 1–35 (2007)
5. Davidson, L.: Large eddy simulations: how to evaluate resolution. International Journal of Heat and Fluid Flow **30**(5), 1016–1025 (2009)
6. Davidson, L., Peng, S.H.: Hybrid LES-RANS: A one-equation SGS model combined with a $k-\omega$ model for predicting recirculating flows. International Journal for Numerical Methods in Fluids **43**, 1003–1018 (2003)
7. Emvin, P.: The full multigrid method applied to turbulent flow in ventilated enclosures using structured and unstructured grids. Ph.D. thesis, Dept. of Thermo and Fluid Dynamics, Chalmers University of Technology, Göteborg (1997)
8. Hinze, J.: Turbulence, 2nd edn. McGraw-Hill, New York (1975)
9. Klein, M., Sadiki, A., Janicka, J.: A digital filter based generation of inflow data for spatially developing direct numerical or large eddy simulations. Journal of Computational Physics **186**(2), 652–665 (2003)
10. Schlatter, P., Stolz, S., Kleiser, L.: Evaluation of high-pass filtered eddy-viscosity models for large-eddy simulation of turbulent flows. Journal of Turbulence **6**(5), 1–21 (2005)
11. Wu, X., Schlüter, J., Moin, P., Pitsch, H., Iaccorino, G., Ham, F.: Computational study on the internal layer in a diffuser. Journal of Fluid Mechanics **550**, 391–412 (2006)

Quality assessment of Dynamic Finite Difference schemes on the Taylor-Green Vortex

Dieter Fauconnier, Chris De Langhe, and Erik Dick

Department of Flow, Heat and Combustion Mechanics, Faculty of Engineering, Ghent University, St. Pietersnieuwstraat 41, 9000 Ghent, Belgium
dieter.fauconnier@ugent.be

Summary. The performance of a class of explicit and implicit *dynamic* finite difference schemes [3, 2] is investigated for the Large-Eddy Simulation of the three-dimensional Taylor-Green Vortex flow [1], in which the dynamic Smagorinsky model and the small-small multiscale Smagorinsky model are used. The numerical errors and the modeling errors and their interactions are investigated.

Key words: Dynamic Finite Difference schemes, Dispersion-Relation Preserving schemes, LES, Taylor-Green Vortex

1 Dynamic Finite Difference Approximations

1.1 Construction

Assume a vector field $\mathbf{u}(\mathbf{x},t)$ defined in continuum space $\mathbb{R}^q$, $q \in \{1,2,3\}$. To simplify the notation, we restrict the formulas to one spatial dimension $q = 1$ and we do not write explicitly the dimension in time, such that $u(x,t) = u(x)$, $x \in \mathbb{R}$. Consider further a one-dimensional uniform Cartesian node distribution with spacing Δ. Consider the Taylor series expansion of the kth-order implicit *central* finite difference approximation for the nth derivative of the continuous and infinitely differentiable field $u(x)$, $x \in \mathbb{R}$, in a node $x = x_i$.

$$\sum_{l=-q}^{q} \alpha_l \overline{\frac{\partial^n u}{\partial x^n}}(x_{i+l}) = \sum_{j=-r}^{r} \frac{\beta_j}{\Delta^n} \overline{u}(x_{i+j}) + \sum_{k'=k}^{\infty} c^*_{k',n} \Delta^{k'} \overline{\frac{\partial^{k'+n} u}{\partial x^{k'+n}}}(x_i), \tag{1}$$

in which α_l and β_j represent the implicit and explicit weighting coefficients, and $c^*_{k',n}$ denote the truncation series coefficients. Note that explicit schemes form a subclass, for which $\alpha_l = 0$, $\forall l \neq 0$. Consider further a similar k^{th}-order implicit finite difference approximation for the nth partial derivative on the *same* computational grid with grid spacing Δ, but expressed as if the grid resolution were $\alpha\Delta$, $\alpha \in \mathbb{N}$ such that the stencil width for the explicit part is $2\alpha r + 1$ whereas that for the implicit part is $2\alpha q + 1$. The Taylor series expansion then reads

M.V. Salvetti et al. (eds.), *Quality and Reliability of Large-Eddy Simulations II*, ERCOFTAC Series 16, DOI 10.1007/978-94-007-0231-8_26,

$$\sum_{l=-q}^{q} \alpha_l \overline{\frac{\partial^n u}{\partial x^n}} (x_{i+\alpha l}) = \sum_{j=-r}^{r} \frac{\beta_j}{(\alpha\Delta)^n} \overline{u}(x_{i+\alpha j}) + \sum_{k'=k}^{\infty} c^*_{k',n} (\alpha\Delta)^{k'} \overline{\frac{\partial^{k'+n} u}{\partial x^{k'+n}}} (x_i). \tag{2}$$

We proceed by dicretisation of the leading order truncation term in expressions (1) and (2), and replacing the theoretical coefficient $c^*_{k,n}$ with a new coefficient $c_{k,n}$, which we determine by comparing the Taylor expansions on both resolutions. Introducing further a blending factor f in the discretized leading order truncation term of the coarse resolution equation (as done in [3, 2]), the finite difference schemes are written as

$$\sum_{l=-q}^{q} \alpha_l \overline{\frac{\partial^n u}{\partial x^n}} (x_{i+l}) \approx \sum_{j=-r}^{r} \frac{\beta_j}{\Delta^n} \overline{u}(x_{i+j}) + c_{k,n}\Delta^k \left. \frac{\delta^{k+n}\overline{u}}{\delta x^{k+n}} \right|^{\Delta} \tag{3}$$

$$\sum_{l=-q}^{q} \alpha_l \overline{\frac{\partial^n u}{\partial x^n}} (x_{i+\alpha l}) \approx \sum_{j=-r}^{r} \frac{\beta_j}{(\alpha\Delta)^n} \overline{u}(x_{i+\alpha_j})$$
$$+ c_{k,n} (\alpha\Delta)^k \left\{ f \left. \frac{\delta^{k+n}\overline{u}}{\delta x^{k+n}} \right|^{\alpha\Delta} + (1-f) \left. \frac{\delta^{k+n}\overline{u}}{\delta x^{k+n}} \right|^{\Delta} \right\}. \tag{4}$$

The purpose of the blending factor f will be explained later. The truncation error $\mathcal{O}(\Delta^k)$, related to expression (3), is given by

$$\sum_{k'=k}^{\infty} c^*_{k',n} \Delta^{k'} \overline{\frac{\partial^{k'+n} u}{\partial x^{k'+n}}} - c_{k,n}\Delta^k \left\{ \overline{\frac{\partial^{k+n} u}{\partial x^{k+n}}} - \sum_{k'=2}^{\infty} c^*_{k',k+n} \Delta^{k'} \overline{\frac{\partial^{k'+k+n} u}{\partial x^{k'+k+n}}} \right\}. \tag{5}$$

A similar expression can be found for the truncation error of (4). Assume now that the discrete non-smooth field $\overline{u}(x)$ is characterized by a cutoff wavenumber $\kappa_c > 0$, such that the spectral content on the computational grid (related to the smoothness of $\overline{u}$) is limited to the range $[0, \kappa_c/\kappa_{max}]$. Then, one can understand that an optimal value for $c_{k,n}$ may be found, for which a norm of the magnitude of (5) is minimal in the spatial domain. This is equivalent of minimizing the dispersion error of the resulting central finite difference scheme, weighted with the energy spectrum of $\overline{u}(x)$, over the interval $[0, \kappa_c/\kappa_{max}]$. Such a value of $c_{k,n}$, generally results in a finite difference approximation of order $\mathcal{O}(\Delta^k)$, unless $c_{k,n} = c^*_{k,n}$, for which it yields $\mathcal{O}(\Delta^{k+2})$. This approach, which is equivalent to the Dispersion-Relation Preserving approach, is more advantageous than increasing the order of accuracy a priori [6]. It was shown in [3] that the optimal coefficient $c_{k,n}$ can be determined by minimizing the difference between the truncated equations (3) and (4). Subtracting (3) and (4), the following symbolic expression for the difference between both approximations is obtained

$$\mathcal{E} = \mathcal{L} + c_{k,n}\mathcal{M} = \mathcal{O}\left((\alpha\Delta)^k\right) - \mathcal{O}\left(\Delta^k\right). \tag{6}$$

The terms $\mathcal{L}$ and $\mathcal{M}$, which are readily found by identification in the former subtraction, can be further simplified [2], yielding

$$\mathscr{L} = c_{k,n}^{*}\left(\alpha^{k}-1\right)\Delta^{k}\,\frac{\delta^{k+n}\overline{u}}{\delta x^{k+n}}\bigg|^{\Delta} \tag{7}$$

$$\mathscr{M} = \left(1-\alpha^{k}\right)\Delta^{k}\,\frac{\delta^{k+n}\overline{u}}{\delta x^{k+n}}\bigg|^{\Delta} - \alpha^{k}\Delta^{k} f\left\{c_{2,k+n}^{*}\left(1-\alpha^{2}\right)\Delta^{2}\,\frac{\delta^{k+n+2}\overline{u}}{\delta x^{k+n+2}}\bigg|^{\Delta}\right\} \tag{8}$$

where $c_{2,k+n}^{*}$ is a constant coefficient known from the Taylor series expansion. The optimal coefficient $c_{k,n}$ can then be determined from (6) by minimizing $\mathscr{E}$ using a least squares approach, i.e. $\frac{\partial}{\partial c_{k,n}}\langle\mathscr{E}^{2}\rangle = 0$, where $\langle\cdot\rangle$ denotes an averaging operator, resulting in the *dynamic coefficient*

$$c_{k,n}^{dyn} = -\frac{\langle\mathscr{L}\mathscr{M}\rangle}{\langle\mathscr{M}\mathscr{M}\rangle}. \tag{9}$$

In the current work, only global spatial averaging is considered for the finite difference schemes. Once $c_{k,n}^{dyn}$ is calculated, its value can be used in the finite difference approximation (3). However, since an explicit finite difference approximation for the $(k+n)^{th}$ derivative is used, the stencil of (3) is then strictly required for this implicit scheme. This is remedied by substituting the explicit $(k+n)^{th}$ derivative by an implicit formulation, which is equivalent of writing (3) immediately in its most compact formulation. We finally obtain the compact implicit discretization

$$\sum_{l=-q}^{q}\left[\alpha_{l}-\frac{c_{k,n}^{dyn}}{c_{k,n}^{*}}\left(\alpha_{l}-\alpha_{l}'\right)\right]\frac{\partial^{n}u}{\partial x^{n}}\left(x_{i+l}\right) \approx \sum_{j=-r}^{r}\frac{\beta_{j}}{\Delta^{n}}\left[1-\frac{c_{k,n}^{dyn}}{c_{k,n}^{*}}\right]\overline{u}\left(x_{i+j}\right) + \frac{c_{k,n}^{dyn}}{c_{k,n}^{*}}\sum_{j=-r-1}^{r+1}\frac{\beta_{j}'}{\Delta^{n}}\overline{u}\left(x_{i+j}\right), \tag{10}$$

where α_{l}' and β_{j}' denote the weighting coefficients of the $k+2^{nd}$-order implicit finite difference approximation. Although for the higher derivatives in (7) and (8) again compact Padé schemes may be used, in this work they are obtained using explicit approximations for reasons of simplicity. The resulting dynamic scheme (10) has a formal order of accuracy k unless $c_{k,n}^{dyn} = c_{k,n}^{*}$, which would lead to order $k+2$.

1.2 High-Reynolds Calibration.

In the following, the role of f and its impact on the resulting coefficient $c_{k,n}^{dyn}$ are briefly discussed

1. If $f=0$, one notices that $c_{k,n}^{dyn} = c_{k,n}^{*}$, regardless the specific spectral characteristics of $\overline{u}$. Then, expression (10) is an approximation with formal asymptotic order of accuracy $\mathscr{O}\left(\Delta^{k+2}\right)$, which makes the dynamic procedure equivalent to Richardson's Extrapolation.
2. Assuming a very smooth field $\overline{u}(x)$ for which $\kappa_c/\kappa_{max} \to 0$, the difference (6) may be considered negligible, i.e. $\mathscr{E}\approx 0$, such that the dynamic coefficient can

be obtained directly from (6). Applying a small perturbation analysis on this expression [2], showed that if $\kappa_c/\kappa_{max} \to 0$, the blending factor approaches its asymptotic value

$$f^* = \frac{1-\alpha^k}{\alpha^k\left(1-\alpha^2\right)} \frac{c^*_{k+2,n}+c^*_{k,n}c^*_{2,k+n}}{c^*_{k,n}c^*_{2,k+n}} \neq 0, \tag{11}$$

3. If $\kappa_c/\kappa_{max} \gg 0$, one may determine the optimal value of the blending factor f by calibrating the modified wavenumber $\kappa_n'^n(\kappa)$ of scheme (10), in which the ratio (9) is transformed in Fourier space [2], for a turbulent spectrum at $\mathrm{Re} \to \infty$ with fixed filter-to-grid cutoff ratio κ_c/κ_{max}, such that the dispersion errors are minimal in the range $[0, \kappa_c/\kappa_{max}]$. The obtained value of f guarantees that the dynamic scheme reaches maximum performance for a Large-Eddy Simulation at high Reynolds numbers with a maximum filter-to-grid cutoff ratio κ_c/κ_{max}. Following [2], the optimal value for the blending factor f is calculated by solving

$$\frac{\partial}{\partial f}\int_0^{\frac{\pi}{\Delta}} \left(\kappa^n - \kappa_n'^n(\kappa, f)\right)^2 E_u(\kappa)\, d\kappa = 0. \tag{12}$$

where $E_u(\kappa)$ represents the energy spectrum of the flow field $\overline{u}(x)$. An idealized and simplified inertial range spectrum for homogeneous isotropic turbulence is introduced [2], defined as

$$E_u(\kappa) = \left[1 - H(\kappa - \kappa_c)\right]\kappa^{-\beta} = \begin{cases} \kappa^{-\beta} & \kappa < \kappa_c \\ 0 & \kappa > \kappa_c \end{cases}, \tag{13}$$

where β determines the slope of the inertial range and the cutoff wavenumber κ_c indicates the highest appearing wavenumber in the (resolved) field $\overline{u}(x)$. Since the dynamic finite difference schemes are applied for the pressure derivatives as well as the velocity derivatives in the Navier-Stokes equations, they are calibrated using the appropriate values of β that correspond to the inertial range of respectively the turbulent velocity $\overline{u}(\mathbf{x})$, i.e. $\beta = -5/3$, and the turbulent pressure field $\overline{p}(\mathbf{x})$, i.e. $\beta = -7/3$. Table 1, gives on overview of the blending factors, as well as the corresponding value of the dynamic coefficient at $\kappa_c = \frac{2}{3}\kappa_{max}$, for the 2nd - and 4th -order explicit and the 4th -order implicit dynamic finite difference schemes which are investigated in this work. Note that for $\beta = 0$, the coefficients $c_{k,n}$, used in the 2nd - and 4th -order explicit and the 4th -order implicit DRP schemes, are obtained directly.

2 Taylor-Green Vortex Setup

The Taylor-Green Vortex, considered as a challenging prototype system that describes the production of small-scale eddies due to the mechanism of vortex-line stretching in homogeneous isotropic turbulence [1], was selected to examine the

Slope	Scheme	$n=1$ f	$n=1$ $c_{k,1}^{dyn}$	$n=2$ f	$n=2$ $c_{k,2}^{dyn}$
	Explicit $k=2$	0.2555	−0.3088	0.2339	−0.1310
$\beta=-\frac{5}{3}$	Explicit $k=4$	0.2298	0.0740	0.2241	0.0203
	Implicit $k=4$	0.1241	0.0121	0.1363	0.0069
	Explicit $k=2$	0.2654	−0.2966	0.2353	−0.1293
$\beta=-\frac{7}{3}$	Explicit $k=4$	0.2329	0.0724	0.2248	0.0201
	Implicit $k=4$	0.1261	0.0120	0.1370	0.0069
	Explicit $k=2$		−0.3344		−0.1345
$\beta=0$	Explicit $k=4$		0.0775		0.0206
	Implicit $k=4$		0.0119		0.0069

Table 1. Numerically obtained optimal blending factors f and corresponding values of $c_{k,n}^{dyn}$ for the inertial range model spectrum at $\kappa_c=\frac{2}{3}\kappa_{max}$.

performance of the developed dynamic finite difference approximations. According to Brachet *et al.* [1], the Taylor-Green Vortex is defined as the periodic three-dimensional incompressible flow, governed by the Navier-Stokes equations and the continuity equation which develops from a single initial vortex structure. The Reynolds number is set to Re $=1500$, which corresponds to a Reynolds number based on the transversal Taylor micro-scale $\mathrm{Re}_\lambda \approx 55$.

The DNS-solution will serve here as a reference solution against which the various LES-solutions are compared. The system of equations is directly solved on a uniform computational grid with 256^3 nodes (128^3 Fourier modes). These settings compare well to those of Brachet *et al.* [1], who used a uniform grid with $N^3=256^3$ for DNS with Reynolds numbers up to Re $=3000$.

The Large-Eddy Simulation is performed on a uniform computational grid with 64^3 nodes and with grid cutoff wavenumber $\kappa_{max}=\pi/\Delta=32$. The LES equations that describe the filtered variables $\overline{u}(\mathbf{x},t)$ and $\overline{p}(\mathbf{x},t)$ are solved in the double-decomposition framework. This implies that the nonlinear term and the *residual-stress* tensor $\overline{\tau}_{ij}=\overline{u_iu_j}-\overline{\overline{u}_i\overline{u}_j}$ are filtered explicitely with a sharp cutoff filter. This approach is motivated by two arguments. First, explicit filtering up to $\kappa_c=\frac{2}{3}\kappa_{max}$ allows to rigorously preclude aliasing. Secondly, the explicit filtering procedure eliminates numerical discretization errors in the high-wavenumber region $[\kappa_c,\kappa_{max}]$. Two residual stress models for $\overline{\tau}_{ij}$ are considered in this work. First, the dynamic Smagorinsky model is used, in which $\overline{\tau}_{ij}=-2\nu_e\overline{S}_{ij}=-2C_s^2\Delta_c^2\overline{\overline{S}\,\overline{S}_{ij}}$ and C_s^2 is determined by the dynamic Germano procedure. Secondly, the small-small multiscale Smagorinsky model is used $\overline{\tau}_{ij}=-2\left(\nu_eS_{ij}''\right)''=-2C_{s,m}^2\Delta_c^2\left(S''S_{ij}''\right)''$ in which $(.)''$, denotes the band-pass filter $[\lambda_c,\kappa_c]$ and $C_{s,m}=C_s\gamma\left[\gamma^{4/3}-1\right]^{-3/4}$, with $\gamma=\kappa_c/\lambda_c$ and $C_s\approx 0.17$. The cutoff wavenumber λ_c that determines the secondary sharp cutoff filter in the dynamic procedure or the sharp cutoff scale-separation filter in the multiscale model is determined as $\lambda_c=\kappa_c/2=\kappa_{max}/3$.

For the DNS and the spectral LES, the partial derivatives in the equations are evaluated in Fourier space by a *pseudo-spectral* method in order to exclude finite difference errors from the solution. Further, the skew-symmetric formulation is adopted for the discretization of the nonlinear term such that it conserves the kinetic energy. The Navier-Stokes equations are solved by means of the pressure-correction algorithm, in which the pressure is obtained from a Poisson equation. The time stepping is performed with the explicit low-storage 4-stage Runge-Kutta method with standard coefficients $\left[\frac{1}{4},\frac{1}{3},\frac{1}{2},1\right]$. In order to guarantee the numerical accuracy of the adopted Runge-Kutta time stepping method, a sufficiently small time step $\Delta t = 0.005$ was chosen for both the DNS and the LES, such that the dispersion and dissipation errors related to this method remain sufficiently low. For the Large-Eddy Simulations with the dynamic finite difference schemes, each partial derivative in the Navier-Stokes equations or the Poisson equation is discretized by the appropriate dynamic finite difference approximation. The implementation involves the calculation of 36 dynamic coefficients which can be evaluated at each Runge-Kutta step, that is 4 times per time step. Since this would lead inevitably to a significant computational overhead, the coefficients are evaluated every 10th time step, which is expected to be sufficient. The computational overhead is only 1.7% in comparison with DRP schemes of comparable accuracy.

3 Quality-Assessment

3.1 Error definitions

In order to separate modeling and numerical errors, the *error decomposition method* of Vreman *et al.* [7] and Meyers *et al.* [4, 5] is adopted. For a specific flow variable ϕ, the total error on ϕ, ε_ϕ, is decomposed into a modeling error contribution $\varepsilon_{\phi,model}$, defined as the difference between ϕ obtained in the DNS and that of the spectral LES, and a numerical error contribution $\varepsilon_{\phi,num}$, defined as the difference between ϕ obtained in the spectral LES and the finite difference LES. The *mathematics-based* error norms are then defined as

$$k_{\varepsilon_\phi}(t) = \iiint_0^{\kappa_{\max}} E_{\varepsilon_\phi}(\boldsymbol{\kappa},t)\,d\boldsymbol{\kappa} = \iiint_0^{\kappa_{\max}} \widehat{\varepsilon_\phi}(\boldsymbol{\kappa},t)\,\widehat{\varepsilon_\phi}^*(-\boldsymbol{\kappa},t)\,d\boldsymbol{\kappa}. \quad (14)$$

whereas the *physics-based* error norms are given by

$$\varepsilon_\phi(t) = \Delta\phi = \iiint_0^{\kappa_{\max}} \boldsymbol{\kappa}^{-q}\Delta E_u(\boldsymbol{\kappa},t)\,d\boldsymbol{\kappa} \quad (15)$$

in which the parameter $q = -1, 0, 2$ determines the specific physics-related quantity ϕ, i.e. integral length scale L_{11}, kinetic energy k or dissipation rate ε. Remark that the sign of ε_ϕ could be either positive or negative, enabling to see interactions between different error sources.

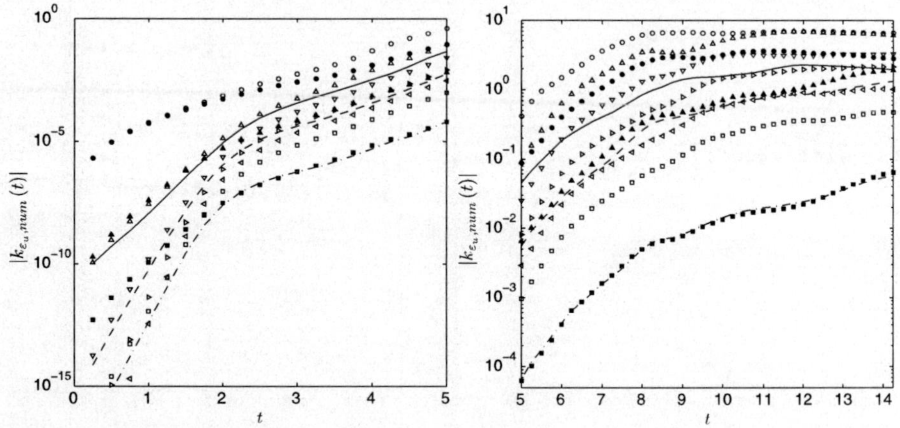

Fig. 1. The global magnitude $k_{\varepsilon_u,num}$ of the numerical errors on the resolved velocity field $\overline{\mathbf{u}}(\mathbf{x},t)$, obtained from the Large-Eddy Simulation of the Taylor-Green Vortex with the dynamic Smagorinsky model: the laminar stages of the Taylor-Green flow at times $0 \leq t \leq 5$ (**left**) and the transitional and turbulent stages at times $5 \leq t \leq 14.25$ (**right**). (○) 2nd -order; (△) 4th -order; (▽) 6th -order; (▷) 8th -order; (◁) 10th -order; (□) 6th -order Padé; (•) 2nd -order explicit DRP; (▲) 4th -order explicit DRP; (■) 4th -order DRP Padé; (———) 2nd -order explicit dynamic; (– – –) 4th -order explicit dynamic; (–·–·) 4th -order dynamic Padé.

3.2 Modeling errors, numerical errors and their interactions

Figure 1 displays the global magnitude of the numerical errors (related to L_2-norm) on the resolved velocity field $\overline{\mathbf{u}}(\mathbf{x},t)$, obtained from the Large-Eddy Simulation with the dynamic Smagorinsky model. Analogous results were obtained with the Multi-scale model. It is observed that the dynamic schemes recover the asymptotic order of accuracy ($\mathscr{O}\left(\Delta^{k+2}\right)$) in the early stages of the simulation, i.e $t \leq 2$, where the flow is still smooth and resolved with DNS-resolution. This is in contrast to the Dispersion-Relation Preserving schemes ($\mathscr{O}\left(\Delta^{k}\right)$), which are suboptimal in these situations, since they have an a priori optimization to a fully developed uniform spectrum. As soon as the resolution of the computational grid becomes inadequate to resolve all scales in the flow (the simulation shifts from DNS-resolution to LES-resolution at $t \approx 2$), the dynamic schemes adapt to the instantaneous solution and achieve an accuracy which is at least as good, or better than that of the Dispersion-Relation Preserving schemes at all times $2 \leq t \leq 14.25$. Studying the results in more detail, the 2nd -order explicit dynamic scheme is observed to achieve the accuracy of the 8th -order central scheme, which is better than that of the corresponding 2nd -order DRP scheme. The 4th -order explicit dynamic scheme obtains slightly better quality than the corresponding DRP scheme, which fluctuates between that of the 8th - and 10th -order standard scheme, although the difference becomes very small. Further, the performance of the 4th -order dynamic implicit finite difference approximation almost collapses with that of the 4th -implicit DRP Padé scheme and both clearly outperform all other schemes, including the standard tridiagonal Padé scheme.

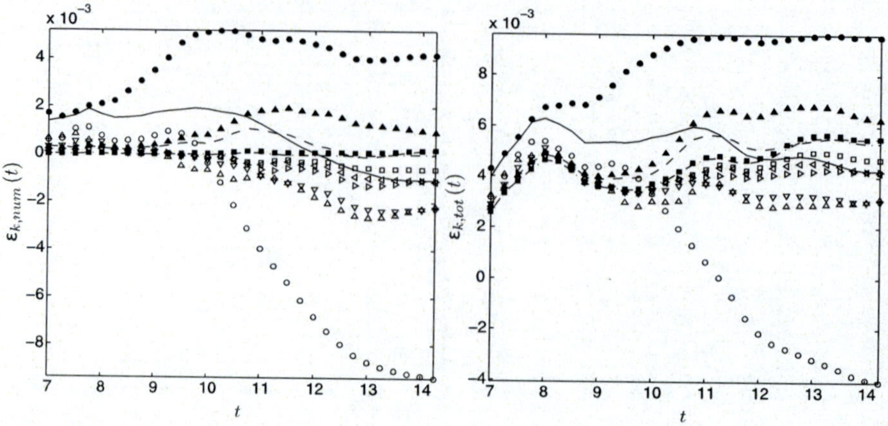

Fig. 2. Numerical error (**left**) and total error (**right**) on the kinetic energy k, in the Large-Eddy Simulation of the Taylor-Green Vortex flow with the multiscale Smagorinsky model model at $t \geq 7$. (○) 2nd -, (△) 4th -, (▽) 6th -, (▷) 8th - and (◁) 10th -order standard schemes; (□) 6th -order Padé; (●) 2nd - and (▲) 4th -order explicit DRP schemes; (■) 4th -order DRP Padé; (———) 2nd - and (− − −) 4th -order explicit dynamic schemes; (−·−·) 4th -order dynamic Padé.

The impact of the numerics on the kinetic energy $k(t)$ is demonstrated in Figure 2. Results on $L_{11}(t)$ and $\varepsilon(t)$ are very similar. Results are only shown for the turbulent period of the flow at $7 \leq t \leq 14.25$, since at earlier times the dynamic finite difference schemes reduce to their asymptotic counterparts. Hence the behaviour of the dynamic schemes in the early stages is no different from that in the previous discussion despite the different error measures. It is observed that the numerical errors of the standard asymptotic finite difference schemes are negative, which indicates a significant reduction in the dissipation due to numerics. In contrast, the numerical errors related to the dynamic finite difference schemes and the Dispersion-Relation Preserving schemes remain positive during the simulation, indicating an increased dissipation due to the numerics. However, the magnitudes of the errors are significantly smaller for the dynamic and DRP schemes than for the standard schemes. Adapting the finite difference schemes for the smallest resolved scales such that the global dispersion error is minimized remains therefore advantageous. Hence, the general conclusions that applied to the mathematics-based errors remain valid for the physics-based errors. One observes that the dynamic finite difference schemes generally perform better than their Dispersion-Relation Preserving counterparts. This good performance is mainly attributed to the ability of the dynamic schemes to adapt to changing flow properties and to their high-Reynolds calibration.

However, a better numerical accuracy does not necessarily lead to a better overall performance, as demonstrated by the total error in Figure 2. We observe that both eddy-viscosity models are too dissipative, despite the use of a multiscale technique or a dynamic procedure, resulting in a positive sign for the modeling error. As a consequence, cancellation of numerical errors and modeling errors are witnessed for the standard finite difference schemes, due to their opposite signs. This confirms the results in [4]. It was shown in [4, 2] that these cancellations depend on the specific quantity that is examined. For instance, the 2nd -order standard scheme leads to the smallest errors for the kinetic energy, followed by the 4th - and 6th -order standard scheme. In contrast, the higher-order standard schemes, the Dispersion-Relation Preserving schemes and the dynamic finite difference schemes, which are believed to have better spectral characteristics, do not perform as well. Since the numerical errors of the dynamic and DRP schemes have the same sign as the modeling errors, they reinforce each other. We note that for subgrid models which have the opposite sign for the modeling error than of those of eddy-viscosity models, the modeling errors and numerical errors of the standard schemes would probably reinforce each other, whereas the modeling errors would cancel out the numerical errors related to the optimized schemes. Hence, cancellation of different errors depends not only on the type of numerical schemes, but also on the type of subgrid model. Finally, it was observed that the multiscale model leads to smaller total errors than the dynamic model due to the smaller modeling error contribution. Despite this, only minor evidence was found that the multiscale model benefits from using more accurate numerical methods, since the modeling errors are still dominant in comparison with the numerical errors. Hence, the total error will only decrease with increasing numerical accuracy if the numerical error contributions are dominant and the modeling error contributions are negligible.

4 Conclusions

In the present work, the performance of a class of explicit and implicit *dynamic* finite difference schemes, developed in [3], was assessed for the Large-Eddy Simulations of the Taylor-Green Vortex at $Re = 1500$, using the dynamic and multiscale Smagorinsky models. The dynamic schemes succeeded in achieving an optimal accuracy for all resolved scales of motion in the flow at any time, rather than focusing only on the asymptotic order of accuracy for the largest resolved scales. In contrast to the DRP schemes, the dynamic schemes systematically recover their potential asymptotic order of accuracy, provided that all scales of motion in the flow field are very well resolved on the computational grid. If the flow contains marginally resolved scales, the dynamic schemes adapt themselves and act similarly to the DRP schemes. The results showed clearly that the tested dynamic finite difference schemes obtain lower numerical errors than the standard asymptotic finite difference schemes, and even the corresponding DRP-schemes, with the same stencil support. Despite the substantial improvement in numerical accuracy, the dynamic schemes and other high-order schemes do not necessarily provide a more accurate solution of the Large-Eddy Simulation, due to advantageous cancellation between numerical errors and modeling errors for less accurate schemes, although this effect depends on the type of subgrid model being used.

Acknowledgments

This research was funded by a Ph.D grant of the Institute for the Promotion of Innovation through Science and Technology in Flanders (IWT-Vlaanderen).

References

1. M.E. Brachet, D.I. Meiron, S.A. Orszag, B.G. Nickel, R.H. Morf, and U. Frisch. Small-scale structure of the Taylor-Green Vortex. *J. Fluid. Mech.*, 130:411–452, 1983.
2. D. Fauconnier, C. De Langhe, and E. Dick. Construction of Implicit Dynamic Finite Difference Schemes and Application to the Large-Eddy Simulation of the Taylor-Green Vortex. *J. Comput. Phys.*, 2009, Accepted.
3. D. Fauconnier, C. De Langhe, and E. Dick. A family of dynamic finite difference schemes for large-eddy simulation. *J. Comput. Phys.*, 228(6):1830–1861, 2009.
4. J. Meyers, B. Geurts, and M. Baelmans. Database analysis of errors in large-eddy simulation. *Phys. Fluids*, 15(9):2740–2755, 2003.
5. J. Meyers, B. Geurts, and M. Baelmans. Optimality of the dynamic procedure for large-eddy simulations. *Phys. Fluids*, 17(4):045–108, 2005.
6. C. K. W. Tam and J. C. Webb. Dispersion-Relation Preserving Finite Difference Schemes for Computational Acoustics. *J. Comp. Phys.*, 107:262–281, 1993.
7. B. Vreman, B. Geurts, and H. Kuerten. Comparison of numerical schemes in large-eddy simulations of the temporal mixing layer. *Int. J. Num. Methods in Fluids*, 22:297–311, 1996.

Stochastic Coherent Adaptive Large-Eddy Simulation with explicit filtering

Giuliano De Stefano[1] and Oleg V. Vasilyev[2]

[1] Dipartimento di Ingegneria Aerospaziale e Meccanica, Seconda Università di Napoli, via Roma 29, I 81031 Aversa, Italy giuliano.destefano@unina2.it

[2] Department of Mechanical Engineering, University of Colorado, Engineering Center, ECME 126, CO 80309 Boulder, USA oleg.vasilyev@colorado.edu

Summary. Stochastic Coherent Adaptive Large-Eddy Simulation is a novel approach to the numerical simulation of turbulence, based upon the wavelet thresholding filter, where the coherent energetic eddies are solved while modelling the influence of the less energetic background flow. In this study, in order to examine the quality and reliability of the method, additional explicit wavelet filtering is introduced by considering two different filtering levels: the physical level, which controls the turbulence model, and the numerical level that is responsible for the accuracy of the numerical simulations. The theoretical basis for explicit filtering and consistent dynamic modelling is given, and some preliminary numerical experiments are presented.

Key words: Adaptive Large-Eddy Simulation, Explicit/Implicit Wavelet Filtering, Dynamic modelling

1 Introduction

The stochastic coherent adaptive large-eddy simulation (SCALES) method is a novel approach to the numerical simulation of turbulence, where the more energetic coherent eddies are solved, while modelling the effect of the less energetic background flow [1]. The formal decomposition between resolved coherent and residual coherent/incoherent motions is obtained through the application of wavelet-based filtering. The space-time evolution of the resolved coherent velocity field is governed by the wavelet-filtered Navier-Stokes equations, where – similarly to any other LES approach – the effect of the unknown residual stresses is modelled.

In order to solve the SCALES governing equations in a computationally efficient manner, the dynamically adaptive wavelet collocation method (AWCM) is used, e.g. [2]. The AWCM procedure is a variable high-order finite-difference method that exploits the same wavelet-based filter to automatically adapt the computational grid to the numerical solution, in both location and scale.

To date, the filtering effect induced by the use of the AWCM has been exploited to implicitly define the filtered-velocity in the SCALES approach, without performing

M.V. Salvetti et al. (eds.), *Quality and Reliability of Large-Eddy Simulations II*, ERCOFTAC Series 16, DOI 10.1007/978-94-007-0231-8_27,

any additional explicit filtering operation. Along this line, the residual stresses have been referred to and treated as subgrid-scale (SGS) stresses, e.g. [3]. It is worth noting that the fundamental issue regarding the interaction between LES filtering, SGS modelling, and numerical errors becomes even more important in the SCALES approach, where the numerical solver allows for the automatic mesh refinement in flow regions with inadequate SGS dissipation.

In this study, in order to examine the quality and reliability of the SCALES methodology, the additional explicit wavelet-filtering procedure is introduced. That is, two different thresholding levels are clearly considered: the physical level, which controls the turbulence model, and the numerical level that is solely responsible for the accuracy of the numerical method. The theoretical basis for SCALES with explicit filtering is given and some preliminary numerical experiments are carried out for decaying homogeneous turbulence at moderate Reynolds-number.

2 Explicit wavelet-filtering approach

2.1 Wavelet-filtered velocity

The formal separation between resolved and unresolved flow structures is obtained through wavelet threshold filtering (WTF), which is performed by applying the wavelet-transform to the unfiltered velocity field, zeroing the wavelet coefficients below a given threshold, and transforming back to the physical space, e.g. [1, 4]. This way, the turbulent velocity field is decomposed into two different parts: a coherent more energetic velocity field and a residual less energetic coherent/incoherent one, i.e., $u_i = \overline{u_i}^{>\varepsilon} + u_i'$, where $\overline{u_i}^{>\varepsilon}$ stands for the wavelet-filtered velocity. Depending on the choice of the WTF level ε that is dictated by the desired turbulence resolution, a relatively small number of wavelets are retained in representing the filtered field $\overline{u_i}^{>\varepsilon}$.

The high compression property of the wavelet-based decomposition is illustrated in Table 1, where the percentage of active wavelets and retained energy/enstrophy are reported as a function of the WTF level for a given turbulent velocity field. The field considered is a realization of a statistically stationary turbulent flow at $Re_\lambda = 126$ (λ being the Taylor microscale) provided by a pseudo-spectral DNS [5]. For instance, by retaining less than 1% of the 512^3 available wavelets, one is able to capture more than 99% of the energy and almost 90% of the DNS enstrophy.

Table 1. Percentage of active wavelets and retained energy/enstrophy for different WTF levels

level ε	wavelets	energy	enstrophy
0.45	0.11%	97.8%	66.0%
0.35	0.26%	99.0%	78.4%
0.25	0.61%	99.5%	87.0%
0.15	1.86%	99.9%	95.1%

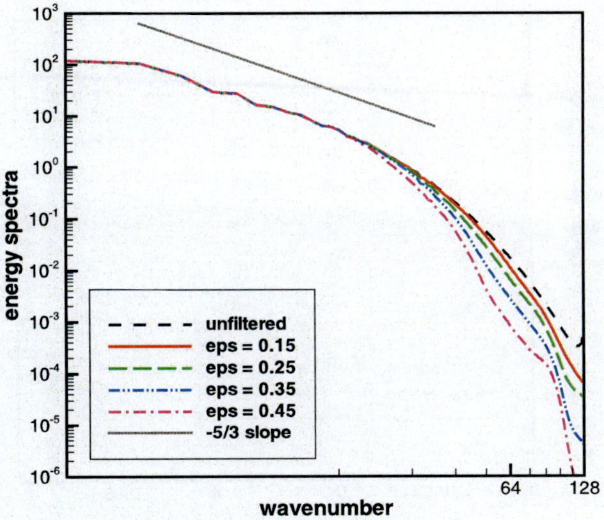

Fig. 1. Energy spectra for wavelet-filtered velocity fields at different levels of resolution (WTF level $\varepsilon = 0.15, 0.25, 0.35$, and 0.45)

Furthermore, one of the distinctive features of WTF stands in the ability to capture coherent energetic eddies of any size. For this reason, the small scale turbulence can be represented – at least partially – by the wavelet-filtered field. This is illustrated in Figure 1, where the energy spectra corresponding to different filtering levels are reported for the above mentioned DNS field.

2.2 Wavelet-filtered equations

The governing equations for SCALES of incompressible turbulent flows are represented by the following wavelet-filtered continuity and Navier-Stokes equations:

$$\frac{\partial \overline{u_i}^{>\varepsilon}}{\partial x_i} = 0, \tag{1}$$

$$\frac{\partial \overline{u_i}^{>\varepsilon}}{\partial t} + \frac{\partial \overline{u_i}^{>\varepsilon}\,\overline{u_j}^{>\varepsilon}}{\partial x_j} = -\frac{1}{\rho}\frac{\partial \overline{p}^{>\varepsilon}}{\partial x_i} + \nu\frac{\partial^2 \overline{u_i}^{>\varepsilon}}{\partial x_j \partial x_j} - \frac{\partial \tau_{ij}}{\partial x_j}, \tag{2}$$

where the unknown subgrid-scale (SGS) stresses

$$\tau_{ij} = \overline{u_i u_j}^{>\varepsilon} - \overline{u}_i^{>\varepsilon}\,\overline{u}_j^{>\varepsilon} \tag{3}$$

need modelling.

From the mathematical point of view, once the SGS stress tensor is given as a function of the resolved velocity $\overline{u_i}^{>\varepsilon}$ and suitable initial conditions are provided,

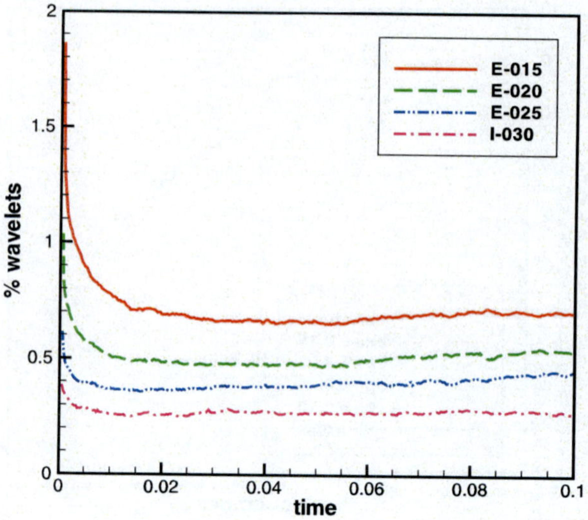

Fig. 2. Percent of retained wavelets for different numerical thresholds (labels as in Table 2)

the SCALES governing equations can be solved using any numerical method. In practice, equation (2) is solved using the dynamically adaptive wavelet collocation method, where the WTF procedure is exploited to automatically adapt the computational grid to the numerical solution, in both location and scale, e.g. [2]. The use of the AWCM procedure involves an unavoidable built-in filtering effect associated to the wavelet numerical threshold ε_g that is used to control the numerical errors. When this induced filtering operation is exploited to unambiguously define the wavelet-filtered velocity assuming $\varepsilon = \varepsilon_g$, there is no reason to discern between explicit and implicit wavelet-filtering, e.g. [3]. However, since the choice of a relatively high threshold level ε_g could affect the accuracy of the numerical simulations, alternatively, one can consider two different superimposed levels of filtering: explicit-filtering at ε and implicit-filtering at $\varepsilon_g < \varepsilon$. The use of such an approach results in adding extra computational modes beyond the modes that are strictly necessary for the desired SCALES solution, but less than in simulations with smaller ε, since the energy cascade is broken by the model.

Considering SCALES with explicit filtering, the momentum equation can be written in the following explicit-filtered form

$$\frac{\partial \overline{u_i}^{>\varepsilon}}{\partial t} + \frac{\partial \overline{\overline{u_i}^{>\varepsilon}\,\overline{u_j}^{>\varepsilon}}^{>\varepsilon}}{\partial x_j} = -\frac{1}{\rho}\frac{\partial \overline{p}^{>\varepsilon}}{\partial x_i} + \nu \frac{\partial^2 \overline{u_i}^{>\varepsilon}}{\partial x_j \partial x_j} - \frac{\partial \tau_{ij}^{>\varepsilon}}{\partial x_j}, \tag{4}$$

where

$$\tau_{ij}^{>\varepsilon} = \overline{u_i u_j}^{>\varepsilon} - \overline{\overline{u_i}^{>\varepsilon}\,\overline{u_j}^{>\varepsilon}}^{>\varepsilon} \tag{5}$$

can be referred to as subfilter-scale (SFS) stresses.

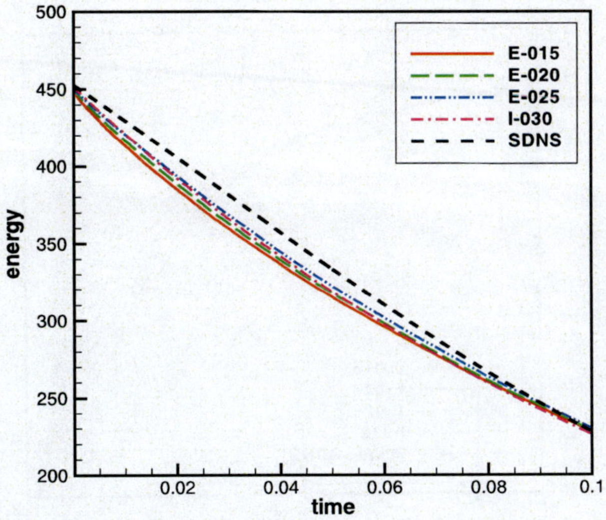

Fig. 3. Resolved kinetic energy for different numerical thresholds (labels as in Table 2)

2.3 Subfilter-scale model

The closure models used in the past for SCALES of turbulent flows should be consistently revised in order to be applied in conjunction with explicit-filtering. For instance, following the eddy-viscosity dynamic Smagorinsky modelling approach proposed in [6], the deviatoric part of the SFS turbulent stress tensor is approximated as

$$\overline{\tau}_{ij}^{>\varepsilon *} \cong -\overline{2C_S\Delta^2\varepsilon^2\left|\overline{S}^{>\varepsilon}\right|\overline{S}_{ij}^{>\varepsilon}}^{>\varepsilon}, \tag{6}$$

where $\overline{S_{ij}}^{>\varepsilon} = 1/2\,(\partial\overline{u_i}^{>\varepsilon}/\partial x_j + \partial\overline{u_j}^{>\varepsilon}/\partial x_i)$ is the filtered rate-of-strain tensor, $\left|\overline{S}^{>\varepsilon}\right| = \left(2\overline{S_{ij}}^{>\varepsilon}\overline{S_{ij}}^{>\varepsilon}\right)^{1/2}$, and Δ is the characteristic explicit-filter length-scale. The residual stress tensor at the test-filter level is defined as the following analog of (5)

$$\overline{T}_{ij}^{>2\varepsilon} = \overline{u_i u_j}^{>2\varepsilon} - \overline{\overline{u}_i^{>2\varepsilon}\,\overline{u}_j^{>2\varepsilon}}^{>2\varepsilon}, \tag{7}$$

where $\overline{(\cdot)}^{>2\varepsilon}$ corresponds to the wavelet test filter at twice the threshold. Since the wavelet filter is a projection operator it satisfies $\overline{\overline{(\cdot)}^{>\varepsilon}}^{>2\varepsilon} \equiv \overline{(\cdot)}^{>2\varepsilon}$.

Therefore, by filtering (5) at the test filter level and combining it with (7), the following modified Germano identity for the known Leonard stresses is obtained:

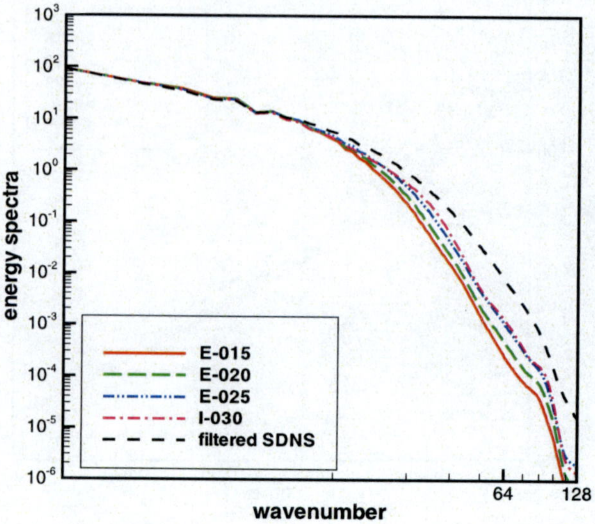

Fig. 4. Energy spectra at $t = 0.05$ for different numerical thresholds (labels as in Table 2)

$$\overline{L}_{ij}^{>2\varepsilon} \equiv \overline{T}_{ij}^{>2\varepsilon} - \overline{\tau_{ij}}^{>2\varepsilon} = \overline{\overline{u_i}^{>\varepsilon}\,\overline{u_j}^{>\varepsilon}}^{>2\varepsilon} - \overline{\overline{u_i}^{>2\varepsilon}\,\overline{u_j}^{>2\varepsilon}}^{>2\varepsilon} . \tag{8}$$

Exploiting the model (6) and the analogous relation for the test filtered SFS stresses that is

$$\overline{T}_{ij}^{>2\varepsilon *} \cong -\overline{2C_S\Delta^2\,(2\varepsilon)^2\left|\overline{S}^{>2\varepsilon}\right|\overline{S}_{ij}^{>2\varepsilon}}^{>2\varepsilon} , \tag{9}$$

one obtains

$$\overline{2C_S\Delta^2\varepsilon^2\left|\overline{S}^{>\varepsilon}\right|\overline{S}_{ij}^{>\varepsilon}}^{>2\varepsilon} - \overline{2C_S\Delta^2\,(2\varepsilon)^2\left|\overline{S}^{>2\varepsilon}\right|\overline{S}_{ij}^{>2\varepsilon}}^{>2\varepsilon} = \overline{L}_{ij}^{>2\varepsilon *} . \tag{10}$$

Finally, a least square solution to (10) leads to the following equation for determining the Smagorinsky model coefficient:

$$2\varepsilon^2\Delta^2 C_S = \frac{\langle \overline{L}_{ij}^{>2\varepsilon *}\,\overline{M}_{ij}^{>2\varepsilon}\rangle}{\langle \overline{M}_{hk}^{>2\varepsilon}\,\overline{M}_{hk}^{>2\varepsilon}\rangle} , \tag{11}$$

where

$$\overline{M}_{ij}^{>2\varepsilon} \equiv \overline{\left|\overline{S}^{>\varepsilon}\right|\overline{S}_{ij}^{>\varepsilon}}^{>2\varepsilon} - 4\left|\overline{S}^{>2\varepsilon}\right|\overline{S}_{ij}^{>2\varepsilon} , \tag{12}$$

and $\langle\cdot\rangle$ denotes volume averaging that is performed to make the numerical solution stable.

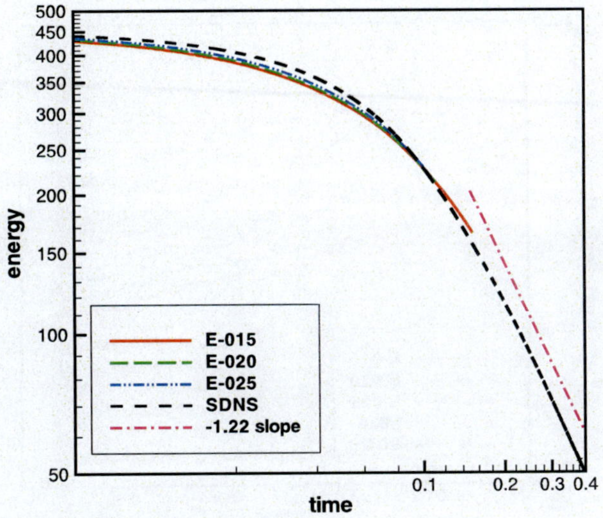

Fig. 5. Resolved kinetic energy for different numerical thresholds in log-log scale (labels as in Table 2)

3 Numerical experiments

In order to make some experiments for SCALES with explicit-filtering, the numerical simulation of incompressible homogeneous decaying turbulence is considered. The initial velocity field is provided by the above-mentioned statistically steady pseudo-spectral DNS solution at $Re_\lambda = 126$ that is obtained by solving the unfiltered Navier-Stokes equations, supplied with the random forcing scheme of Eswaran & Pope [8], with 256^3 Fourier modes. The same pseudo-spectral code with the same resolution is used to produce a reference DNS solution for the present decaying case (for discussion SDNS) as in [5].

It is worth noting that the initial Reynolds-number is too low to allow for an extended inertial range in the energy spectrum. Nevertheless, as illustrated in Figure 1, the theoretical $-5/3$ slope for the inertial scaling regime is well represented by the SDNS solution.

Due to the finite difference nature of the wavelet-based solver, the initial SCALES resolution is doubled in each direction with respect to SDNS, in order to retain approximately the same initial energy content. For this reason, SCALES is run using a maximum resolution corresponding to 512^3 grid points (or, equivalently, wavelets). However, the actual number of wavelets used in the simulation is very low with respect to the above maximum value, owing to the high compression property of the wavelet-transform-based method discussed in Section 2.1. In addition, due to the decaying nature of the turbulent flow considered, a great number of wavelets is only required during the initial period, with a

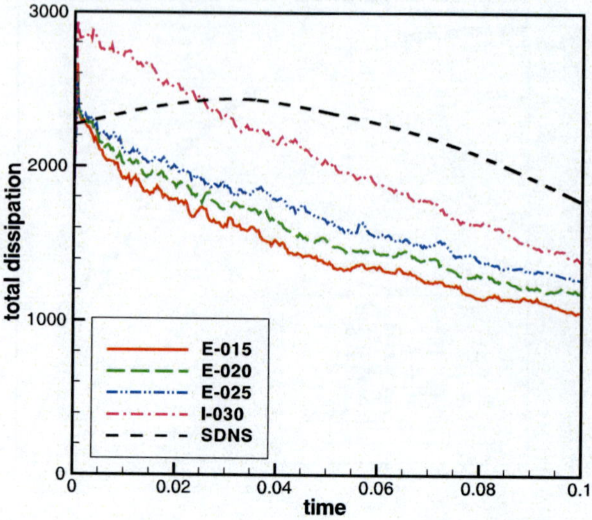

Fig. 6. Total dissipation for different numerical thresholds (labels as in Table 2)

gradual decrease of the necessary level of numerical resolution as turbulence decays.

In this preliminary study, a number of calculations are performed for a given explicit WTF level, which is $\varepsilon = 0.30$, and different numerical thresholds, ranging from $\varepsilon_g = 0.25$ down to $\varepsilon_g = 0.15$. The wavelet-filter at threshold ε is explicitly applied upon the resolved velocity field step-by-step during the simulation, which is practically equivalent to solving (4). Given ε, with the progressive improvement of the numerical accuracy, a grid-independent SCALES solution is approached for decreasing ε_g. This way, one can examine the pure combined effect of "physical" filtering and SGS modelling.

Some interesting results are illustrated for a time interval corresponding to approximately one SDNS initial eddy-turnover time. This short time is however sufficient to have a significant decay in the Reynolds-number that becomes $Re_\lambda \cong 72$. As further reference, the SCALES solution corresponding to $\varepsilon_g = 0.30$ without any explicit filtering is considered in the following (for discussion I-030). All the solutions presented are summarized in Table 2.

Table 2. SCALES calculations performed

label	ε_g	ε	explicit
E-015	0.15	0.30	yes
E-020	0.20	0.30	yes
E-025	0.25	0.30	yes
I-030	0.30	0.30	no

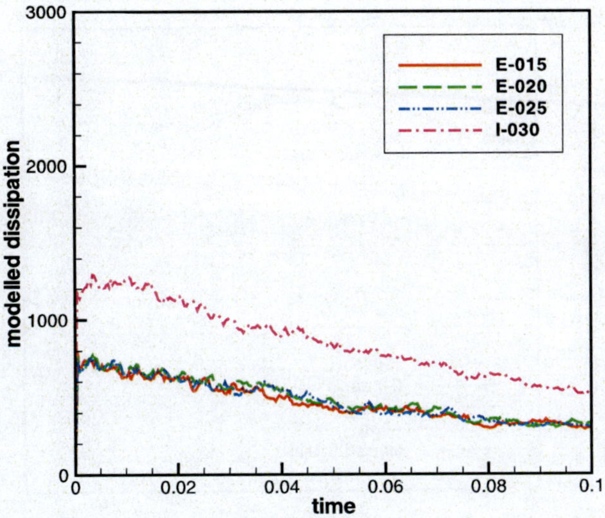

Fig. 7. Modelled dissipation for different numerical thresholds (labels as in Table 2)

As expected, the number of wavelets retained in the calculation increases with the decrease of ε_g. However, even for the smallest value of ε_g, the percent of active wavelets is less than 1% of the total number of available wavelets, as illustrated in Figure 2. The energy decay for the different solutions is reported in Figure 3, while the energy spectra at a given time instant ($t = 0.05$) are illustrated in Figure 4.

Owing to the preliminary character of the present experiments, the time interval for the SCALES solutions is too short to capture the self-similarity regime that exists for homogeneous isotropic decaying turbulence, e.g. [9]. In fact, the decay exponent varies in time during the first period of the decay, while it is expected to be constant at large time instants. By plotting the energy decay in log-log scale, as illustrated in Figure 5, the constant decay exponent can be empirically estimated to be about -1.22 for the SDNS solution.

The time histories of total dissipation (resolved plus modelled) and modelled dissipation alone are depicted in Figures 6 and 7, respectively. The spectral distribution of resolved dissipation is plotted in Figure 8. As discussed for the energy decay, similarly, the self-similarity regime for the energy dissipation rate can not be observed. By plotting the total dissipation in log-log scale, as illustrated in Figure 9, the corresponding constant decay exponent can be empirically estimated at large time instants to be about -1.86 for the SDNS solution.

By making a comparison with the implicit-filtering approach, it appears that the additional wavelet-filtering applied upon the SCALES solution induces additional dissipation of kinetic energy. This built-in extra dissipation should be taken into account when considering the energy budget, as well as the flow statistics that involve

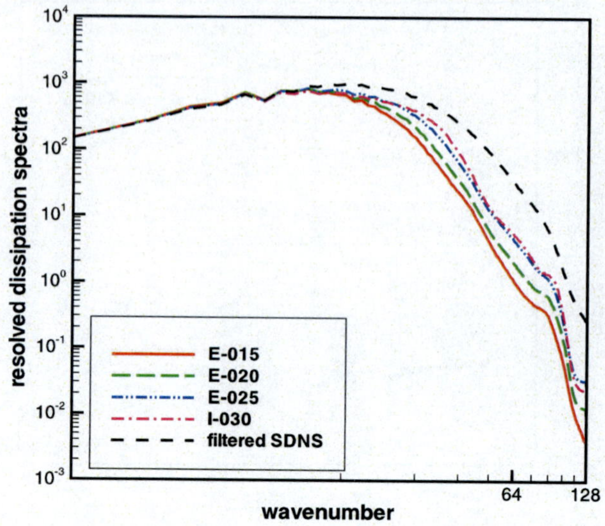

Fig. 8. Resolved dissipation spectra at $t = 0.05$ for different numerical thresholds (labels as in Table 2)

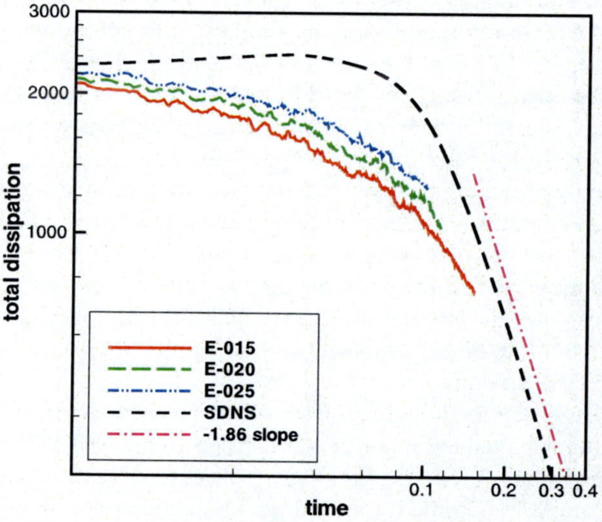

Fig. 9. Total dissipation for different numerical thresholds in log-log scale (labels as in Table 2)

dissipation scales like, for instance, the Taylor-scale Reynolds-number reported in Figure 10.

For decreasing ε_g, the SCALES solution appears to depart from SDNS, approaching a grid-independent solution that represents the ideal evolution of the wavelet-filtered velocity field corresponding to the most energetic coherent ed-

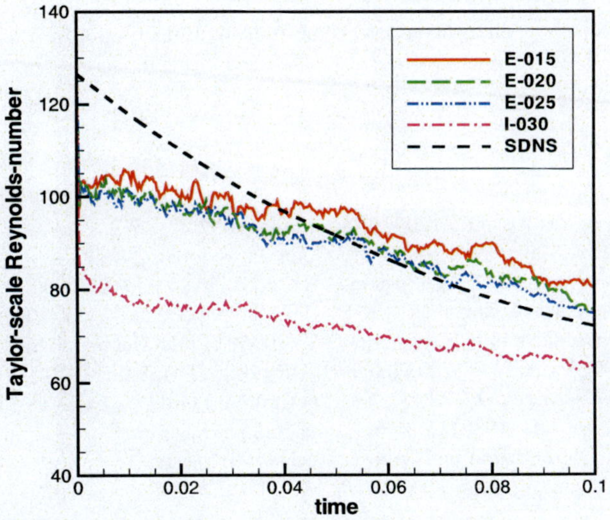

Fig. 10. Taylor-scale Reynolds-number for different numerical thresholds (labels as in Table 2)

dies (as defined by the prescribed WTF level). An even lower numerical threshold should be considered to make this behavior more evident. Given ε, the SFS dissipation provided by the modelling procedure does not depend upon the grid resolution determined by the choice of ε_g, while the resolved dissipation tends to be lower, corresponding to the reduced enstrophy contained in the explicitly-filtered field. In the coarse-grid case (I-030), in absence of explicit-filtering ($\varepsilon = \varepsilon_g$), the model automatically compensates for the lack of dissipation owing to the adaptive nature of the SCALES approach. However, in this case the picture becomes very complex and, in order to make it possible to better discern between the filtering and the numerical issues, the explicit-filtering procedure seems preferable.

4 Concluding remarks

In this work, the SCALES method is applied with superimposed explicit filtering. The study of the effect of numerical thresholding level on the accuracy and computational efficiency of SCALES is carried out. The explicit filtering allows the analysis of the quality and reliability of SCALES solutions with respect to ideal grid-independent calculations, thus enhancing our knowledge about the strong interactions between wavelet-compression and modelled turbulent dissipation in the wavelet-based numerical simulations of turbulence. Another possibility that will be examined in the future consists in varying the SFS threshold ε for a given very small value of ε_g. The use of a smaller SFS thresholding level would increase the number of resolved coherent structures, and the influence of the model would become less important, since the coherent eddies dynamics would be captured by the resolved modes.

References

1. Goldstein D E, Vasilyev O V (2004) Phys Fluids 16:2497–2513
2. Vasilyev O V, Kevlahan N K-R (2005) J Comp Phys 206:412–431
3. De Stefano G, Vasilyev O V, Goldstein D E (2008) Phys Fluids 20:045102.1–14
4. Farge M, Schneider K, Kevlahan N K-R (1999) Phys Fluids 11:2187–2201
5. De Stefano G, Goldstein D E, Vasilyev O V (2005) J Fluid Mech 525:263–274
6. Goldstein D E, Vasilyev O V, Kevlahan N K-R (2005) J Turbul 6:1–20
7. De Stefano G, Vasilyev O V (2004) Theoret Comput Fluid Dynamics 18:27–41
8. Eswaran V, Pope S B (1988) Comput Fluids 16:257–278
9. George W K (1992) Phys Fluids A 4(7):1492–1509

Error reduction in LES via adaptive moving grids

Claudia Hertel[1] and Jochen Fröhlich[2]

[1] Institut für Strömungsmechanik, TU Dresden, 01062 Dresden, Germany
claudia.hertel@tu-dresden.de
[2] Institut für Strömungsmechanik, TU Dresden, 01062 Dresden, Germany
jochen.froehlich@tu-dresden.de

Summary. In complex turbulent flows the length scale varies substantially over the computational domain. When modelling such flows with large eddy simulation (LES) this must be accounted for. In the present paper a self-adaptive method is presented to adjust the step size of the computational grid to the local resolution requirements of an LES. An $r-$adaptive method is used involving a moving mesh PDE. Different physically motivated monitor functions are proposed, most of them closely related to LES, and applied to the turbulent flow over periodic hills.

Key words: large eddy simulation, moving mesh, adaptivity, separated flow, periodic hills

1 Introduction

Comparing the results of an LES to those of a DNS, very good agreement can be obtained in some cases, in other cases deviations are observed. These depend on the details of the LES and on the quantity considered. The LES error essentially contains two components, the discretization error and the modelling error, which are intermingled but can be separated in tests [5]. While their ratio grossly depends on the ratio of the filer size to the grid size, both decrease if the grid is refined. This leads to the idea of reducing the error of an LES by optimally refining the grid, i.e. positioning a given number of grid points in an optimal way using a self-adaptive method. Adaptive methods comprise three ingredients, a PDE solver, a method to accomplish local changes in grid resolution and a measure or indicator for the need of refinement. Different algorithms to generate local refinement have been developed. These comprise h-refinement (usually by locally inserting grid points), p-refinement (by increasing the local order of the discretization), and r-refinement (by redistributing the given points in space). The latter is employed here as it leaves the data structure unchanged during the simulation.

The basic element of each r-adaptive method is the strategy according to which the grid points are moved. The localization-based method employed in the present

M.V. Salvetti et al. (eds.), *Quality and Reliability of Large-Eddy Simulations II*,
ERCOFTAC Series 16, DOI 10.1007/978-94-007-0231-8_28,

paper defines directly the grid points by using the concept of the moving mesh PDE by Russell et al. [6] which links the distribution of grid points to the so-called monitor function.

In contrast to laminar flows or statistical turbulence modelling, changes in local resolution do not only affect the discretization error but also the equation to be solved itself, via the subgrid-scale modelling term. Refinement criteria therefore have to be LES-specific. In the present work several physically motivated criteria are proposed which are particularly easy to evaluate.

Two attitudes can be followed when adapting the grid in an LES: One is to optimize the grid for the instantaneous solution, the other is to optimize in a statistical sense. The latter leads to a two-phase procedure consisting of a first phase in which the LES is conducted and the grid adapted according to statistical data and a second phase where the optimized grid is frozen and the final statistics are accumulated. This attitude is followed here, but the former can be obtained with exactly the same method by just a change in parameters (τ in (4) below).

2 Numerical framework and LES modelling

The filtered Navier-Stokes equations on curvilinear coordinates are solved using a collocated variable arrangement on block-structured grids. Time integration is accomplished by a fractional step method, consisting of a Runge Kutta scheme as predictor and a pressure-correction equation as corrector. In space, a Finite Volume method with central interpolation is employed to discretize the governing equations. To account for moving grids, the ALE (arbitrary Lagrangian Eulerian) framework has been implemented. The momentum equation then reads [3]

$$\begin{aligned} \frac{\mathrm{d}}{\mathrm{d}t} \int_{V(t)} \rho\, u_i \,\mathrm{d}V \;+\; \int_{V(t)} \rho \frac{\partial\left[u_i\left(u_j - u_{N,j}\right)\right]}{\partial x_j} \mathrm{d}V \\ = -\int_{V(t)} \frac{\partial p}{\partial x_i} \mathrm{d}V + \int_{V(t)} \frac{\partial \tau_{ij}}{\partial x_j} \mathrm{d}V + \int_{V(t)} f_i \,\mathrm{d}V \,, \end{aligned} \tag{1}$$

where $u_{N,i}$ specifies the grid velocity and $V(t)$ the time dependent volume of the computational cells. When dicretizing (1) the Space Conservation Law has to be fulfilled to guarantee that the discrete values of the grid velocity $u_{N,i}$ are calculated in a way consistent with the change of volume over time of each cell [2].

Subgrid-scale modelling is accomplished using the Smagorinsky model [8] with a constant of $C_S = 0.1$. Since the grid is intentionally coarse, a wall function is employed [10].

3 Moving Mesh PDE

The r-adaptive method employed controls the positions $\mathbf{x}$ of the computational grid in physical space by minimizing a mesh adaption functional [7]

$$I[\boldsymbol{\xi},t] = \frac{1}{2}\int_{\mathbf{S}_p}\sum_{i=1}^{3}(\nabla\xi^i)^T G^{-1}\nabla\xi^i \mathrm{d}\mathbf{x} \tag{2}$$

where $\boldsymbol{\xi} = \boldsymbol{\xi}(\mathbf{x},t)$ is the uniformly discretized computational space and $\mathbf{x} = \mathbf{x}(\boldsymbol{\xi},t)$ the inverse mapping. The monitor function G is a 3×3 symmetric positive definite matrix, proportional to the density of grid points. To achieve equally distributed and regular grids with respect to G a Moving Mesh Partial Differential Equation (MMPDE), resulting from minimizing the functional (2), can be devised [7]

$$\tau\frac{\partial\mathbf{x}}{\partial t} = P\left[\sum_{i=1}^{3}\sum_{j=1}^{3}\underbrace{(\mathbf{a}^i\cdot G^{-1}\mathbf{a}^j)}_{A_{ij}}\frac{\partial^2\mathbf{x}}{\partial\xi^i\partial\xi^j} - \sum_{i=1}^{3}\underbrace{\left(\mathbf{a}^i\cdot\frac{\partial G^{-1}}{\partial\xi^i}\mathbf{a}^i\right)}_{B_i}\frac{\partial\mathbf{x}}{\partial\xi^i}\right]. \tag{3}$$

In this equation, $\mathbf{a}^i = \triangledown\xi^i$ identifies the contravariant basis vectors ($\triangledown$ is the gradient with respect to $\mathbf{x}$), τ the time scaling parameter for the grid velocity and A_{ij}, B_i coefficients depending on both, the physical information G, and geometrical quantities $\mathbf{a}^i$.

Choosing $G = \omega\mathscr{I}$, with $\mathscr{I}$ the identity matrix, reduces the adaption criterion to the choice of the scalar monitor function ω [11]. For the local scaling parameter P in (3) a combination of the coefficients A_{ii} and B_i is used to guarantee an equal order of magnitude of the coefficients of the MMPDE over the domain. With these choices (3) becomes

$$\tau\frac{\partial\mathbf{x}}{\partial t} = \frac{1}{\sqrt{\sum_i(A_{ii}^2+B_i^2)}}\frac{1}{\omega^2}\sum_{i=1}^{3}\sum_{j=1}^{3}(\mathbf{a}^i\cdot\mathbf{a}^j)\frac{\partial}{\partial\xi^i}\left(\omega\frac{\partial\mathbf{x}}{\partial\xi^j}\right). \tag{4}$$

The shape of the boundary is fixed throughout the simulation but the grid points are able to move on the boundary. The grid movement in tangential direction at the walls is linked directly to the movement of the wall nearest points to achieve orthogonal grids at the wall.

Equation (4) is discretized by a Finite Difference formulation for the cell corner coordinates. Using cell centers as unknowns as employed for velocities and pressure is not possible because given these, a valid grid can only be computed in exceptional cases. For the spatial discretization central differences are used and the Euler implicit scheme is employed in time.

4 Choice of the monitor function

The heart of each moving mesh method is the monitor function and hence the *Quantity of Interest* ψ (QoI) according to which the MMPDE tries to equidistribute the grid. In the present paper the monitor function ω has the form

$$\omega(\xi,\eta) = \sqrt{1+\alpha\left(\frac{\psi}{\psi_{max}}\right)^2}. \tag{5}$$

In each time step the QoI ψ is scaled with the maximum value in the domain ψ_{max} to achieve $\omega \in \left[1; \sqrt{1+\alpha}\right]$. The parameter α is chosen once and for all to scale the maximum value of ω. Different choices of ψ were tested.

Gradient of velocity:

$$\psi = \nabla \langle u \rangle \,. \tag{6}$$

Here, $\langle u \rangle$ is the time- and z-averaged streamwise velocity component. The grid refinement hence should take place in flow regions where the gradient is high.

Turbulent viscosity:

$$\psi = \frac{\langle \nu_t \rangle}{\langle \nu_t \rangle + \langle \nu \rangle} \tag{7}$$

Here, ν is the molecular viscosity and ν_t the modelled viscosity, predicted by the Smagorinsky model. This criterion aims to equidistribute the ratio of modelled to total dissipation, very closely related to the ratio of SGS dissipation ε_t to total dissipation ε_μ, the subgrid-activity parameter defined in [5]

$$s = \frac{\langle \varepsilon_t \rangle}{\langle \varepsilon_t \rangle + \langle \varepsilon_\mu \rangle} \,. \tag{8}$$

Modelled turbulent kinetic energy (TKE):

$$\psi = \frac{\langle k_{sgs} \rangle}{\langle k_{res} \rangle + \langle k_{sgs} \rangle} \tag{9}$$

According to the basic idea of LES this criterion tries to achieve a uniform distribution of the percentage of modeled TKE over the domain. To evaluate the modelled TKE of the subgrid-scales an approach from Berselli, Iliescu and Layton [1] is used by setting

$$k_{sgs} \approx \left(2^{1/3} - 1\right) 0.5 \left|\bar{\mathbf{u}} - \bar{\bar{\mathbf{u}}}\right|^2 \,. \tag{10}$$

Here, $\bar{\mathbf{u}}$ is the resolved velocity of the LES, while $\bar{\bar{\mathbf{u}}}$ is obtained from an explicit filtering operation using weights of $1/8$ for the point itself and $1/16, 1/32, 1/64$ for the neighbour points. The resolved TKE is determined as $k_{res} = (u_i'u_i' + v_i'v_i' + w_i'w_i')/2$. This model for k_{sgs} leads to unwanted, non physically high values of ψ in regions where k_{res} is vanishing. Two remedies of ψ were tested. The first approach is designed to avoid the division by very low values by adding a constant to the local value:

$$\psi = \frac{\langle k_{sgs} \rangle}{\langle k_{tot} \rangle + C\, k_{tot,max}} \,. \tag{11}$$

Two values of C were tested here: $C = 0.1$ and $C = 0.001$.

The second variant replaces the local TKE by the maximum in the domain:

$$\psi = \frac{\langle k_{sgs} \rangle}{k_{tot,max}} \,. \tag{12}$$

5 Application to the flow over periodic hills

The method proposed above was implemented in the in-house code LESOCC2 and applied to the turbulent flow over periodic hills [4, 9]. The configuration is a channel with hill-shaped constrictions as depicted in Fig. 1 with $L_x = 9h$, $L_y = 3.035h$. The flow is supposed to be periodic in streamwise and spanwise direction ($L_z = 4.5h$).

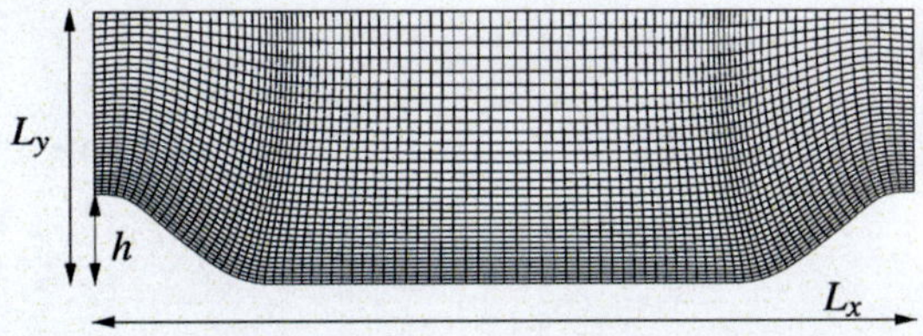

Fig. 1. Two-dimensional slice of the geometry investigated and initial grid.

To demonstrate the improvements achievable by using an MMPDE, a coarse grid is chosen on purpose. It consists of $89 \times 33 \times 49$ grid points (135168 cells) in $x-$, $y-$ and $z-$direction, respectively. The simulations were conducted with $Re_h = 10595$, based on the bulk velocity at the hill crest and the hill height h. The same physical and numerical parameters are used throughout. The adaption was run with $\alpha = 50$, $\tau = 1.0$ and only the QoI changed. All simulations started with the same well converged solution on the stationary grid displayed in Fig. 1. They were performed over 10000 time steps, with the grid being adapted every 100 time steps. Towards the end of these simulations the increments of the grid points became very small. Grid adaption was then turned off and the statistics were accumulated over 50 flow-through-times. In the simulations below k_{sgs} in (9)-(12) was determined by averaging in spanwise direction.

Final grid and monitor function

The corner grid points of the initial grid of each QoI tested are shown in Fig. 2(a-e) together with the monitor function ω at the first time step. The final grids of the adaption are shown on the right side of Fig. 2.

The gradient of the streamwise velocity $\langle u \rangle$ generates a monitor function with peak values close to the surface of the hill crest. The grid is hence refined substantially in this region so that the separation region is well resolved. Low values of ω elsewhere in the domain redistribute the grid points almost equally (compare Fig. 2(a) and Fig. 2(f)).

The monitor function (7) using the turbulent viscosity shows a smoother distribution of ω as can be seen in Fig. 2(b). The grid is refined near the hill crest, but with the turbulent viscosity being zero at a solid wall no refinement in normal direction at the wall can be achieved. The grid there is coarsed instead as displayed in Fig. 2(g). For the monitor function based on the modelled TKE different results are obtained for the variants introduced in Sec. 4. The results in Fig. 2(i) and 2(k), for (11), where a constant is added to the resolved TKE in the denominator, show only slight differences in the grids obtained. In both cases the maximum of the monitor function,

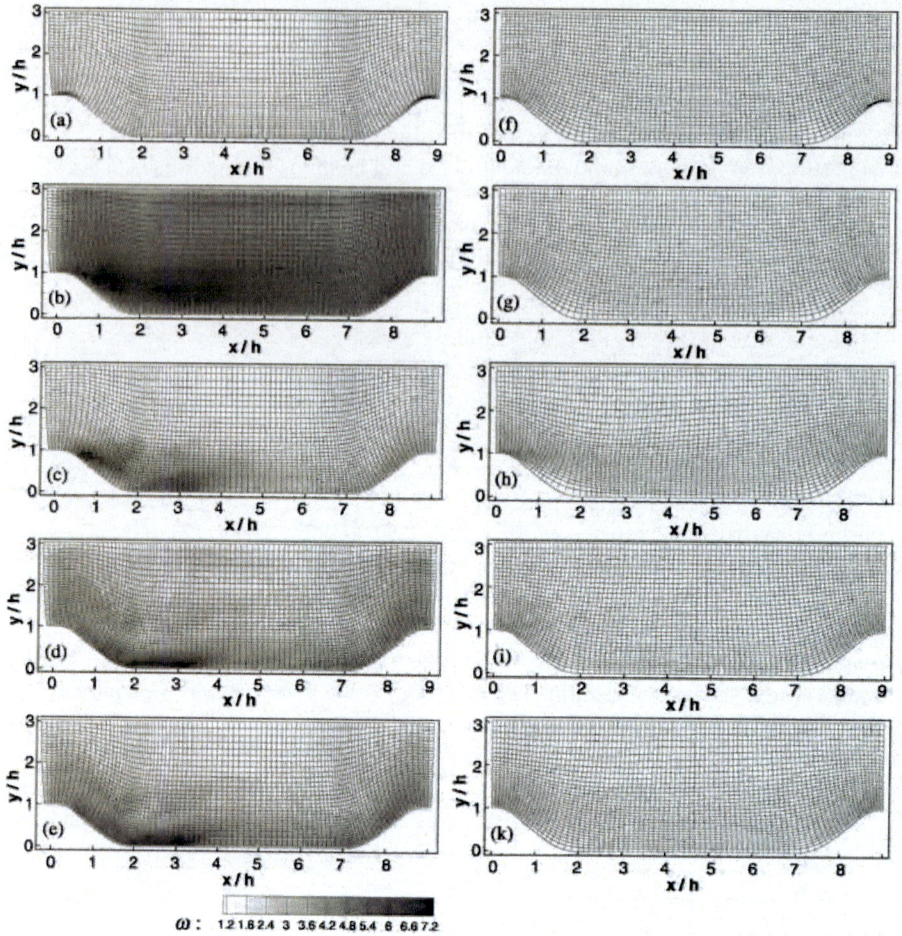

Fig. 2. Initial grid together with monitor function ω at the first time step (left) and the final grid after 100 adaptions (right). QoI: (a,f) gradient of streamwise velocity; (b,g) turbulent viscosity; (c,h) modelled TKE (12); (d,i) modelled TKE (11) with $C = 0.1$; (e,k) modelled TKE (11) with $C = 0.001$.

and hence the maximum for grid refinement, is situated downstream of the separation point and in the recirculation area. The second variant related to the modelled TKE, (12), leads to grid refinement in the shear layer resulting from a maximum of the monitor function ω in that region. Similar to (7), the criteria (11), (12) yield low values of the monitor function at walls and hence large cells at the walls.

Properties of mean flow on adapted grids

The correct determination of the separation and reattachement point of the flow has a strong influence on the overall quality of the LES of the turbulent hill flow [9]. The values obtained with the present simulations are reported in Table 1.

Table 1. Separation and reattachement points for simulations on final adapted grids compared to the data of Fröhlich et al. [4]

	fine grid [4]	initial grid	$\nabla\langle u\rangle$	$\frac{\langle \nu_t\rangle}{\langle \nu_{tot}\rangle}$	(12)	(11) $C=0.1$	(11) $C=1e-3$
x_{sep}/h	0.2	0.5	0.3	0.65	0.45	0.55	0.55
x_{rea}/h	4.6	3.1	4.7	2.6	3.4	3.1	2.85

Comparing separation and reattachement point for the simulation using the initial grid without any adaption to the reference values shows, that the separation point from the curved surface is calculated too far downstream and hence the recirculation area is too small. The flow reattaches too early. All QoIs tested except the gradient of streamwise velocity according to (6) do not improve the prediction of the separation point substantially in the present setup. For the monitor function (6) a separation and reattachement point close to the results of Fröhlich et al. [4] is achieved, due to the fine grid resolution in wall normal and tangential direction at the hill crest.

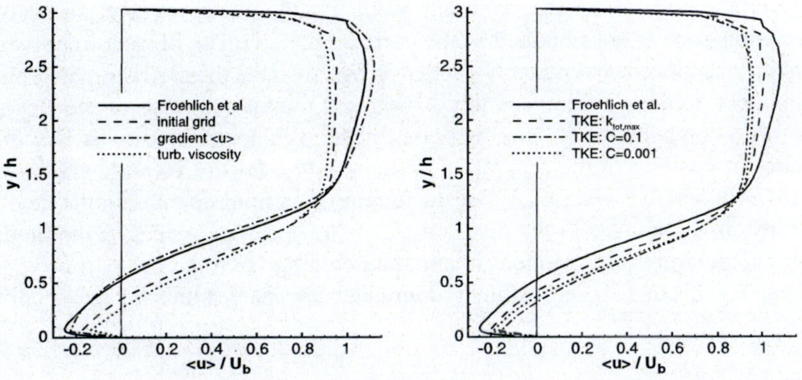

Fig. 3. Averaged streamwise velocity at position $x/h = 2.0$ in the middle of the recirculation area.

This trend can also be seen in the streamwise velocity at $x/h = 2.0$ plotted in Fig. 3 as an example. This position is located in the middle of the recirculation area so that the effect of the separation point on the downstream flow can be seen. The well predicted separation point obtained with the gradient monitor function (6) results in a well calculated velocity profile in the recirculation area. As for the initial grid, the profiles obtained with the QoIs containing the modelled TKE or the turbulent viscosity show too small recirculation areas. The worst separation point was obtained with (7), which is in line with the profile of $\langle u\rangle$ showing no improvement compared to the initial grid in that case. For the modelled TKE (12) a slight improvement of the

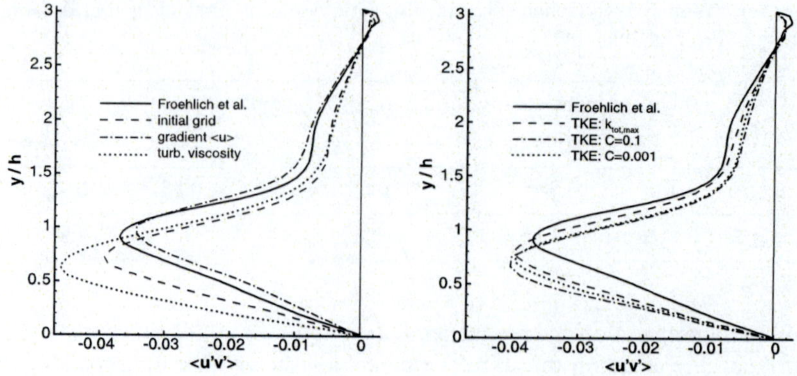

Fig. 4. Reynolds stress $\langle u'v' \rangle$ at position $x/h = 2.0$ in the middle of the recirculation area.

separation point can be achieved (Table 1), also visible in the velocity profile where the grid refinement in the shear layer brings further improvement.

In Fig. 4, the Reynolds shear stress $\langle u'v' \rangle$ is shown for the same position as the velocity profiles. The solution obtained with the initial grid and no adaption shows a wrong vertical position for the maximum value, i.e. the free shear layer on top of the recirculation zone is not obtained at the correct position. The biggest improvement can again be achieved by the monitor function based on the mean streamwise velocity gradient. The biggest differences in position and maximum value of the Reynolds stress $\langle u'v' \rangle$ compared to [4] are obtained for the turbulent viscosity as QoI due to the deficient position of the separation point and, like for the velocity gradient, too few grid points in the shear layer. For the monitor functions containing the modelled TKE only small changes in the distribution of $\langle u'v' \rangle$ can be seen. For the modified variant (12) a slight improvement is obtained coming from the better resolution of the shear layer, but still the result is dominated by the position of the separation point.

Computational costs

When evaluating the improvements obtained with the *r*-adaptive method the computational costs are of interest. All simulations were performed on one processor of the SGI Altix at ZIH, Dresden (ccNUMA architecture, dual core Itanium 2, 1.6 GHz). Two issues determine the computational efficiency, the convergence behaviour of the Poisson solver in each time step and the size of the time step. Already during the first part of the simulations, where the adaption was performed, differences in the computation time were noticed. since the grid is distorted differently by the used criteria, the convergence behaviour is different in the simulations. Hence, although the same number of time steps was executed, the computation time differs. It is longest for the adaption using the gradient of the averaged velocity (6), needing 6.5 hours, while the shortest was the one using the turbulent viscosity (7) with 5.3 hours. The time used to solve the MMPDE was less than 3% of the simulation time for all criteria.

Table 2. Time step size and computation time in the second phase of the simulations.

	initial grid	$\nabla\langle u\rangle$	$\frac{\langle \nu_t\rangle}{\langle \nu_{tot}\rangle}$	(12)	(11) $C=0.1$	(11) $C=1e-3$
time step $[\times 10^{-2}]$	2.069	1.512	1.736	1.760	1.717	1.786
CPU time [*hours*]	13.0	19.0	13.6	15.2	13.75	14.0

Table 2 provides information about the second phase of the simulations where the grid resulting from the respective adaption was frozen and averages accumulated over 50 flow-through times. The size of the time step was adjusted automatically to comply with the CFL-limit and the diffusion limit of the time scheme. The local clustering of points in specific regions led to a reduction in all cases considered. Although criteria (7) and (11) yield a time step about 20% smaller than for the initial grid, the CPU time requirement is only increased by about 5% due to the different convergence behaviour. The grid obtained with (6) yields the largest CPU cost due to the small cells at the crest reducing Δt and the somewhat slower convergence.

6 Conclusions

In the present study, the grid of an LES was adapted to the statistical properties of the flow according to several quantities of interest. In the applications it was observed that the LES-specific quantities, like those based on k_{res}, ν_t etc. were not as successful as the criterion based on the gradient of the mean streamwise velocity as the latter substantially clusters the points in the separation region. Adaption for the core and the near-wall region hence seem to have different requirements. Simulations with blended criteria are under way.

Acknowledgments

This work is sponsored by the German Research Foundation (DFG) in the framework of Priority Programme SPP 1276, MetStröm. The authors would like to thank their colleagues A. Dörnbrack, J. Lang, C. Kühnlein and S. Löbig for instructive discussions and a passionating cooperation.

References

1. Berselli LC, Iliescu W, Layton WJ (2006) Mathematics of Large Eddy Simulation of Turbulent Flows. Springer, Berlin Heidelberg New York
2. Demirdžić I, Perić M (1988), Int. J. Num. Meth. Fluids 8:1037–1050
3. Ferziger JH, Perić M (2002) Computational Methods for Fluid Dynamics. Springer, Berlin Heidelberg New York

4. Fröhlich J, Mellen CP, Rodi W, Temmerman L, Leschziner MA (2005), J. Fluid Mech. 526:19–66
5. Geurts, Fröhlich J (2002), Phys. Fluids 14:L41–L44
6. Huang W, Ren Y, Russell RD (1994), SIAM J. Numer. Anal. 31:709–730
7. Huang W (2001), J. Comp. Phys. 171:753–775
8. Smagorinsky J (1963), Mon. Weather Rev. 91:99–165
9. Temmerman L, Leschziner M, Mellen CP, Fröhlich J (2003), Int. J. Heat Fluid Flow 24:157–180
10. Werner H, Wengle H (1993), Selected Papers from the 8th Symposium on Turbulent Shear Flows (ed F Durst, R Friedrich, B Launder, F Schmidt, U Schumann & J Whitelaw) 155–168
11. Winslow A (1967), J. Comput. Phys. 1:149–172

Influence of Reynolds number and grid resolution on large-eddy simulations of self-similar jets based on relaxation filtering

Christophe Bogey[1] and Christophe Bailly[2]

[1] LMFA, UMR CNRS 5509, Ecole Centrale de Lyon, 69134 Ecully, France
christophe.bogey@ec-lyon.fr
[2] Same address christophe.bailly@ec-lyon.fr, Institut Universitaire de France

Summary. Large-eddy simulations are performed using low-dissipation numerical schemes combined with a relaxation filtering as subgrid dissipation to investigate the influence of the Reynolds number and the grid resolution on self-similar turbulent circular jets. Three jets with the same initial parameters except for the diameters yielding Reynolds numbers of 1800, 3600 and 11000 are first considered. Then two additional jets at Reynolds number 3600 are calculated using coarser grids in the turbulent flow regions. Energy dissipation and filtering activity are examined in the different simulations, and mean and turbulent properties of the jets are compared.

Key words: Large-Eddy Simulation, relaxation filtering, jet, Reynolds number

1 Introduction

This work deals with the influence of the Reynolds number and grid resolution on self-similar jets computed by large-eddy simulations based on relaxation filtering (LES-RF) [1, 2], in which a high-order filtering is applied to the flow variables to relaxe energy. It is the continuation of an LES of a jet at Reynolds number $\mathrm{Re}_D = 11000$, which provided results in good agreement with measurements by Panchapakesan and Lumley [3] for a jet at the same Reynolds number [2].

The first motivation of this work is to assess the LES methodology by examining the way in which the filtering activity adjusts to the physical and numerical parameters of the simulations. The LES method, or similar methods, have been applied successfully by different research teams [1, 2, 4, 5, 6, 7], but its limitations are still to be clearly evidenced. One important question when using LES is in particular whether it is possible to reproduce and investigate Reynolds number effects [1, 8] with the subgrid modelling that is implemented. The second motivation of the present work is therefore to study the influence of the Reynolds number on the self-similar turbulent jets using the LES-RF approach. The impact of

M.V. Salvetti et al. (eds.), *Quality and Reliability of Large-Eddy Simulations II*,
ERCOFTAC Series 16, DOI 10.1007/978-94-007-0231-8_29,

the Reynolds number on jet flow development has indeed been shown, experimentally [9, 10, 11, 12, 13, 14], as well as numerically [1], to be significant up to Reynolds numbers around $\mathrm{Re}_D = u_j D/\nu \simeq 10^4$, where u_j and D are the jet inlet velocity and diameter, and ν is the kinematic molecular viscosity.

Thus, in order to deal with the effects of the Reynolds number, two jets with the same conditions as the jet at $\mathrm{Re}_D = 11000$, but with different diameters yielding lower Reynolds numbers of 1800 and 3600, are considered. Then, two other jets at $\mathrm{Re}_D = 3600$, in which coarser grids are used in the turbulent flow regions, are calculated. In this way, the variations of the results with the Reynolds number are explored at a fixed grid resolution, those with the grid resolution at a fixed Reynolds number.

In this paper, the main parameters of the simulations are first defined. Then vorticity fields are shown, the filtering activity in the different LES is discussed, and some preliminary comparisons between the mean and turbulent properties of the jets are reported.

2 Methods and parameters

2.1 LES methodology

The flow governing equations are the 3-D Cartesian filtered compressible Navier-Stokes equations, rearranged by Vreman *et al.* [15, 16] in the following form

$$\frac{\partial \overline{\rho}}{\partial t} + \frac{\partial \overline{\rho}\widetilde{u}_j}{\partial x_j} = 0 \tag{1}$$

$$\frac{\partial \overline{\rho}\widetilde{u}_i}{\partial t} + \frac{\partial \overline{\rho}\widetilde{u}_i\widetilde{u}_j}{\partial x_j} = -\frac{\partial \overline{p}}{\partial x_i} + \frac{\partial \widetilde{\tau}_{ij}}{\partial x_j} \tag{2}$$

$$\frac{\partial \overline{\rho}\widetilde{e}_t}{\partial t} + \frac{\partial ((\overline{\rho}\widetilde{e}_t + \overline{p})\widetilde{u}_j)}{\partial x_j} = -\frac{\partial \widetilde{q}_j}{\partial x_j} + \frac{\partial \widetilde{\tau}_{ij}\widetilde{u}_i}{\partial x_j} \tag{3}$$

where ρ represents the density, u_i the velocity, p the pressure, τ_{ij} the viscous stress tensor, e_t the total energy density, and q_j the heat flux. The overbar denotes a filtered quantity, and the filtering is assumed to commute with time and spatial derivatives. The tilde denotes a quantity calculated from the filtered variables $\overline{\rho}$, $\overline{\rho u_i}$ and $\overline{p}$. Thus the calculated velocity is $\widetilde{u}_i = \overline{\rho u_i}/\overline{\rho}$ (Favre filtering), and the calculated total energy is $\overline{\rho}\widetilde{e}_t = \overline{p}/(\gamma-1) + \overline{\rho}\widetilde{u}_i\widetilde{u}_i/2$ for a perfect gas, where γ is the specific heat ratio. The viscous stress tensor is defined by $\widetilde{\tau}_{ij} = 2\widetilde{\mu}(\widetilde{s}_{ij} - \widetilde{s}_{kk}\delta_{ij}/3)$ with $\widetilde{s}_{ij} = (\partial\widetilde{u}_i/\partial x_j + \partial\widetilde{u}_j/\partial x_i)/2$. The viscosity $\widetilde{\mu} = \mu(\widetilde{T})$ is provided by Sutherland's law, and the temperature $\widetilde{T}$ is obtained using the state equation $\overline{p} = \overline{\rho} r \widetilde{T}$. The heat flux is given by $\widetilde{q}_j = -\lambda \partial \widetilde{T}/\partial x_j$ where $\lambda = \widetilde{\mu} c_p/\sigma$ is the thermal conductivity, σ is the Prandtl number and c_p the specific heat at constant pressure.

In the present LES methodology, the effects of the energy-dissipating subgrid scales are taken into account by the application of a high-order/selective filtering to the flow variables $\overline{\rho}$, $\overline{\rho u_i}$ and $\overline{p}$, to relaxe energy through the smaller scales discretized, without affecting significantly the scales accurately resolved [1]. This LES methodology based on relaxation filtering (LES-RF) was in particular developed to avoid an artificial decrease of the effective Reynolds number of the flow, as it might be the case with modellings involving eddy viscosity [8].

During the simulations, the terms in the budgets for the turbulent kinetic energy, including the dissipation induced by the relaxation filtering, are all computed explicitly from the LES equations. The application of the filtering to the momentum variables can indeed be integrated into the right-hand side of equation (2), in the following way

$$\frac{\partial \overline{\rho}\widetilde{u}_i}{\partial t} + \frac{\partial \overline{\rho}\widetilde{u}_i\widetilde{u}_j}{\partial x_j} = -\frac{\partial \overline{p}}{\partial x_i} + \frac{\partial \widetilde{\tau}_{ij}}{\partial x_j} + D^j_{sf}\left(\overline{\rho u_i}\right) \tag{4}$$

where D^j_{sf} is the filtering operator in the j-th direction. More details can be found in a recent paper [2].

2.2 Numerical algorithm

The LES equations (1), (2) and (3) are solved using low-dissipation and low-dispersion explicit schemes [17]. Fourth-order 11-point finite differences are implemented for spatial discretization, and a second-order 6-stage low-storage Runge-Kutta algorithm is applied for time integration. The relaxation filtering is carried out using an explicit second-order 11-point filter designed to damp only the shortest waves resolved by the grid, discretized by fewer than four points per wavelength. It is applied with a constant strength, every second iteration, which was found to be sufficient in previous jet simulations [8] because of the very small time steps due to the explicit time integration.

2.3 Simulation definition

Round jets with the same inflow parameters, including shear-layer thickness and forcing, but with different diameters yielding $\mathrm{Re}_D = 1800$, $\mathrm{Re}_D = 3600$ and $\mathrm{Re}_D = 11000$, are simulated. Computational domains extending up to 90, 120 and 150 jet radii r_0 in the downstream direction are used, respectively, so that a part of the self-similarity region of the jets is calculated.

In the three jets referred to as jetRe1800, jetRe3600 and jetRe11000 in table 1, only the Reynolds number varies, and the grid properties are the same. A uniform axial grid spacing $\Delta x = r_0/4$ is in particular specified in the downstream direction for $x \geq 25r_0$, see for instance the exhaustive description of jetRe11000 in [2].

In the two additional cases jetRe3600dx2 and jetRe3600dx4 reported in table 1, the Reynolds number and conditions are the same as in jetRe3600, but the axial grid spacing is stretched at a rate of 2% from $x = 45r_0$ to reach, farther downstream, $\Delta x = r_0/2$ for $x \geq 58r_0$ in jetRe3600dx2 and $\Delta x = r_0$ for $x \geq 84r_0$ in jetRe3600dx4.

Table 1. Simulation parameters: jet diameter-based Reynolds numbers Re_D, mesh spacings Δx in the self-similarity flow regions, numbers of grid points (n_x, n_y, n_z) and of time steps n_t, simulation times T normalized by D/u_j.

Reference	Re_D	Δx	$n_x \times n_y \times n_z$	n_t	Tu_j/D
jetRe1800	1800	$r_0/4$	$411 \times 211 \times 211$	1.6×10^6	0.79×10^5
jetRe3600	3600	$r_0/4$	$531 \times 261 \times 261$	1.6×10^6	0.79×10^5
jetRe11000	11000	$r_0/4$	$651 \times 261 \times 261$	2.8×10^6	1.34×10^5
jetRe3600dx2	3600	$r_0/2$	$383 \times 261 \times 261$	1.4×10^6	0.69×10^5
jetRe3600dx4	3600	r_0	$321 \times 261 \times 261$	1.3×10^6	0.64×10^5

3 Results

3.1 Vorticity snapshots

Mean flow and turbulence properties, including the second-order and third-order velocity moments, and the energy budgets, have been evaluated from the LES flow fields. First illustrations are given here by vorticity snapshots.

In figure 1 displaying vorticity obtained for jetRe11000 and jetRe1800, the effects of the Reynolds number on the turbulent transition, and on the turbulent scales can be seen. At higher Reynolds number, the transition occurs more rapidly, and more fine scales are visible.

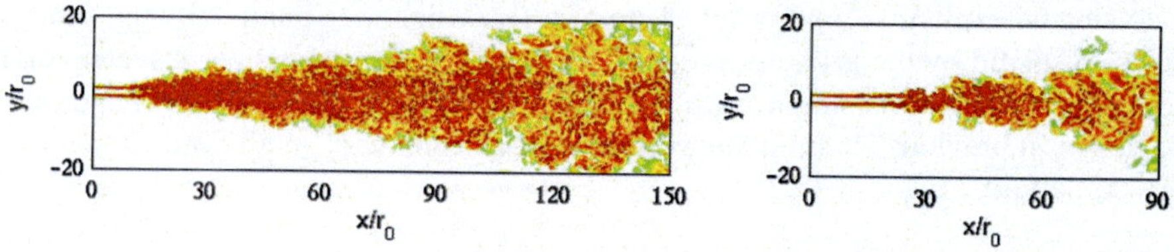

Fig. 1. Snapshots of vorticity norm normalized by the jet diameter and the centerline mean velocity, $|\omega|D/u_c$, in the plane $z = 0$, for jetRe11000 (left) and jetRe1800 (right). The color scale ranges for levels from 0.4 to 2.

In figure 2 showing vorticity for jetRe3600, jetRe3600dx2 and jetRe3600dx4, the impact of the grid resolution on the downstream turbulence appears to be significant. The use of coarser grids progressively removes the smaller turbulent structures for $x \geq 120r_0$. These scales are especially damped in jetRe3600dx4, for which $\Delta x = r_0$ in the downstream region.

3.2 Energy dissipation and filtering activity

In the present LES, dissipation is the sum of the viscous and filtering dissipations, whose contributions vary with the Reynolds number and the grid resolution. To study this point, the variations along the jet axis of the subgrid-activity parameter [18], that

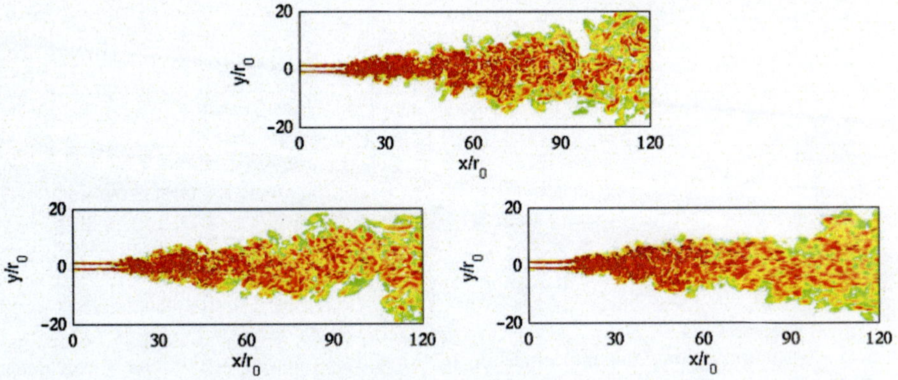

Fig. 2. Snapshots of vorticity norm $|\omega|D/u_c$ in the plane $z = 0$, for jetRe3600 (top), jetRe3600dx2 (bottom left) and jetRe3600dx4 (bottom right). Same color scale as in figure 1.

is the filtering-activity parameter here, are presented in figure 3. They are calculated from the budgets for the turbulent energy as the ratio between the filtering dissipation and the total dissipation.

On the left graph, as expected from previous results [1], the filtering activity is found to increase with the Reynolds number. At $x = 90r_0$ for example, filtering is responsible for about 40% of the energy dissipation in jetRe11000, but only for about 10% in jetRe1800. In the three jets with different Reynolds numbers, the filtering activity also becomes lower as the jets develop in the downstream direction with growing turbulent length scales.

The right graph shows the effects of the grid coarsening on the filtering activity. The mesh stretching, beginning at $x = 45r_0$, leads to a strengthening of the contribution of the relaxation filtering to the energy dissipation. In jetRe3600, jetRe3600dx2 and jetRe3600dx4, one thus finds filtering activities of 0.47, 0.26 and 0.18 at $x = 120r_0$, as the axial mesh spacings are respectively $\Delta x = r_0/4$, $\Delta x = r_0/2$ and $\Delta x = r_0$ in the downstream flow region.

To illustrate the spatial resolution in the present LES, the ratios between typical turbulence length scales and the axial mesh spacing Δx are represented in figure 4. The axial integral length scale is evaluated from the scaling law $L_f = 0.0385x$ observed experimentally in [19] as well as numerically in [20], while the transverse Taylor scale and the Kolmogorov scale are calculated using the relations of isotropic turbulence, $\lambda_g = (15L_f\nu/[u'u']^{1/2})^{1/2}$ and $\eta = (L_f^{1/4}(\nu/[u'u']^{1/2})^{3/4}$, respectively, where the brackets denote time averaging.

The length scales increase as the jet develops. This is shown in the left figure dealing with the three simulations jetRe1800, jetRe3600 and jetRe11000 in which the same uniform axial mesh spacing is used. It is moreover interesting to observe that, far downstream in the self-similarity regions of these three jets, the integral length scales are well resolved ($L_f \geq 10\Delta x$), and the Taylor scales are discretized ($\lambda_g \geq 2\Delta x$).

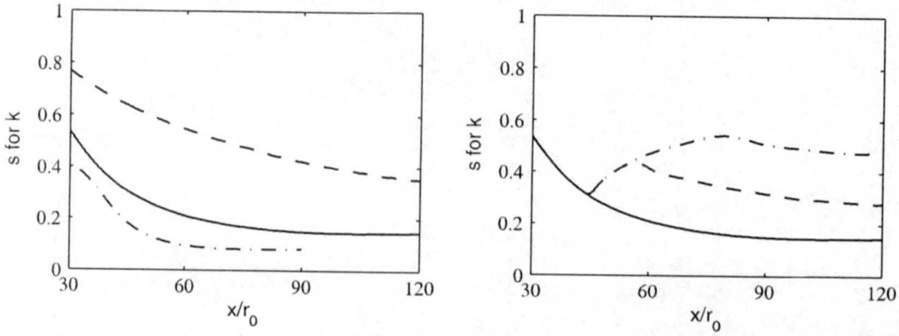

Fig. 3. Variations along the jet centerline of the filtering-activity parameter s calculated from the turbulent kinetic energy. Left: for — — — jetRe11000, ——— jetRe3600, — · — · jetRe1800; and right: for ——— jetRe3600, — — — jetRe3600dx2, — · — · jetRe3600dx4.

This is unfortunately not the case in the right figure with coarser mesh spacings. More precisely, the LES resolution is poor but acceptable in jetRe3600dx2, and insufficient in jetRe3600dx4. At $x = 90r_0$, one can indeed notice that $L_f = 6.9\Delta x$ and $\lambda_g = 3.7\Delta x$ in jetRe3600dx2, and $L_f = 3.5\Delta x$ and $\lambda_g = \Delta x$ in jetRe3600dx4. In the latter simulation, the integral length scale is therefore not accurately resolved, and might be affected by the relaxation filtering.

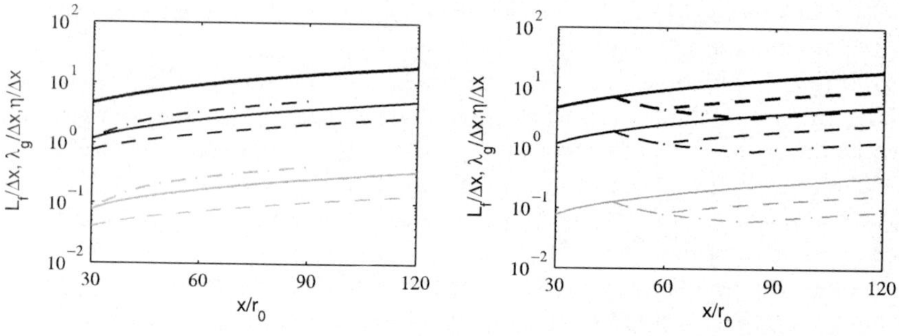

Fig. 4. Variations along the jet centerline of the ratios between the axial integral length scale, the transverse Taylor and Kolmogorov scales, and the axial mesh spacing: thick lines $L_f/\Delta x$, thin lines $\lambda_g/\Delta x$, grey lines $\eta/\Delta x$. Left: for — — — jetRe11000, ——— jetRe3600, — · — · jetRe1800; and right: for ——— jetRe3600, — — — jetRe3600dx2, — · — · jetRe3600dx4.

3.3 Mean and turbulent jet development

To characterize the jet mean flows, the variations of the centerline mean axial velocity u_c are considered. In self-preserving jets, it has indeed been evidenced that

$$\frac{u_c}{u_j} = B \times \frac{D}{(x - x_0)} \tag{5}$$

where B is the decay constant, and x_0 denotes a virtual origin. The variations of the jet half-width are not displayed here because of the limited size of the paper, but they provide similar results.

The profiles obtained in the present jets for u_j/u_c are plotted in figure 5. In all cases, linear increases are observed sufficiently far in the downstream direction, suggesting that the mean flow is self-similar. In addition, the self-similar mean flow appears in the left figure to develop at a lower rate at higher Reynolds number, in agreement with the experimental data obtained for plane jets by Deo *et al.* [13]. The influence of the grid resolution in the right figure finally seems quite weak.

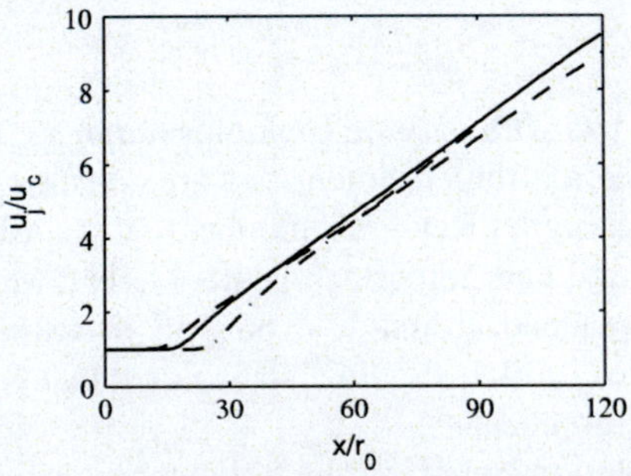

Fig. 5. Variations of the inverse of centerline mean axial velocity u_j/u_c. Left: for — — — jetRe11000, ——— jetRe3600, — · — · jetRe1800; and right: for ——— jetRe3600, — — — jetRe3600dx2, — · — · jetRe3600dx4.

To quantify properties of the mean flow self-similarity, the local decay constant B' is evaluated from the profiles of figure 5 in the following way

$$\frac{1}{B'} = \frac{\mathrm{d}(u_j/u_c)}{\mathrm{d}(x/D)} \tag{6}$$

and they are represented in figure 6. In all jets, after a transitional period, the local decay constant tends to an asymptotic value in the downstream direction, which indicates self-similarity. Constants $B = 5.8$, $B = 6.1$ and $B = 6.4$ are thus determined in the three jets of the left graph, jetRe1800, jetRe3600, and jetRe11000, with varying Reynolds numbers. The similar variations of B' in the right graph for the three jets at same Reynolds number also support that the self-similar jet mean flow depend only weakly on the LES resolution.

The turbulent jet flow features are now examined by presenting in figure 7 and 8 the centerline profiles of the axial and radial turbulent intensities $[u'u']^{1/2}/u_c$ and $[v'v']^{1/2}/u_c$. In this way the axial locations at which the turbulent jets achieve self-similarity can in particular be determined.

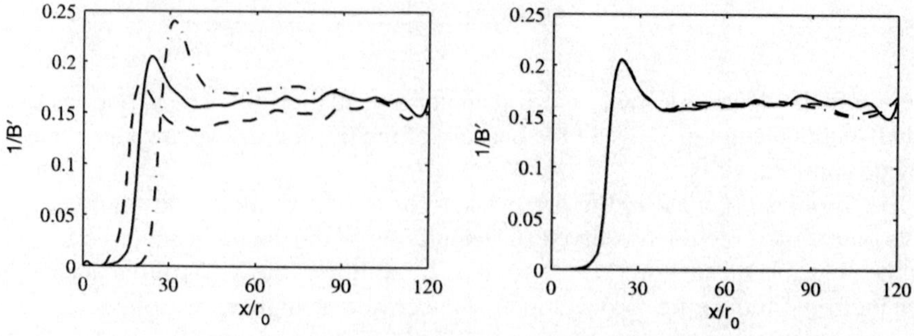

Fig. 6. Axial variations of the inverse of local decay constant $1/B'$. Left: for — — — jetRe11000, ——— jetRe3600, — · — · jetRe1800; and right: for ——— jetRe3600, — — — jetRe3600dx2, — · — · jetRe3600dx4.

In the left figures dealing with the three jets with varying Reynolds numbers, the establishment of self-similarity, obtained when turbulent intensities are constant on the jet axis, is shown to occur more slowly at higher Reynolds number, *i.e.* at farther axial distance, in agreement with experimental and numerical results [1, 11]. More quantitatively, self-similarity seems to be reached around $x = 60r_0$ in jetRe1800, $x = 70r_0$ in jetRe3600 and $x = 120r_0$ in jetRe11000. It should also be noted that very close asymptotic values are observed in the three jets.

The right figures show that the grid resolution can affect the turbulent intensities in the jet self-similarity region. With respect to jetRe3600, the alterations are moderate in jetRe3600dx2, but appreciable in jetRe3600dx4. The self-similarity of the turbulent flow in the latter jet is even not well established because, instead of remaining constant, both axial and radial components of the turbulent intensities decrease on the jet axis far from the inflow. This could be expected because of the insufficient resolution of the turbulent length scales in jetRe3600dx4, already discussed in previous section.

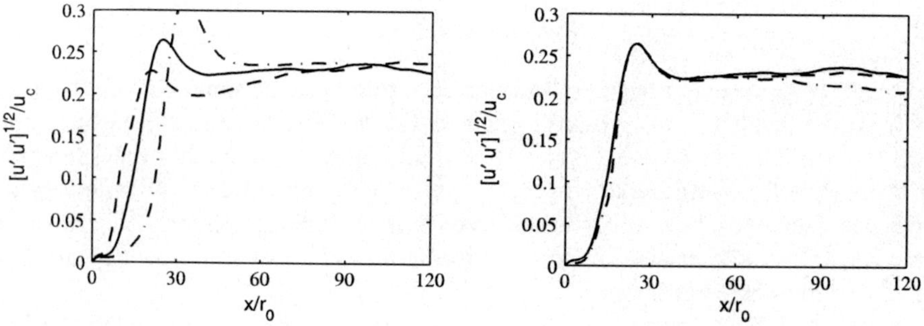

Fig. 7. Variations of centerline axial turbulence intensities $[u'u']^{1/2}/u_c$. Left: for — — — jetRe11000, ——— jetRe3600, — · — · jetRe1800; and right: for ——— jetRe3600, — — — jetRe3600dx2, — · — · jetRe3600dx4.

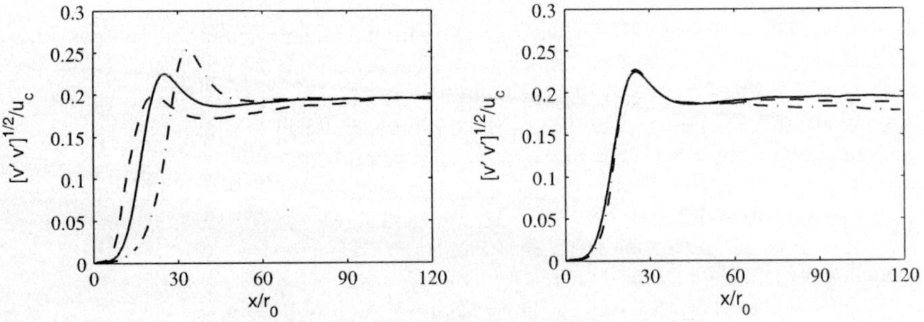

Fig. 8. Variations of centerline radial turbulence intensities $[v'v']^{1/2}/u_c$. Left: for — — — jetRe11000, ——— jetRe3600, — · — · jetRe1800; and right: for ——— jetRe3600, — — — jetRe3600dx2, — · — · jetRe3600dx4.

4 Concluding remarks

The present paper provides preliminary results on the influence of the Reynolds number and grid resolution on round turbulent jets simulated by LES based on relaxation filtering. The subgrid or filtering activity is first shown to adjust by itself to the Reynolds number, to the grid resolution and to the development of the turbulent length scales of the flow in the different computations. The effects of the Reynolds number are noted to be especially significant on the characteristics of the self-preserving mean flows, and on the distance required to achieve self-similarity of the turbulent flows, in agreement with experimental findings. The effects of the grid resolution are however rather weak on the mean flow features, but appreciable on the turbulence features. A decrease of the turbulent intensities is in particular observed for insufficient grid resolution.

Acknowledgments

The authors gratefully acknowledge the *Institut du Développement et des Ressources en Informatique Scientifique* of the CNRS and the *Centre de Calcul Recherche et Technologie* of the CEA for providing CPU time on Nec computers and for technical assistance.

References

1. Bogey C, Bailly C (2006) Phys Fluids 18(6): 065101
2. Bogey C, Bailly C (2009) J Fluid Mech 627: 129–160
3. Panchapakesan NR, Lumley JL (1993) J Fluid Mech 246: 197–223
4. Stolz S, Adams NA, Kleiser L (2001) Phys. Fluids 13(4):997–1015
5. Mathew J, Lechner R, Foysi H, Sesterhenn J, Friedrich R (2003) Phys. Fluids 15(8):2279–2289

6. Rizzetta DP, Visbal MR, Blaisdell GA (2003) Int. J. Numer. Meth. Fluids 42:665–693
7. Schlatter P, Stolz S, Kleiser L (2006) Int. J. Heat Fluid Flow 27:549–558
8. Bogey C, Bailly C (2006) Int J Heat Fluid Flow 27(4):603–610
9. Lemieux GP, Oosthuizen PH (1985) AIAA J 23: 1845–1847
10. Namer I, Ötügen MV (1988) Exp Fluids 6: 387–399
11. Pitts WM (1991) Exp Fluids (11): 135–141
12. Kwon SJ, Seo IW (2005) Exp Fluids 38: 801–812
13. Deo RC, Mi J, Nathan GJ (2008) Phys Fluids (20): 075108
14. Fellouah H, Ball CG, Pollard A (2009) Int J Heat Mass Transfer 52:3943–3954
15. Vreman B, Geurts B, Kuerten H (1995) Appl Sci Res 54:191–203
16. Vreman B, Geurts B, Kuerten H (1997) J Fluid Mech. 339:357–390
17. Bogey C, Bailly C (2004) J Comput Phys 194(1): 194–214
18. Geurts BJ, Fröhlich J (2002) Phys Fluids 14(6):41–44
19. Wygnanski I, Fiedler H (1969) J Fluid Mech 38(3): 577–612
20. Bogey C, Bailly C (2006) Computers & Fluids 35(10): 1344–1358

An Examination of the Spatial Resolution Requirements for LES of a Compressible Jet

James R. DeBonis[1]

NASA Glenn Research Center, Cleveland, Ohio 44135 james.r.debonis@nasa.gov

Summary. This work examines the grid requirements necessary for a properly resolved large-eddy simulation (LES) of a compressible jet. The numerical scheme used for the analysis and its corresponding computational grid are used to estimate, *a priori*, the resolution of the simulation. This estimated resolution, expressed in terms of wave number, is compared to the resolution in the turbulent spectra obtained from the simulation. Two levels of grid resolution are examined. The solution yields good agreement with experimental data for mean flow and first-order turbulent statistics. The estimated resolution of the analysis properly predicts the trends with respect to the computed turbulent spectra.

Key words: spatial discretization, grid resolution, jet flow

1 Introduction

Large-eddy simulation (LES) is becoming a widely used tool in the fluid dynamics community. The technique offers the promise of improved predictions and increased information for turbulent flows over the well established Reynolds Averaged Navier-Stokes (RANS) methods. The key to an accurate LES is the proper resolution of the large-scale turbulent structures which dominate the flow. The resolution of these structures is determined by a combination of the numerical scheme and computational grid used. Determining how much grid is necessary for an accurate analysis is done by trial and error, a computationally expensive undertaking.

Fourier analysis has been used to analyze the properties of numerical schemes [1]. This technique has been useful in both developing and identifying high fidelity numerical schemes for LES. The present work extends this idea by mapping the resolution of the numerical scheme onto the computational grid to give an *a priori* estimate of the simulation's resolution at every point in the domain. This estimated resolution is then compared to the actual resolution of the turbulent spectra extracted from the simulation. If the required resolution of the turbulent spectra necessary for an accurate LES simulation can be established for a given type of flow, this technique offers a simple tool to construct quality grids.

M.V. Salvetti et al. (eds.), *Quality and Reliability of Large-Eddy Simulations II*, ERCOFTAC Series 16, DOI 10.1007/978-94-007-0231-8_30,

The flow considered here is a Mach 0.9 jet. The prediction of turbulent jet flows is of great interest to the aerospace community. Researchers throughout the world are attempting to reduce the noise emitted by jet engines. Understanding and modeling the noise producing mechanism of the jet is critical to this effort. Reynolds Averaged Navier-Stokes (RANS) methods fail to accurately predict the turbulent structure of the jet [2]. LES has the potential to improve the accuracy of the predictions and can provide additional turbulent information. In this work, a high-resolution numerical scheme is used to model the jet flow. The resolution of the scheme is determined and mapped onto the computational grids. Solutions for two computational grids are obtained and compared to experiment. Finally the estimated resolution is compared to the resolution computed from the turbulent spectra.

2 Numerical Method

The code used in this study, WRLES (Wave Resolving Large-Eddy Simulation), is a special purpose large-eddy simulation code that uses high-resolution temporal and spatial discretization schemes to accurately simulate the convection of turbulent structures. The code solves the compressible Favre-filtered Navier-Stokes equations. It was developed at the NASA Glenn Research Center for the study of turbulent jets, but can be applied to other flows.

The code uses a family of explicit Runge-Kutta time stepping schemes written in a general M-stage 2-N storage formulation[3]. One-stage is required for each order of accuracy desired. Additional stages can be used to increase accuracy or alternatively, to reduce dispersion error. The 4-stage, 3rd-order scheme of Carpenter and Kennedy is used in this study[4].

Central differencing is used for the spatial discretization because of its non-dissipative properties. This helps ensure the accurate convection of turbulence. Both standard nth-order central difference stencils, designed to minimize truncation error, and Dispersion Relation Preserving (DRP) stencils are available in the code. The code can accommodate stencils up to 13 points. DRP schemes, originally developed by Tam[5], are designed to minimize the dispersion error in the simulation. They do this by sacrificing the scheme's formal order of accuracy in favor of lower dispersion at smaller wave numbers. The 13 point DRP scheme of Bogey and Bailly[6] is used here.

This lack of dissipation in central differencing makes the schemes unstable. In order to ensure a stable solution without adversely affecting the resolving properties of the scheme, solution filtering is used. This is a low-pass filter that leaves the low wavenumber structures, well resolved by the simulation, untouched. It removes the high wavenumber structures that are not properly resolved and can cause instability. The filter must be properly matched to the differencing scheme, so that the filter removes only those waves that are not properly computed. Bogey and Bailly developed filters to match their DRP stencils. For the standard central difference schemes the filters of Kennedy and Carpenter [7] are implemented in the code.

3 Estimating Spatial Resolution

Fourier analysis has been used to evaluate the ability of numerical schemes to resolve wave motion [1]. The technique analyzes the numerical scheme on the one-dimensional convection equation and can provide the scheme's behavior in terms of dissipative and dispersive errors. The effect of the temporal discretization is not considered here.

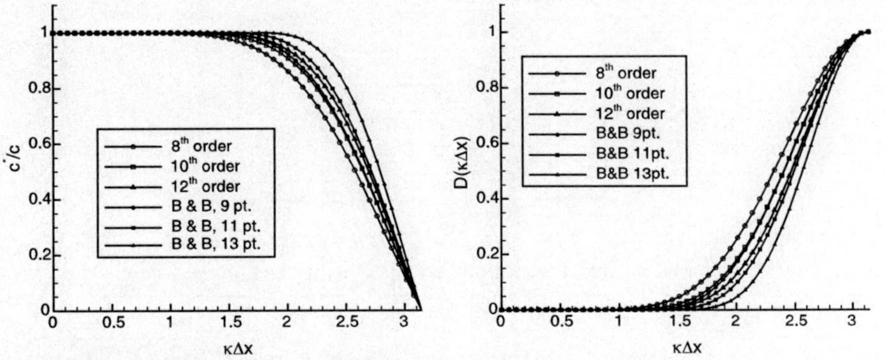

(a) Error in phase speed for the finite difference stencils

(b) Damping functions of the explict filters

Fig. 1. Properties of the numerical scheme

Fourier analysis shows that central difference schemes have no inherent dissipation, and thus are ideally suited for LES calculations. They do however produce dispersive errors. Figure 1(a) shows the phase error, written as the ratio of numerical phase speed to actual phase speed, c^*/c, versus wavenumber per grid spacing, $(\kappa\Delta x)$, for both standard 8th, 10th and 12th order schemes and the equivalent stencil width DRP schemes from Bogey and Bailly. They chose $5 \cdot 10^{-4}$ as the maximum phase error level corresponding to a "properly resolved" wave. Using this error level, the maximum resolvable wave number per grid spacing, $(\kappa\Delta x)_{max}$, is determined for each stencil.

The damping functions, $D(\kappa\Delta x)$, are shown in figure 1(b) for the standard filters of Kennedy and Carpenter [7] and the DRP filters derived by Bogey and Bailly to match their difference stencils. The damping is zero until it reaches a "cutoff" wave number, where it rapidly increases, effectively removing all the structures of higher waves numbers. Since filters typically also utilize a coefficient, σ, to regulate the amount of damping, the total amount of dissipation is $\sigma D(\kappa\Delta x)$. The cutoff wave number, $(\kappa\Delta x)_{cut}$ is the wave number where the total amount of dissipation exceeds $5 \cdot 10^{-4}$, the same error level used for the differencing schemes. A typical damping coefficient of 0.2 is used here.

The maximum resolvable wave number for each scheme is plotted against the stencil size as solid lines in figure 2. The cutoff wave number of the corresponding filter is plotted as a dashed line. For a stable solution, the cutoff wave number of

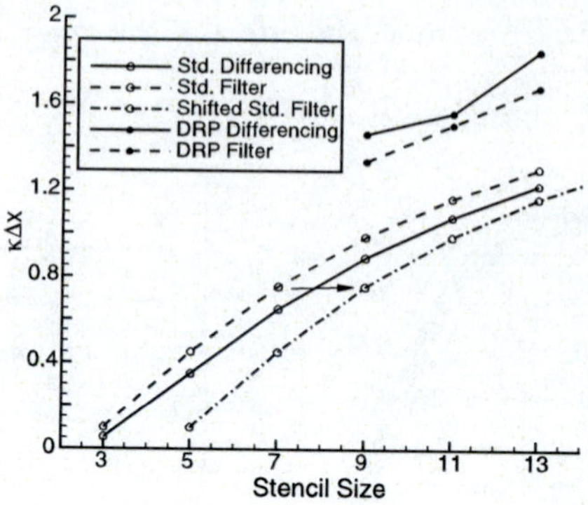

Fig. 2. Cutoff wave numbers for schemes and filters

the filter should be less than the maximum resolvable wave number of the numerical scheme. This ensures that all of the structures within the domain are properly resolved. For the standard schemes, the filter's cutoff is too high, leaving improperly resolved structures in the flowfield. For these schemes, shifting the order of the filter down to the next level should insure a stable solution. Figure 2 clearly illustrates superior resolution offered by the DRP schemes.

For the numerical schemes used here, the filter limits the resolution of the simulation and the cutoff wave number, $(\kappa\Delta x)_{cut}$, should be used in the estimation procedure. For schemes without an explicit filter, $(\kappa\Delta x)_{max}$ would be used. For each grid cell, the spacing in each direction (Δx, Δy, and Δz) is computed and the limiting, maximum, spacing is chosen. The cutoff wave number per grid spacing can then be mapped on the grid to provide the maximum resolvable wave number at every point in the computational domain. This technique provides an *a priori* estimate of resolution for a given analysis. As a better understanding of the resolution required for a successful LES is gained, this technique may prove to be a useful tool in the construction of computational grids and the selection of numerical schemes.

4 Modeling the Mach 0.9 Jet

The geometry and corresponding experimental data used for this study was from the 2 inch diameter Acoustic Reference Nozzle (ARN) tested at the NASA Glenn Research Center by Bridges and Wernet [8]. The jet Reynolds number, $Re_j = \rho_j U_j D_j/\mu_j$, from the experiment was approximately 2 million. In order to reduce the range of turbulent scales in the simulation, the Reynolds number of the simulation was reduced to 50,000.

The computational grid (figure 3) models the internal and external nozzle geometry from plenum to exit. The majority of grid points are placed in the area of interest, downstream of the nozzle exit in the plume region. The grid extends 40 jet diameters, D_j, downstream of the nozzle exit and 30 diameters radially from the centerline. Two different grid densities were examined (table 1). The baseline grid contained 3.5 million grid points. The smallest grid spacing was in the radial direction due to the the clustering near solid walls. The ability to resolve a three-dimensional turbulent eddy is limited by the largest grid spacing in each of the three directions. Therefore the refined grid maintained the radial grid spacing and added additional points in both the axial and azimuthal directions to improve overall resolution.

Grid	Interior of Nozzle	Exterior of Nozzle	Nozzle Plume	Total
baseline	45 x 55 x 102	45 x 65 x 102	196 x 148 x 102	3,509,616
refined	45 x 55 x 132	45 x 65 x 132	294 x 148 x 132	6,456,384

Table 1. Dimensions of the computational grid

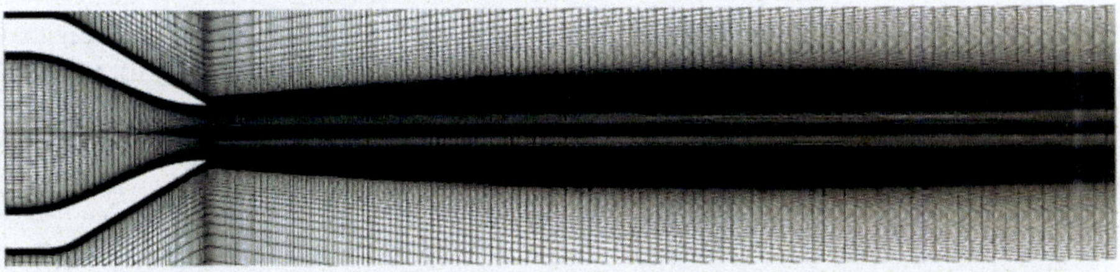

Fig. 3. Baseline computational grid

The resolution of the simulation was estimated using the technique outlined in section 3. Contours of the estimated resolvable wave number in a streamwise plane through the center of the jet is shown in figure 4. The wavenumber is nondimensionalized by the jet diameter, a reasonable estimate of the maximum turbulent length scale. In the jet plume, the resolution varies with axial distance from the jet exit, and is constant in the radial and azimuthal directions, indicating that the axial grid spacing is the limiting factor.

Total pressure and total temperature, consistent with a Mach 0.9 cold jet, were specified at the inflow of the nozzle plenum. Freestream conditions were specified using a farfield characteristic boundary condition. The experiment had no freestream velocity, but a very low Mach number, 0.05, was specified at the computation's farfield boundary to maintain stability. Static pressure was specified at the outflow and at $x/D_j > 30$, the grid spacing is gradually increased and a sixth-order filter is applied to remove spurious reflections. The centerline of the computation is treated

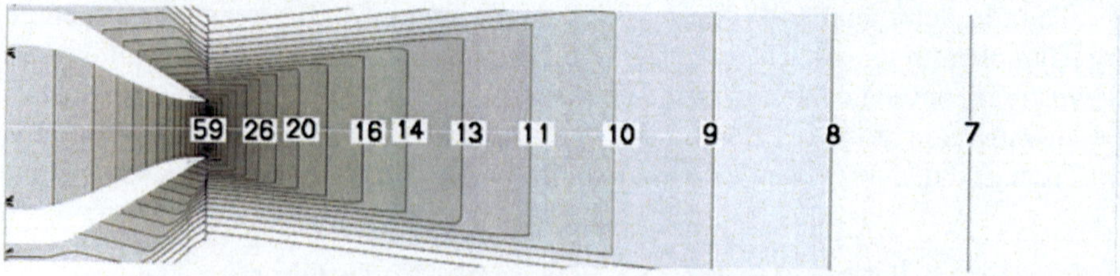

Fig. 4. Contours of estimated resolvable wave number, $(\kappa D_j)_{est}$

using Hixon's approach [9]. This method avoids stability problems by creating a difference stencil that spans the singularity at the jet centerline.

5 Results

Solutions for both grids were computed using hybrid MPI/OpenMP parallel processing on a Linux cluster. No sub-grid scale model was used and the solution relies on the dissipation of the filter to represent the sub-grid scales.

A time step of approximately $2.45 \cdot 10^{-8}$ seconds was used. An instantaneous flowfield was saved every $2.5 \cdot 10^{-5}$ seconds, yielding frequency resolution up to 20kHz. A total of 2,048 instantaneous flowfields were saved, representing 0.0512 seconds of physical time. The saved solution files were post processed to obtain, mean flow, turbulent statistics and turbulent spectra.

5.1 Prediction of the Flowfield

Instantaneous Mach number contours for the refined grid are shown in figure 5. The flow is dominated by turbulent mixing. The turbulent structures range in size from the order of the nozzle lip thickness to larger than the nozzle diameter. In general the size of the structures increases with distance from the nozzle exit. The thin void along the centerline of the domain is an artifact of the centerline boundary condition.

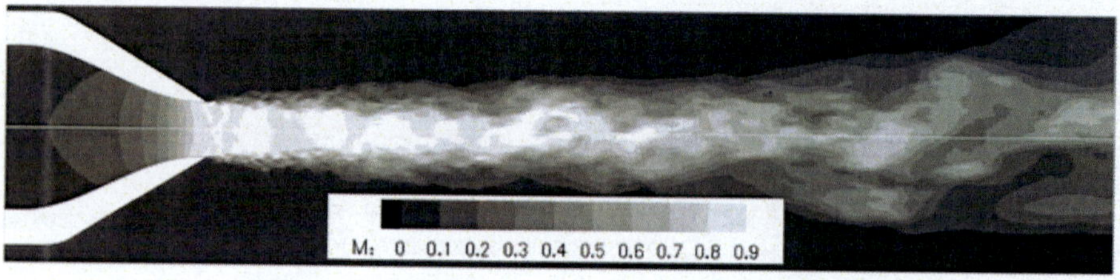

Fig. 5. Contours of Instantaneous Mach number, M

The prediction of the mean velocity along the jet centerline for both grids is compared to experiment in figure 6. Agreement with experiment is excellent for both grid levels. For the baseline grid, the end of the potential core (the point where the velocity begins to decay) is predicted upstream of the experimental position. The refined grid prediction of the end of the potential core compares well with experiment. Downstream of the potential core, both grids do a good job of predicting the centerline velocity decay up to $x/D_j = 15$. Beyond this point irregularities in the experimental data make comparison difficult.

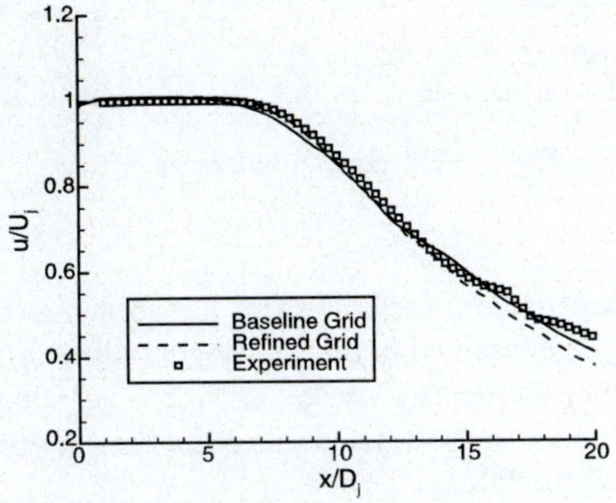

Fig. 6. Centerline velocity profiles

Axial, u'/U_j, and radial, v'/U_j, turbulent intensities on the jet centerline, for both grids, are shown in figure 7. On the baseline grid, the axial intensity levels are over predicted and the radial intensity levels are under predicted. Grid refinement improves both predictions, decreasing the axial intensity and increasing the radial intensity. Note that the prediction of radial intensity was improved by refining the axial and azimuthal grid spacing. This illustrates the importance of resolving the turbulent structures in all three computational directions and suggests that isotropic grid cells are ideal for LES.

5.2 Resolution of the Turbulent Structures

Turbulent spectra were computed at three locations in the jet mixing layer. The points were located on the jet lip line, $r/D_j = 0.5$ at three axial locations, $x/D_j = 4$, 7, and 10. At each of these points, the three velocity components for each saved solution were extracted, yielding a time history of the velocity. The data was converted to instantaneous turbulent kinetic energy and was processed using a Fast Fourier Transform. This yielded a turbulent spectrum in the frequency domain. The flow was assumed to be statistically stationary and Taylor's hypothesis [10] was used to convert the frequency domain to the spatial domain.

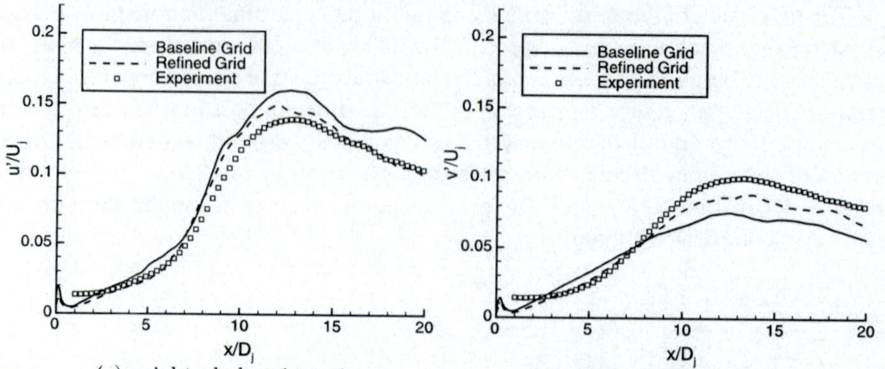

(a) axial turbulent intensity
(b) radial turbulent intensity

Fig. 7. Centerline turbulent intensities

Plots of the turbulent spectra for both grids are shown in figure 8. The majority of the turbulent energy is contained in the small wave number/large-scale structures. At larger wave numbers the energy decays with a $-5/3$ slope for a short period, consistent with turbulence theory. Beyond this point the energy decays more rapidly, most likely the effect of the filter.

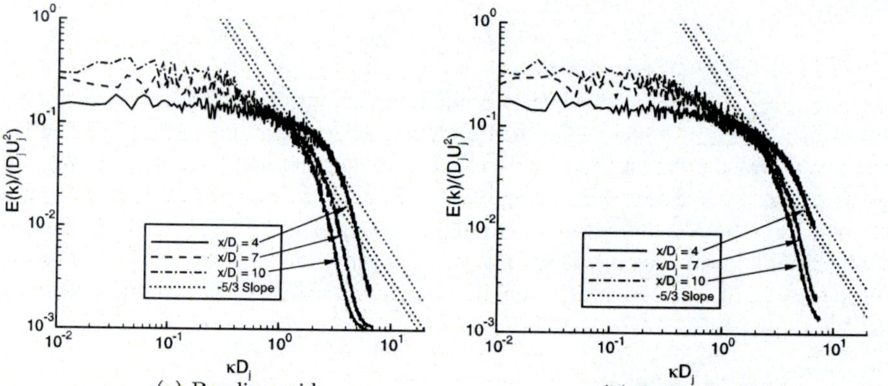

(a) Baseline grid
(b) Refined grid

Fig. 8. Turbulent spectra

The spectra was used to determine the maximum resolved wave number in the flow at each position. This wave number was chosen at the point where the spectra deviated from the theoretical $-5/3$ decay. Table 2 contains the estimated resolvable wave number based on the grid and numerical scheme and the actual resolved wave number computed from the spectra. The computed results do not match the estimates. However, for the six cases considered the ratio between the estimated and computed numbers is very similar, ranging from 4.23 to 4.70. This fact lends credence to the estimating technique. There are several assumptions inherent in the estimates that would alter the overall level of the estimated wave number. Foremost is the *ad hoc* selection of the "properly resolved" error level. It is also important to note that the resolution scales properly with grid refinement. At a given axial position, the ratio between refined and baseline grid estimated wave number is nearly identical to that same ratio computed from the spectra.

	Baseline Grid			Refined Grid		
x/D_j	4	7	10	4	7	10
Estimate	14.1	10.6	8.70	20.0	15.4	13.1
From Spectra	3.33	2.31	1.85	4.58	3.33	2.81
Ratio	4.23	4.59	4.70	4.37	4.62	4.66

Table 2. Resolved wave number, κD_j

6 Summary and Conclusions

A high resolution numerical scheme was used to perform a large-eddy simulation of a Mach 0.9 jet. A four-stage third-order low-dispersion Runge-Kutta scheme was used for the temporal discretization and the 13-point Dispersion Relation Preserving scheme of Bogey and Bailley [6] was used for the spatial discretization. A filter designed to match the spatial discretization scheme was used to maintain stability and to represent the sub-grid scale dissipation. Fourier analysis of the spatial discretization scheme was used to characterize the ability of the scheme to resolve waves on a computational grid. This information was applied to the computational grid used in the study to obtain an *a priori* estimate of the resolution of the turbulent structures.

Two different grids were examined, a baseline grid (3.5 million points) and a refined grid (6.5 million points), which increased the resolution in the axial and azimuthal directions. The results of the computation were compared to the experimental data of Bridges and Wernet [8]. Excellent agreement was obtained for the mean flow properties at both grid resolutions. Good agreement was achieved for the turbulent intensities for the baseline grid and the agreement improved significantly with grid refinement. The grid refinement suggests that grid cells that tend toward isotropic are ideal for predicting turbulent quantities.

Turbulent energy spectra were obtained at three points in the flowfield for both grids. The spectra appear reasonable and reproduce the theoretical $-5/3$ slope turbulent decay for a small range of wave numbers. The spectra was used to determine the actual resolution of the computation. This data was compared to the estimated resolution. The absolute values of the resolved wave numbers did not agree, but the estimates appear to properly predict the trends. This technique of estimating the resolution appears to be a reasonable method to help construct computational grids and estimate resources needed for a large-eddy simulation.

References

1. Vichnevetsky, R. and Bowles, J. B., *Fourier Analysis of Numerical Approximations of Hyperbolic Equations*, Society for Applied and Industrial Mathematics, 1982.
2. Georgiadis, N. J. and DeBonis, J. R., "Navier-Stokes Analysis Methods for Turbulent Jet Flows with Application to Aircraft Exhaust Nozzles," *Progress in the Aerospace Sciences*, Vol. 42, 2006, pp. 377–418.
3. Williamson, J. H., "Low-Storage Runge-Kutta Schemes," *Journal of Computational Physics*, Vol. 35, 1980, pp. 48.
4. Carpenter, M. H. and Kennedy, C. A., "Fourth-Order 2N-Storage Runge-Kutta Schemes," NASA TM 109112, 1994.
5. Tam, C. K. W. and Webb, J. C., "Dispersion Relation-Preserving Finite Difference Schemes for Computational Aeroacoustics," *Journal of Computational Physics*, Vol. 107, 1993, pp. 262–281.
6. Bogey, C. and Bailly, C., "A Family of Low Dispersive and Low Dissipative Explicit Schemes for Flow and Noise Computations," *Journal of Computational Physics*, Vol. 194, 2004, pp. 194–214.
7. Kennedy, C. A. and Carpenter, M. H., "Comparison of Several Numerical Methods for Simulation of Compressible Shear Layers," NASA TP 3484, 1997.
8. Bridges, J. E. and Wernet, M. P., "Measurements of the Aeroacoustic Sound Source in Hot Jets," AIAA Paper 2003-3130, 2003.
9. Hixon, R., Shih, S.-H., and Mankbadi, R. R., "Numerical treatment of cylindrical coordinate centerline singularities," AIAA Paper 1999-2391, 1999.
10. Pope, S. B., *Turbulent Flows*, Cambridge University Press, 2000, pp. 223–224.

A Computational Uncertainty Analysis of LES/DNS: towards building a reliable engineering turbulence prediction capability

Hongyi Xu[1]

Rotary-Wing Aerodynamics, Aerodynamics Laboratory, Institute for Aerospace Research, National Research Council Canada, Hongyi.Xu@nrc-cnrc.gc.ca

Summary. A computational uncertainty analysis is conducted for turbulence simulation. The major sources of error, the model error and the numerical errors, are analyzed strictly based on the physical nature of turbulence. Through these analyses, a clear technical roadmap is laid out towards building a reliable engineering turbulence simulation capability.

Key words: Computational Uncertainty Analysis, Direct Numerical Simulation, Large-eddy Simulation, Multi-grid Technique

1 Introduction

To develop a reliable engineering turbulence prediction capability, the identification and quantification of uncertainty in a turbulence simulation are critically important. The current analysis is aimed at a precise evaluation of the major sources of error and the development of relevant technologies to minimize or to eliminate these errors. The major sources of error in a turbulence simulation are classified as the physical, or model, errors and the numerical errors. The physical error is inherently associated with the physical model governing the turbulence phenomenon, which are broadly classified as Reynolds Averaged Navier-Stokes (RANS), Large Eddy Simulation (LES) and Direct Numerical Simulation (DNS). These physical models, based on which a flow solver is developed, govern the accuracy and quality of the simulation. For numerical errors, three sources can be identified: (1) the numerical discretization; (2) the solution procedure and (3) the initial and boundary conditions.

In the current paper, the errors from RANS, LES and DNS models are analyzed based on their physical nature and are qualitatively evaluated by DNS results in [1]. An analysis of the turbulence isotropy properties led to reevaluating the validity of the turbulent eddy-viscosity, a well-known fundamental assumption in both RANS closure models and LES sub-grid scale (SGS) models. The importance of eliminating discretization errors is demonstrated through the grid-dependence study conducted in [1]. Verification and validation of the DNS results led to the discovery of the

M.V. Salvetti et al. (eds.), *Quality and Reliability of Large-Eddy Simulations II*, ERCOFTAC Series 16, DOI 10.1007/978-94-007-0231-8_31,

'law-of-the-corners' formulation for the streamwise velocity, as opposed to the 'law-of-the-wall' [2].

As indicated in Xu et al. [3], when solving the incompressible unsteady N-S equations, solution convergence for the momentum equations is relatively easy and the majority of the CPU time is spent solving the pressure Poisson equation. A robust convergence acceleration technique, i.e. the flexible-cycle additive-correction multigrid (FCAC-MG) method, was developed to drive the residuals of pressure Poisson equation down to a satisfactory level. This solution technology was originally developed for simple geometry flows and the current research extends this technique to complex geometry flows, including flow past a rectangular block and flow past an airfoil.

Also, the current research addresses the importance of obtaining realistic turbulent inflow conditions. Towards this end, the temporal LES/DNS simulation capability, developed based on the FCAC-MG method, provides a powerful tool to generate realistic turbulent inflow conditions. The spatial simulation of an airfoil in a fully-developed turbulent channel demonstrated the importance and uniqueness of applying the temporal approach to obtain realistic turbulent inflow conditions.

In summary, the aforementioned technology development enabled the establishment of a reliable turbulence simulation capability, including the generation of realistic turbulent inflow for a relevant flow geometry and accurate spatial simulation capturing the turbulence in a complex flow configuration.

2 Model Error Analysis for RANS, LES

The RANS model is based on the averaging procedure proposed by [4], leading to the RANS equations to solve the mean flow field. The effects of turbulence are lumped into the Reynolds stresses, giving rise to the well-known turbulence closure problem. Kolmogorov [5] proposed a generalized form of Boussinesq's hypothesis, $\rho\overline{u_i'u_j'} = A\delta_{ij} + \mu_T(\partial\overline{u}_i/\partial x_j + \partial\overline{u}_j/\partial x_i)$, relating the Reynolds stress to the strain of mean velocity based on an analogy of the motions between fluid particles and turbulent eddies. The turbulent eddy viscosity, μ_T, then became the foundation of the RANS model. The accuracy of RANS is primarily dependent on μ_T in a turbulence closure model, while the grid resolution plays only a limited role. Based on DNS data [1], the validity of μ_T can be directly interrogated so that the errors from a RANS or LES model can be more clearly understood. The major postulation in the constitutive equation for a Newtonian fluid is the isotropy assumption [6] resulting in the reduction of the stress-strain coefficient tensor E_{ijkm} in $\tau_{ij} = E_{ijkm}s_{km}$ to two scalar invariants (μ, λ) known as the first and second coefficients of viscosity. The isotropy assumption of turbulent eddy motions implicitly contained in the Boussinesq hypothesis becomes a disputable issue. DNS data [1] allow a direct check of the validity of this assumption. Figure 1(1) gives the flow configuration of a square annular duct that contains three representative regions: the flat-wall along the wall-bisector and the concave and convex corners along corner-bisector. Figure 1(2) shows two groups of probing points where the turbulent energy spectra (TES) were collected. Figures 2 and 3 provide the TES in three spatial directions at these points.

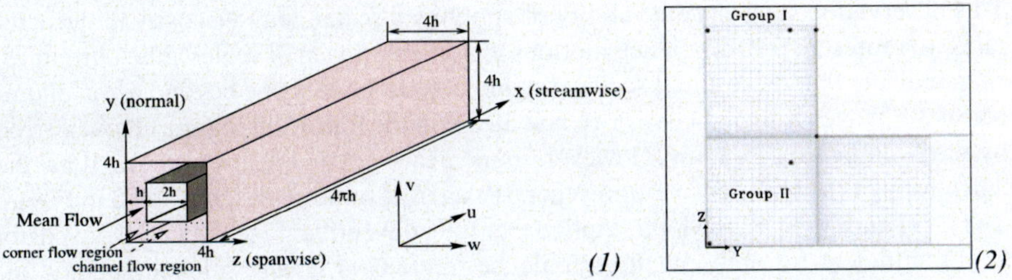

Fig. 1. (1) Flow configuration of square annular duct; (2) TES probing points

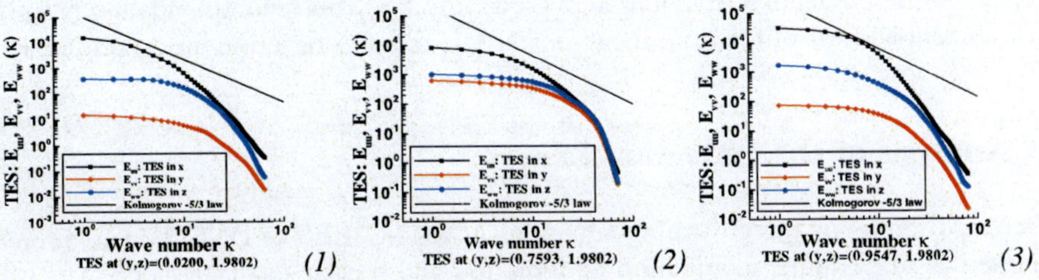

Fig. 2. TES along the wall-bisector (Group I)

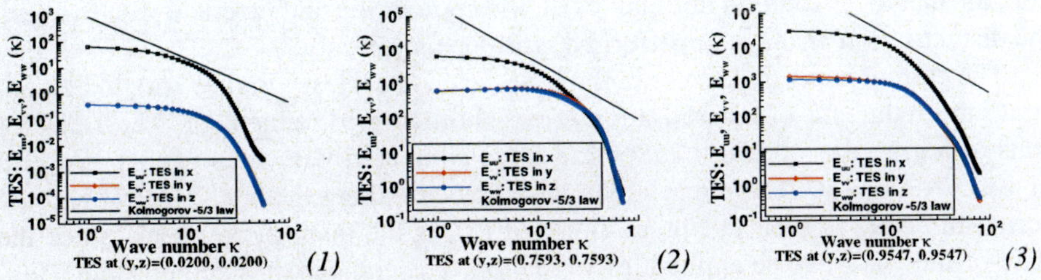

Fig. 3. TES along the corner-bisector (Group II)

Strong anisotropies are the dominant features in the entire flow domain for the large-scale motions, with wavenumbers ranging from 1 to 10. Therefore, the turbulent eddy viscosity concept is not valid for these motion scales, which is the major cause of failure for RANS models. As demonstrated in Figure 2-(2) and 3-(2), the homogeneous turbulence away from the wall and corners is characterized by strong anisotropy and flow-configuration dependence in large-scale motions. As the energy cascades down to smaller scales, the TES unanimously approach an isotropic state. This observation suggests the validity of applying a scalar eddy viscosity in these regions to represent the turbulence motions at sufficiently small scales. The current results indicate that these small scales roughly consist of motions beyond a wavenumber of 30. Wavenumbers between 10 and 30 are in the transition regime characterized by the Kolmogorov -5/3 law. However, in the near-wall and near-corner regions, the anisotropy of the turbulence motions persists even at small scales, as shown in Figure 2-(1), (3) and 3-(1), (3), which implies that the conventional (scalar) eddy viscosity is not sufficient to represent the turbulence motions at small scales in specific regions, such as the near-wall and near-corners. A more general form of Boussinesq's hypothesis, $\rho\overline{u_i'u_j'} = A\delta_{ij} + \mu_{Tijkm}(\partial\bar{u}_k/\partial x_m + \partial\bar{u}_m/\partial x_k)$, could be a solution, where the repeated sub-indices, k and m, imply the Einstein summation.

Based on the above analysis, a more general form of turbulence eddy viscosity, i.e. the tensor form of eddy viscosity, is recommended for future research of turbulence closure model in RANS and SGS model in LES. This recommendation permits a sufficient address of the prominent anisotropic features of turbulence motions.

3 Analysis of Discretisation Errors

Before discretising the governing equations of RANS, LES or DNS models, proper numerical algorithms, including both temporal and spatial schemes, need to be selected and these schemes must be capable of correctly reflecting the physical nature of the flow phenomenon, typically represented by an elliptic, parabolic or hyperbolic equation. If the numerical algorithms are properly selected and correctly implemented, the discretisation errors can be minimized using high-order schemes, and can be driven asymptotically to zero as the grid becomes more refined. Since, in general, an analytical solution does not exist, the appropriate and practical way to check the discretisation errors is to perform a grid-dependence study.

The current paper makes use of the DNS results in [1], see Figure 1(a), to demonstrate the reduction of discretisation errors through grid refinement. The DNS results at a grid size of $256 \times 258 \times 258$ were compared with LES results at a grid of $130 \times 130 \times 130$. The mean streamwise velocity along the wall-bisector and the corner-bisector, as seen in Figure 4(1) and (2), is the quantity to check, since the 'law-of-the-wall' can be applied for validation. The comparisons demonstrate a tangible amount of grid dependence between the solutions of the LES and DNS grids. Also, the velocity profiles exhibit the salient feature of boundary-layer type flows along both the wall and corner bisectors, i.e. high velocity gradients in the near-wall and near-corner (both concave and convex) regions and mild gradients away from these regions. To validate that the current grid refinement made the solution asymptotically approach the right physics, the 'law-of-the-wall' from von Karman [2] is applied to the velocity profile along the wall bisector near both the inner and outer walls, shown in Figure 4(3). The comparison shows the grid-diffusion effects of the coarse (LES) grid, which cause an over-prediction of the streamwise velocity compared to the logarithmic distribution set by the 'law-of-the-wall'. The solution with the DNS grid, through effective reducing of the discretisation errors, exhibits a perfect match with the 'law-of-the-wall' relation in both the viscous sub-layer and the turbulent outer-layer of the flat walls.

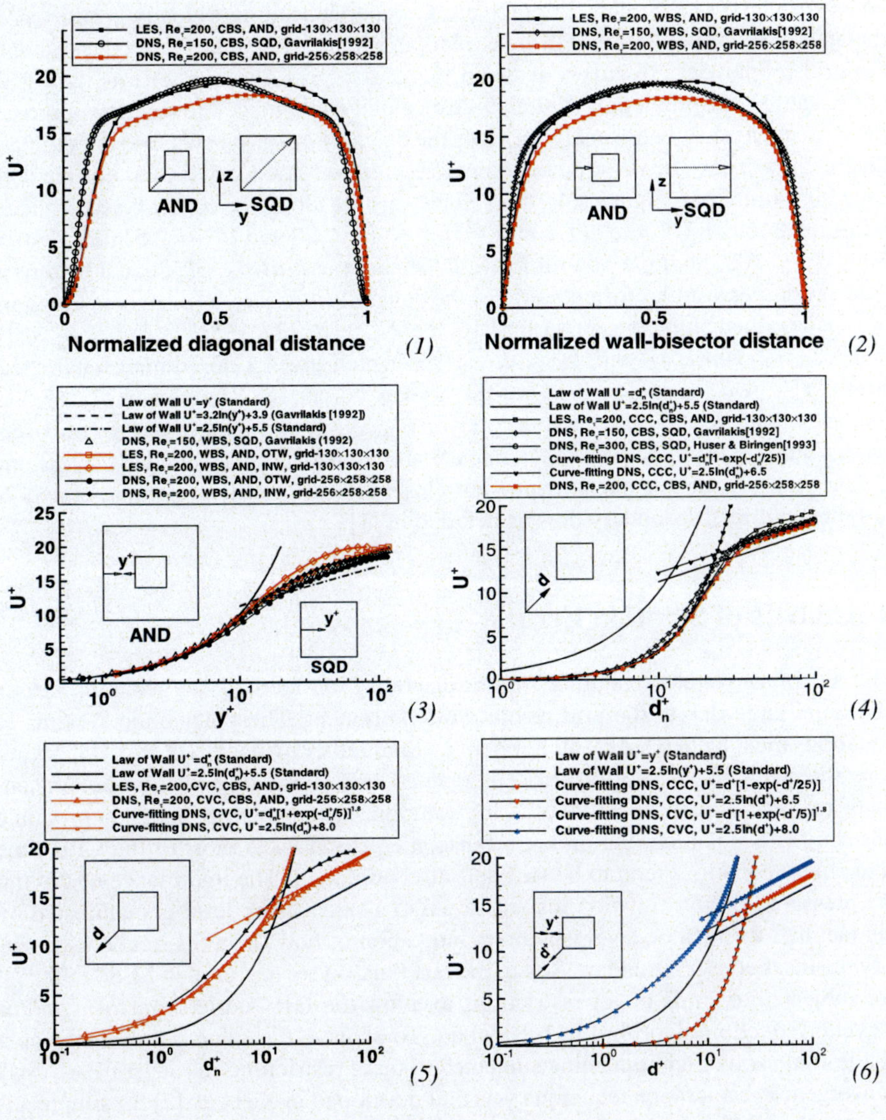

Fig. 4. Streamwise velocity profiles along the (1) corner bisector,(2) wall bisector, (3) validation against 'law-of-the-wall' along wall bisector, (4) 'law-of-the concave corner', (5) 'law-of-the convex corner' and (6) Comparison of 'law-of-the-wall' and 'law-of-the-corner'

For reference, these abbreviations are used in the legends of the following figures: SQD: square duct; AND: annular duct; WBS: wall bisector; CBS: corner bisector; CCC: concave corner; CVC: convex corner; INW: inner wall; OTW: outer wall.

By plotting the velocity profiles on the logarithmic scale along the corner bisector near the concave and convex 90° corners, 'law-of-the-corner' type of formulations can be derived, which reveal the prominent damping and enhancing effects of the concave and convex corners, respectively. Comparisons of the velocity profiles, near the wall in Figure 4(3) and near the corners in Figure 4(4) and (5), demonstrate that the grid refinement effectively reduced the coarse grid diffusion effects, that generally cause over-prediction of streamwise velocity, and made the solution approach the correct physics. These analyses led to the discovery of two new types of boundary layers, the concave 90° corner and convex 90° corner boundary layers. For the concave 90° corner, see Figure 4(4), the velocity profile along the corner bisector yields a formulation, i.e. $U^+ = d_n^+(1-e^{-d_n^+/25})$, $0 \leq d_n^+ \leq 20$ and $U^+ = 2.5\ln(d_n^+)+6.5$, $30 \leq d_n^+ \leq 100$, giving a non-dimensional shear stress, $\partial U^+/\partial d_n^+$, equal to zero at concave corner. Similar formulation for the convex 90° corner, as displayed in Figure 4(5), is obtained along the corner bisector, i.e. $U^+ = d_n^+(1+e^{-d_n^+/5})^{1.8}, 0 \leq d_n^+ \leq 10$ and $U^+ = 2.5\ln(d_n^+)+8.0$, $20 \leq d_n^+ \leq 100$ which gives a non-dimensional shear stress, $\partial U^+/\partial d_n^+$, equal to 3.257 at the convex-corner tip. The plot in Figure 4(6) indicates that the flat-plate boundary layer is actually the neutral curve (zero secondary flow case) in between the concave and convex corner boundary layers and the mechanisms of this phenomenon are closely related to the prominent effects of turbulence-driven secondary flow as detailed in [1].

4 Analysis of Solution Errors

The algebraic systems resulting from a discretisation method are generally sparse and quite large due to the grid resolution requirement. These algebraic systems of equations must be tackled by an iterative method rather than a direct method. Therefore, the unavoidable solution errors associated with an iterative method become another major source of uncertainty in the solution. When solving the incompressible unsteady N-S equation, the pressure Poisson equation takes most of the CPU time since these equations tend to be stiff and ill-conditioned. The importance of driving the pressure Poisson equation residue down to a satisfactory level is manifested by the fact that the residue level is a direct indication of how the numerical results satisfy the mass conservation law. So far, the fast Poisson solvers used in LES/DNS, [7], are subjected to some major restrictions to allow the fast Fourier transform and/or a cyclic reduction algorithm to be applied. To achieve the same performance as a fast Poisson solver without being subjected to these restrictions, a robust FCAC-MG convergence acceleration technique was first developed in Xu et al. [3] for simple geometric flows. The major advantage of a flexible-cycle over the conventional V or W cycles is that the computations on a given grid level always have the opportunity to move up or down one grid level, depending on whether the residual reduction on the current grid level is satisfied. This flexibility makes the flexible-cycle scheme more efficient in the corrections between the fine and coarse grids. The additive-correction scheme preserves the integral conservation property on all multi-grid levels. This conservation property is particularly important in truthfully reflecting the turbulence physics based on conservation laws. Therefore, as a combination of these two solution strategies, the FCAC-MG acceleration technique is highly efficient, reliable and robust, making it feasible for CPU-intensive computations in LES/DNS.

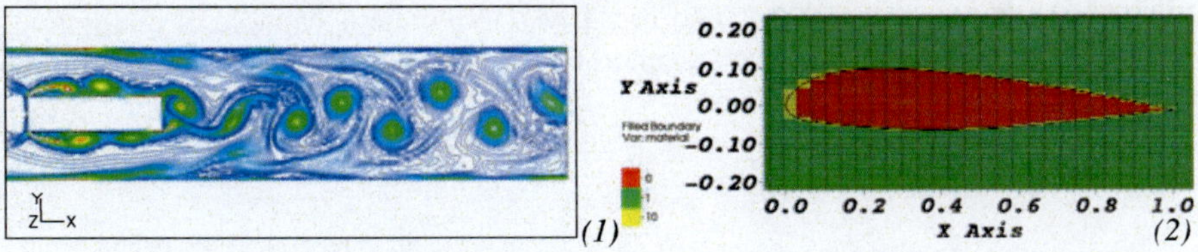

Fig. 5. Flow configurations around (1) 2-D rectangular block; (2) 2-D airfoil

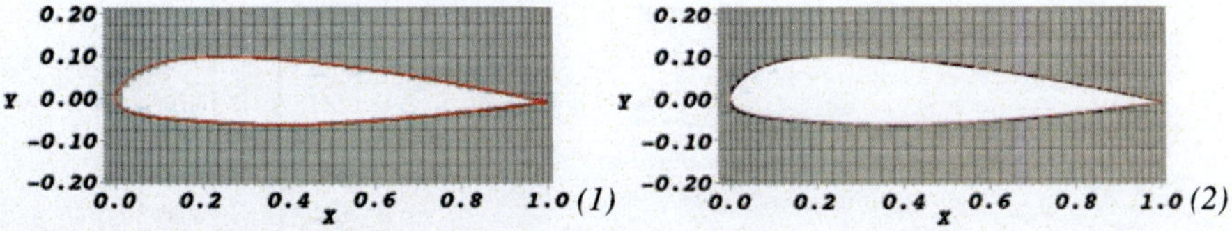

Fig. 6. (1) The initial immersed boundary grid around an airfoil without AMR cutting; (2) The grid with four levels of AMR cutting

To extend the FCAC-MG technique to complex geometric flows, the flow solver was first developed to handle the immersed Cartesian geometry, such as the flow past rectangular block in a confined chamber, see Figure 5(1). To further develop the solution technology to immersed curvilinear geometry, as illustrated by an airfoil in Figure 5(2), the Immersed Boundary (IB) method and Adaptive Mesh Refinement (AMR) technology, first reported in [8], was implemented into the LES/DNS solver and coupled with the FCAC-MG method. To achieve the geometry adaptation, an algorithm was developed to identify the cutting cells that intersect the flow geometry and the flow domain is then divided into three group of meshes, namely, (1) fluid cells, (2) cutting cells and (3) solid cells, as seen in Figure 5(2) represented by the green, yellow and red colors. An AMR method was developed to refine progressively the cutting cells so that the airfoil geometry in Figure 6(1) could be more accurately represented by the AMR grid. Figure 6(1) presents the initial immersed boundary grid without AMR cutting and Figure 6(2) provides the grid with four levels of AMR refinement that capture more accurately the airfoil geometry. The solution technology based on this AMR refinement grid is currently under development. Towards this end, based on the discovery that no solution errors should be tolerable at coarsest multigrid level, a fast direct solver at the coarsest grid was developed for a general immersed flow-structure topology represented by the cutting cells. Figure 7 illustrates a general sparse matrix structure at the coarsest multigrid level with a mesh of $N_x \times N_y \times N_z = 5 \times 5 \times 4$ including boundary points. This development guarantees the FCAC-MG method is capable of driving the residues of both mass and momentum equations down to the machine error. Figure 8(1) and (2) present a typical convergence history and one flexible-cycle of multigrid iteration. As an initial approximation, the simulation was conducted on the grid without AMR grid cutting. Figure 9(1) demonstrates that the current LES/DNS code is capable of qualitatively capturing the unsteady vortex wake generated even by a rough grid representation of airfoil geometry.

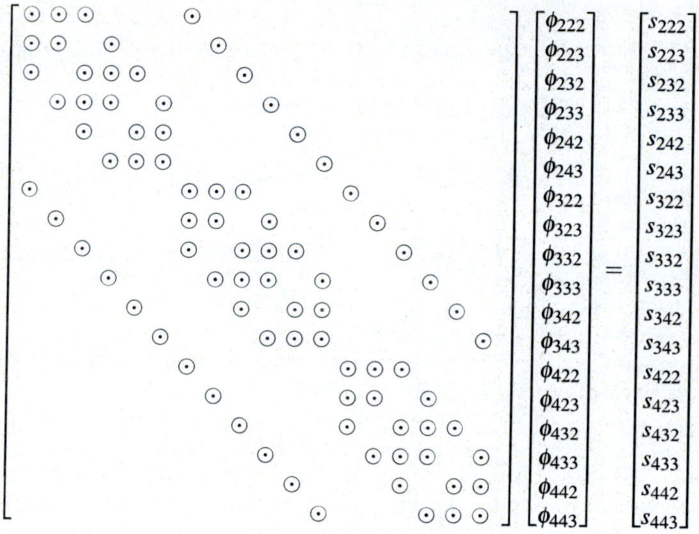

Fig. 7. Sparse matrix pattern at the coarsest grid level in multigrid

5 Importance of the initial and boundary conditions in LES/DNS

One of the most challenging issues in computational fluid dynamics (CFD) is prescribing accurate initial and boundary conditions that carry the correct physics and reflect realistic flow conditions ([9], [10]). This issue in conventional RANS simulations has not received adequate attention primarily for two reasons: (1) most RANS calculations are aimed at resolving flows in a steady state so that an uniform flow assumption for the initial and inflow boundary conditions was sufficient, and (2) the errors introduced by a RANS model often prevail over the incorrectness of the initial and boundary conditions and, therefore, a higher accuracy in these conditions

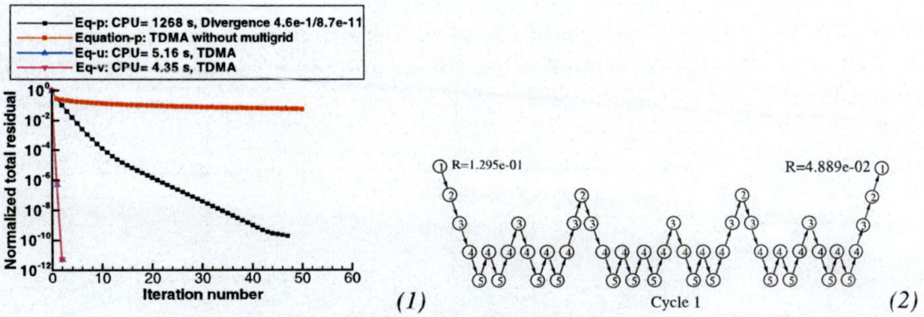

Fig. 8. (1) Convergence histories for the momentum and pressure Poisson equations and (2) a cycle pattern of the FCAC-MG for the flow past rectangular block

was not justified. However, the issue must be unequivocally resolved in LES/DNS since the target of these simulations is to resolve the detailed spectra of the unsteady turbulence motions.

In this regard, the temporal simulation, as reported in [14] and [15], can be used to generate realistic turbulent inflow which is imposed onto the inlet surface of a relevant spatial simulation. This strategy was applied in the current paper to simulate the turbulent flow past an airfoil in a plane channel, as depicted in Figure 9(2). The current simulation made use of the results obtained in a separate DNS of plane channel flow to generate the inlet conditions for the flow past airfoil.

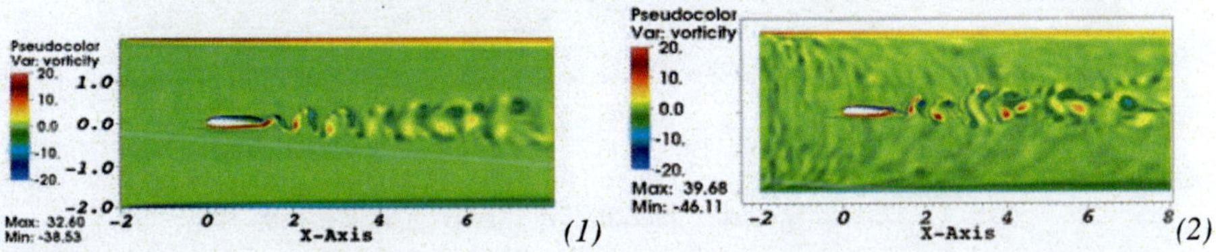

Fig. 9. Unsteady vortex shedding of airfoil (1) uniform inflow; (2) fully-developed turbulent inflow

Figure 9(2) presents the airfoil in the environment of fully developed turbulence in the confined channel. Comparing with the airfoil flow in uniform inflow condition in Figure 9(1), the inflow turbulence has a significant impact on the flow patterns both on airfoil surfaces and on the airfoil wake shedding. Due to the incoming turbulence, the flow on the surface of the airfoil tends to have an early transition to turbulence and causes a significant shortening the length of the airfoil wake. For the uniform inflow case, the vortex shedding occurred beyond the trailing edge of the airfoil. However, under the turbulent inflow condition, the vortex shedding took place before the trailing edge, giving rise to a lot of flow separations, reattachment and subsequently vortices rolling on the surface of the airfoil. All these observations can be seen vividly by the animation of the vortices evolution, as illustrated in Figures 10 and 11.

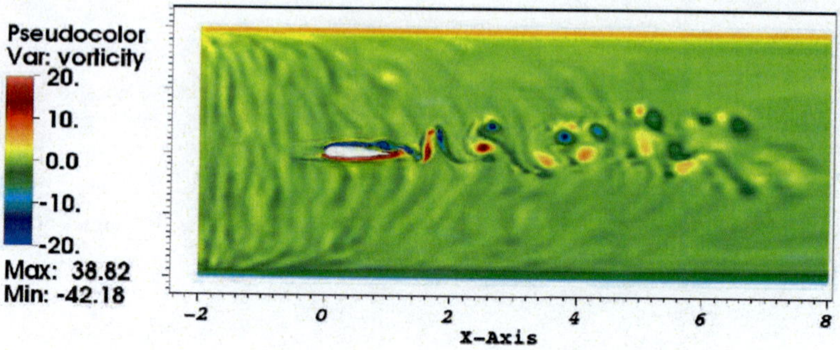

Fig. 10. Unsteady vortex shedding of airfoil at time instant I

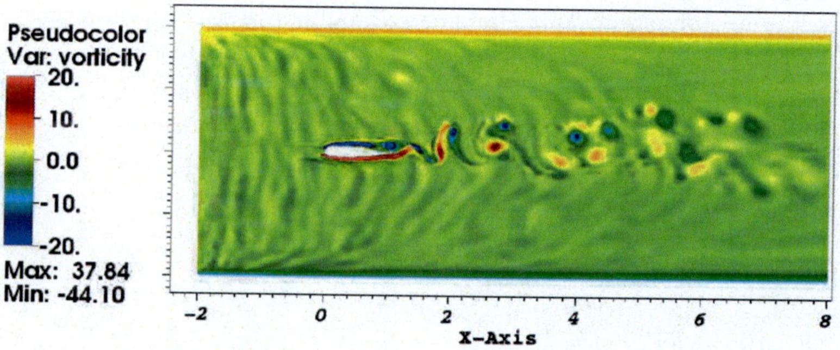

Fig. 11. Unsteady vortex shedding of airfoil at time instant II

6 Conclusions

A computational uncertainty analysis was conducted for generic turbulence simulation using RANS, LES and DNS. The sources of errors, namely, the model error and the numerical errors, were identified and analyzed. Technologies were developed to minimize or to eliminate these errors. To reduce model error, a more general form of eddy viscosity, i.e. the tensor form eddy viscosity, is recommended to address the anisotropic nature of turbulence motions. The adaptive mesh refinement (AMR) technology, both geometric and solution adaptive methods, is proposed to smartly reduce the discretization error. To eliminate the solution error, particularly for complex geometry flows, FCAC-MG solution technology is applied to immersed boundary method coupled with the AMR technique. The importance of the realistic turbulent inflow is qualitatively demonstrated through the flow past an airfoil in a channel. These research and development laid out a technical road map for the future LES/DNS practice, which is expected to promote the quality of LES/DNS. More quantitative comparisons for specific flow will be provided in future work.

References

1. Xu H. (2009) J. Fluid Mech., 621:23-57
2. von Karman T. (1930) 3rd Proc. Int. Congr. Appl. Mech.
3. Xu H., Yuan W. and Khalid M. (2005) J. Comp. Phys., 209:504-540
4. Reynolds O. (1895) Phil. Trans. Roy. Soc. London, A 186:123-164
5. Kolmogorov A.N. (1942) Izv. Akad. Nauk U.S.S.R. Physics, 6:56-58
6. Warsi Z. U. A. (1993) CRC Press, 45-50
7. Wesseling P. (2001) Series in Computational Mathematics, Springer-Verlag
8. Berger M.J. and Oliger J. (1984) J. Comp. Phys., 53, 484
9. Rogallo R.S. and Moin P. (1984) Ann. Rev. Fluid Mech., 16:99-137
10. Lund T. S. (1998) J. Comp. Phys., 140:233-258
11. Lopes A.S. and Palma J.M.L.M. (2002) J. Comp. Phys., 175:713-738
12. Spalart P.R. (1988) J. Fluid Mech., 187:61-98
13. Kim J., Moin P. and Moser R. (1987) J. Fluid Mech., 177:133-166
14. Akselvoll K. and Moin P. (1995) Report TF-63, Dept. of Mech. Eng., Stanford University
15. Xu H., Khalid M., and Pollard A. (2003) Intl. J. Computational Fluid Dynamics, 17(5):339-356

Computational error-minimization for LES of non-premixed turbulent combustion

Bernard J. Geurts[1,2] and Andreas Kempf[3]

[1] Multiscale Modeling and Simulation, Faculty EEMCS, University of Twente, P.O. Box 217, 7500 AE Enschede, The Netherlands b.j.geurts@utwente.nl
[2] Anisotropic Turbulence, Faculty Applied Physics, Eindhoven University of Technology, P.O. Box 513, 5600 MB Eindhoven, The Netherlands
[3] Department of Mechanical Engineering, Imperial College London, Exhibition Road, South Kensington, London SW7 2AZ, United Kingdom a.kempf@imperial.ac.uk

Summary. We apply the error-landscape analysis to turbulent non-premixed combustion in a bluff-body flame and investigate the error-reduction that can be achieved by adopting the SIPI algorithm (successive inverse polynomial interpolation) for direct optimization of the combined effect of discretization and modeling errors. Small scale turbulent flow aspects are modeled using the Smagorinsky model and a flamelet formulation is adopted for the combustion process. A systematic study of numerical predictions at various resolutions and different levels of subgrid dissipation is conducted, providing an overview of partial error-cancellation. The general structure of the error-landscape is similar to that found for single phase homogeneous isotropic flow - the application of SIPI results in a considerable reduction of the total error (15-50 % improvement in relative error) after a small number of iterations. The SIPI approach provides an impression of the sensitivity of predictions on numerical and modeling parameters.

Key words: turbulence, non-premixed combustion, error-landscape, successive inverse polynomial interpolation

1 Introduction

An assessment of the reliability with which primary turbulent flow phenomena and process steps in non-premixed combustion can be predicted using large-eddy simulation (LES) is a central issue that needs to be addressed in order to properly support the application of LES to flow problems of realistic complexity. The overall reliability is influenced, on the one hand, by the accuracy with which the small-scale turbulence and combustion are modeled, and, on the other hand, by the error that arises from the spatial and temporal discretization. Resolving this issue of reliability of LES is a central pacing item in the development of LES toward a design tool of industrial interest.

M.V. Salvetti et al. (eds.), *Quality and Reliability of Large-Eddy Simulations II*,
ERCOFTAC Series 16, DOI 10.1007/978-94-007-0231-8_32,

In this paper, we present a computational error-analysis of large-eddy simulation applied to turbulent non-premixed combustion. We extend the error-landscape approach [1] to heterogeneous turbulence. The geometry of the Sydney bluff-body flame is adopted, while the Smagorinsky model [2] and a flamelet formulation [3] are used for the turbulent stresses and the combustion contributions respectively. We quantify the total simulation error relative to experimental data [4]. The total simulation error displays a general structure that is in many respects comparable to that found for homogeneous decaying turbulence - it is indicative of a partial cancellation of errors due to modeling and due to discretization. This property can be adopted to computationally reduce error levels using an automated algorithmic approach. We present results based on the SIPI method (successive inverse polynomial interpolation) and illustrate the level of error-reduction that can be achieved and the computational overhead that is required.

Turbulent combustion presents major challenges to computational modeling strategies. Compared to the turbulent flow of a simple fluid such as air, combustion considerably increases the complexity of the problem. A variety of modeling strategies has been developed, addressing *(i)* the dynamic consequences of small-scale turbulent motions, *(ii)* the resulting multitude of chemical transformations that constitute the total combustion, and *(iii)* the complete coupling between these nonlinear processes. Such modeling is aimed at reducing the complexity of the underlying dynamical system to allow simulation at a strongly reduced spatial resolution, i.e., at manageable computational costs. However, these modeling steps also introduce sources of error into the computations. In addition, the spatial discretization that is used to represent the governing equations will be a source of simulation error, which is particularly relevant at modest resolution of the smaller resolved scales, as is often the case in realistic LES [5]. Finally, these sources of error combine nonlinearly and may result in significant error-accumulation under certain simulation conditions.

The organization of this paper is as follows. A brief introduction to the simulation of a turbulent non-premixed Sydney flame is provided in Section 2. The error-landscape analysis of this flow problem is presented in Section 3. The errors due to modeling and discretization were found to partially cancel. This provides the opportunity to adopt the SIPI algorithm (successive inverse polynomial interpolation) to iteratively optimize the total error that occurs. The application of SIPI is illustrated in Section 4. Concluding remarks are collected in Section 5.

2 Simulation of a turbulent bluff-body flame

In this section we introduce the mathematical model and the numerical method used for simulation of a turbulent non-premixed Sydney flame. Some flow visualization is given of the flame structure that is obtained.

A smoothed description of the turbulent flow associated with a combustion process arises from the application of a spatial filter, denoted by an overbar. Associated with this the density weighted Favre averaging is also employed, e.g., to obtain the smoothed velocity field; this is denoted by a tilde. The spatially filtered Navier Stokes equations are solved together with the continuity equation:

$$\frac{\partial}{\partial t}(\overline{\rho}\widetilde{u_i}) + \frac{\partial}{\partial x_j}(\overline{\rho}\widetilde{u_i u_j}) = \tag{1}$$

$$\frac{\partial}{\partial x_j}\left[\overline{\rho}\left(\nu\widetilde{\frac{\partial u_j}{\partial x_i}} + \nu\widetilde{\frac{\partial u_i}{\partial x_j}}\right) - \frac{2}{3}\overline{\rho}\nu\widetilde{\frac{\partial u_k}{\partial x_k}}\delta_{ij}\right] - \frac{\partial \overline{p}}{\partial x_i}$$

$$\frac{\partial \overline{\rho}}{\partial t} + \frac{\partial \overline{\rho}\widetilde{u}_j}{\partial x_j} = 0 \tag{2}$$

The flow is described by velocity u_i, pressure p and density ρ. The kinematic viscosity is denoted by ν. We will restrict ourselves to low Mach numbers - in these conditions the density ρ is variable but independent of the pressure p. The momentum equation as given in (1) contains a closure problem in view of the occurrence of contributions $\widetilde{u_i u_j}$. These can not be expressed in terms of the Favre averaged velocities. Instead the approximate form $\widetilde{u}_i\widetilde{u}_j$ is identified next to the unresolved sub-grid stresses τ_{ij}^{sgs} according to $\widetilde{u_i u_j} = \widetilde{u}_i\widetilde{u}_j + \tau_{ij}^{sgs}$. The unknown subgrid stress tensor τ_{ij}^{sgs} is modeled through Smagorinsky's eddy viscosity model [2]:

$$\tau_{ij}^{sgs} = \nu_t\left(\frac{\partial \widetilde{u}_j}{x_i} + \frac{\partial \widetilde{u}_i}{\partial x_j} - \frac{2}{3}\frac{\partial \widetilde{u}_k}{\partial x_k}\delta_{ij}\right) \tag{3}$$

The turbulent viscosity is obtained from the deformation rate tensor

$$\nu_t = (C_S\Delta)^2\sqrt{2\widetilde{S}_{ij}\widetilde{S}_{ij}} \qquad \text{with} \qquad \widetilde{S}_{ij} = \frac{1}{2}\left(\frac{\partial \widetilde{u}_j}{\partial x_i} + \frac{\partial \widetilde{u}_i}{\partial x_j}\right) \tag{4}$$

Here we introduced the Smagorinsky coefficient C_S next to the filter-width parameter Δ. Often Δ is taken equal to the grid-spacing. Various values have been suggested for the Smagorinsky constant C_S - a typical reference value is 0.17. In this paper we will treat C_S as an adjustable parameter and use this degree of freedom to obtain improved accuracy of predictions. Various more refined sub-grid models have been proposed in literature [6]. For the Sydney flame the use of a fine resolution in combination with the Smagorinsky model was found to be a good point of departure for the purpose of this paper, i.e., the assessment of the reliability of a particular computational modeling of a turbulent non-premixed combustion process.

We consider non-premixed combustion that is limited by the rate at which fuel and oxidizer mix. For such processes we may assume these species to react as soon as they come together. We incorporate combustion using a flamelet approach according to Williams [11] and Peters [3], based on Bilger's mixture fraction [12]. The mixture fraction f represents the local ratio of atomic mass from the fuel nozzle to the atomic mass from the air-coflow. The evolution of the mixture fraction is obtained from solving

$$\frac{\partial}{\partial t}\left(\overline{\rho}\widetilde{f}\right) + \frac{\partial}{\partial x_j}\left(\overline{\rho}\widetilde{f}\widetilde{u}_j\right) = \frac{\partial}{\partial x_j}\left(\overline{\rho}\left(\frac{\nu_t}{\sigma_t} + \frac{\nu}{\sigma}\right)\frac{\partial \widetilde{f}}{\partial x_j}\right) \tag{5}$$

This equation includes an eddy-diffusivity term to represent the unresolved scalar fluxes from the turbulent viscosity ν_t and the turbulent Schmidt-number $\sigma_t = D_t/\nu_t \approx 0.7$ [13] which relates the turbulent diffusivity to the turbulent viscosity. The mixture fraction is important for the evaluation of the tabulated flamelet description.

The numerical simulations are based on the *PsiPhi* LES code for general transported scalars Φ and derived quantities Ψ, using algorithms similar to the flowsi code [14, 15]. The code uses a Low-Mach assumption to solve the Favre Filtered Navier Stokes equations on a collocated Cartesian grid which consists of cubical cells. The *PsiPhi* code applies a low storage Runge-Kutta scheme that is third order accurate for linear problems. In *PsiPhi*, diffusion discretization schemes of both 2nd and 4th order accuracy are available. For convection, a 2nd order central scheme is used. For scalar transport, numerical oscillations must be prevented through non-oscillatory schemes. We employ a TVD (Total Variation Diminishing) scheme that blends a second order central scheme with a first order upwind scheme. Avoiding numerical oscillation is particularly important for the mixture-fraction.

The test-case considered here is the turbulent non-premixed bluff-body burner designed at Sydney university [4]. For this burner, both flow-field and scalar data are available. The Sydney burner consists of an axial bluff-body located in an open wind-tunnel, causing air recirculation down-stream of the bluff-body. A circular hole in the center of the bluff-body ejects gaseous fuel into the recirculation zone, leading to a stabilized flame. We focus on the test-case 'HM1e' with a co-flow velocity of $U_{co} = 35$ m/s, a jet velocity of $U_j = 108$ m/s, and a fuel composition of 50% [vol.] methane and 50% [vol.] hydrogen.

The Sydney Bluff Body Flame has been discretized on a computational domain of 200 mm in axial direction and 160 mm in each of the lateral directions, using cubic cells on grids of [40.0, 14.1, 5.0, 1.8, 0.63] million cells. Several simulations were performed on each of these grids for different values of the Smagorinsky parameter C_S. Fig. 1 shows snapshots of the flame region at a resolution of 5 M cells and different Smagorinsky parameters. We observe that a considerable increase in C_S relative to the reference value of 0.173 induces a significant smoothing of the smaller scales in the flame. More small scales appear in the snapshots at lower values of C_S. However, from these snapshots it is not possible to identify which of these simulations is actually the more accurate one. Intuitively, too much smoothing will imply a problem that is not hindered much by discretization errors. Conversely, a problem in which many small scales are kept may rely less on the quality of the sub-grid model and hence be also less affected by possible errors in the small-scale modeling. However, a very smooth flow will rely more on proper representation of the small scales by the sub-grid model while a lively small-scale turbulent flow may be affected by discretization errors. Where to strike an optimal balance between these sources of error is not easy to answer. However, following the error-landscape approach[1] a systematic framework is offered with which the total simulation error can be assessed via numerical experimentation. We performed this analysis for the Sydney flame and briefly sketch some findings in the next section.

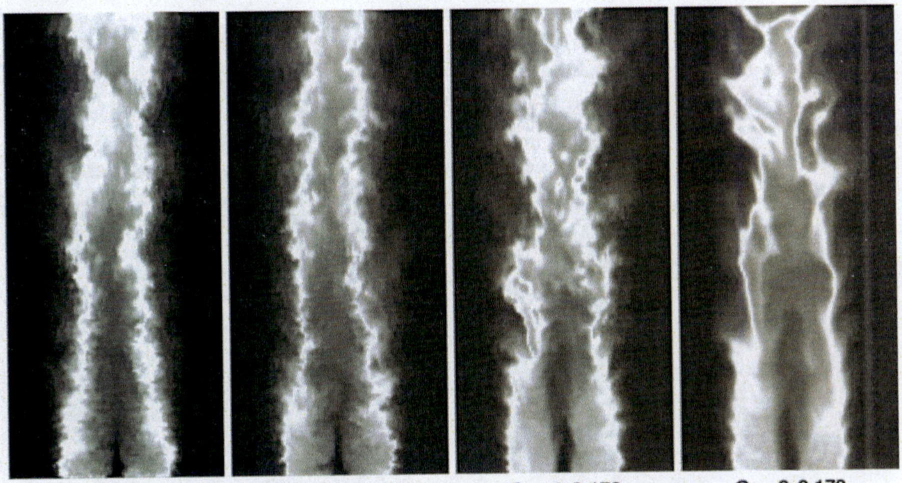

Fig. 1. Instantaneous temperature fields of the Sydney bluff-body flame for four different model constants C_S using 5 M cells. The figure shows how larger model coefficients suppress the smallest structures through overemphasized diffusion. With the smallest model coefficient, cell-sized structures which induce large numerical errors become important.

3 Error landscape analysis of a turbulent combustion

In this section we introduce the error-landscape approach and apply this to the Sydney bluff body flame. We confront the simulations with experimental data [4, 16]. The basic structure of the error-landscape will be sketched afterwards in combination with the possibility for computational error-reduction.

The introduction of a sub-grid model to represent the small-scale turbulent motions and a spatial discretization method to evaluate the derivatives in the governing equations constitute sources of error. Their theoretical assessment in LES is hampered by the fact that practical LES is routinely executed at marginal sub-grid resolutions and at fairly large filter-widths. Correspondingly, both numerical and modeling errors are expected to be large and do not lend themselves for an analysis based on Taylor expansions and Richardson extrapolation. Rather, a direct simulation approach is adhered to. From a systematic variation of important system parameters associated with numerical and modeling errors a precise assessment of the total simulation error can be obtained.

In any large-eddy simulation the numerical flow prediction will be influenced by the interplay between sources arising from flaws in the sub-filter model and errors due to the numerical treatment at marginal sub-filter resolution. For strongly turbulent flow it is unclear how these sources of error will accumulate. This introduces the issue of reliability of a given LES. For the Smagorinsky model as adopted in this paper, a detailed impression of the total error behavior in a given flow may be obtained by considering the total error as a function of the numerical resolution $N \sim 1/h$ and the Smagorinsky parameter C_S, taking the filter-width equal to the grid-spacing. Any simulation in particular is then characterized by its combination (C_S,N). Correspondingly, one may label the total error E in this way. From contour-plots of E as function of C_S and N one may infer under what conditions on the model parameter and/or the numerical resolution one may expect a given level of error. Such a contour-plot of $E(C_S,N)$ will be referred to as an error-landscape. It requires a systematic collection of individual LES on a grid of (C_S,N) points. From such a study one may, afterwards, identify the 'optimal refinement' strategy, which specifies the value of $\widehat{C_S}$ at which the total error is minimal at given N.

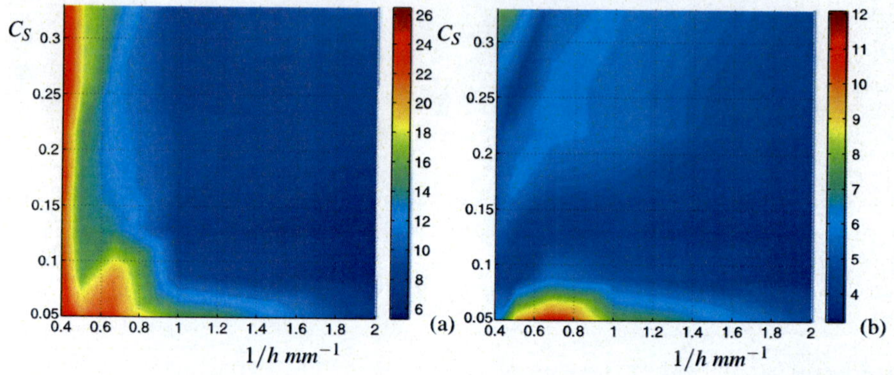

Fig. 2. Error-landscape obtained for the error in the streamwise velocity $\overline{u}$ (a) and $rms(\overline{u})$. Simulations were performed at grid-spacings of $0.5, 1.0, 1.5, 2.0, 2.5$ (mm) and a range of C_S values. The errors are defined relative to experimental data obtained at a number of spatial locations.

A crucial element in the error-landscape approach is the measure that is adopted to quantify the total error. This aspect is dependent on the particular flow and application that is being investigated. In this paper we investigate the total simulation error in terms of differences that occur in mean and root-mean-square properties of the flow, such as velocity components, temperature and mixture fraction, relative to the experimental findings. We can express the error in terms of individual properties or as a weighted average of several of them simultaneously. In Fig. 2 we present the error-landscape evaluated for the mean and rms streamwise velocity component. Experimental data for these velocity profiles are available at a number of locations - taking the discrete L_2-norm of the difference between these profiles and the LES results yields an impression of the total simulation error as function of spatial resolution and Smagorinsky constant. We observe that the error increases very rapidly in the mean in case the spatial resolution drops below a certain level. In addition, the use of very low and very high values of C_S induces an increase in the total simulation error. In between, a striking valley structure develops, which is more pronounced at coarser resolutions. This valley structure in the error landscape is indicative of partial error cancellation [1].

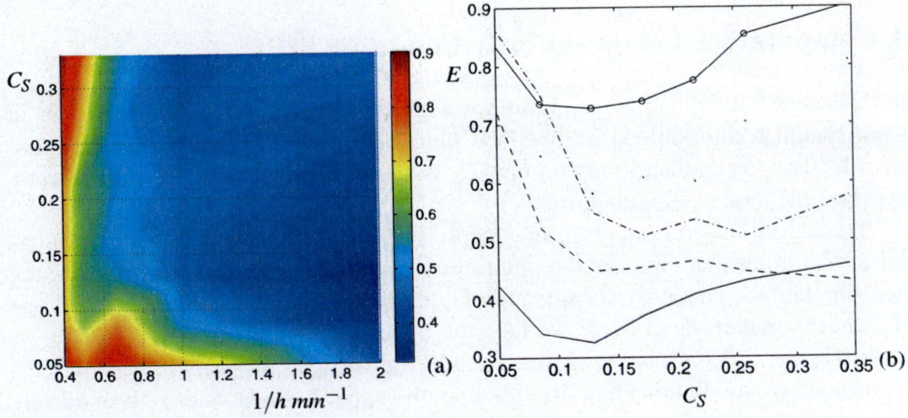

Fig. 3. Error-landscape obtained for the combined error in the streamwise velocity $\overline{u}$, the spanwise velocity $\overline{v}$, the temperature $\widetilde{T}$ and the mixture fraction $\widetilde{f}$. Each error-component was normalized by its respective maximum and subsequently added with equal weight (a). A collection of cross-sections of the error E at constant spatial resolution is shown in (b): $h = 0.5, 1.0, 1.5, 2.0, 2.5$ mm plotted solid, dash, dash-dot, dot and solid with circle respectively.

A number of flow properties is available experimentally for error-analysis, including temperature, mixture fraction and spanwise velocity. Each of these quantities yields quite different error-landscapes. A more complete assessment arises from error measures that include an appropriate mixture of such physical quantities. This also provides flexibility in assigning different weights to different flow properties to reflect application specific emphasis. As an illustration we show the combined error in streamwise and spanwise velocity, temperature and mixture fraction in Fig. 3. In order to include errors in various quantities that may be of different magnitudes, we combine basic error-landscapes in the individual flow properties, normalized by their maximal value. An equal weighing of these normalized error-landscapes shows a 'valley-structure'. The error-behavior is expressed accurately in terms of the graphs of error versus C_S at a number of resolutions in Fig. 3(b). We observe that the overall error-levels increase with decreasing resolution. Moreover, a local minimum of the total error can be appreciated for values of the Smagorinsky coefficient of about 0.1-0.2, depending on the spatial resolution that was adopted. In some cases the reduction in relative error compared to the lower value of $C_S = 0.05$ is as high as 40-50 % and quite localized around a specific C_S-value. At other resolutions the benefit of adapting C_S is less clear and a more shallow minimum is observed. This error-structure is reminiscent to that established for homogeneous, isotropic, decaying turbulence [7, 8]. To automatically obtain a reduced error-level, we may employ an iterative successive inverse polynomial interpolation (SIPI) optimization procedure [9]. We turn to this in the next section.

4 Computational error optimization using SIPI

In this section we illustrate an algorithm with which optimal model coefficients at fixed spatial resolution can be obtained iteratively. Instead of recording an entire error-landscape and identifying optimal C_S values only afterwards, one may directly optimize the total simulation error.

The algorithm proceeds in a few steps. The first task is to determine an interval of C_S values that contains the optimum. This can be determined by performing two simulations, one at C_S considerably higher than suggested in literature and one at a much smaller value. Here we take the upper boundary to be $C_{S,up} = 2C_{S,Lilly}$ where $C_{S,Lilly} = 0.173$, and the lower boundary at $C_{S,low} = C_{S,Lilly}/4$. These settings correspond to simulations that are too smooth (upper boundary) or only modestly smoothed. The first iteration is then performed at $C_S = (C_{S,up} + C_{S,low})/2$, after which we have obtained three measurements of the error at their corresponding C_S values. Further refinements can be obtained using the SIPI minimization algorithm - this approach is preferred as it does not require derivatives of the error E [10]. At fixed resolution N^3 locally around its minimum $E(C_S)$ can be approximated by a parabola in C_S. Based on the three initial iterations we may construct such an interpolating parabola and perform a new iteration at a C_S value that corresponds to the minimum of this parabola. On an interval around the minimum where the error function is strictly convex, this algorithm will be guaranteed to converge - in practice 5-6 iterations are quite sufficient to approximate the optimum with fair accuracy.

In Fig. 4 we show the iteration process at different spatial resolutions. The iteration process rapidly converges to a value that is near a local minimum of the error-landscape. If the error-landscape is very flat the iterations suggest quite large values of C_S. In such cases the precise value of C_S is not very critical and very dissipative sub-grid modeling is still acceptable. The SIPI procedure automatically yields suggestions for an optimal C_S at which the error is (much) smaller compared to $C_S = C_{S,Lilly}/4$. The benefit of SIPI as measured in terms of the error-reduction relative to the lower bound for C_S varies from about 25% at the highest resolution to about 50 % for intermediate resolutions and about 15 % at the coarsest resolutions. Interestingly, the optimal C_S increases for $h = 2.5, 2.0, 1.5, 1.0$ mm and jumps to a value around 0.1 for $h = 0.5$ mm. It appears that only at $h = 0.5$ mm all dynamically relevant scales are represented - a full DNS may be hypothesized to require another factor 3-6 more grid-points at which the optimal $C_S \approx 0$.

5 Concluding remarks

In this paper we presented large-eddy simulations of turbulent, non-premixed combustion that arises in a bluff-body burner known in literature as the Sydney bluff-

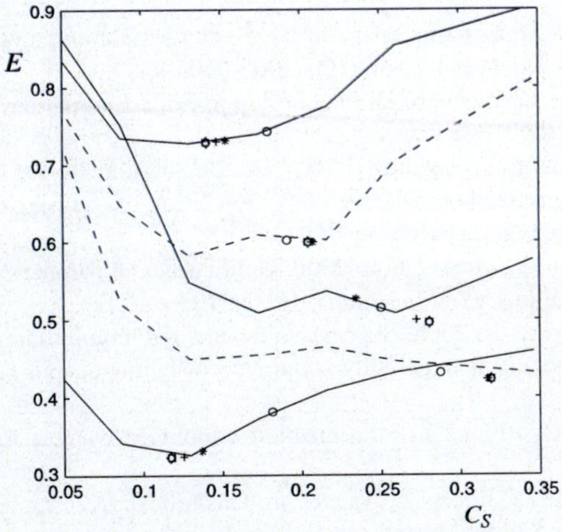

Fig. 4. SIPI convergence for the total simulation error E shown using simulated SIPI based on error-data obtained at grid-spacing $0.5, 1.0, 1.5, 2.0, 2.5$ mm from bottom to top, labeled with solid and dashed curves consecutively. After the first three iterations, subsequent iterands are labeled with $\circ$, $*$, $+$, diamond and square.

body flame [4]. The Smagorinsky model was adopted to represent the small turbulent motions and a flamelet model was adopted to describe the combustion process. The influence of a Smagorinsky coefficient that is either too high or too low was illustrated on the basis of selected snapshots. This provides a first impression of the effect of errors due to the discretization and errors due to shortcomings in the sub-grid model. In order to quantify the errors in more detail the error-landscape approach was used [1]. Using a definition of the total simulation error relative to experimental data, a concise overview of the reliability of LES of this problem can be arrived at. The error-landscape that was obtained showed a qualitative similarity with the 'valley-structure' that was seen in an analysis of LES of homogeneous isotropic turbulence. This also provides the opportunity to directly reduce the error-level by a carefully generated sequence of simulations. Such a SIPI approach was found to reduce the error by 15-50 % depending on the spatial resolution that was adopted.

References

1. Meyers, J., Geurts, B.J., Baelmans, M.: 2003. Database-analysis of errors in large-eddy simulations. *Phys. of Fluids*, **15**, 2740
2. Smagorinsky, J.: 1963. General circulation experiments with the primitive equations. *Mon. Weather Rev.*, **91**, 99
3. Peters, N.: 1984. Laminar Diffusion Flamelet Models in Non-Premixed Turbulent Combustion. *Prog. Energy Combust. Sci.*, **10**, pp. 319-339
4. Dally, B., Masri, A., Barlow, S., Fiechtner, G.: 1998. Instantaneous Mean Compositional Structure of Bluff-Body Stabilized Nonpremixed Flames. *Combust. Flame*, **114**, pp. 119-148

5. Geurts, B.J.: 2006. Interacting errors in large-eddy simulation: a review of recent developments. *J. of Turbulence*, 179596 (JOT-2005-0106.R1)
6. Geurts B.J.: 2003. Elements of Direct and Large-Eddy Simulation. Edwards Publishing. ISBN: 1-930217-07-2, 388 pages
7. Meyers, J., Geurts, B.J., Sagaut, P.: 2007. A computational error assessment of central finite-volume discretizations in large-eddy simulation of a Smagorinsky fluid. *J. Comp. Phys.* http://dx.doi.org/10.1016/j.jcp.2007.07.012.
8. Meyers, J., Sagaut, P., Geurts, B.J.: 2006. Optimal model parameters for multi-objective large-eddy simulations. *Phys. of Fluids*, **18**, 095103
9. Geurts, B.J., Meyers, J.: 2006. Successive inverse polynomial interpolation to optimize Smagorinsky's model for large-eddy simulation of homogeneous turbulence. *Phys. Fluids*, **18**, 118102
10. Brent, R.: 1973. Algorithms for Minimization without Derivatives. Englewood Cliffs N.J.: Prentice-Hall.
11. Williams, F.A.: 1975. Recent Advances in Theoretical Descriptions of Turbulent Diffusion Flames, in S.N.B. Murthy (ed.), Turbulent Mixing in Nonreactive and Reactive Flows, pp. 189-208, Plenum Press, New York, US
12. Bilger, R., Starner, S., Kee, R.: 1990. On Reduced Mechanisms for Methane-Air Combustion in Nonpremixed Flames. *Combust. Flame*, **80**, pp. 135-149
13. Forkel, H., Janicka, J.: 2000. Large Eddy Simulation of a Turbulent Hydrogen Diffusion Flame. *Flow, Turbulence and Combustion*, **65**, pp. 163-175
14. Kempf, A., Lindstedt, R.P., Janicka, J.: 2006. Large-eddy simulation of a bluff-body stabilized non-premixed flame. *Combust. Flame*, **144**, pp. 170-189
15. Stein, O., Kempf, A.: 2007. LES of the Sydney swirl flame series: A study of vortex breakdown in isothermal and reacting flows. *Proc. Combust. Inst.*, **31**, pp. 1755-1763
16. Dally, B., Fletcher, D., Masri, A.: 1998. *Combust. Theory and Modeling*, **2**, pp. 193-219

Part III

Mathematical analysis and foundation for SGS modeling

Assessment of eddy resolving techniques for the flow over periodically arranged hills up to Re=37,000

Michael Manhart[1], Christoph Rapp[1], Nikolaus Peller[1], Michael Breuer[2], Orhan Aybay[2], Jordan A. Denev[3], Carlos J. Falconi[3]

[1] Fachgebiet Hydromechanik, Technische Universität München
80333 München, Germany m.manhart;ch.rapp;n.peller@bv.tum.de
[2] Professur für Strömungsmechanik, Helmut-Schmidt-Universität Hamburg
22043 Hamburg, Germany breuer@hsu-hh.de
[3] Institut für Technische Chemie und Polymerchemie, Universität Karlsruhe
76128 Karlsruhe, Germany denev;falconi@ict.uni-karlsruhe.de

Summary. The turbulent flow over periodically arranged geometrically two-dimensional hills in a channel at a Reynolds number of $Re = 37,000$ has been considered as benchmark case for various eddy-resolving methods. The aim of this study is to assess various LES models and numerical approaches in a turbulent flow detaching from a curved surface. We compare results of a Cartesian grid solver using the immersed boundary method with various curvilinear approaches ranging from standard eddy-viscosity subgrid-scale models to hybrid LES-RANS models. The results are validated by a recent experiment conducted in a water channel by particle image velocimetry and laser-Doppler anemometry.

Key words: Quality, Reliability, Large-Eddy Simulation, Hybrid LES-RANS

1 Introduction

Large-Eddy Simulation of complex aerodynamic configurations heavily stresses available computer hardware and codes through excessive grid refinement necessary to resolve the immediate wall layer. The scaling of the viscous layer with Reynolds number leads to a severe bottleneck if flows at high Reynolds numbers are to be predicted. Furthermore, complex high Reynolds number flows are difficult to predict because of the representation of the geometric details. Concerning the representation of the geometry, several grid systems have been developed, Cartesian Immersed Boundary (IBM), Curvilinear Body Fitted and unstructured grids. In this study, we consider the former two by comparing results from three different flow solvers, a Cartesian IBM code (MGLET, [11]) a curvilinear code that allows for a hybrid method (LESOCC, [2]) and a curvilinear code with a recently developed wavelet-based subgrid-scale model (LESOCC2, [7]).

While IBM methods are able to represent highly complex shapes with many geometric details, they have problems to refine the grid close to a wall if the wall is not

M.V. Salvetti et al. (eds.), *Quality and Reliability of Large-Eddy Simulations II*, ERCOFTAC Series 16, DOI 10.1007/978-94-007-0231-8_33,

aligned with the grid. This problem results in poor wall shear stress predictions if not two (or even three in the general case) directions are refined at the wall. Consequently, it has been found in a previous study at a low Reynolds number that an IBM method requires a larger total number of grid cells than a curvilinear one. We expect that with increasingly thinner viscous sublayers, i.e. with higher Reynolds numbers, this problem would be more severe. The performance of an IBM method together with three different sub-grid scale models is checked by the flow solver MGLET. We compare standard Smagorinsky [13], the WALE subrid-scale model [10] and a Lagrangian dynamic model [9].

As an alternative to explicit wall functions, hybrid methods have become popular in recent years. Hybrid methods, such as Detached Eddy Simulation (DES) promise to save resources in the wall layer by applying Reynolds Averaged (RANS) models in the wall layer and LES outside. Since at high Reynolds numbers, wall units can not be resolved by a grid that resolves the large scales of the outer flow, the spatial filter performed by this grid tends towards a Reynolds filter thus justifying RANS modeling close to the wall. Issues of these so-called hybrid methods are the location and definition of the interface between RANS and LES zone and the low-Reynolds modification of the terms in the applied model equations. A recent development of Jaffrezic and Breuer [6] showed excellent results for the considered flow case at a lower Reynolds number. We assess this approach, a combination of Schumann's one-equation subgrid-scale model in the LES zone with an explicit algebraic Reynolds stress model in the RANS zone in comparison with a pure LES with Schumann's model by the curvilinear code LESOCC.

A recently proposed wavelet-based eddy-viscosity subgrid-scale model [4] is checked in comparison to the standard Smagorinsky model with the curvilinear code LESOCC2. In this model the eddy-viscosity is set proportional to the wavelet-details from a two-level biorthonormal wavelet decomposition of the velocity vector.

2 Test case

We consider in this paper the flow over a periodic arrangement of smoothly contoured two-dimensional hills featuring important details of flow separation from a curved surface with subsequent reattachment (see Figure 1). It goes back to the experiments of Almeida *et al.* [1], but has been modified by Mellen *et al.* [8] to be more suitable for numerical simulations. Consequently, it has been extensively studied over the past few years e.g. [3, 5] and used as basis for workshops on turbulence modeling. Up to now, highly resolved LES and DNS are available for Reynolds numbers $Re \leq 10,595$ that can be regarded as reference solutions [3]. In addition, an experiment corresponding to this new setup was designed at the Technische Universität München in order to provide experimental reference data up to $Re = 37,000$ [12]. In this experiment, careful checks have been made to validated the measurement techniques (PIV and LDA). Two-dimensionality of the flow in the middle of the channel has been proven for $Re \geq 10,595$.

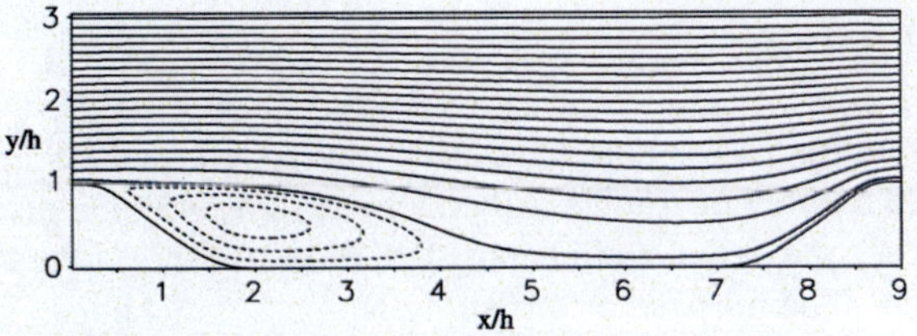

Fig. 1. Geometry of periodic hills in a channel.

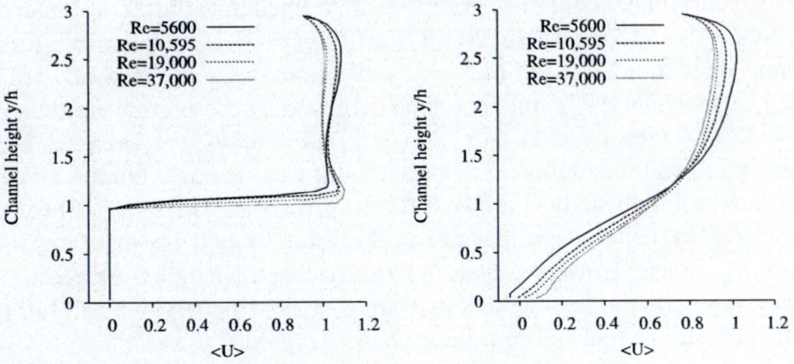

Fig. 2. Dependence of averaged velocity profiles on Reynolds number. Experiment by Rapp [12]. $x/h = 0.05$ (left) and $x/h = 4.0$ (right).

The flow field reveals all structural features of separating and reattaching flows including curved streamlines, highly unsteady separation and reattachment lines, a fluctuating shear layer above the separation zone giving rise to Kelvin-Helmholtz vortices and strongly amplified fluctuations. On the windward side of the hill the flow is accelerated with a maximal wall shear stress occurring approximately in the upper third of the hill slope. Two secondary separation zones were observed at $Re = 10,595$, one in front of the hill and one at the hill crest. A detailed analysis of the flow field was reported by Fröhlich *et al.* [5] and Breuer *et al.* [3].

Of special interest for this study is the Reynolds number dependence of the flow field. As can be seen in Figure 2, the mean streamwise velocity profiles develop an overshoot directly above the hill crest ($x/h = 0.05$) with increasing Reynolds number. Above $Re = 19,000$ the profile is flat towards the upper wall over the hill. The separation point moves upwards with increasing Reynolds number. Close to the reattachment point ($x/h = 4.0$) we see again the flatter velocity profiles at the higher Reynolds numbers. The two positions shown in Figure 2 reveal the most pronounced Reynolds number dependent features of the averaged velocity field. These are also the positions at which the strongest deviations between different simulations have been observed in previous studies.

3 Numerical details

The codes used in this assessment have been developed for LES of turbulent flows and validated in a number of studies. All codes use finite-volume discretizations with standard central second-order spatial approximations for this study and Runge-Kutta time advancement. The purpose of this study is to assess what can be achieved with relatively coarse grids. In this context, it is of major interest, how the performance of the schemes changes with increasing Reynolds number.

The two curvilinear codes use grids that are refined towards the wall and result in 10^6 (LESOCC2) and $1.5 \cdot 10^6$ (LESOCC) cells, respectively. The most critical point concerning resolution is at the maximal wall shear stress at the windward side of the hill. The spacings Δy^+_{max} that could be achieved by these grids are rather coarse for $Re = 37,000$ (see table 1). For the Cartesian method it is even more difficult to achieve a full wall resolution at that point with a comparable number of cells. We performed a resolution study with two different grids, one with $2 \cdot 10^6$ (2M), the other with $3.8 \cdot 10^6$ (4M) cells. The achievable wall resolutions at the point of maximal wall stress are very coarse. However, it has to be noted that the length over which this wall stress is acting upon the flow field is very short, approximately one half hill heights. In all other positions, the wall resolutions can be regarded as sufficient.

Table 1. Main parameters of grids used in this study. X: streamwise; Y: wall normal; Z: spanwise direction. Wall resolution Δy^+_{max} at point of maximal wall shear stress.

Code	$NX \times NY \times NZ$	$\Delta y^+_{max}(10,595)$	$\Delta y^+_{max}(37,000)$
MGLET (4M)	$216 \times 168 \times 104$	11	30
MGLET (2M)	$176 \times 120 \times 96$	32	88
LESOCC	$160 \times 160 \times 60$	1	3.4
LESOCC2	$160 \times 100 \times 60$	4.3	11.8

4 Results

The focus of this study lies on the capability of the different modeling approaches to accurately predict the size and shape of the separation downstream of the hill at a medium Reynolds number. It can be stated that all methods succeed in qualitatively predicting the separation zone. The main difference between the various models and approaches can be seen in the time averaged streamwise velocity profiles at the hill crest $x/h = 0.05$ and the reattachment region $x/h = 4.0$ which are discussed in the remainder.

The performance of the IBM method is discussed for $Re = 10,595$ and $Re = 37,000$ (Figures 3 and 4). The overall accordance is satisfying. The peak over the hill crest that was measured in the experiment is not reached by the simulations at $Re = 10,595$. Especially the Smagorinsky model (SMA) seems to underpredict this feature. This underprediction goes in hand with a delayed reattachment (see $x/h = 4.0$). The Lagrangian subgrid-scale model (LAG) performs best at the lower Reynolds number. At the higher Reynolds number, the WALE model (WAL) performs a little better than the Lagrangian one but both are fully satisfying over the hill crest and at the point of reattachment. The Lagrangian model with $2 \cdot 10^6$ (2M) cells even outperforms the Smagorinsky model at twice the number of cells. But both reattach too late. The comparison between experiment and simulation is better at the higher Reynolds number than at the lower one. This is surprising as the resolution of the viscous layer is much worse at the higher Reynolds number and no wall model was used.

The performance of the curvilinear codes is checked at $Re = 37,000$ (Figure 5). Concerning the overshoot over the hill crest there are two different solutions, the finer grid (1.5M) comes closer to the experiment than the coarser one (1M). It seems that a too small overshoot over the hill crest goes in hand with a too long reattachment length. The hybrid model (HYB) performs similar to Schumann's model (SCH) with a slightly longer recirculation zone. The wavelet-based model (WAV) is slightly better than the Smagorinsky model. To check the performance of the curvilinear schemes at $Re = 10,595$, we plot profiles from Smagorinsky (LESOCC2, 1M-SMA), wavelet-based (LESOCC2, 1M-WAV) and dynamic Smagorinsky (LESOCC, 12M-DSM). Note, that the latter one was computed on a $12 \cdot 10^6$-cells mesh that is fine enough so that the effect of SGS models disappears [3]. This simulation can be regarded as reference for LES at this Reynolds number as no significant change of the results can be expected at higher resolutions. At this Reynolds number, the standard Smagorinsky model performs nearly as well on one million as the dynamic one on 12 million cells. The wavelet-based model is a little too dissipative at the wall as no damping of the SGS-viscosity was undertaken. This issue has already been identified for further improvements of this model. Note that Van Driest damping was used for the Smagorinsky model.

The assessment of the averaged velocity profiles over the hill crest and close to reattachment reveals the observation that an underprediction of the overshoot above the hill crest results in a longer recirculation zone in the lee of the hill. This behavior can not fully be explained by the values of the total shear stresses (resolved plus modeled) at the hill crest (Figure 7). The higher Reynolds shear stress values come together with a longer separation bubble. This is counterintuitive but can be explained by the fact that the momentum transport over the shear layer bounding the separation zone is due to the fluctuations in that shear layer which start over the hill crest in a very thin layer. This layer is represented by the sharp shear stress peak immediately above the wall that was not measured in the experiment, presumably due to resolution problems. Thus the large peak in the middle of the channel is a remainder of upstream processes. It can be concluded that it is responsible for the shape of the velocity profile in the bulk of the flow which was observed to be more or less flat in the various simulations. If the shear stresses are strong, the overshoot just over the crest of the hill is damped due to increased momentum transfer from the bulk to the location of the overshoot. This conclusion is supported by the fact that all results with lower peak (overshoot) values directly over the hill crest show an increased level of Reynolds shear stress in the bulk.

Now let's examine the behavior of the individual methods upstream of the hill which is supposed to be important for the shape of the velocity profile above the crest. The windward side of the hill is characterized by a strong acceleration of the bulk velocity by a factor of 3/2. An examination of the Reynolds shear stresses at the middle of the hill slope at $x/h = 8.0$ reveals a nearly perfect match of all methods with the measurements at $y/h \geq 1.5$ (not shown). In this region the streamlines are nearly parallel (compare Figure 1). Below, the numerical results deviate from the experiment. Reaching towards the lower wall, the streamlines are more and more aligned with the hill slope. The momentum transfer perpendicular to the streamlines can therefore only be examined in a coordinate system that is aligned with the wall which is done in the following.

The Reynolds number dependence of the shear stress that is evident from the experiment is hardly predicted by the numerical schemes (Figures 8 and 9). Note that in these figures, on the left hand side the numerical results should coincide with the lower Reynolds number (hollow symbols) and on the right hand side, they should match the higher Reynolds number (full symbols). While in the experiment the shear stress drops by a factor of $3/2$ from $Re = 10,595$ to $Re = 37,000$, this drop is not observed in most of the models. In the Cartesian code, the Lagrangian and the WALE models drop by some amount but not like the experimental values. At the lower Reynolds number, the values are mainly too small, at the higher one too large. The results from the Smagorinsky model seem to be independent of the Reynolds number, an observation that can also be made in the curvilinear codes. Those results lie generally above the experimental ones except for the high resolution simulation with $12 \cdot 10^6$ cells (12M-DSM) at $Re = 10,595$. The main trend is that the higher the momentum transport perpendicular to the streamlines is, the lower is the overshoot just above the hill crest. This is explained by the fact that this overshoot is an inviscid effect that is attenuated by strong shear stresses upstream. The excessive near-wall bump apparent in the IBM method is a consequence of the poor near-wall resolution of this method. Its effect on the velocity profiles remains limited as the overall level remains small.

5 Conclusions

The present study compares several eddy resolving approaches for numerical simulation of turbulent flow with experimental data at low to medium Reynolds number flow over periodically arranged hills. Special attention is placed on the Reynolds number dependence of the results. The following observations could be done.

The acceleration and streamline curvature at the windward side of the hill generates an overshoot in the velocity profile above the hill crest. This is an inviscid effect. Viscous and turbulent stresses attenuate this overshoot. Both, viscous and turbulent stresses decrease with increasing Reynolds number, consequently the overshoot increases with Reynolds number. Various methods were not able to reproduce

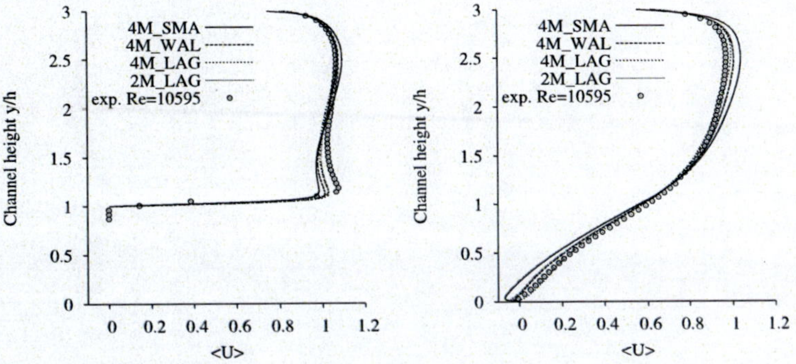

Fig. 3. Comparison of averaged velocities from MGLET with experimental data at $Re = 10,595$. Experiment by Rapp [12]. $x/h = 0.05$ (left) and $x/h = 4.0$ (right).

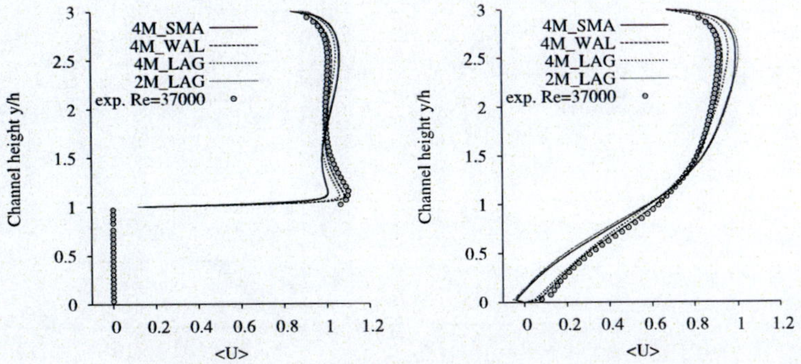

Fig. 4. Comparison of averaged velocities from MGLET with experimental data at $Re = 37,000$. Experiment by Rapp [12]. $x/h = 0.05$ (left) and $x/h = 4.0$ (right).

this Reynolds number dependence, especially the classical Smagorinsky model. The successful models include the Lagrangian dynamic, the WALE and the newly developed wavelet-based model as well as Schumann's model with or without hybrid wall modeling. It has to be mentioned that the hybrid wall modeling tested in this study represents a promising way as the interface between RANS and LES zone does not have to be defined a priori and a smooth transition is achieved at the interface. Van Driest damping is required when the wall resolution is in the range of ten wall units. The newly proposed wavelet-based model still has to be tested with Van Driest damping although it's results are fully satisfying.

A correct prediction of the maximum wall shear stress, which is difficult for IBM methods seems to have less impact on the global results at higher Reynolds numbers than at lower ones. This can be explained by the ratio of wall shear stress to wall pressure which shifts in favor of the wall pressure at higher Reynolds numbers. As a consequence, results at higher Reynolds number are less prone to immediate wall resolution than to overall grid resolution in the bulk of the flow. In this respect, the penalty of Cartesian IBM methods concerning wall resolution is less severe for this flow when Reynolds number is high.

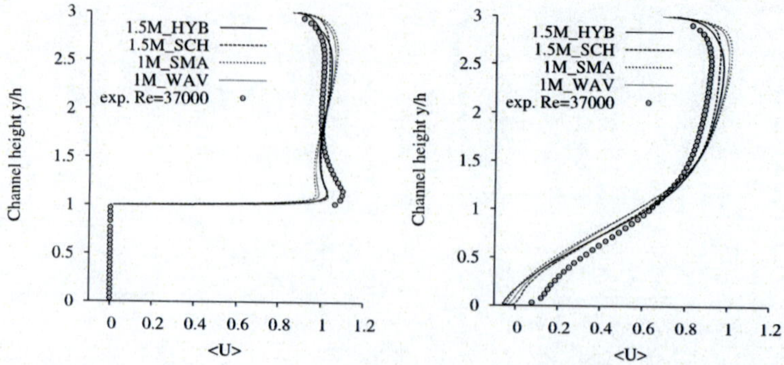

Fig. 5. Comparison of averaged velocities from LESOCC and LESOCC2 with experimental data at $Re = 37,000$. Experiment by Rapp [12]. $x/h = 0.05$ (left) and $x/h = 4.0$ (right).

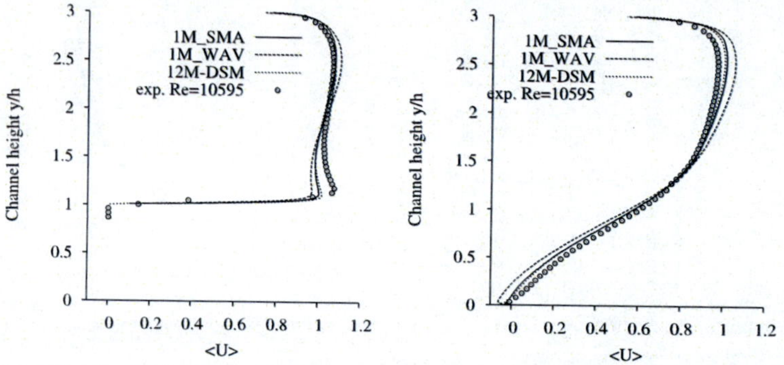

Fig. 6. Comparison of averaged velocities from LESOCC and LESOCC2 with experimental data at $Re = 10,595$. Experiment by Rapp [12]. $x/h = 0.05$ (left) and $x/h = 4.0$ (right).

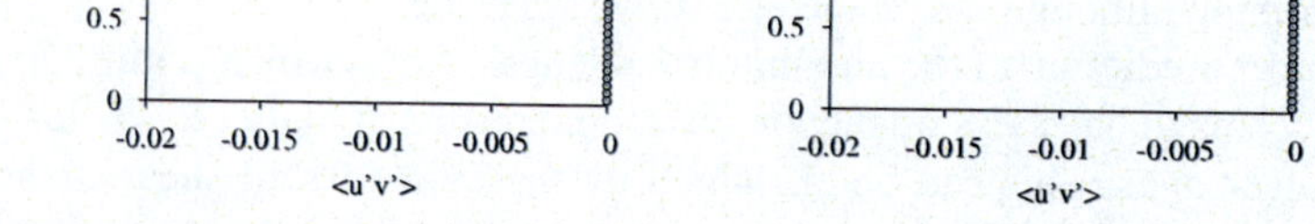

Fig. 7. Comparison of Reynolds shear stress from MGLET (left) and LESOCC and LESOCC2 (right) with experimental data at $Re = 37,000$. Experiment by Rapp [12]. $x/h = 0.05$.

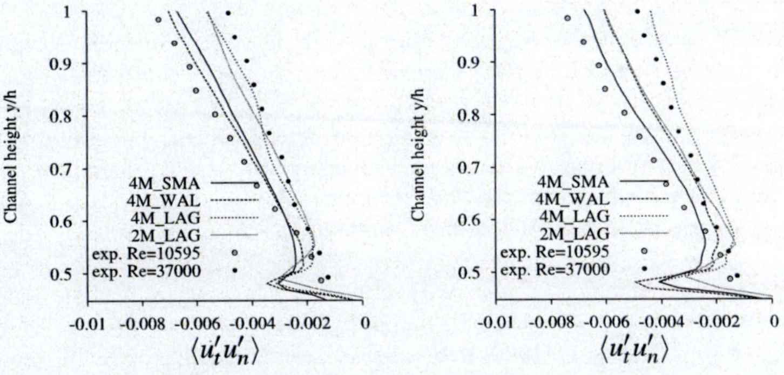

Fig. 8. Comparison of Reynolds shear stresses from MGLET at $x/h = 8.0$ in a coordinate system aligned with the wall. Experiment by Rapp [12]. $Re = 10,595$ (left) and $Re = 37,000$ (right).

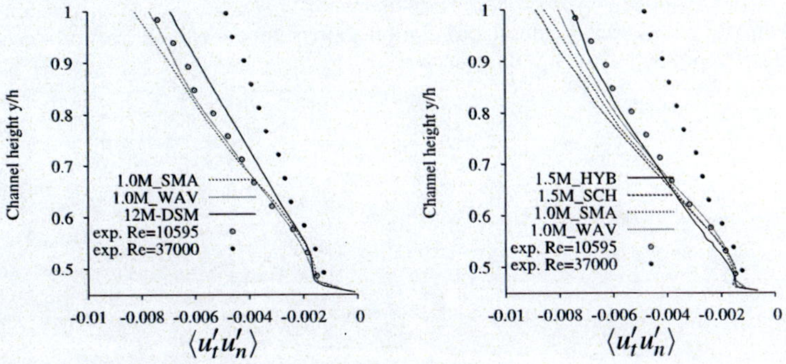

Fig. 9. Comparison of Reynolds shear stresses from LESOCC and LESOCC2 at $x/h = 8.0$ in a coordinate system aligned with the wall. Experiment by Rapp [12]. $Re = 10,595$ (left) and $Re = 37,000$ (right).

References

1. Almeida, G.P., Heitor, M.V. (1993): Wake flows behind two-dimensional model hills, *Expl. Thermal Fluid Sci.* 7, pp. 87-101.
2. Breuer, M. (1998) Large-eddy simulation of the sub-critical flow past a circular cylinder: numerical and modeling aspects *Int. J. Numer. Methods Fluids* 28, pp. 1281-1302
3. Breuer, M., Peller, N., Rapp, Ch., Manhart, M. (2009) Flow over periodic hills - numerical and experimental study over a wide range of Reynolds numbers. *Computers and Fluids* 38, pp. 433-457
4. Denev, J., Falconi, C., Fröhlich, J. and Bockhorn, H. Wavelet-adapted subgrid-scale models for LES *Second Int. Conf. on Turbulence and Interaction*, May 31 - June 5, 2009, Sainte-Luce, Martinique
5. Fröhlich, J., Mellen, C.P., Rodi, W., Temmerman, L., Leschziner, M.A. (2005): *Highly resolved large-eddy simulation of separated flow in a channel with streamwise periodic constrictions*, *J. Fluid Mech.* 526, pp. 19-66.

6. Jaffrézic B, and Breuer, M. (2008) Application of an explicit algebraic Reynolds stress model within a hybrid LES-RANS method *Flow, Turbulence and Combustion* 81, pp. 415-448
7. Hinterberger, C. (2004) Dreidimensionale und tiefengemittelte Large-Eddy-Simulation von Flachwasserströmungen *PhD thesis*, University of Karlsruhe, Germany
8. Mellen, C.P, Fröhlich, J., Rodi, W. (2000): Large-eddy simulation of the flow over periodic hills, Proc. of 16th IMACS World Congress, Lausanne, Switzerland, (eds.) Deville, M., Owens, R.
9. Meneveau, C., Lund, T.S., Cabot, W.H. (1996) A Langrangian dynamic subgrid-scale model of turbulence. *J. Fluid Mech.* 319, pp. 353-385.
10. Nicoud, F. and Ducros, F. (1999) Subgrid-scale stress modelling based on the square of the velocity gradient tensor, *Flow, Turb. and Comb.* 62, pp. 183-200.
11. Peller, N., Le Duc, A., Tremblay, F., Manhart, M. (2006) High-order stable interpolations for immersed boundary methods. *Int. J. Numer. Methods Fluids* 52, pp. 1175-1193
12. Rapp, C. (2009) Experimentelle Studie der turbulenten Strömung über periodische Hügel, *PhD thesis*, Technische Universität München, Germany
13. Smagorinsky, J. (1963) General circulation experiments with the primitive equations. *Monthly Weather Rev.* 91, pp. 99-164.

From suitable weak solutions to entropy viscosity

Jean-Luc Guermond[1]*, Richard Pasquetti[2], Bojan Popov[1]

[1] Department of Mathematics, Texas A&M University, College Station, Texas 77843, USA
guermond@math.tamu.edu

[2] Lab. J.A. Dieudonné, UMR CNRS 6621, Université de Nice-Sophia Antipolis, Nice, France Richard.Pasquetti@unice.fr

Summary. This paper focuses on the notion of suitable weak solutions for the three-dimensional incompressible Navier–Stokes equations and discusses the relevance of this notion to Computational Fluid Dynamics. The purpose of the paper is twofold (i) to recall basic mathematical properties of the three-dimensional incompressible Navier-Stokes equations and to show how they relate to LES (ii) to introduce an entropy viscosity technique based on the notion of suitable weak solution and to illustrate numerically this concept.

Key words: Quality, Reliability, Large-Eddy Simulation, Suitable weak solutions, Entropy viscosity

1 Introduction

The question addressed in this paper is that of constructing approximate solutions to the three-dimensional incompressible Navier-Stokes equations using under-resolved meshes. The use of under-resolved meshes cannot be avoided when the Reynolds number is large, which is very often the case in engineering situations. At the present time, simulating time-dependent flows at Reynolds numbers greater than a few thousands is a challenging task due to the heuristic Kolmogorov estimate $\mathcal{O}(R_e^{9/4})$ for the total number of degrees of freedom which is required to simulate flows at a given value of R_e.

In the wake of [6, 12, 9], the objective of this paper is to show that the notion of suitable weak solutions introduced by Scheffer [22] is a sound, firm, mathematical ground which could be useful to LES modelers to build energetically coherent theories.

* This material is based upon work supported by the National Science Foundation grants DMS-07138229 and DMS-0811041 and partially supported by Award No. KUS-C1-016-04, made by King Abdullah University of Science and Technology (KAUST). This work was also supported by Lawrence Livermore National Security, LLC, under Task Order B575366 and Master Task Agreement B575363

M.V. Salvetti et al. (eds.), *Quality and Reliability of Large-Eddy Simulations II*, ERCOFTAC Series 16, DOI 10.1007/978-94-007-0231-8_34,

This paper is organized in four parts. In the first part (Section 2) we recall basic mathematical properties of suitable weak solutions and we mention some new results regarding the approximation of these solutions. We want to draw the attention of the community on the fact that, contrary to finite elements/volumes/differences and wavelets, at the time of this writing it is still unknown whether Fourier-based DNS in the periodic cube produce weak solutions that are suitable. In other words there is still no mathematical proof that Fourier-based DNS is energetically consistent. Considering the importance of Fourier-based DNS in CFD, this theoretical gap is worrisome.

In the second part of this paper (Section 3) we propose a model that aims at controlling the energy balance at the grid scale in a way which is consistent with the notion of suitable solutions, i.e., suitable solution are locally dissipative. The key idea consists of adding a numerical viscosity proportional to the default to equilibrium in the local energy equation. This model henceforth referred to as LES or entropy viscosity has been proposed in [10, 9].

In the third part (Section 4) we test the entropy viscosity technique on some scalar nonlinear conservations laws using various discrete settings comprising Fourier expansions and finite elements.

The entropy viscosity technique is adapted to the compressible Euler equations in the fourth part of this paper (Section 5).

The test reported in Section 4 and Section 5 should convince the reader that computing a numerical viscosity proportional to the entropy residual is a very efficient stabilization technique. We conjecture that this method should be a good candidate for LES and could provide a reasonable mathematical background for LES.

2 Suitable weak solutions

2.1 The Navier-Stokes problem

Let Ω be a connected, open, bounded domain in $\mathbb{R}^3$ and consider the time-dependent incompressible Navier–Stokes equations in Ω

$$\begin{cases} \partial_t \mathbf{u} + \mathbf{u}\cdot\nabla\mathbf{u} + \nabla p - R_e^{-1}\Delta\mathbf{u} = \mathbf{f} \quad \text{in } Q_T, \\ \nabla\cdot\mathbf{u} = 0 \quad \text{in } Q_T, \\ \mathbf{u}|_{t=0} = \mathbf{u}_0, \quad \mathbf{u} \text{ periodic or } \mathbf{u}|_\Gamma = 0, \end{cases} \tag{1}$$

where $Q_T = \Omega \times (0,T)$ is the space-time domain, Γ is the boundary of Ω, and R_e is the Reynolds number.

Spaces of $\mathbb{R}^3$-valued functions on Ω and $\mathbb{R}^3$-valued functions are denoted in bold fonts. The Euclidean norm in $\mathbb{R}^3$ is denoted $|\cdot|$. In the following c is a generic constant which may depend on the data $\mathbf{f}$, $\mathbf{u}_0$, R_e, Ω, T. The value of c may vary at each occurrence. Whenever E is a normed space, $\|\cdot\|_E$ denotes a norm in E.

2.2 Suitable weak solutions

It is known since Leray [18] and Hopf [13] that weak solutions to (1) exist, but the question of uniqueness of these solutions is still open. The major obstacle in the way is that the a priori energy estimates obtained so far do not preclude the occurrence of so-called vorticity bursts reaching scales smaller than the Kolmogorov scale.

Uniqueness is intimately related to smoothness. A very interesting approached to the smoothness question has been developed by Scheffer [22]. The idea is to study the Hausdorff measure of the singular set of weak solutions (the singular set is composed of those points in time and space where the solution is not essentially bounded in any neighborhood of these points). Proving that the measure of the singular set is zero would amount to proving that there is no singularity. To carry out this program Scheffer introduced the notion of suitable weak solutions which boils down to the following

Definiton 1 *Let* $(\mathbf{u},p)$, $\mathbf{u} \in L^2((0,T);\mathbf{H}^1(\Omega)) \cap L^\infty((0,T);\mathbf{L}^2(\Omega))$, $p \in \mathscr{D}'((0,T);L^2(\Omega))$, *be a weak solution to the Navier-Stokes equation (1). The pair* $(\mathbf{u},p)$ *is said to be suitable if the local energy balance*

$$\partial_t(\tfrac{1}{2}\mathbf{u}^2) + \nabla\cdot((\tfrac{1}{2}\mathbf{u}^2 + p)\mathbf{u}) - R_e^{-1}\Delta(\tfrac{1}{2}\mathbf{u}^2) + R_e^{-1}(\nabla\mathbf{u})^2 - \mathbf{f}\cdot\mathbf{u} \leq 0 \tag{2}$$

is satisfied in the distributional sense, i.e., in $\mathscr{D}'(Q_T;\mathbb{R}^+)$.

It is remarkable that the above inequality is similar to entropy conditions for conservation laws. Think of it as an entropy inequality where the kinetic energy would play the role of an entropy.

Suitable weak solutions are known to exist always. They can be constructed by regularizing the nonlinear term (i.e., Leray regularization) and passing to the limit. With this notion Scheffer was able to derive a bound from above of some Hausdorff measure of the singular set. The remarkable fact about this result is that it cannot (yet) be obtained without invoking suitability, i.e., it is not known if every weak solution satisfies (2). The result of Scheffer has been improved by Caffarelli-Kohn-Nirenberg and is now referred as the Caffarelli-Kohn-Nirenberg Theorem [3, 20] in the literature. In a nutshell, this result asserts that the one-dimensional Hausdorff measure of the set of singularities of a suitable weak solution is zero. In other words, if singularities exist, they must lie on a space-time set whose dimension is smaller than that of a space-time line. To the present time, this is the best partial regularity result available for the Navier–Stokes equations. For any practical purpose, this theorem asserts that suitable weak solutions are almost classical. The word "almost" is important here; although suitable weak solutions are the most regular solutions known to exist, they may still have singular points, i.e., be not classical.

2.3 Direct Numerical Simulations (DNS)

Since DNS is the highest court in LES-land, one is certainly entitled to ask whether limits of DNS solutions (as the mesh-size go to zero) are suitable. This may seem to be a dumb question to ask, but surprisingly the answer is not yet totally clear.

To formalize an answer to the above question let us assume that the velocity and pressure are approximated in some discrete spaces $\mathbf{X}_h$ and M_h, respectively (h denoting the mesh-size). We now introduce a notion of discrete commutator.

Definiton 2 *The space $\mathbf{X}_h$ (resp. M_h) is said to have the discrete commutator property if there is an operator $P_h \in \mathscr{L}(\mathbf{H}^1_0(\Omega);\mathbf{X}_h)$ (resp. $Q_h \in \mathscr{L}(H^1(\Omega);M_h)$) such that the following holds for all ϕ in $W^{2,\infty}_0(\Omega)$ and all $v_h \in \mathbf{X}_h$ (resp. all $q_h \in M_h$)*

$$\|\phi v_h - P_h(\phi v_h)\|_{\mathbf{H}^l(\Omega)} \leq c h^{1+m-l} \|v_h\|_{\mathbf{H}^m(\Omega)} \|\phi\|_{W^{m+1,\infty}(\Omega)}, \qquad 0 \leq l \leq m \leq 1$$
$$\|\phi q_h - Q_h(\phi q_h)\|_{H^l(\Omega)} \leq c h^{1+m-l} \|q_h\|_{H^m(\Omega)} \|\phi\|_{W^{m+1,\infty}(\Omega)}.$$

When P_h (resp. Q_h) is a projector, the above definition is an estimate of the operator norm of the commutator $[\Phi, P_h] := \Phi \circ P_h - P_h \circ \Phi$ where $\Phi(v) := \phi v$. This property is also called 'super-approximation' in the finite element literature [12, 14, 15]. The discrete commutator property is known to hold in discrete spaces where there exist projectors that have local approximation properties, see Bertoluzza [2]. It is known to hold for finite elements and wavelets.

The best available results so far concerning the construction of suitable weak solution as limits of DNS solution is summarized in the following

Theorem 1 ([7, 8]) *If $\mathbf{X}_h$ and M_h have the discrete commutator property (in addition to having the usual reasonable approximation properties), the pair (u_h, p_h) convergences, up to subsequences, to a suitable weak solution to (1).*

In a nutshell this theorem says that DNS solutions based on finite elements and wavelets (and very likely finite volumes and finite differences, splines, etc.) converge to suitable weak solutions.

One may then wonder if DNS solutions based on Fourier approximation in the periodic cube also converge to suitable solutions. Surprisingly enough, this is still unknown at the time of this writing. The main obstacle in the way is that the discrete commutator property does not hold for Fourier approximations. Actually, counter-examples to the discrete commutator property can be constructed. We are then led to seriously consider the following

Open question 1 *Do Fourier-based DNS solutions converge to suitable weak solution as the degree of the approximation goes to infinity?*

I think this question should equally attract the interest of mathematicians and DNS specialists. For mathematicians, investigating this problem might be a way to set a wedge that could separate the class of suitable weak solutions from that of those that are weak only. For CFD specialists, it would certainly be re-assuring to know that Fourier-based DNS solutions locally dissipate energy correctly.

One way to interpret the above results is that Finite Elements, Wavelets, Finite Differences, etc. have enough built-in numerical dissipation to help the energy cascade to go in the right direction, i.e., the energy at extremely fine scales is always dissipated when using approximation methods having local interpolation properties. Contrary to Finite Elements, wavelets, etc. the Fourier technique is so accurate that it does not induce enough numerical diffusion to counteract the Gibbs-Wilbraham phenomenon. Whether energy is correctly dissipated locally for Fourier approximations is still a mystery. The key here is the lack of localization.

2.4 More open questions for DNS

Let us finish this section by bringing the following open question to the attention of the DNS community

Open question 2 *Do weak solutions to the Navier-Stokes equation satisfy the global energy balance?*

Again, this question may seem ridiculous, but global energy balance has not yet been proved for large data (initial or forcing term). The DNS community could contribute to solving this question by verifying the global energy balance of DNS solutions as the mesh size is refined.

Of course, calling $\mathbf{u}_h$ the DNS solution, it is clear that any DNS algorithm can be tweaked so that global energy balance is exact

$$\frac{1}{2}\|\mathbf{u}_h(T)\|^2_{\mathbf{L}^2(\Omega)} + \int_0^T R_e^{-1}\|\nabla\mathbf{u}_h(t)\|_{\mathbf{L}^2(\Omega)}\,\mathrm{d}t = \frac{1}{2}\|\mathbf{u}_0\|^2_{\mathbf{L}^2(\Omega)}. \tag{3}$$

This is not the point. Unless it can be proved someday that $\mathbf{u}_h$ converges strongly to some $\mathbf{u}$ in $L^2((0,T);\mathbf{H}^1(\Omega))$, there is no reason for the limit solution, say $\mathbf{u}$, to satisfy the equality in (3). The best that can be deduced is the inequality

$$\frac{1}{2}\|\mathbf{u}(T)\|^2_{\mathbf{L}^2(\Omega)} + \int_0^T R_e^{-1}\|\nabla\mathbf{u}(t)\|_{\mathbf{L}^2(\Omega)}\,dt \le \frac{1}{2}\|\mathbf{u}_0\|^2_{\mathbf{L}^2(\Omega)}. \tag{4}$$

More precisely, although $\nabla\mathbf{u}_h$ converges weakly to $\nabla\mathbf{u}$ in $\mathbf{L}^2(\Omega)$, it is possible that $\lim_{h\to 0}\|\nabla\mathbf{u}_h\|_{\mathbf{L}^2(\Omega)} \not\to \|\nabla\mathbf{u}\|_{\mathbf{L}^2(\Omega)}$. To understand the difficulty, set $h = 1/N$ and consider of the function

$$v_h(x) = \sin(2\pi x/h), \qquad x \in (0,1). \tag{5}$$

Clearly $v_h \to 0$ weakly in $L^2(0,1)$ as $h \to 0$, but $\|v_h\|_{L^2(0,1)} = 1$ and $\|\lim_{h\to 0} v_h\|_{L^2(0,1)} = 0$. In other words $\lim_{h\to 0}\|v_h\|_{L^2(0,1)} \neq \|\lim_{h\to 0} v_h\|_{L^2(0,1)}$. It is remarkable though that any amount of smoothing is sufficient to transform weak convergence into strong convergence. Hence, extending v_h by zero over $\mathbb{R}$ it can be shown that $\lim_{h\to 0}\|\varphi_\varepsilon * v_h\|_{L^2(0,1)} = \|\lim_{h\to 0}\varphi_\varepsilon * v_h\|_{L^2(0,1)}$ for any reasonable smoothing kernel φ and any $\varepsilon > 0$.

Upon setting $\overline{\mathbf{u}}_h(t) := \varphi_\varepsilon * \mathbf{u}_h(t)$, DNS simulations could help solve the above question by verifying whether the following holds for all times T

$$\frac{1}{2}\|\overline{\mathbf{u}}_h(T)\|^2_{\mathbf{L}^2(\Omega)} + \int_0^T R_e^{-1}\|\nabla\overline{\mathbf{u}}_h(t)\|_{\mathbf{L}^2(\Omega)}\,\mathrm{d}t \approx \frac{1}{2}\|\mathbf{u}_0\|^2_{\mathbf{L}^2(\Omega)} \tag{6}$$

for small resolutions (i.e., $h \to 0$) and any $\varepsilon \gg h$, say $\varepsilon = h^\alpha$ with $\alpha \in (0,1)$.

3 Proposal for a LES model based on suitability

The goal of this section is to explore some implications the notion of suitable solutions may have when it comes to approximate the Navier-Stokes equations on a finite grid. In other words, since $\lim_{h\to 0}$ is a mathematical dream which is unachievable with the computing power currently available, can we anyway draw something useful from the existence of suitable solutions?

3.1 Practical interpretation of the notion of suitable solution

At high Reynolds numbers CFD is always under-resolved. In other words, even if one uses a discrete setting admitting a discrete commutator property, the results of Theorem 1 is useless for practical purposes since the approximate solution thus calculated may be far from a (the?) suitable solution. The limit $h \to 0$ is an ideal situation from which practical CFD simulations are usually far. Then, one may ask oneself what is the use of the notion of suitable solutions? Is it a notion that we should care about in CFD?

To answer the above question, let us rephrase the definition of suitability. Let $\mathbf{u}$, p be a weak solution of the Navier-Stokes equations in the Leray class. Let us define the residual of the momentum equation

$$\mathbf{R}(x,t) := \partial_t \mathbf{u} - R_e^{-1}\Delta\mathbf{u} + \mathbf{u}\cdot\nabla\mathbf{u} + \nabla p - \mathbf{f}. \tag{7}$$

$\mathbf{u}$, p being a weak solution means that the residual $\mathbf{R}(x,t)$ is zero in the distribution sense. Is it then clear that the power of the residual, $\mathbf{R}(x,t)\cdot\mathbf{u}$, is zero? Well, no, since it is not known whether $\mathbf{u}$ is smooth enough to be tested against $\mathbf{R}(x,t)$, i.e., it is not clear whether the integral $\int_0^T\int_\Omega \mathbf{R}(x,t)\cdot\mathbf{u}(\mathbf{x},t)\,\mathrm{d}\mathbf{x}\,\mathrm{d}t$ makes sense.

Consider the one-dimensional inviscid Burgers equation $\partial_t u + \frac{1}{2}\partial_x(u^2) = 0$ for instance. In the distribution sense $0 = R(x,t) := \partial_t u + \frac{1}{2}\partial_x(u^2)$, but the unique entropy equation is, among all the weak solutions, the only one that satisfies $R(x,t)u = \frac{1}{2}\partial_t u^2 + \frac{1}{3}\partial_x(u^3) \le 0$. It is indeed true that the product $R(x,t)u = 0$ at points (x,t) where u is smooth, but in shocks $R(x,t)u$ is a negative Dirac measure. More precisely consider the following solution $u(x,t) = 1 - H(x - \frac{1}{2}t)$ where H is the Heaviside function, $x \in (-\infty,+\infty)$ and $t \ge 0$. One easily verifies that u solves the Burgers equations with initial data $u_0(x) = 1 - H(x)$, i.e., $R(x,t) = 0$, but $\frac{1}{2}\partial_t u^2 + \frac{1}{3}\partial_x(u^3) = -\frac{1}{12}\delta(x - \frac{1}{2}t) \neq 0$, where δ is the Dirac measure. This example may help the reader to understand why the open question (2) is still open and might not have an obvious answer. When approximating the solutions of the Burgers equation, one can certainly come up with algorithms that are energy preserving (i.e., $\int_{-\infty}^{\infty} \frac{1}{2}u_h^2(x,t)\,\mathrm{d}x = \int_{-\infty}^{\infty} \frac{1}{2}u_0^2(x)\,\mathrm{d}x$). Using such a technique would be a disastrous idea, since the energy preserving solution to the Burgers equation is not the correct one.

The definition of a suitable solution can be rephrased as follows: A suitable solution is one for which the power of the residual is negative in the distribution sense in Q_{T^*}, i.e.,

$$\partial_t(\tfrac{1}{2}\mathbf{u}^2) + \nabla\cdot((\tfrac{1}{2}\mathbf{u}^2 + p)\mathbf{u}) - R_e^{-1}\Delta(\tfrac{1}{2}\mathbf{u}^2) + R_e^{-1}(\nabla\mathbf{u})^2 - \mathbf{f}\cdot\mathbf{u} \le 0, \tag{8}$$

The reader may verify by himself that indeed, (8) is formally equivalent to $\mathbf{R}(x,t)\cdot\mathbf{u} \leq 0$, (the term "formally" meaning: in the optimistic hypothesis that $\mathbf{u}$ and p are smooth functions). In other words, if singularities occur, suitable solutions are such that these singularities dissipate energy.

3.2 What happens in under-resolved simulations?

Let us now focus our attention on under-resolved numerical simulations. Being under-resolved in a space-time region means that the numerical solution experiences large gradients that cannot be correctly represented by the mesh in the region in question. In other words, for all practical purpose, the numerical solution is singular at the considered mesh scale (i.e., behaves like a singular one on the available mesh). As time progresses the large unresolved gradients are likely to produce even larger gradients through nonlinear interactions, i.e., we have to deal with energy accumulation at the grid scale. The question is non-longer to determine whether the solution(s) to the Navier-Stokes equation is (are) classical or not (a debate that a pragmatic reader may think to be of remote academic interest), it just now amounts to deciding what to do with a quasi-singular numerical solution.

Let us rephrase the situation in mathematical terms. Let $(\mathbf{u}_h, p_h)$ be the approximate velocity and the approximate pressure, the subscript h representing the typical mesh-size. Let $D_h(x,t)$ be the numerical residual of the energy (entropy) equation

$$D_h(x,t) := \partial_t(\tfrac{1}{2}\mathbf{u}_h^2) + \nabla\cdot((\tfrac{1}{2}\mathbf{u}_h^2 + p_h)\mathbf{u}_h) - R_e^{-1}\Delta(\tfrac{1}{2}\mathbf{u}_h^2) + R_e^{-1}(\nabla\mathbf{u}_h)^2 - \mathbf{f}\cdot\mathbf{u}_h. \quad (9)$$

Being under-resolved in a neighborhood of (x_0,t_0) means that $D_h(x_0,t_0)$ is significantly larger than the consistency error of the method. If locally the power of the numerical singularity is negative, i.e., $D_h(x_0,t_0) \leq 0$, we do not have anything to fear since energy is cascading down and is eventually lost in the subgrid scales, a scenario in agreement with the Kolmogorov cascade. On the other hand if the numerical singularity produces energy, i.e., $D_h(x_0,t_0) > 0$, all the bets are off since the situation is out of control and, by analogy with a shock that would produce energy, is unphysical.

In conclusion ensuring that $D_h(x_0,t_0) \leq 0$ is a highly desirable feature. If it could be enforced everywhere in the domain, that would mean that the energy gently cascades down in the subgrid scales and is eventually dissipated. Rephrased in eddy terms, this condition would guaranty that every eddy of size similar to the mesh-size would eventually be dissipated. Hence in under-resolved situations, one should wish the approximate solution to be suitable in the discrete sense, i.e.,

$$D_h(x,t) \leq 0, \qquad \forall(x,t) \in Q_T. \quad (10)$$

The above discussion leads us to propose the following tentative definition for LES

Definiton 3 *A LES solution is an approximation of the Navier-Stokes equation that satisfies* (10).

3.3 A LES model based on suitability

Of course (10) cannot be enforced in addition to the discrete momentum conservation and the discrete mass conservation. But, similarly to the entropy condition for nonlinear conservation laws, (10) can be incorporated in the algorithm that calculates the pair $(\mathbf{u}_h, p_h)$.

Possibilities are numerous, but the technique that we propose is to use (10) to construct an artificial viscosity by setting

$$\nu_h(x,t) := \min\left(c_{\max}|\mathbf{u}_h(x,t)|h(x), c h^2(x)\frac{|D_h(x,t))|}{\|\mathbf{u}_h^2\|_{\mathbf{L}^\infty(\Omega)}}\right), \tag{11}$$

where $h(x)$ is the local mesh size in the neighborhood of x, $\|\mathbf{u}_h^2\|_{\mathbf{L}^\infty(\Omega)}$ is the maximum norm of $\mathbf{u}_h^2$ (this is just a normalizing term), $c_{\max} \approx 0.5$ and c are adjustable constants. The momentum equation is then be modified by adding the term $-\nabla\cdot(\nu_h(x,t)\nabla\mathbf{u}_h)$. The quantity $c_{\max}|\mathbf{u}_h(x,t)|h(x)$ is the viscosity that would be induced by first-order up-winding on a uniform Cartesian grid. Definition (11) implies that the LES viscosity never exceeds the first-order up-wind viscosity. When the mesh is fine enough to resolve all the scales, the quantity $|D_h(x,t))|$ is of the same order as the consistency error of the numerical method which is used and $h(x)^2|D_h(x,t))|$ is far smaller than the first-order up-wind viscosity. This observation implies that $\nu_h(x,t)$ is a consistent viscosity, i.e., it vanishes when all the scales are resolved. The LES viscosity is active only in under-resolved region if spurious energy is generated at the mesh scale, i.e., when energy seems to be coming up from subgrid scales. Note that the LES viscosity as defined in (11) may be oscillatory, so that a smoothing (local averaging) may be required.

4 Numerical illustrations for scalar conservation laws

Our goal in this section is to describe how the LES viscosity model proposed above can be used to solve nonlinear scalar conservations laws. We change the terminology by renaming the LES viscosity an entropy viscosity.

4.1 Scalar conservation equations

We consider the equation

$$\partial_t u + \nabla\cdot\mathbf{f}(u) = 0 \tag{12}$$

subject to the initial condition $u|_{t=0} = u_0$ and the appropriate boundary conditions. In some cases we will solve the Cauchy problem (restricted to a bounded domain) and in other cases we will specify the corresponding boundary conditions. It is well known that the Cauchy or the initial boundary value problem has a unique entropy solution (see [16, 1]) which satisfies an additional set of differential inequalities

$$\partial_t E(u) + \nabla\cdot\mathbf{F}(u) \le 0 \tag{13}$$

for any pair of functions $E(u)$ and $\mathbf{F}(u)$ such that E is convex and $\mathbf{F}(u) = \int E'(u)\mathbf{f}'(u)\,du$. The function E is called entropy and $\mathbf{F}$ is the associated entropy flux. The most well known pairs are the Kružkov's pairs generated by $E_c(u) = |u-c|$, where c is any arbitrary constant. It is known that $E(u) = \frac{1}{2}u^2$ is enough to select the unique entropy solution when $\mathbf{f}$ is convex.

4.2 The algorithm

Assume that we have at hand a finite element mesh $\mathscr{T}_h$ and that the local approximation is done using polynomials of degree at most k. We first assume that the time is continuous. The entropy viscosity method proceeds as follows:

- Compute the entropy residual, $D_h(u) := \partial_t E(u) + \nabla\cdot\mathbf{F}(u)$
- For each cell $K \in \mathscr{T}_h$ compute the local mesh size: $h_K = \mathrm{diam}(K)/k$.
- Let $\overline{E}(u_h)$ be the average entropy over the domain. On each cell, construct a viscosity associated with the entropy residual:

$$\nu_E := c_E\, h_K^2 \|D_h\|_{L^\infty(K)} / \|E(u_h) - \overline{E}(u_h)\|_{L^\infty(\Omega)}$$

- On each cell, compute an upper bound of the viscosity based on the maximum local wave speed: $\beta_K = \||\mathbf{f}'(u)|\|_{\infty,K}$, where $|\cdot|$ is the Euclidean norm:

$$\nu_{\max} := c_{\max} h_K \beta_K$$

- Define the entropy viscosity on each mesh cell K:

$$\nu_h := \min(\nu_{\max}, \nu_E) \tag{14}$$

  If required, one may smooth the thus obtained entropy viscosity. This is especially useful when high order approximations are concerned, i.e. if $k > 2$ (see the companion paper [11]).
- Solution method: Galerkin + entropy viscosity:

$$\int_\Omega (\partial_t u_h + \nabla\cdot\mathbf{f}(u_h) v_h \,d\mathbf{x} + \sum_K \int_K \nu_h \nabla u_h \cdot \nabla v_h \,d\mathbf{x} = 0, \quad \forall v_h$$

The time marching can be done with an explicit Runge-Kutta method (RK3 or RK4). Denoting u_h^n, u_h^{n-1} and u_h^{n-2}, the approximations of u at time t_n, t_{n-1} and t_{n-2}, one simple possibility to evaluate the entropy residual consists of setting

$$D_h = \frac{1}{2\Delta t}(3E(u_h^n) - 4E(u_h^{n-1}) + E(u_h^{n-2})) + \nabla\cdot\mathbf{F}(u_h^n), \tag{15}$$

which is formally second-order accurate in time.

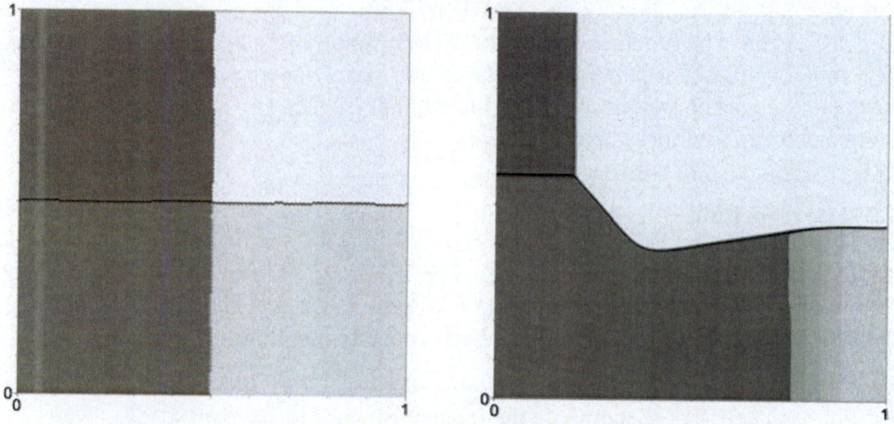

Fig. 1. Burgers. Initial data (left); $\mathbb{P}_1$ approximation at $t = 0.5$, $3\,10^4$ nodes (right)

4.3 Inviscid Burgers equation

To numerically illustrate the above algorithm we first consider the inviscid Burgers equation in two space dimensions with $\mathbf{f}(u) = \frac{1}{2}(u^2, u^2)$ and $E(u) = \frac{1}{2}u^2$. The initial data is piecewise constant in the four quadrants of $\mathbb{R}^2 + (\frac{1}{2}, \frac{1}{2})$; $u_0 = -1, -0.2, 0.5, 0.8$ in the top right, top left, bottom left, and bottom right quadrant, respectively. The field u_0 is shown in the left panel of Figure 1. The solution computed with $\mathbb{P}_1$ elements ($3\,10^4$ $\mathbb{P}_1$ nodes) is shown in the right panel of Figure 1.

We perform convergence tests in the square $(0,1)^2$. The tests are done with $\mathbb{P}_1$ and $\mathbb{P}_2$ Lagrange finite elements on unstructured Delaunay meshes. The solution is computed at $t = 0.5$ and the error is measured in the L^1- and L^2-norm. We report in Table 1 the results obtained with $\mathbb{P}_1$ and $\mathbb{P}_2$ Lagrange finite elements on various meshes. The quantity h refers to the typical mesh-size for each mesh. The time stepping is done using the SSP RK3 method, see [5]. The entropy residual is computed with the explicit BDF2 formula (15) based on the three previous time levels. The coefficients $c_{\max}$ and c in (14) are $c_{\max} = 0.4/k$ and $c_E = 1$. As expected, the convergence rates in the L^1- and L^2-norms are close to the theoretical $\mathscr{O}(h)$ and $\mathscr{O}(h^{\frac{1}{2}})$ orders, respectively. These orders are optimal since the solution is in BV only.

We now redo the above convergence tests with Fourier expansions, see [11] for details on the algorithm. The results are reported in Table 2. The computation is done by extending the computational domain by symmetry about the axes $\{x = 1\}$ and $\{y = 1\}$ and the initial data is extended so as to make the extension periodic. The time marching is done by using the standard Runge-Kutta scheme (RK4). The nonlinearity is de-aliased using the $\frac{3}{2}$-padding rule. The entropy viscosity is made explicit and computed by using the BDF2 formula (15) based on the three previous time levels. The entropy viscosity is computed in the physical space at the Fourier nodes. The coefficients $c_{\max}$ and c in (14) are $c_{\max} = 0.5$ and $c_E = 8$. Here again we observe quasi-optimal convergence rates both in L^1- and L^2-norms.

h	$\mathbb{P}_1$				$\mathbb{P}_2$			
	L^2	rate	L^1	rate	L^2	rate	L^1	rate
5.00E-2	2.3651E-1	–	9.3661E-2	–	1.8068E-1	–	5.2531E-2	–
2.50E-2	1.7653E-1	0.422	4.9934E-2	0.907	1.2956E-1	0.480	2.7212E-2	0.949
1.25E-2	1.2788E-1	0.465	2.5990E-2	0.942	9.5508E-2	0.440	1.4588E-2	0.899
6.25E-3	9.3631E-2	0.449	1.3583E-2	0.936	6.8806E-2	0.473	7.6435E-3	0.932
3.12E-3	6.7498E-2	0.472	6.9797E-3	0.961	–	–	–	–

Table 1. Convergence tests for the inviscid Burgers equation. $\mathbb{P}_1$ approximation (left), $\mathbb{P}_2$ approximation (right).

N	h=1/N	L^1	rate	L^2	rate
36	2.78E-2	1.92E-2	–	1.02E-1	–
72	1.39E-2	9.99E-3	0.94	7.28E-2	0.49
144	6.94E-3	5.34E-3	0.89	5.41E-2	0.43
288	3.47E-3	2.79E-3	0.95	3.80E-2	0.51

Table 2. Convergence tests for Burgers. Fourier approximation.

4.4 KPP rotating wave

We now illustrate the capability of the entropy viscosity method to deal with non-convex fluxes by solving the two dimensional scalar conservation law

$$\partial_t u + \nabla \cdot \mathbf{f}(u) = 0, \quad \mathbf{f}(u) = (\sin u, \cos u) \tag{16}$$

subject to the following initial condition

$$u(x,y,0) = u_0(x,y) = \begin{cases} 3.5\pi, & x^2+y^2 < 1; \\ 0.25\pi, & \text{otherwise.} \end{cases} \tag{17}$$

This test proposed in [17] is a challenging exercise to many high-order numerical schemes because the exact solution has a two-dimensional composite wave structure. For example the central-upwind schemes based on WENO5, Minmod 2 and SuperBee reconstructions fail this test case, see [17] for details.

The solution is computed at $t = 1.0$ and is shown in Figure 2. The left panel shows the reference solution from [17] on a Cartesian grid 400×400, i.e., $h = 0.01$. This solution is computed using an adaptive WENO5/Minmod 1 reconstruction; the Minmod 1 reconstruction is used only in the transition zones where the flux convexity changes and WENO5 is used everywhere else. The center panel shows the $\mathbb{P}_2$ approximation using the entropy viscosity on a quasi-uniform mesh of mesh size $h = 0.0125$. The composite wave is captured well and the accuracy of the $\mathbb{P}_2$ approximation is similar to that of the reference solution. The right panel shows the ratio of the entropy viscosity to the first-order upwind viscosity, which is very small outside the shock and saturates to 1 in the shock, as expected. This panel illustrates quite well the auto-adaption property of the entropy viscosity.

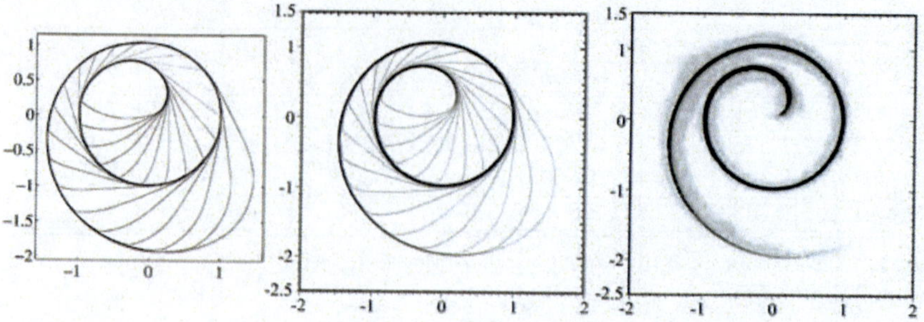

Fig. 2. KPP rotating wave. Left: adaptive WENO5/Minmod 1 from [17] on Cartesian grid $\Delta x = \Delta y = \frac{1}{100}$; Center: entropy-viscosity $\mathbb{P}_2$ approximation, $h = 0.0125$; Right: Ratio of the entropy viscosity to its maximum value.

5 Numerical illustration for the Euler equations

We extend in this section the entropy viscosity method to the compressible gas dynamics for perfect gases.

5.1 The Euler equations

We consider the Euler equations for a perfect gas. These equations state the conservation of mass, momentum and energy (see, e.g. [19] for an overview) and can be put into the following conservative form:

$$\partial_t \mathbf{c} + \nabla \cdot (\mathbf{f}(\mathbf{c})) = 0, \quad \mathbf{c} = \begin{pmatrix} \rho \\ \mathbf{m} \\ E \end{pmatrix}, \quad \mathbf{f}(\mathbf{c}) = \begin{pmatrix} \mathbf{m} \\ \mathbf{m} \otimes \frac{\mathbf{m}}{\rho} + p\mathbb{I} \\ \frac{\mathbf{m}}{\rho}(E+p) \end{pmatrix} \tag{18}$$

where the independent variables are the density ρ, the momentum vector field $\mathbf{m}$ and the total energy E. The velocity vector field $\mathbf{u}$ is defined by $\mathbf{u} := \mathbf{m}/\rho$. The symbol $\mathbb{I}$ denotes the identity matrix in $\mathbb{R}^d$. The pressure is expressed via the equation of state of ideal gases:

$$p = \rho T, \quad \text{with} \quad T = (\gamma - 1)\left(\frac{E}{\rho} - \frac{1}{2}\mathbf{u}^2\right) \tag{19}$$

where γ is the adiabatic constant and T is the temperature. We also introduce the entropy functional

$$S(p,\rho) = \frac{\rho}{\gamma - 1}\log(p/\rho^\gamma). \tag{20}$$

This quantity satisfies the following inequality

$$\partial_t S + \nabla \cdot (\mathbf{u}S) \geq 0, \tag{21}$$

with equality if all the fields are smooth.

5.2 Description of the algorithm for finite elements

The main idea of the algorithm consists of introducing an artificial viscosity and an artificial thermal diffusivity in the spirit of the compressible Navier-Stokes equations, i.e., we augment the Euler system with additional viscous fluxes. This idea has been investigated in [10] in one space dimension using Fourier expansions for space approximation. The algorithm that we presently use is slightly simplified.

This algorithm proceeds similarly to what is described in §4.2 for scalar conservation laws. At each time step we do the following: (i) we evaluate the residual of the entropy equation, (ii) we compute the associated artificial viscosities, (iii) then we update the mass, momentum, and total energy, the fluxes being augmented with the following viscous flux

$$\mathbf{f}_{visc}(\mathbf{c}_h) = \begin{pmatrix} 0 \\ -\mu_h \nabla \mathbf{u}_h \\ -\mu_h \nabla(\frac{1}{2}\mathbf{u}_h^2) - \kappa_h \nabla T_h \end{pmatrix}. \tag{22}$$

Let us now be more specific. Let Δt be the time step and let $\mathbf{c}_h^n$, $\mathbf{c}_h^{n-1}$, etc. be the approximations of the solution at times t_n, t_{n-1}, etc. We define the physical entropy $S_h^n = \frac{\rho_h^n}{\gamma-1}\log(p_h^n/(\rho_h^n)^\gamma)$. Then the task consists of evaluating the residual for the entropy conservation equation. One possible option consists of setting,

$$D_h = \tfrac{1}{2\Delta t}(3S_h^n - 4S_h^{n-1} + S_h^{n-2}) + \nabla\cdot(\mathbf{u}_h^n S_h^n). \tag{23}$$

which is formally second-order accurate. For each mesh cell K in $\mathscr{T}_h$ we first compute the entropy viscosity associated with the residual:

$$\mu_S = c_E\, h_K^2 \|\rho_h\|_{L^\infty(K)} \|D_h\|_{L^\infty(K)}. \tag{24}$$

Note that μ_S has the dimension of a transport coefficient times a density and that no normalization is needed, since the log coming in the definition of the physical entropy is dimensionless.

For the maximum value of the viscosity we use:

$$\mu_{\max} = c_{\max} h_K^2 \|\rho_h^n((\mathbf{u}_h^n)^2 + \gamma T_h^n)^{\frac{1}{2}}\|_{L^\infty(K)} \tag{25}$$

Thus, the first-order upwind viscosity is roughly estimated to be $h_K c_{\max}\|\rho_h^n((\mathbf{u}_h^n)^2 + \gamma T_h^n)^{\frac{1}{2}}\|_{L^\infty(K)}$, where $((\mathbf{u}_h^n)^2 + \gamma T_h^n)^{\frac{1}{2}}$ is an estimation of the local wave speed. Taking $c_{\max} = \frac{1}{2}$ would roughly amount to limit the viscosity with the first-order upwind viscosity on uniform Cartesian grids.

Finally, we set:

$$\mu_h = \min(\mu_{\max}, \mu_S) \tag{26}$$

$$\kappa_h = \mathscr{P}\frac{\mu_h}{\gamma - 1}. \tag{27}$$

In the previous expressions the tunable coefficients $c_{\max}$ and c_E depend on the time-marching technique and on the space approximation method, but are independent of the time step Δt and the mesh-size h. The coefficient $\mathscr{P}$ is an artificial Prandtl number which can be chosen to of order 1.

This approach has gives satisfactory results, but still with some small spurious oscillations of the solution. We have checked that one way to improve the situation is to add an artificial diffusion term for the density. Improvements may indeed be obtained by augmenting the continuity equation with a dissipation term, which viscosity, say ν_h, is again linked to μ_h, i.e., $\nu_h = \mathscr{P}_\rho \mu_h / \|\rho_h\|_{L^\infty(K)}$.

5.3 Mach 3 step

We illustrate the algorithm described above by considering the Mach 3 flow in a wind tunnel with a forward facing step. This benchmark test has been proposed by Emery [4]. The geometry of the domain is shown in Figure 3. The initial conditions are specified in terms of the primitive variables

$$(\rho, \mathbf{u}, p)^T(x,y,0) = (1.4, (3.0, 0.0), 1.0)^T. \tag{28}$$

These initial conditions are also prescribed as inflow boundary conditions along the $\{x=0\}$ axis. The outflow boundary at $\{x=3\}$ is free. The slip condition $\mathbf{u}\cdot\mathbf{n} = 0$ is specified on the solid wall of the tunnel where $\mathbf{n}$ is the unit outward normal on $\partial\Omega$. This problem was popularized by Woodward and Colella's extensive study [23] of the performance of various numerical methods in the presence of strong shocks.

We show in Figure 3 the density field at $t = 4$ on two different meshes with $\mathbb{P}_1$ Lagrange finite elements. The results shown in the left panels have been obtained on a mesh composed of 4813 $\mathbb{P}_1$ nodes and the results shown in the right panels have been obtained on a mesh composed of 893468 $\mathbb{P}_1$ nodes. These computations have been done with $c_{\max} = 0.25$, $c_E = 1$, $\mathscr{P} = 0.1$ and $\mathscr{P}_\rho = 0.1$. The tests have been run with CFL $= 0.5$. Our solutions agree, at least in the eye-ball norm, with other reference solutions that can be found in the literature. The contact discontinuity emerging from the three-shock interaction point is present in both simulations and is captured quite accurately. A Kelvin-Helmholtz instability develops along the contact discontinuity on the refined mesh.

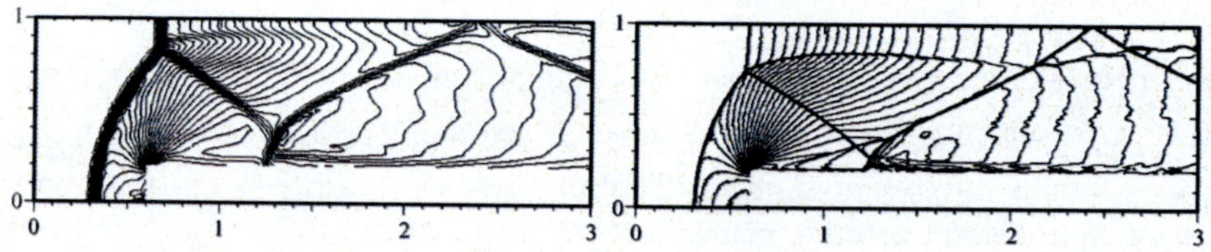

Fig. 3. Mach 3 step, density, $t = 4$, density, $\mathbb{P}_1$ approximation. Left: $h = 0.25$, 4813 $\mathbb{P}_1$ nodes. Right: $h = 0.003$, 893468 $\mathbb{P}_1$ nodes.

As reported in [23] we have observed that the way the velocity boundary condition is implemented in the vicinity of the corner of the step somewhat influences the quality of the solution. We do not enforce any boundary condition at the node at the corner of the step in the computations shown in Figure 3; enforcing the slip condition at this point implies $\mathbf{u} = 0$, which is too strong a constraint.

5.4 Double Mach reflection

We now solve the so-called double Mach reflection problem at Mach 10. This problem, popularized by Woodward and Colella (see [23] for complete description), involves a Mach 10 shock in air ($\gamma = 1.4$) that impinges a wall with a 60 degree angle. The undisturbed air ahead of the shock has density 1.4 and pressure 1. The computational domain is $\Omega = (0,4)\times(0,1)$. The reflecting wall lies at the bottom of the domain and starts at $x = \frac{1}{6}$, i.e., free slip boundary condition is enforced on $\{x \geq \frac{1}{6}, y = 0\}$. The shock makes a 60 degree angle with the x-axis. Outflow boundary conditions are enforced at $\{0 \leq x < \frac{1}{6}, y = 0\}$ and $\{x = 4\}$. The values along the top boundary $\{y = 1\}$) are set to describe the motion of the initial Mach 10 shock. The flow is computed at time $t = 0.2$

The control parameters of the entropy viscosity are $c_{\max} = 0.25$, $c_E = 0.25$, $\mathscr{P} = 0.075$ and $\mathscr{P}_\rho = 0$. The tests have been run with CFL $= 0.5$

We show in Figure 4 the solution computed with $\mathbb{P}_1$ Lagrange polynomials on a mesh composed of 453969 nodes. The left panel displays the density field in the region $0 \leq x \leq 3$. The right panel shows a close up view of the density in the region of the three-shock interaction point. To evaluate the influence of the control parameter c_E we show in the bottom panel the density field computed with $c_E = 1$. The overall features are unchanged but using $c_E = 1$ slightly smeared the roll-up of the front jet and removed small oscillations.

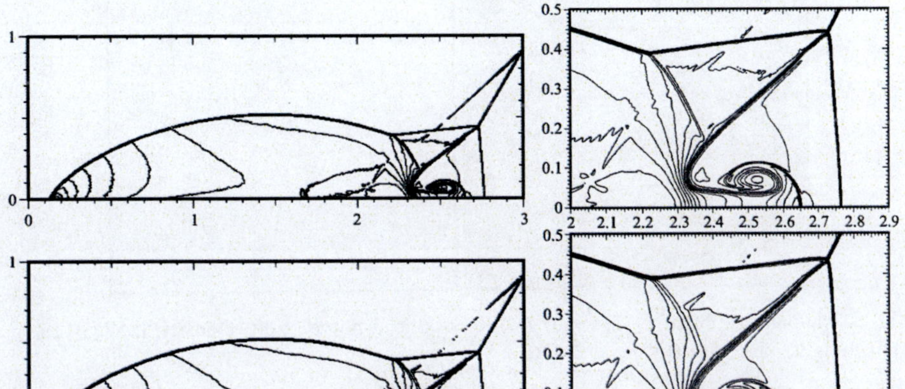

Fig. 4. Double Mach reflection at $t = 0.2$, $M = 10$, density field, 453969 $\mathbb{P}_1$ nodes. Left: global view. Right: close up view in the region of three-shock interaction point. Top: $c_E = 0.25$. Bottom: $c_E = 1$

5.5 A Riemann problem with Fourier approximation

We finish this series of tests by showing how the method performs with the Fourier approximation. The algorithm is the same as that described in Section 5.2, without using any stabilization term for the density, see [11].

The technique is validated by solving the benchmark problem number 12 from [21]. It is a two-dimensional Riemann problem set in $\mathbb{R}^2$. In the restricted computational domain $(0,1)^2$ the initial set of data is defined as follows:

$$\begin{aligned}
&p=1, && \rho=4/5, && \mathbf{u}=(0,0) && 0<x<1/2, && 0<y<1/2,\\
&p=1, && \rho=1, && \mathbf{u}=(3/\sqrt{17},0) && 0<x<1/2, && 1/2<y<1,\\
&p=1, && \rho=1, && \mathbf{u}=(0.,3/\sqrt{17}), && 1/2<x<1, && 0<y<1/2,\\
&p=2/5, && \rho=17/32, && \mathbf{u}=(0,0), && 1/2<x<1, && 0.5<y<1.
\end{aligned} \tag{29}$$

Proceeding as in Section 4.3, the problem is first made periodic by extending the computational domain to $(0,2)^2$, and the initial data are extended by symmetry about the axes $\{x=1\}$ and $\{y=1\}$. The solution is computed at time $t=0.2$.

The time marching algorithm is the same as in Section 4.3. The nonlinear terms are de-aliased. The control parameters for the entropy viscosity are $c_{\max}=\frac{1}{2}$, $c_E=20$, and $\mathscr{P}=1$.

We show in Figure 5 results obtained with 400 Fourier modes in each direction, *i.e.* with 400 grid-points in $(0,1)^2$. They compare well with those obtained with other more sophisticated shock capturing methods, see [21].

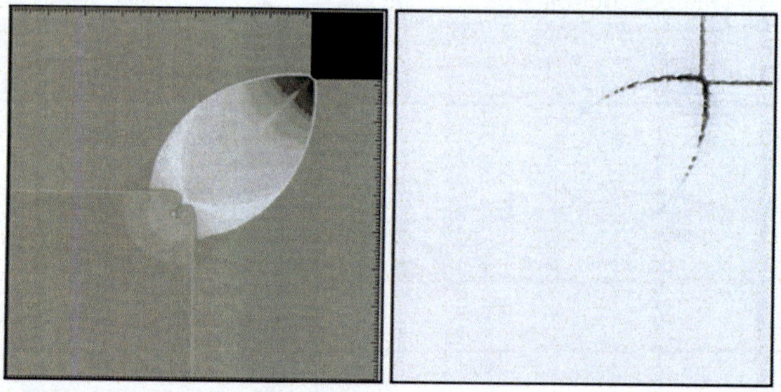

Fig. 5. Riemann problem # 12, $t=0.2$, 400×400 grid-points. Left: Density, $0.528\le\rho_h\le 1.707$. Right: Viscosity, $0\le\mu_h<3.410^{-3}$.

6 Conclusions

We have recalled basic mathematical properties of the three-dimensional incompressible Navier-Stokes equations and showed how they might relate to LES. Some fundamental questions regarding Fourier-based DNS have been raised. The notion of suitable solution lead us to introduce the new concept of entropy viscosity. The key idea consists of adding a numerical viscosity proportional to the default to equilibrium in the local energy equation. To evaluate this idea we have applied it to nonlinear conservation laws and showed that it is very efficient and very simple to implement. The entropy viscosity concept seems promising and is a potential candidate for LES.

References

1. C. Bardos, A.Y. le Roux, and J.-C. Nédélec. First order quasilinear equations with boundary conditions. *Comm. Partial Differential Equations*, 4(9):1017–1034, 1979.
2. S. Bertoluzza. The discrete commutator property of approximation spaces. *C. R. Acad. Sci. Paris, Sér. I*, 329(12):1097–1102, 1999.
3. L. Caffarelli, R. Kohn, and L. Nirenberg. Partial regularity of suitable weak solutions of the Navier-Stokes equations. *Comm. Pure Appl. Math.*, 35(6):771–831, 1982.
4. A.F. Emery. An evaluation of several differencing methods for inviscid fluid flow problems. *J. Computational Phys.*, 2:306–331, 1968.
5. S. Gottlieb, C.-W. Shu, and E. Tadmor. Strong stability-preserving high-order time discretization methods. *SIAM Rev.*, 43(1):89–112 (electronic), 2001.
6. J.-L. Guermond, and S. Prudhomme. On the construction of suitable solutions to the navier–stokes equations and questions regarding the definition of large eddy simulation. *Physica D*, 207:64–78, 2005.
7. J.-L. Guermond. Finite-element-based Faedo-Galerkin weak solutions to the Navier-Stokes equations in the three-dimensional torus are suitable. *J. Math. Pures Appl. (9)*, 85(3):451–464, 2006.
8. J.-L. Guermond. Faedo-Galerkin weak solutions of the Navier–Stokes equations with Dirichlet boundary conditions are suitable. *J. Math. Pures Appl.*, 88:87–106, 2007.
9. J.-L. Guermond. On the use of the notion of suitable weak solutions in CFD. *Int. J. Numer. Methods Fluids*, 57:1153–1170, 2008.
10. J.-L. Guermond and R. Pasquetti. Entropy-based nonlinear viscosity for fourier approximations of conservation laws. *C. R. Math. Acad. Sci. Paris*, 346:801–806, 2008.
11. J.-L. Guermond and R. Pasquetti. Entropy viscosity method for high-order approximations of conservation laws. In E. Ronquist, editor, *ICOSAHOM09*, Lecture Notes in computational Science and Engineering, Berlin, 2010. Springer-Verlag.
12. J. Hoffman and C. Johnson. *Computational turbulent incompressible flow*, volume 4 of *Applied Mathematics: Body and Soul*. Springer, Berlin, 2007.
13. E. Hopf. Über die Anfangswertaufgabe für die hydrodynamischen Grundgleichungen. *Math. Nachr.*, 4:213–231, 1951.
14. C. Johnson and A. Szepessy. On the convergence of a finite element method for a nonlinear hyperbolic conservation law. *Math. Comp.*, 49(180):427–444, 1987.
15. C. Johnson, A. Szepessy, and P. Hansbo. On the convergence of shock-capturing streamline diffusion finite element methods for hyperbolic conservation laws. *Math. Comp.*, 54(189):107–129, 1990.
16. S. N. Kružkov. First order quasilinear equations with several independent variables. *Mat. Sb. (N.S.)*, 81 (123):228–255, 1970.
17. A. Kurganov, G. Petrova, and B. Popov. Adaptive semidiscrete central-upwind schemes for nonconvex hyperbolic conservation laws. *SIAM J. Sci. Comput.*, 29(6):2381–2401 (electronic), 2007.

18. J. Leray. Essai sur le mouvement d'un fluide visqueux emplissant l'espace. *Acta Math.*, 63:193–248, 1934.
19. R. J. LeVeque. *Finite volume methods for hyperbolic problems.* Cambridge Texts in Applied Mathematics. Cambridge University Press, Cambridge, 2002.
20. F. Lin. A new proof of the Caffarelli-Kohn-Nirenberg theorem. *Comm. Pure Appl. Math.*, 51(3):241–257, 1998.
21. R. Liska and B. Wendroff. Comparison of several difference schemes on 1D and 2D test problems for the Euler equations. *SIAM J. Sci. Comput.*, 25(3):995–1017 (electronic), 2003.
22. V. Scheffer. Hausdorff measure and the Navier-Stokes equations. *Comm. Math. Phys.*, 55(2):97–112, 1977.
23. P. Woodward and P. Colella. The numerical simulation of two-dimensional fluid flow with strong shocks. *J. Comput. Phys.*, 54(1):115–173, 1984.

A new deconvolution approach

Massimo Germano

Dip. di Ing. Aeronautica e Spaziale, Politecnico di Torino, Italy
massimo.germano@polito.it

Summary. A new deconvolution method recently proposed by author [1] is developed and applied to the differential filters [2, 3], in the context of the large eddy simulation of turbulent flows and the related subgrid scale modeling.

Key words: Large-Eddy Simulation, Subgrid Scale Modeling, Deconvolution Methods

1 Introduction

One basic problem of the Large Eddy Simulation of turbulent flows is to express the filtered value $\langle f(a)\rangle_g$ of a function $f(a)$

$$\langle f(a)\rangle_g = \mathscr{G}[f(a)] \tag{1}$$

in terms of the computed set of the filtered quantities $\langle a\rangle_g$

$$\langle a\rangle_g = \mathscr{G}[a] \tag{2}$$

where $\mathscr{G}$ is a generic filtering operator that we assume linear and constant preserving

$$\langle a+b\rangle_g = \langle a\rangle_g + \langle b\rangle_g \quad ; \quad \langle k\rangle_g = k \tag{3}$$

where k is a constant. If the filter $\mathscr{G}$ is explicitly given and is provided with an inverse operator $\mathscr{D} = \mathscr{G}^{-1}$, formally we can write

$$f(a) = f(a^*) \tag{4}$$

where

$$a^* = \mathscr{D}[\langle a\rangle_g] \tag{5}$$

and the problem is formally solved by the explicit relation

$$\langle f(a)\rangle_g = \mathscr{G}[f(\mathscr{D}[\langle a\rangle_g])] \tag{6}$$

M.V. Salvetti et al. (eds.), *Quality and Reliability of Large-Eddy Simulations II*, ERCOFTAC Series 16, DOI 10.1007/978-94-007-0231-8_35,

We remark that due to the numerical discretization usually only some part of the original field can be recovered, and approximate deconvolution operators $\mathscr{D}^* \sim \mathscr{G}^{-1}$ coupled with some additional regularization [4] are usually adopted. Functional contributions [5] like the Smagorinsky model or associated dynamic versions have been proposed and validated [6, 7]. Here we will not enter in such kind of problems and we refer to recent books and papers devoted to the mathematical and the computational aspects of the deconvolution methods applied to the large eddy simulation of turbulent flows [8, 9]. The main interest of this paper is to develop a different deconvolution approach recently proposed by the author [1]. We remark that in the large eddy simulation we are not particularly interested to the reconstruction of the original quantity a from the filtered quantities $\langle a \rangle_g$ but mainly to derive from $\langle a \rangle_g$ the filtered function $\langle f(a) \rangle_g$. As a consequence the proposed new deconvolution approach is based on the direct application of $\mathscr{D}$ both to $\langle a \rangle_g$ and to the filtered function $\langle f(a) \rangle_g$. We can write

$$f^*(a) = f(a^*) \tag{7}$$

where

$$f^*(a) = \mathscr{D}[\langle f(a) \rangle_g] \tag{8}$$

and the new explicit final relation is the following

$$\mathscr{D}[\langle f(a) \rangle_g] = f(\mathscr{D}[\langle a \rangle_g]) \tag{9}$$

In particular we will consider the following quantities

$$\begin{aligned} q &= \overline{f(a)} - f(\overline{a}) \\ s &= f(\mathscr{D}[\overline{a}]) - \mathscr{D}[f(\overline{a})] \end{aligned} \tag{10}$$

where we have introduced the equivalent notation

$$\overline{\cdots} = \langle \cdots \rangle_g = \mathscr{G}[\cdots] \tag{11}$$

If we apply the usual deconvolution we have

$$q = \overline{s} \tag{12}$$

while if we apply the new deconvolution method expressed by (9) we obtain

$$\mathscr{D}[q] = s \tag{13}$$

and we will call *filtered* deconvolution the first one given by (12), and *defiltered* deconvolution the second one given by (13).

2 Application of the new deconvolution method to LES

In the following we will examine the peculiarities of such new approach in the case of a generic inverse filter given by

$$\mathscr{D} = \mathscr{I} + \varepsilon \mathscr{L} \tag{14}$$

where $\mathscr{I}$ is the identity operator, ε a constant and $\mathscr{L}$ a generic linear *difference* operator such that

$$\langle a+b \rangle_l = \langle a \rangle_l + \langle b \rangle_l \quad ; \quad \langle k \rangle_l = 0 \tag{15}$$

where k is a constant. In this case if we apply to the quantity q

$$q = \overline{f(a)} - f(\bar{a})$$

the *defiltered* deconvolution expressed by (13) we have

$$s = \mathscr{D}[q] = q + \varepsilon \mathscr{L}[q] \tag{16}$$

and we can explicitly write

$$s = f(\mathscr{D}[\bar{a}]) - \mathscr{D}[f(\bar{a})] = f(\bar{a} + \varepsilon \mathscr{L}[\bar{a}]) - (f(\bar{a}) + \varepsilon \mathscr{L}[f(\bar{a})]) \tag{17}$$

Let us now consider in more detail the proposed new defiltered deconvolution (16) and let us remark first of all that in the trivial case of $f(a) = a$ we obviously have $q = s = 0$. More important in the context of the nonlinear terms related to the fluid flows is the case $f(a) = u_i u_k$, where u_i are the cartesian components of the velocity field. In this case we have

$$\begin{aligned} f(a) &= u_i u_k \\ q_{ik} &= \overline{u_i u_k} - \bar{u}_i \bar{u}_k \\ s_{ik} &= \mathscr{D}[q_{ik}] = \varepsilon \left(\bar{u}_i \mathscr{L}[\bar{u}_k] + \bar{u}_k \mathscr{L}[\bar{u}_i] - \mathscr{L}[\bar{u}_i \bar{u}_k] \right) + \varepsilon^2 \mathscr{L}[\bar{u}_i] \mathscr{L}[\bar{u}_k] \end{aligned} \tag{18}$$

and more generally we can apply this new deconvolution approach directly to the Navier-Stokes equations for an incompressible flow. In the case of the continuity operator

$$f(a) = \mathscr{C}[u_k] = \frac{\partial u_k}{\partial x_k} \tag{19}$$

we have

$$\overline{\frac{\partial u_k}{\partial x_k}} - \frac{\partial \bar{u}_k}{\partial x_k} = q$$

$$s = \mathscr{D}[q] = \varepsilon \left(\frac{\partial (\mathscr{L}[\bar{u}_k])}{\partial x_k} - \mathscr{L}\left[\frac{\partial \bar{u}_k}{\partial x_k} \right] \right) \tag{20}$$

and if we consider the momentum operator

$$f(a) = \mathscr{N}_i[u_k, p] = \frac{\partial u_i}{\partial t} + \frac{\partial (u_i u_k)}{\partial x_k} + \frac{\partial p}{\partial x_i} - \nu \frac{\partial^2 u_i}{\partial x_k \partial x_k} \tag{21}$$

we can write

$$\overline{\mathscr{N}_i[u_k, p]} - \mathscr{N}_i[\bar{u}_k, \bar{p}] = \frac{\partial q_{ik}}{\partial x_k} + q_i$$

where

$$q_{ik} = \overline{u_i u_k} - \bar{u}_i \bar{u}_k$$
$$s_{ik} = \mathscr{D}[q_{ik}] = \varepsilon\left(\bar{u}_i \mathscr{L}[\bar{u}_k] + \bar{u}_k \mathscr{L}[\bar{u}_i] - \mathscr{L}[\bar{u}_i \bar{u}_k]\right) + \varepsilon^2 \mathscr{L}[\bar{u}_i]\mathscr{L}[\bar{u}_k] \qquad (22)$$

and where

$$q_i = \overline{\frac{\partial u_i}{\partial t}} - \frac{\partial \bar{u}_i}{\partial t} + \overline{\frac{\partial (u_i u_k)}{\partial x_k}} - \frac{\partial \overline{u_i u_k}}{\partial x_k} + \\ + \overline{\frac{\partial p}{\partial x_i}} - \frac{\partial \bar{p}}{\partial x_i} - \nu \overline{\frac{\partial^2 u_i}{\partial x_k \partial x_k}} - \nu \frac{\partial^2 \bar{u}_i}{\partial x_k \partial x_k}$$

$$s_i = \mathscr{D}[q_i] = \varepsilon\left(\frac{\partial(\mathscr{L}[\bar{u}_i])}{\partial t} - \mathscr{L}\left[\frac{\partial \bar{u}_i}{\partial t}\right]\right) + \varepsilon\left(\frac{\partial(\mathscr{L}[\overline{u_i u_k}])}{\partial x_k} - \mathscr{L}\left[\frac{\partial(\overline{u_i u_k})}{\partial x_k}\right]\right) + \\ + \varepsilon\left(\frac{\partial(\mathscr{L}[\bar{p}])}{\partial x_i} - \mathscr{L}\left[\frac{\partial \bar{p}}{\partial x_i}\right]\right) - \varepsilon\nu\left(\frac{\partial^2(\mathscr{L}[\bar{u}_i])}{\partial x_k \partial x_k} - \mathscr{L}\left[\frac{\partial^2 \bar{u}_k}{\partial x_k \partial x_i}\right]\right) \qquad (23)$$

We remark that for incompressible flows we have

$$\overline{\frac{\partial u_k}{\partial x_k}} = 0 \quad ; \quad \overline{\mathscr{N}_i[u_k, p]} = 0 \qquad (24)$$

so that as regards the continuity equation we can formally write

$$\frac{\partial \bar{u}_k}{\partial x_k} = -q$$
$$q + \varepsilon \mathscr{L}[q] = s \qquad (25)$$

and as regards the momentum equation we can formally write

$$\frac{\partial \bar{u}_i}{\partial t} + \frac{\partial(\bar{u}_i \bar{u}_k)}{\partial x_k} + \frac{\partial \bar{p}}{\partial x_i} - \nu \frac{\partial^2 \bar{u}_i}{\partial x_k \partial x_k} = -\frac{\partial q_{ik}}{\partial x_k} - q_i$$
$$q_i + \varepsilon \mathscr{L}[q_i] = s_i$$
$$q_{ik} + \varepsilon \mathscr{L}[q_{ik}] = s_{ik} \qquad (26)$$

It is worth noting that q and q_i are subgrid contributions due to the non homogeneity of the filter, and in particular they are null in the case the filter commutes with the time and the space derivatives. The subgrid contributions expressed by q_{ij} are due to the nonlinearity of the convective term, and they can be read as usual as the subgrid stresses associated to the particular filtering operator.

3 Application of the new deconvolution method to the differential filters

Let us now consider particular difference operators $\mathscr{L}$ given by a linear differential form [2] in space and time expressed by the relation

$$\mathscr{L} = \frac{\partial}{\partial t} + v_m \frac{\partial}{\partial x_m} + \alpha_{mn} \frac{\partial^2}{\partial x_m \partial x_n} + \cdots \quad (27)$$

and we recall that a differential filter $\mathscr{D}$

$$\mathscr{D} = \mathscr{I} + \varepsilon \mathscr{L}$$

is formally defined [3] as the inverse $\mathscr{D} = \mathscr{G}^{-1}$ of a filtering operator $\mathscr{G}$

$$\langle a \rangle_g = \mathscr{G}[a] = \int G(\mathbf{x} - \mathbf{x}', t - t')\, a(\mathbf{x}', t')\, d\mathbf{x}'dt' \quad (28)$$

where $G(\mathbf{x} - \mathbf{x}', t - t')$ is the Green's function associated to the differential operator $\mathscr{D}$. We remark that in this case the constant ε has the dimension of a time and the relations (25,26) become evolutionary equations for q, q_i and q_{ik} given by

$$\frac{\partial q}{\partial t} + v_m \frac{\partial q}{\partial x_m} + \alpha_{mn} \frac{\partial^2 q}{\partial x_m \partial x_n} + \cdots = \frac{s - q}{\varepsilon}$$

$$\frac{\partial q_i}{\partial t} + v_m \frac{\partial q_i}{\partial x_m} + \alpha_{mn} \frac{\partial q_i}{\partial x_m \partial x_n} + \cdots = \frac{s_i - q_i}{\varepsilon}$$

$$\frac{\partial q_{ik}}{\partial t} + v_m \frac{\partial q_{ik}}{\partial x_m} + \alpha_{mn} \frac{\partial q_{ik}}{\partial x_m \partial x_n} + \cdots = \frac{s_{ik} - q_{ik}}{\varepsilon} \quad (29)$$

If the filter is non homogeneous the coefficients v_m and α_{mn} depends on space and time, and we have

$$\frac{\partial \mathscr{L}}{\partial t} - \mathscr{L} \frac{\partial}{\partial t} = \frac{\partial v_m}{\partial t} \frac{\partial}{\partial x_m} + \frac{\partial \alpha_{mn}}{\partial t} \frac{\partial^2}{\partial x_m \partial x_n} + \cdots$$

$$\frac{\partial \mathscr{L}}{\partial x_i} - \mathscr{L} \frac{\partial}{\partial x_i} = \frac{\partial v_m}{\partial x_i} \frac{\partial}{\partial x_m} + \frac{\partial \alpha_{mn}}{\partial x_i} \frac{\partial^2}{\partial x_m \partial x_n} + \cdots \quad (30)$$

In the following we will only consider the homogeneous case, with constant coefficients. In this case s, q, s_i and q_i are identically null.

4 The new deconvolution approach applied to the low-pass filter

The simplest differential filter is the low pass filter in time [10]. Its inverse is given by

$$\mathscr{D} = \mathscr{I} + \varepsilon \frac{\partial}{\partial t} \quad (31)$$

so that in this case we simply have

$$\mathscr{L} = \frac{\partial}{\partial t} \tag{32}$$

and it is easy to derive the associated equation for the subgrid stresses q_{ij}

$$\frac{\partial q_{ij}}{\partial t} = \frac{s_{ij} - q_{ij}}{\varepsilon} \tag{33}$$

where

$$s_{ij} = \varepsilon^2 \frac{\partial \bar{u}_i}{\partial t} \frac{\partial \bar{u}_j}{\partial t} \tag{34}$$

It is interesting to derive from (33) the equation for the subgrid kinetic energy

$$K = \frac{\overline{u_k u_k} - \bar{u}_k \bar{u}_k}{2} = \frac{q_{kk}}{2}$$

We have

$$\frac{\partial K}{\partial t} = -\frac{K}{\varepsilon} + \varepsilon \frac{\partial \bar{u}_k}{\partial t} \frac{\partial \bar{u}_k}{\partial t} \tag{35}$$

and we see that the dissipation of the subgrid kinetic energy is based on a relaxation term, while the production is related to the time derivatives of the velocity. We remark that in the *equilibrium* conditions, $\varepsilon \to 0$, we have

$$q_{ij} \sim s_{ij} = \varepsilon^2 \frac{\partial \bar{u}_k}{\partial t} \frac{\partial \bar{u}_k}{\partial t} \tag{36}$$

5 The new deconvolution approach applied to the parabolic filter

Another interesting differential filter is the parabolic one [10]. Its inverse is given by

$$\mathscr{D} = \mathscr{I} + \varepsilon \left(\frac{\partial}{\partial t} - \alpha \frac{\partial^2}{\partial x_m^2} \right) \tag{37}$$

so that in this case we have

$$\mathscr{L} = \frac{\partial}{\partial t} - \alpha \frac{\partial^2}{\partial x_m^2} \tag{38}$$

The associated equation for the subgrid stresses q_{ij} is given by

$$\frac{\partial q_{ij}}{\partial t} - \alpha \frac{\partial^2 q_{ij}}{\partial x_m^2} = \frac{s_{ij} - q_{ij}}{\varepsilon} \tag{39}$$

where now

$$\begin{aligned} s_{ij} = {} & 2\varepsilon\alpha \frac{\partial \bar{u}_i}{\partial x_m} \frac{\partial \bar{u}_j}{\partial x_m} + \\ & + \varepsilon^2 \left(\frac{\partial \bar{u}_i}{\partial t} \frac{\partial \bar{u}_j}{\partial t} - \alpha \frac{\partial \bar{u}_i}{\partial t} \frac{\partial^2 \bar{u}_j}{\partial x_m^2} - \alpha \frac{\partial \bar{u}_j}{\partial t} \frac{\partial^2 \bar{u}_i}{\partial x_m^2} + \alpha^2 \frac{\partial \bar{u}_i}{\partial x_m^2} \frac{\partial^2 \bar{u}_j}{\partial x_m^2} \right) \end{aligned} \tag{40}$$

We remark that in the limit $\varepsilon \to 0$ we simply have

$$s_{ij} = 2\varepsilon\alpha \frac{\partial \bar{u}_i}{\partial x_m}\frac{\partial \bar{u}_j}{\partial x_m} \tag{41}$$

so that we can write the equation for the subgrid kinetic energy

$$K = \frac{\overline{u_k u_k} - \bar{u}_k\bar{u}_k}{2} = \frac{q_{kk}}{2}$$

in the simple form

$$\frac{\partial K}{\partial t} - \alpha\frac{\partial^2 K}{\partial x_m^2} = -\frac{K}{\varepsilon} + \alpha\frac{\partial \bar{u}_k}{\partial x_m}\frac{\partial \bar{u}_k}{\partial x_m} \tag{42}$$

We see that in the case of the parabolic filter the dissipation of the subgrid kinetic energy is based on a relaxation term, while the production is related to the velocity gradient. Finally it is worth noticing that in the *equilibrium* conditions, $\varepsilon \to 0$, we have

$$q_{ij} \sim s_{ij} = 2\varepsilon\alpha \frac{\partial \bar{u}_i}{\partial x_m}\frac{\partial \bar{u}_j}{\partial x_m} \tag{43}$$

that corresponds to the well known gradient model.

6 The new deconvolution approach applied to the convective filter

Let us now consider a convective-diffusive filter [10] whose inverse is given by the differential equation

$$\mathscr{D} = \mathscr{I} + \varepsilon\left(\frac{\partial}{\partial t} + v_m\frac{\partial}{\partial x_m} - \alpha\frac{\partial^2}{\partial x_m^2}\right) \tag{44}$$

so that in this case we have

$$\mathscr{L} = \frac{\partial}{\partial t} + v_m\frac{\partial}{\partial x_m} - \alpha\frac{\partial^2}{\partial x_m^2} \tag{45}$$

The associated equation for the subgrid stresses q_{ij} is given by

$$\frac{\partial q_{ij}}{\partial t} + v_m\frac{\partial}{\partial x_m} - \alpha\frac{\partial^2 q_{ij}}{\partial x_m^2} = \frac{s_{ij} - q_{ij}}{\varepsilon} \tag{46}$$

and it is interesting to notice that in the limit $\varepsilon \to 0$ the source terms s_{ij} are the same of the parabolic filter, given by (40), so that we can finally write

$$\frac{\partial K}{\partial t} + v_m\frac{\partial K}{\partial x_m} - \alpha\frac{\partial^2 K}{\partial x_m^2} = -\frac{K}{\varepsilon} + \alpha\frac{\partial \bar{u}_k}{\partial x_m}\frac{\partial \bar{u}_k}{\partial x_m} \tag{47}$$

7 Conclusions

A new deconvolution approach [1] recently proposed by the author has been applied to the differential filters. It results that the associated subgrid stresses are in this case given by evolutionary equations in time. In particular have been examined in detail the low-pass filter in time, the parabolic filter and a general convective-diffusive filter. The evolutionary equations of the subgrid kinetic energy associated to these filters are particularly interesting. They show that for these filters the dissipation of the subgrid kinetic energy is given by a relaxation term, while the production term is related to the velocity gradient and to the velocity derivative in time.

References

1. Germano M (2009) Phys Fluids 21:045107/3
2. Germano M (1986) Phys Fluids 29:1755–1757
3. Germano M (1986) Phys Fluids 29:1757–1758
4. Stolz S, Adams N A (1999) Phys Fluids 11:1699–1701
5. Sagaut P (2005) Large Eddy Simulation for Incompressible Flows. Springer, Berlin Heidelberg New York
6. Habisreutinger M A, Bouffanais R, Leriche E, Deville M O (2007) J Comput Phys 224:241–266
7. Kuerten J G M, Geurts B J, Vreman A W, Germano M (1999) Phys. Fluids 11:3778–3785
8. Berselli L C, Iliescu T, Layton W (2006) Mathematics of Large Eddy Simulation of Turbulent Flows, Springer, Berlin Heidelberg New York
9. Dunca A, Epshteyn Y (2006) SIAM J. Math. Anal. 1890–1902.
10. Germano M (2000) Fundamentals of Large Eddy Simulation. In: Peyret R, Krause E (eds) Advanced Turbulent Flows Computations, CISM Courses and Lectures 395:81–130 Springer, Berlin Heidelberg New York

Horizontal Approximate Deconvolution for Stratified Flows: Analysis and Computations

Luigi C. Berselli[1], Traian Iliescu[2], and Tamay Özgökmen[3]

[1] Dipartimento di Matematica Applicata "U. Dini", Via F. Buonarroti 1/c, I-56127, Pisa ITALY, Phone:+390502213846, Fax: +390502213802, berselli@dma.unipi.it

[2] Department of Mathematics, Virginia Tech, Blacksburg, VA, United States, iliescu@vt.edu

[3] RSMAS/MPO, University of Miami, Miami, FL, United States, tozgokmen@rsmas.miami.edu

Summary. In this paper we propose a new Large Eddy Simulation model derived by approximate deconvolution obtained by means of wave-number asymptotic expansions. This LES model is designed for oceanic flows and in particular to simulate mixing of fluids with different temperatures, density or salinity. The model -which exploits some ideas well diffused in the community- is based on a suitable horizontal filtering of the equations. We prove a couple of *a-priori* estimates, showing certain mathematical properties and we present also the results of some preliminary numerical experiments.

Key words: Stratified flows, Rational/Clark LES model, Boussinesq equations, LES in domains with boundary

1 Introduction

Mixing is one of the most important processes to understand transport of pollutants as well as the details of thermohaline circulation. Despite the increase in computational power, the scales in the ocean circulation can not be all resolved simultaneously. Basin models are configured for $\mathcal{O}$(1000km) to $\mathcal{O}$(10km), and regional or coastal models from $\mathcal{O}$(100km) to $\mathcal{O}$(1km), requiring both sub-grid-scale parametrization. However, there exist small-scale ocean flows (which take place below this inherently coarse numerical resolution) that often play a significant role in an accurate representation of the large ocean scales. More precise motivations for the study of this physical problem and the requirement of suitable numerical methods to handle all scales is explained for instance in [24, 26]. To a first approximation, the mixing phenomena can be described mathematically by means of the Boussinesq system of partial differential equations:

M.V. Salvetti et al. (eds.), *Quality and Reliability of Large-Eddy Simulations II*, ERCOFTAC Series 16, DOI 10.1007/978-94-007-0231-8_36,

$$\begin{aligned} &\partial_t \mathbf{u} + \nabla \cdot (\mathbf{u} \otimes \mathbf{u}) - \frac{1}{Re} \Delta \mathbf{u} + \nabla p = -\frac{1}{Fr^2} \rho' \mathbf{e}_3, \quad \text{with } \nabla \cdot \mathbf{u} = 0, \\ &\partial_t \rho' + \nabla \cdot (\rho' \mathbf{u}) - \frac{1}{Re\,Pr} \Delta \rho' = 0. \end{aligned} \tag{1}$$

The unknowns $(\mathbf{u}, p, \rho')$ are velocity, pressure, and "salinity perturbation," respectively, and $\mathbf{e}_3 = (0,0,1)$. The non-dimensional parameters are the Reynolds number Re, the Prandtl number Pr, and the Froude number Fr. The problem we consider takes naturally place in an *elongated domain* (We model the small portion of the ocean we consider as a parallelepiped)

$$D := \left\{ x \in \mathbb{R}^3 : \ -\pi < x_1, x_2 < \pi, \quad -d < x_3 < d \right\} \qquad \text{with} \qquad d << 1,$$

hence the boundary is flat and we enforce periodic boundary conditions on the "*horizontal variables*" $x_h := (x_1, x_2)$. We use the subscript "$_h$" to denote differential operators acting only on the horizontal variables. In particular, we use the following notation

$$\Delta_h := \partial^2_{x_1} + \partial^2_{x_2} \qquad \text{and} \qquad \nabla_h := (\partial_{x_1}, \partial_{x_2}).$$

One of the features (which is well-know by practitioners) of this problem is that filtering seems to be required only in the horizontal directions, because the mixing takes place mainly along these two directions. In a previous work [3] we started analyzing it by means of stochastic parametrization and Itô's calculus. Since the problem we have in mind is that of stratified fluids, another justification for the use of horizontal viscosities/filtering comes from the study of Ekman boundary layers for rotating fluids (see [9] and references therein) and in particular the system of the Navier-Stokes equations with partial viscosity:

$$\partial_t \mathbf{u} + \nabla \cdot (\mathbf{u} \otimes \mathbf{u}) - \frac{1}{Re} \Delta_h \mathbf{u} + \nabla p = \mathbf{f}, \quad \text{with } \nabla \cdot \mathbf{u} = 0.$$

These are the main modeling motivations for the anisotropic LES method (2) we will introduce. Other motivations, which are consequences of the mathematical properties, will be explained in the sequel, but another peculiarity of the new method we propose is that we do not need artificial boundary conditions. The objective of the present paper is to propose a new LES method, to prove existence of suitable classes of solutions, and to compare the results with some 3D simulations performed with other LES methods.

2 An anisotropic Large Eddy Simulation model

One of the earliest LES models is the "Gradient" method (known also as Taylor or Clark method and introduced in [10, 21]) which is based on an approximation of the subgrid-scale term by means of asymptotic expansion in wave-numbers. In the Gradient model the turbulent stress-tensor is $\boldsymbol{\tau}_{\mathrm{G}}(\mathbf{w}, \mathbf{w}) = \alpha^2 \nabla \mathbf{w} \nabla \mathbf{w}^T$, where $\left[\nabla \mathbf{w} \nabla \mathbf{w}^T\right]_{ij} := \sum_{k=1}^3 \partial_{x_k} w^i \, \partial_{x_k} w^j$. The derivation of this model and the basic results regarding the mathematical analysis are collected in [5]. It is well-known that instabilities occur in the numerical implementation and that some kind of smoothing must be added in order to have effective simulations, see [17]. In particular, a filtered version of the Gradient method is called Rational or[4] Clark-α method, and the stress-tensor reads as

$$\boldsymbol{\tau}_{\text{R-C}}(\mathbf{w},\mathbf{w}) = (\text{I}-\alpha^2\Delta)^{-1}\alpha^2\nabla\mathbf{w}\nabla\mathbf{w}^T,$$

cf. [31, 32, 33, 34]. The Rational method has been derived [16] by a rational approximation of wave-numbers, while the Clark-α method is based on analogies with the Helmholtz filtering. The mathematical analysis of these models can be found in [4, 8]. Following the approach described in the introduction we consider a smoothing acting only in the horizontal variables, which can be performed by means of convolution with the following kernel: $g_\alpha(x) = \frac{1}{\pi\alpha^2}\text{Exp}(-\frac{x_1^2+x_2^2}{\alpha^2})$. By taking the 2D Fourier Transform, by performing the Taylor series expansion, and by neglecting terms which are formally of order of α^4 one gets (apart multiplicative constants) the following expression for the subgrid-scale term $\boldsymbol{\tau}_{\text{hG}}(\mathbf{w},\mathbf{w}) = \alpha^2\nabla_h\mathbf{w}\nabla_h\mathbf{w}^T$, where $\left[\nabla_h\mathbf{w}\nabla_h\mathbf{w}^T\right]_{ij} := \alpha^2\sum_{k=1}^2 \partial_{x_k}w^i\,\partial_{x_k}w^j$. We will mainly consider the "horizontal version" of the Rational-Clark-α method,

$$\boldsymbol{\tau}_{\text{hR-C}}(\mathbf{w},\mathbf{w}) = (\text{I}-\alpha^2\Delta_h)^{-1}\alpha^2\nabla_h\mathbf{w}\nabla_h\mathbf{w}^T.$$

From the computational point of view the inversion of a horizontal Laplacian is less time/memory consuming than the usual one, but the main advantage of the use of the horizontal Laplacian is that *there is no need* to introduce extra/artificial boundary conditions for the Helmholtz operator and that the value $\mathbf{w}$ on the boundary of the domain D can be imposed. The issue of the boundary conditions for LES models is generally very complex, cf. [5, Ch. 9]. The challenging property of our approach based on horizontal filtering is that part of the problem (due also to the particular shape of the domain) is overcome. In the sequel, we sketch the main mathematical properties of the space filtered Navier-Stokes equations approximated by the "horizontal Rational/Clark" model

$$\partial_t\mathbf{w} + \nabla\cdot(\mathbf{w}\otimes\mathbf{w}) - \frac{1}{Re}\Delta\mathbf{w} + \nabla\cdot(\text{I}-\alpha^2\Delta_h)^{-1}\alpha^2\nabla_h\mathbf{w}\nabla_h\mathbf{w}^T + \nabla q = \mathbf{f}, \qquad (2)$$

with $\nabla\cdot\mathbf{w} = 0$. It seems that the LES model and the viscous term in equation (2) have different forms. Indeed, the viscous term retains its original 3D form, whereas the LES model displays the horizontal filtering operation. This difference, however, is only in form, since the Laplace operator in the viscous term is applied to $\mathbf{w}$, which is the *horizontally filtered* velocity. Thus, both the viscous term and the LES model in equation (2) include the horizontal filtering operation, the former implicitly (through $\mathbf{w}$), and the latter explicitly. The reason for the dependence on the horizontal filter is displayed explicitly in the LES model and is due to the nonlinear term $\nabla\cdot(\mathbf{w}\otimes\mathbf{w})$;

[4] The two methods differ by a multiplicative factor, which is nevertheless critical for the well-posedness results, see Remark 2 and also Section 5.

since the Laplacian is linear, the differentiation and filtering operators commute, and thus the only dependence on horizontal filtering in the viscous term is through $\mathbf{w}$. Full details on the horizontal filtering, in conjunction with other LES models (with more appealing and neat mathematical properties) will appear in [2].

2.1 Some properties of the horizontal Rational/Clark model

Given a vector field $\mathbf{f}$, an "averaged" field $\mathbf{u}$ is defined by solving the following *horizontal Helmholtz-Stokes* problem:

$$\begin{aligned} \mathbf{u}-\alpha^2\Delta_h\mathbf{u}+\nabla q=\mathbf{f}, &\quad \text{with } \nabla\cdot\mathbf{u}=0 \text{ in } D,\\ \mathbf{u}\cdot\mathbf{n}=0 &\quad \text{on } \partial D, \end{aligned} \tag{3}$$

and throughout the paper $\alpha > 0$ will be a fixed number. We use the notation

$$L^2(D) := \left\{\phi : D\to\mathbb{R},\ x_h\text{-periodic, measurable, with} \int_D |\phi|^2\,dx < +\infty\right\},$$

and $\mathbb{L}^2 := (L^2(D))^3$. We use the same symbol $\|\,.\,\|$ for the norm in both spaces. Next, by setting $\partial D := \left\{x\in\mathbb{R}^3 : -\pi < x_1, x_2 < \pi,\ x_3 = \pm d\right\}$ we can define

$$\mathbb{H}^1_h := \left\{\mathbf{u}\in L^2 : \nabla_h\mathbf{u}\in (L^2(D))^6,\ \nabla\cdot\mathbf{u}\in L^2(D),\ \mathbf{u}\cdot\mathbf{n}=0 \text{ in } H^{-1/2}(\partial D)\right\},$$

where $\mathbf{n}$ denotes the exterior unit normal vector on ∂D. Observe also that since $\mathbf{u}$ has divergence in $L^2(D)$, then $\frac{\partial u^3}{\partial x_3} = -\nabla_h\cdot\mathbf{u}_h = -\frac{\partial u^1}{\partial x_1} - \frac{\partial u^2}{\partial x_2} \in L^2(D)$. This shows that some extra-regularity for the vertical component of the velocity is obtained for free, and this is another interesting mathematical advantage.

Lemma 1 (See [2]). *Let $\mathbf{f}\in\mathbb{L}^2$ and $\alpha>0$. Then, there exists a unique $(\mathbf{u},q)\in\mathbb{H}^1_h\times L^2(D)/\mathbb{R}$ solution of (3) and a constant $c=c(\alpha)$ such that*

$$\|\mathbf{u}\| + \alpha\|\nabla_h\mathbf{u}\| + \alpha\|\nabla u^3\| + \|q\| \le c\|\mathbf{f}\|.$$

Remark 1. A relevant point (for a rigorous proof see [2]) is that if $\nabla\cdot\mathbf{f}=0$ and if $\mathbf{f}$ is tangential to the boundary, one can prove that $q=0$. Consequently, the two operators coincide and one has to solve three uncoupled Helmholtz problem, instead of the Stokes problem, since the field $(\mathrm{I}-\alpha^2\Delta_h)^{-1}\mathbf{f}$, turns out to be divergence-free. This motivates the name Helmholtz-Stokes. In particular, in the sequel the operator $(\mathrm{I}-\alpha^2\Delta_h)^{-1}$ appearing in the various equations is the horizontal*Helmholtz* one.

What is most important for understanding the mathematical properties of solutions of (2) is the following lemma.

Lemma 2. *For all sufficiently smooth vector fields $\mathbf{w}$ on D, x_h-periodic, such that $\nabla\cdot\mathbf{w}=0$ in D and $\mathbf{w}\cdot\mathbf{n}=0$ on ∂D, it holds*

$$\int_D \left[\nabla\cdot(\mathbf{w}\otimes\mathbf{w}) + \nabla\cdot(\mathrm{I}-\alpha^2\Delta_h)^{-1}\alpha^2\nabla_h\mathbf{w}\nabla_h\mathbf{w}^T\right]\cdot(\mathrm{I}-\alpha^2\Delta_h)\mathbf{w}\,dx = 0.$$

Proof. The proof is obtained by direct integration by parts (cf. [4, 8]). In fact, as usual $\int_D \nabla\cdot(\mathbf{w}\otimes\mathbf{w})\cdot\mathbf{w}\,dx = 0$; next, observe that

$$\int_D \nabla\cdot(\mathbf{w}\otimes\mathbf{w})\cdot(-\alpha^2\Delta_h\mathbf{w}) = \alpha^2\sum_{k=1}^{2}\int_D \partial_{x_k}\left[(\mathbf{w}\cdot\nabla)\mathbf{w}\right](\partial_{x_k}\mathbf{w})$$
$$= \alpha^2\sum_{i,j=1}^{3}\sum_{k=1}^{2}\int_D \partial_{x_k}\mathbf{w}^j\partial_{x_j}\mathbf{w}^i\partial_{x_k}\mathbf{w}^i\,dx,$$

since one term vanishes by integration by parts. Moreover, we have

$$\int_D \nabla\cdot(\mathrm{I}-\alpha^2\Delta_h)^{-1}\alpha^2\nabla_h\mathbf{w}\nabla_h\mathbf{w}^T\cdot(\mathrm{I}-\alpha^2\Delta_h)\mathbf{w} = \alpha^2\int_D \nabla\cdot\left(\nabla_h\mathbf{w}\nabla_h\mathbf{w}^T\right)\mathbf{w},$$

since in our setting the operator $(\mathrm{I}-\alpha^2\Delta_h)$ is self-adjoint. Finally, a further integration by parts (possible since $\mathbf{w}\cdot\mathbf{n}=0$ on ∂D) shows that

$$\alpha^2\int_D \nabla\cdot\left(\nabla_h\mathbf{w}\nabla_h\mathbf{w}^T\right)\mathbf{w} = -\alpha^2\int_D\sum_{i,j=1}^{3}\sum_{k=1}^{2}\partial_{x_k}\mathbf{w}^i\partial_{x_k}\mathbf{w}^j\partial_{x_j}\mathbf{w}^i,$$

and adding together the various terms we end the proof. □

Remark 2. Lemma 2 shows that $(\mathrm{I}-\alpha^2\Delta_h)\mathbf{w}$ is a "*good multiplier*" but again it is relevant (concerning the boundary conditions) that it involves only derivatives in the horizontal directions. Moreover, it is important to note that the Rational model corresponds to the following stress-tensor

$$\boldsymbol{\tau}_{\mathrm{RLES}}(\mathbf{w},\mathbf{w}) = (\mathrm{I}-\frac{\alpha^2}{2}\Delta)^{-1}\alpha^2\nabla\mathbf{w}\nabla\mathbf{w}^T$$

In this case the "good multiplier" is $(\mathrm{I}-\frac{\alpha^2}{2}\Delta)\mathbf{w}$ but the factor $\alpha^2/2$ does not imply that the cancellation of the nonlinearities, as in Lemma 2. Results in [1, 4] show that the term $\nabla\cdot(\mathrm{I}-\frac{\alpha^2}{2}\Delta)^{-1}\alpha^2\nabla\mathbf{w}\nabla\mathbf{w}^T$ plays the same role (in terms of Sobolev spaces inequalities) of the convective term, not allowing for global-in-time estimates. This is one of the subtle differences between the Clark-α and the Rational Model. Obviously the same difference occurs also for the horizontal version of both methods.

The main result concerning system (2) is the following.

Theorem 1. *Let be given* $\mathbf{w}_0\in\mathbb{H}^1_h$, *and* $\mathbf{f}\in L^2(0,T;L^2)$. *Then, there exists a solution* $\mathbf{w}$ *(x_h-periodic and vanishing on ∂D) in the sense of distributions to system (2) such that*

$$\mathbf{w},\ \nabla_h\mathbf{w},\ \nabla w^3 \in L^\infty(0,T;L^2)\cap L^2(0,T;H^1). \tag{4}$$

Proof. We sketch the proof, that is based on the usual Galerkin method. We look for an approximate solution $\mathbf{w}^m = \sum_{k=1}^m c^k(t)\,\mathscr{E}_k(x)$, where $\mathscr{E}_k(x)$ are eigenfunction of the Stokes operator with our boundary conditions. The explicit form of the eigenfunctions is know, see [23, 29], but for our purposes it is enough to observe that they are of the type $e^{i(k_1x_1+k_2x_2)}f(x_3)$, for $k_1,k_2 \in \mathbb{Z}$ and for certain smooth f, vanishing at $x_3 = \pm d$. This is relevant, since $\Delta_h \mathscr{E}_k$ still is an eigenfunction. Hence, the *a priori* estimate obtained by testing the equation by $(\mathrm{I} - \alpha^2\Delta_h)\mathbf{w}^m$ is completely justified. By performing standard integrations by parts (possible by x_h-periodicity) we get $\int_D \partial_t \mathbf{w}^m \cdot (\mathrm{I} - \alpha^2\Delta_h)\mathbf{w}^m = \frac{1}{2}\frac{d}{dt}\left(\|\mathbf{w}^m\|^2 + \alpha^2\|\nabla_h \mathbf{w}^m\|^2\right)$ and also $-\int_D \Delta \mathbf{w}^m \cdot (\mathrm{I} - \alpha^2\Delta_h)\mathbf{w}^m = \|\nabla \mathbf{w}^m\|^2 + \alpha^2\|\nabla\nabla_h \mathbf{w}^m\|^2$. Hence, application of Lemma 2 shows the following differential inequality for $\mathbf{w}^m$:

$$\frac{1}{2}\frac{d}{dt}\left(\|\mathbf{w}^m\|^2 + \alpha^2\|\nabla_h \mathbf{w}^m\|^2\right) + \frac{1}{Re}\left(\|\nabla \mathbf{w}^m\|^2 + \alpha^2\|\nabla\nabla_h \mathbf{w}^m\|^2\right) \leq c\|\mathbf{f}\|^2.$$

Uniform bounds in $m \in \mathbb{N}$ and other standard arguments, see *e.g.*. [11], can be used to prove that (a sub-sequence of) $\mathbf{w}^m$ converges (as $m \to +\infty$) towards a solution of (2). Regularity of w^3 follows from $\nabla \cdot \mathbf{w} = 0$. □

Remark 3. The solution $\mathbf{w}$ has regularity properties which lie "in between" those of weak and strong solutions for the NSE (In particular we are missing the control of $\partial_{x_3}\mathbf{w}$ in $L^\infty(0,T;L^2)$.) Consequently, we are not able to prove uniqueness within this class of solutions. In addition, one cannot start a bootstrapping argument to improve the regularity of $\mathbf{w}$: By using as test function $(\mathrm{I} - \alpha^2\Delta_h)^2\mathbf{w}^m$ (the counterpart of the H^2-estimates in [8]) one does not get good *a priori* estimates. Roughly speaking, it seems that multipliers which are good for $\nabla \cdot (\mathbf{w} \otimes \mathbf{w})$ are not good for $\nabla \cdot (\mathrm{I} - \frac{\alpha^2}{2}\Delta)^{-1}\alpha^2\nabla_h \mathbf{w} \nabla_h \mathbf{w}^T$ and *vice-versa*. Based on these observations we think that this is not the best model (at least from the point of view of uniqueness and stability of solutions) to be implemented with the horizontal filtering. Much better theoretical results can be proved in the framework of Approximate Deconvolution Models (ADM), *à la* Stolz & Adams, see [30, 20, 2]). With these models the special expression of the stress-tensor allows us to prove uniqueness, using the same multiplier and just the regularity in (4).

3 On the Boussinesq system

We consider now the Boussinesq system (1) and we study a LES model in which the velocity equation is filtered, while not the equation for the salinity. We use this approach because results of [13, 19] show that in presence of nonzero viscosity (and also with vanishing diffusivity) the classical conditions which ensure regularity of the NSE imply the continuation of smooth solutions of (1). Improved theoretical results on this topic will appear in a forthcoming paper [6]. Here, we consider the following horizontal LES model for the Boussinesq system (still with $\nabla \cdot \mathbf{w} = 0$)

$$\begin{aligned}\partial_t \mathbf{w} + \nabla \cdot (\mathbf{w} \otimes \mathbf{w}) - \frac{1}{Re}\Delta \mathbf{w} + \nabla \cdot (\mathrm{I} - \alpha^2\Delta_h)^{-1}\alpha^2\nabla_h \mathbf{w} \nabla_h \mathbf{w}^T + \nabla q &= -\frac{1}{Fr^2}\rho' \mathbf{e}_3,\\ \partial_t \rho' + \nabla \cdot (\rho' \mathbf{w}) - \frac{1}{Re\,Pr}\Delta \rho' &= 0.\end{aligned} \tag{5}$$

With the same Galerkin approach of Sec. 2.1 one proves the following result.

Theorem 2. *Let be given* $(\mathbf{w}_0, \rho'_0) \in \mathbb{H}^1_h \times L^2$. *Then, there exists a weak solution to* (5) *(x_h-periodic and with* $\mathbf{w}$ *vanishing on* ∂D*) such that*

$$\mathbf{w},\, \nabla_h \mathbf{w},\, \nabla w^3,\, \rho' \in L^\infty(0,T;L^2) \cap L^2(0,T;H^1).$$

The proof, is based on the following *a-priori* estimate (obtained by using as test function $((\mathrm{I} - \alpha^2 \Delta_h)\mathbf{w}, \rho')$ and by performing some integrations by parts)

$$\frac{1}{2}\frac{d}{dt}(\|\mathbf{w}\|^2 + \alpha^2 \|\nabla_h \mathbf{w}\|^2 + \|\rho'\|^2) + \frac{1}{Re}(\|\nabla \mathbf{w}\|^2 + \alpha^2 \|\nabla \nabla_h \mathbf{w}\|^2) + \|\nabla \rho'\|^2 \leq c(\|\rho\|^2 + \|\mathbf{f}\|^2),$$

and by using the same standard tools as before. Also in this case we are not able to prove uniqueness, since the control of the whole $\nabla \mathbf{w}$ is missing.

In the next section we explain how the results can be improved by changing the class of models we consider. In particular, here we are testing the horizontal Rational/Clark method since it is the one for which we have at disposal a numerical code to perform preliminary assessment and we can precisely compare the results with those previously obtained. In addition, we wanted to test the performances of these methods in situations where the theoretical results are not conclusive.

4 Perspectives for future studies

Even if the numerical simulations for (2)-(5) are quite nice, the results are not completely satisfactory from the point of view of functional analysis. Good theoretical results can be surely obtained if the regularization in (5) is done with the Clark-α model instead of its horizontal counterpart, while less trivial is the limit $k := \frac{1}{Re\,Pr} \to 0^+$ for this model.

Moving to ADM we can easily show that the following system is well-posed

$$\begin{aligned} \partial_t \mathbf{w} + \nabla \cdot (\mathrm{I} - \alpha^2 \Delta)^{-1} (\mathbf{w} \otimes \mathbf{w}) - \frac{1}{Re}\Delta \mathbf{w} + \nabla q &= -\frac{1}{Fr^2}\rho' \mathbf{e}_3. \\ \partial_t \rho' + \nabla \cdot (\rho' \mathbf{w}) - \frac{1}{Re\,Pr}\Delta \rho' &= 0. \end{aligned} \tag{6}$$

This is the Boussinesq version of the Layton & Lewandowski model (or simplified Bardina) and good properties follow because with $((\mathrm{I} - \alpha^2 \Delta)\mathbf{w}, \rho')$ as test function, one gets directly $\nabla \mathbf{u} \in L^\infty(0,T;L^2)$, cf. [8, 20]. Observe that *no regularization* is needed in the equation for ρ'. The computational problems with this model concern the boundary conditions, since the inversion of the Laplace operator requires *ad hoc* boundary conditions on ∂D (Essentially it works only in the periodic setting.) This observation is one of the main reasons that led to the introduction of the *horizontal version* of this model in [2]. For the Boussinesq system, this will be naturally generalized by considering the equation

$$\partial_t \mathbf{w} + \nabla \cdot (\mathrm{I} - \alpha^2 \Delta_h)^{-1} (\mathbf{w} \otimes \mathbf{w}) - \frac{1}{Re}\Delta \mathbf{w} + \nabla q = -\frac{1}{Fr^2}\rho' \mathbf{e}_3,$$

coupled with (6)$_2$. From the mathematical point of view the resulting system does not give the sufficient control on $\nabla \mathbf{w}$ to prove uniqueness. Guided by the results on mathematical analysis (we can prove global-in-time existence and uniqueness of strong solutions [2]), we conjecture that a suitable ADM model for the Boussinesq system will be the following one

$$\partial_t \mathbf{w} + \nabla \cdot (\mathrm{I} - \alpha^2 \Delta_h)^{-1} (\mathbf{w} \otimes \mathbf{w}) - \frac{1}{Re} \Delta \mathbf{w} + \nabla q = -\frac{1}{Fr^2} \rho' \mathbf{e}_3,$$
$$\partial_t \rho' + \nabla \cdot (\mathrm{I} - \alpha^2 \Delta_h)^{-1} (\rho' \mathbf{w}) - \frac{1}{Re\, Pr} \Delta \rho' = 0.$$

All theoretical results on the Boussinesq equations will be collected in the forthcoming paper [6], while numerical testing of all different methods will be the object of future research.

5 Numerical Results

In this section, we investigate how the horizontal RLES model and the horizontal Clark-α models perform in the numerical simulation of 3D turbulent stratified flows. To this end, we compare the performance of four LES models: 1) the RLES model, 2) the Clark-α model, 3) the horizontal RLES model, and 4) the horizontal Clark-α model - against DNS results. The comparison criterion is simple: the closer the LES results are to the benchmark DNS results, the better the LES model. To ensure a fair assessment of the performance of the LES models, we also included under-resolved numerical simulations without any LES modeling, which we denoted by DNS*. Thus, it is expected that the LES models produce better results than the DNS*, at the very least.

Although most of the theoretical developments in this paper have been centered around the NSE (see also [2]), the numerical illustrations in this section are for the Boussinesq equations, since -as explained in the introduction- it is one of the commonly used mathematical models in the numerical investigation of oceanic and atmospheric flows. The model setup is similar to that in [25], which contains a detailed discussion of the boundary conditions, initial conditions and parameters used. We now briefly list them; for more details, the reader is referred to [25].

We consider the *lock-exchange* problem, a popular benchmark problem for the numerical investigation of mixing in stratified flows [24, 25, 26]. The computational domain is $-\frac{L}{2} \le x \le \frac{L}{2}$, $0 \le y \le W$, and $0 \le z \le H$, where $L/H = 2$ and $W/H = 1$. At the top, bottom, left and right boundaries, no-flow and free-slip boundary conditions are used for the velocity components (u, v, w), while no-flux (insulation) conditions are used for the density perturbation ρ', i.e., $\dfrac{\partial w}{\partial \mathbf{n}} = 0$; $(u, v, w) \cdot \mathbf{n} = 0$; $\dfrac{\partial \rho'}{\partial \mathbf{n}} = 0$, where $\mathbf{n}$ is the normal to the boundary. In the horizontal directions, periodic boundary conditions are used both for velocity and density perturbation fields: $u(x,0,z) = u(x,W,z)$; $v(x,0,z) = v(x,W,z)$; $\rho'(x,0,z) = \rho'(x,W,z)$. Since $\alpha \to 0^+$

near the boundary, the filtered variables approach the unfiltered variables. Thus, for the LES models, we used the same boundary conditions as those used for the DNS. The non-dimensional parameters have the same values as those in [25]: The Froude number is $Fr = 2^{-\frac{1}{2}}$, the Prandtl number is $Pr = 7$, and the Reynolds number is $Re = 10^4$. The numerical study is conducted using Nek5000, which is a spectral element code developed by Paul Fischer and his group [14, 22, 27]. All experiments were run on Virginia Tech's SystemX, based on 2200 Apple G5 processors with 2.3 GHz and InfiniBand interconnect. The accuracy of the LES models is evaluated through *a posteriori* testing. The main measure used is the background/reference potential energy (RPE), which exactly quantifies mixing in an enclosed system [25]. RPE is the minimum potential energy that can be obtained through an adiabatic redistribution of the water masses. Finally, for the RLES model, we chose $\gamma = 3$ and $\gamma_T = 15$, cf. [24, 25] for the notation.

There are also a couple of significant differences from the previous studies [24, 25]. First, in this study we are integrating much longer in time. As explained next, this has an effect on the conclusions regarding the performance of the LES models. Second, we implemented the horizontal version of the RLES and Clark-α models. Although the gradient tensor has been implemented in its horizontal form, the Helmholtz operator is still in its original isotropic form. We are currently implementing its anisotropic (horizontal) version.

The results of our numerical simulations are presented in figures 1, 2, and 3. In Figure 1, we present snapshots of DNS for the density perturbation ρ' at different times. This time evolution of the density perturbation will represent the benchmark for our LES runs.

In figure 2, we present snapshots of the density perturbation ρ' at $t = 3.0$. Notice that all four models (horizontal RLES, horizontal Clark-α, DNS, and DNS*) produce practically indistinguishable results.

Finally, in figure 3 we investigate the ability of the LES models to reproduce the DNS RPE curve. We also compare the isotropic and horizontal versions of the LES models. The behavior of the LES models depends on the time-interval considered. At the beginning of the numerical simulation, both LES models (horizontal RLES and horizontal Clark-α) produce better results than the under-resolved simulation (DNS*) and the horizontal RLES model is more accurate than the Clark-α model. Towards the end of the simulation, however, the quality of the results produced by the LES models degrades. Indeed, the LES models yield RPE curves that are farther away from the benchmark RPE curve than the RPE curve produced by the under-resolved DNS* runs. We emphasize that this behavior was not displayed in our previous studies [24, 25], since there we did not integrate as long in time as in our present study. The DNS results in figure 3 have not been filtered. From our previous experience with the lock-exchange problem, filtering the DNS data does not change the overall qualitative results. We believe that this is mainly due to the fact that the filtering radius α is not too large. In realistic ocean flows, where computational resources become utterly scarce, however, one needs to consider a very large filtering radius. In this case, one needs to replace the DNS data in figure 3 with its filtered counterpart, where the horizontal filtering is carried out by convolution with the kernel $g_\alpha(\cdot)$ introduced in Section 2. We plan to investigate the effect of replacing DNS data with filtered DNS data in a forthcoming study.

These new results open a new research avenue that we are currently pursuing. On the other hand, the preliminary tests of this report show that the new methods we propose are promising and may be used to improve performances of previous simulations. Further analysis and tests are running, and we hope to give some new insight in the numerical simulation of mixing phenomena involved in ocean flows.

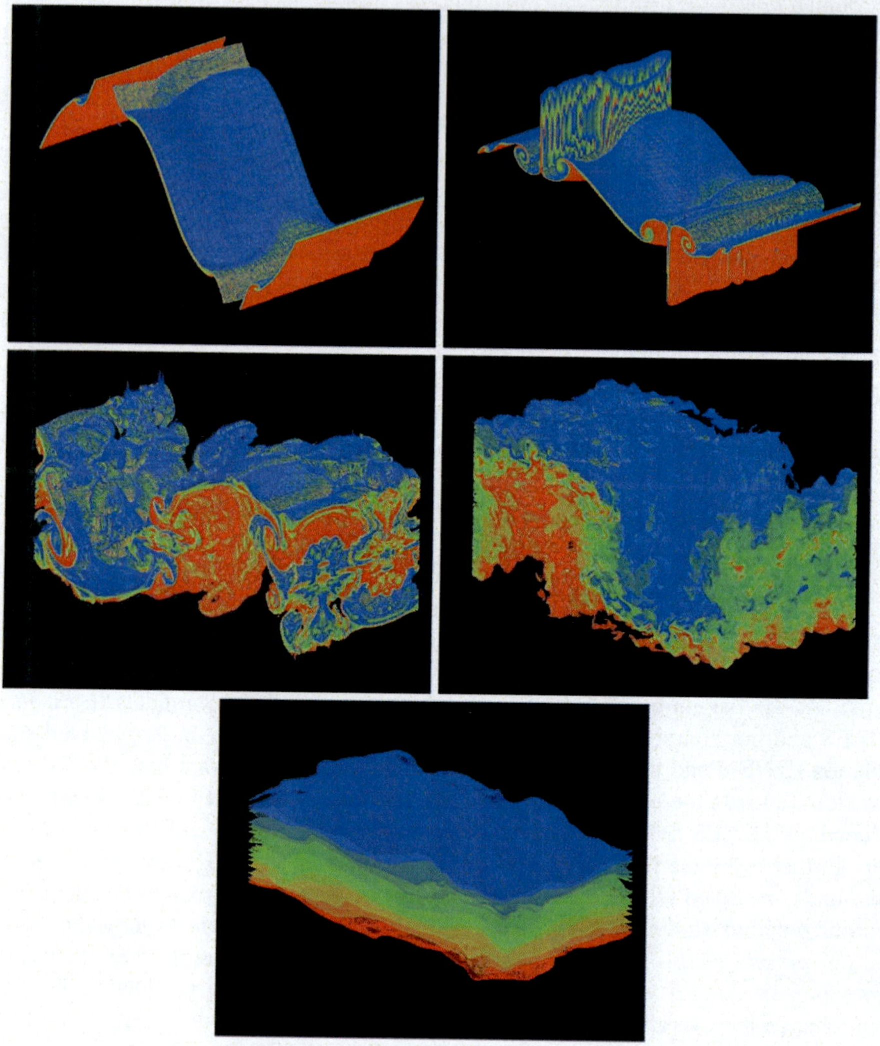

Fig. 1. Density perturbation snapshots. DNS at: (a) t=0.8; (b) t=1.2; (c) t=3.0; (d) t=5.0; and (e) t=45.0.

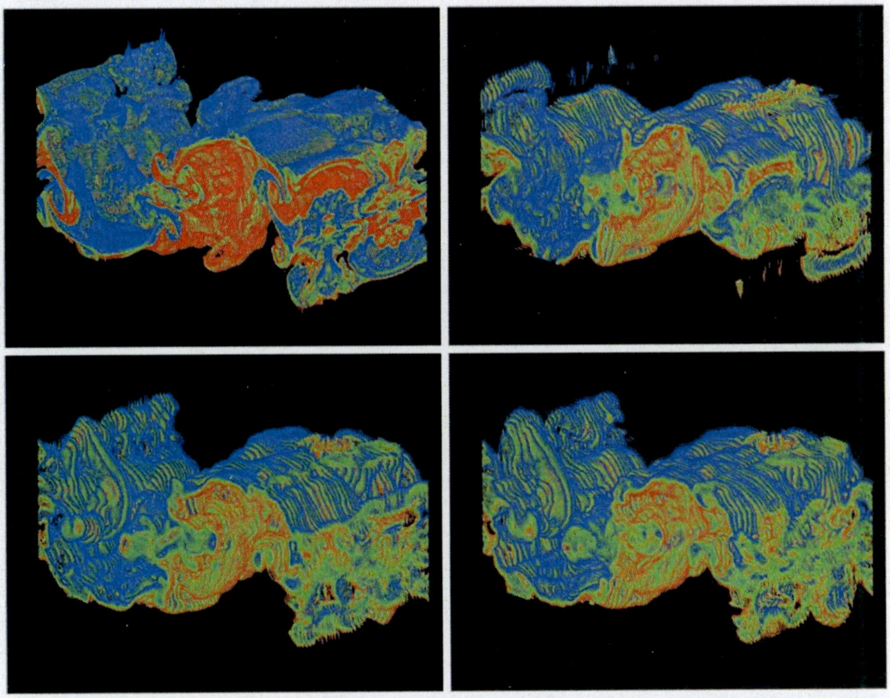

Fig. 2. Density perturbation snapshots. (a) DNS; (b) DNS*; (c) Clark-α horizontal; and (d) RLES horizontal.

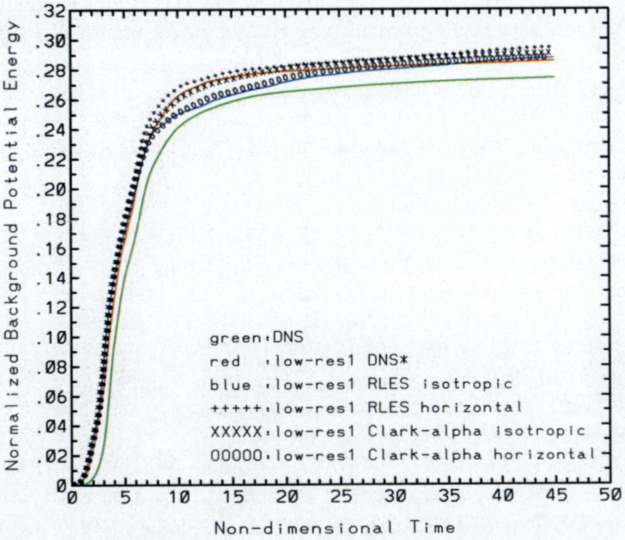

Fig. 3. RPE curves for DNS, DNS*, Clark-α horizontal, RLES horizontal, Clark-α, and RLES.

References

1. Barbato D, Berselli LC, Grisanti CR (2007) J. Math. Fluid Mech. 9:44–74
2. Berselli LC (2009) Analysis of a scale similarity LES model designed for certain stratified flows. Tech. Report., Univ. Pisa
3. Berselli LC, Flandoli F (2009) On a stochastic approach to Eddy Viscosity models for turbulent flows. In: Contributions to current challenges in mathematical fluid mechanics, Adv. Math. Fluid Mech., Birkhäuser, Basel, 57–83.
4. Berselli LC, Galdi GP, Iliescu T, Layton WJ (2002) Math. Models Methods Appl. Sci. 12:1131–1152
5. Berselli LC, Iliescu T, Layton WJ (2006) Mathematics of Large Eddy Simulation of turbulent flows. Springer-Verlag, Berlin
6. Berselli LC, Spirito S (2009) Remarks on the Boussinesq equations. In preparation
7. Cantero MI, Lee JR, Balachandar S, Garcia M (2007) J. Fluid Mech. 586:1–39
8. Cao C, Holm DD, Titi ES (2005) J. Turbul. 6, Paper 20.
9. Chemin JY, Desjardins B, Gallagher I, Grenier E (2000) M2AN Math. Model. Numer. Anal. 34:315–335
10. Clark R, Ferziger J, Reynolds W (1979) J. Fluid Mech. 91:1–16
11. Constantin P, Foias C (1988) Navier-Stokes equations. Univ. of Chicago Press, Chicago
12. Deville M, Fischer P, Mund E (2002) High-order methods for incompressible fluid flow. Cambridge Univ. Press, Cambridge
13. Fan J, Ozawa T (2009) Nonlinearity 22:553–568
14. Fischer P (1997) J. Comp. Phys. 133:84–101
15. Fischer P, Mullen J (2001) C. R. Acad. Sci. Paris Sér. I Math. 332:265–270
16. Galdi GP, Layton WJ (2000) Math. Models Methods Appl. Sci. 10:343–350
17. Geurts BJ, Holm DD (2003) Phys. Fluids 15:L13–L16
18. Härtel C, Meiburg E, Necker F (2000) J. Fluid Mech. 418:189–212
19. Ishimura N, Morimoto H (1999) Math. Models Methods Appl. Sci. 9:1323–1332
20. Layton WJ, Lewandowski R (2006) Discrete Contin. Dyn. Syst. Ser. B 6:111–128
21. Leonard A (1974) Adv. in Geophysics 18A:237–248
22. Maday Y, Patera AT (1989) Spectral element methods for the incompressible Navier-Stokes equations. In: Noor AK and Oden JT (eds) State–of–the–Art Surveys in Computational Mechanics, ASME, N.Y.
23. Nerli A, Camarri S (2006) Meccanica 41:671–680
24. Özgökmen T, Iliescu T, Fischer P (2009) Ocean Modelling 26:134–155
25. Özgökmen T, Iliescu T, Fischer P (2009) Ocean Modelling, 30:190–206.
26. Özgökmen T, Iliescu T, Fischer P, Srinivasan A, Duan J (2007) Ocean Modelling 16:106–140
27. Patera A (1984) J. Comput. Phys. 54:468–488
28. Pope S (2000) Turbulent flows. Cambridge Univ. Press, Cambridge
29. Rummler B (1997) Z. Angew. Math. Mech. 77:669–675
30. Stolz S, Adams NA (1999) Phys. Fluids 11:1699–1701
31. Vreman B, Geurts B, Kuerten H (1996) Theor. Comput. Fluid Dyn. 8:309–324
32. Vreman B, Geurts B, Kuerten H (1997) J. Fluid Mech. 339:357–390
33. Winckelmans GS, Jeanmart H, Carati D (2002) Phys. Fluids 14:1809–1811
34. Winckelmans GS, Wray AA, Vasilyev OV, Jeanmart H (2001) Phys. Fluids 13:1385–1403

The effect of subfilter-scale physics on regularization models

Jonathan Pietarila Graham[1], Darryl Holm[2,3], Pablo Mininni[4,5], and Annick Pouquet[4]

[1]Max-Planck-Institut für Sonnensystemforschung, Katlenburg-Lindau, Germany [2]Department of Mathematics, Imperial College London, London, UK [3]Computer and Computational Science Division, Los Alamos National Laboratory, NM, USA [4]National Center for Atmospheric Research,[†] Boulder, Colorado, USA [5]Departamento de Física, Facultad de Ciencias Exactas y Naturales, Universidad de Buenos Aires, Ciudad Universitaria, Buenos Aires, Argentina

Summary. The subfilter-scale (SFS) physics of regularization models are investigated to understand the regularizations' performance as SFS models. The strong suppression of spectrally local SFS interactions and the conservation of small-scale circulation in the Lagrangian-averaged Navier-Stokes $\alpha-$model (LANS$-\alpha$) is found to lead to the formation of rigid bodies. These contaminate the superfilter-scale energy spectrum with a scaling that approaches k^{+1} as the SFS spectra is resolved. The Clark$-\alpha$ and Leray$-\alpha$ models, truncations of LANS$-\alpha$, do not conserve small-scale circulation and do not develop rigid bodies. LANS$-\alpha$, however, is closest to Navier-Stokes in intermittency properties. For magnetohydrodynamics (MHD), the presence of the Lorentz force as a source (or sink) for circulation and as a facilitator of both spectrally nonlocal large to small scale interactions as well as local SFS interactions prevents the formation of rigid bodies in Lagrangian-averaged MHD (LAMHD$-\alpha$). We find LAMHD$-\alpha$ performs well as a predictor of superfilter-scale energy spectra and of intermittent current sheets at high Reynolds numbers. We expect it may prove to be a generally applicable MHD-LES.

Key words: LES, Subgrid-scale processes, alpha models, MHD, intermittency

1 Introduction

Regularization modeling for Navier-Stokes and magnetohydrodynamics (MHD) promises several advantages. Unlike the situation for Navier-Stokes [16], for a regularization, we are guaranteed the computability of solutions: we can achieve a direct numerical solution (DNS) of the model equations. It is worth noting that we can then achieve a grid-independent model. As only the spectral distribution of energy, not the dissipative processes as in many LES, is modified, a well-defined Reynolds number

[†] The National Center for Atmospheric Research is sponsored by the National Science Foundation

M.V. Salvetti et al. (eds.), *Quality and Reliability of Large-Eddy Simulations II*, ERCOFTAC Series 16, DOI 10.1007/978-94-007-0231-8_37,

(*Re*) is retained (instead of the usual approach of modeling the behavior of the flow in the limit of very large *Re*). Thus, the models may be more applicable to intermittent phenomenon where the length of the inertial range can be important [13]. Since the models do not introduce the effect of the small scales *ad hoc* but rather preserve the mathematical properties of the underlying equations, their application can further our understanding of turbulence and turbulence modeling. The methods are also more easily generalized to other problems (e.g., MHD).

In this paper we address two separate questions. One is the question of the practical applicability of regularization models as subfilter-scale (SFS) models. When addressing this question, the filter width, α, will be placed in the inertial range and the grid spacing, Δ, will be just small enough to achieve a DNS of the regularization. Our aim is then to determine how well the model's DNS reproduces a "DNS" of Navier-Stokes compared *at scales larger than* α. Our second question is "How do the models work?" Here, we aim to understand how the new SFS physics allow reproduction of the superfilter-scale properties. To address this question, we must necessarily spend the majority of our numerical resolution to resolve the subfilter-scale inertial range. The filter width, α, will be a large fraction of the computational domain and no superfilter-scale comparisons can be made with Navier-Stokes. Instead, comparisons for scales smaller than α will elucidate the differences in physics between the model and Navier-Stokes (or MHD). It is these very differences that allow the model to reduce computational cost when employed as a SFS.

2 Navier-Stokes

2.1 LANS$-\alpha$ and rigid body formation

The first model we consider is the Lagrangian-averaged Navier-Stokes (LANS) $\alpha-$model [3, 18]. It is derived by Lagrangian averaging fluid motions followed by application of Taylor's frozen-in turbulence approximation as the model's one and only closure: fluctuations about the Lagrangian mean smaller than α are swept along by the large-scale flow and are not allowed to interact with one another. The model is attractive as it retains the Hamiltonian structure of Navier-Stokes, preserves Kelvin's theorem (conserves small-scale circulation in the absence of dissipation), and conserves both total energy and helicity. These properties are conserved in the H^1_α norm instead of the usual L^2 norm. This is essential when interpreting results of the model as, for example, quantities involving the square velocity, $|\mathbf{v}|^2$, must now be replaced with the dot product $\mathbf{v}\cdot\bar{\mathbf{v}}$ where $\bar{\mathbf{v}}$ is the smoothed velocity. Physically, the model retains nonlocal interactions between the superfilter and subfilter scales while the flux of energy in subfilter scales is reduced by the limit on local small-scale to small-scale interactions.

The LANS$-\alpha$ model is given by, with ν the viscosity:

$$\begin{aligned} \partial_t v_i + \partial_j(\bar{v}_j v_i) + \partial_i \pi + v_j \partial_i \bar{v}_j &= \nu \partial_{jj} v_i \\ \partial_j v_j = \partial_j \bar{v}_j &= 0. \end{aligned} \tag{1}$$

It differs from Navier-Stokes both in advection by the smoothed velocity and the addition of a second nonlinear term associated with the conservation of circulation. Traditionally, LANS$-\alpha$ is used with an inverse Helmholtz operator as the filter: $\bar{v}_i = (1-\alpha^2\partial_{jj})^{-1}v_i$. In this case, LANS$-\alpha$ can be written as a LES with

$$\bar{\tau}^{\alpha}_{ij} = (1-\alpha^2\partial_{jj})^{-1}\alpha^2(\partial_m\bar{v}_i\partial_m\bar{v}_j + \partial_m\bar{v}_i\partial_j\bar{v}_m - \partial_i\bar{v}_m\partial_j\bar{v}_m) \tag{2}$$

as the Reynolds subfilter-scale stress tensor. The model allows for a reduction in resolution without changing the dissipative terms by altering the SFS scaling properties. Near $k = 2\pi/\alpha$, the H^1_α energy spectrum is predicted to transition from k^β with $\beta = -5/3$ at larger scales to $\beta = -1$ at smaller scales [5]. Consequently, dissipation goes as k^1 instead of $k^{1/3}$ and the same amount of power is dissipated in fewer scales. The change in spectral scaling also predicts a resolution requirement in degrees of freedom, dof, for LANS [5],

$$dof_\alpha \sim \alpha^{-1}Re^{3/2}\,, \tag{3}$$

which has been confirmed in numerical experiments [12]. Resolving dof_α allows the identification of LANS$-\alpha$ as a grid-independent SFS model. When compared with the result for Navier-Stokes,

$$dof_{NS} \sim Re^{9/4}\,, \tag{4}$$

we see that LANS$-\alpha$ should improve as a SFS model for larger Re. This is an encouraging prediction as it compared well with dynamic eddy viscosity [27] and dynamic mixed (similarity) eddy viscosity [8] at moderate Re.

We have, however, found that LANS$-\alpha$ develops a problem at large Re: it develops a positive-exponent power-law bump in its small-scale energy spectrum and a contamination of superfilter-scale spectral properties [12]. To investigate the SFS physics responsible for this, we employ a filter 1/3 the size of our 256^3 computational cube in a pseudo-spectral calculation [10, 11] with a Taylor-Green (TG) forcing [30] and $Re \approx 8000$. As shown in Fig. 1, the observed scaling law is k^{+1}. This was shown to be associated with the formation in the flow of passively swept regions, called *rigid bodies* [12]. These form as a consequence of disallowing sub-α-scale fluctuations to interact with each other in the closure approximation. A rigid body cannot support longitudinal velocity increments: $\delta\bar{v}_{\|}(l) \equiv [\bar{\mathbf{v}}(\mathbf{x}) - \bar{\mathbf{v}}(\mathbf{x}+\mathbf{l})]\cdot\mathbf{l}/l = 0$. This predicts a scaling relation, $\delta\bar{v} \sim l^0$, and, with $v \sim \alpha^2k^2\bar{v}$ for $l \ll \alpha$, an energy spectrum of

$$E_\alpha(k) \sim \bar{v}vk^{-1} \sim k^1 \tag{5}$$

which is compatible with the observed SFS energy spectrum. Inside rigid bodies there can be no turbulent cascade of energy to smaller scales (no internal degrees of freedom). From the Kármán-Howarth theorem, we should then expect to be able to detect rigid bodies by visualizing the cubed velocity increments (which are proportional to the energy flux). The regions which correspond to negligible flux are shown as black in the inset of Fig. 1. Filtering these regions out, allows us to obtain a (convolved) energy spectra for the remaining white portion of the flow. This spectrum is shown as a dashed line in Fig. 1 and has a negative spectral slope close to the

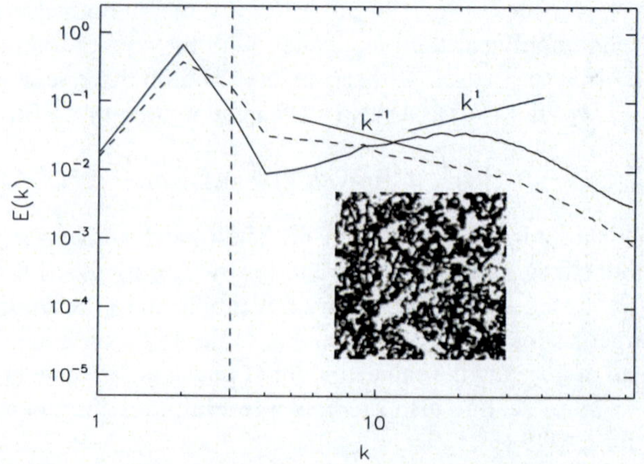

Fig. 1. Energy spectrum for LANS$-\alpha$ (solid line) with filter scale $\alpha = 2\pi/3$ (vertical dashed line). The SFS inertial range is seen to match a k^{+1} power law. Insert shows thresholded cubed velocity increment $|\delta\bar{v}_{\|}(2\pi/10)|^3 < 10^{-2}$ in black. These regions do not contribute to the turbulent cascade of energy to smaller scales and are identified with rigid bodies. A spectrum of only the white regions (dashed line) is consistent with the predicted k^{-1} scaling outside rigid bodies.

predicted k^{-1} energy spectrum. The resulting picture of the model's behavior is to produce two spatially separate scalings. The white portions of the flow possess the predicted LANS$-\alpha$ scaling and are responsible for the observation of the predicted dof_{α}. The black portions are rigid bodies whose k^1 energy spectrum dominates over k^{-1} for large k and are responsible for the observed spectral contamination. Note that suitable spectra can be obtained with very small α [12] or with modified viscous length scale (LANS$-\alpha\beta$ [20]).

2.2 Clark$-\alpha$, Leray$-\alpha$, and influence of circulation on rigid bodies

The formation of rigid bodies in LANS$-\alpha$ limits the reduction of numerical dof saved compared to Navier-Stokes to a factor of $1/12$ regardless of Re [12]. It is desirable, then, to alter the model in such a way to prevent the formation of rigid bodies. Truncation of the Reynolds SFS stress tensor, Eq. (2), to the first term results in the Clark$-\alpha$ model [2] and to the first two terms results in the Leray$-\alpha$ model [7, 8]. Both these models are regularizations and conserve the total energy of the flow. They do not, however, conserve the helicity nor the small-scale circulation. Considering the rotational properties of a rigid body (in the absence of viscous friction), these models' circulation properties may be incompatible with rigid body formation. This is, indeed, borne out in the model's SFS energy spectrum (Fig. 2a). While LANS$-\alpha$ exhibits a positive-exponent power law in this case ($\alpha = 2\pi/3$, $Re \approx 3300$, TG forcing), both Clark$-\alpha$ and Leray$-\alpha$ are free from this signature of rigid body formation. However, LANS$-\alpha$'s intermittency properties are more similar to Navier-Stokes than the other two models (Fig. 2b).

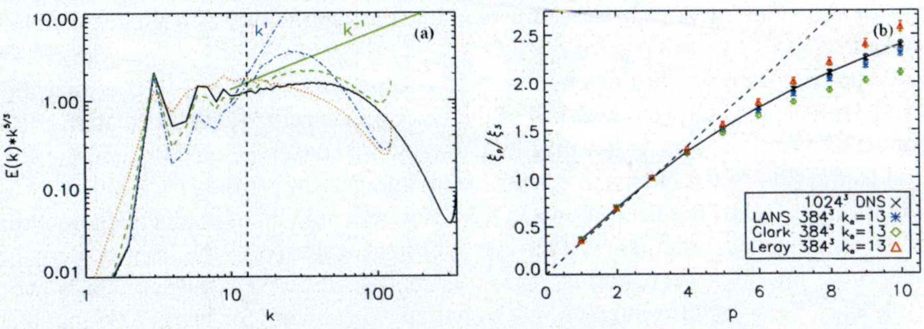

Fig. 2. (a) Compensated energy spectra ($2\pi/\alpha$ vertical dashed line) for Navier-Stokes (solid black), LANS$-\alpha$ (blue dash-dotted), Clark$-\alpha$ (green dashed), and Leray$-\alpha$ (red dotted). (b) Normalized structure function scaling exponent ξ_p/ξ_3 versus order p. Clark$-\alpha$ is the best approximation for the superfilter-scale spectrum, whereas high-order intermittency properties are best reproduced by LANS$-\alpha$ [14].

3 MHD: circulation and outlook for LES

In MHD, the situation is quite different since small-scale circulation is broken by the Lorentz force $\mathbf{j}\times\mathbf{b}$, with $\mathbf{j}=\boldsymbol{\nabla}\times\mathbf{b}$ the current, $\mathbf{b}$ being the induction. This force acts as source (sink) of circulation, Γ, as opposed to the insufficient modeling of Γ in the Leray$-\alpha$ and Clark$-\alpha$ models. This can be seen in Kelvin's theorem,

$$\frac{d}{dt}\Gamma = \frac{d}{dt}\oint_{\mathscr{C}} \mathbf{v}\cdot d\mathbf{r} = \oint_{\mathscr{C}} \mathbf{j}\times\mathbf{b}\cdot d\mathbf{r}\,. \tag{6}$$

This may prevent the formation of rigid bodies even while conserving all the correct physical properties of the flow.

The LES equations for MHD are given by

$$\begin{aligned} \partial_t\bar{\mathbf{v}} + \bar{\boldsymbol{\omega}}\times\bar{\mathbf{v}} &= \bar{\mathbf{j}}\times\bar{\mathbf{b}} - \boldsymbol{\nabla}\bar{\Pi} + \nu\nabla^2\bar{\mathbf{v}} - \nabla\cdot\bar{\tau} \\ \partial_t\bar{\mathbf{b}} &= \boldsymbol{\nabla}\times\left(\bar{\mathbf{v}}\times\bar{\mathbf{b}}\right) + \eta\nabla^2\bar{\mathbf{b}} - \nabla\cdot\bar{\tau}^b\,, \end{aligned} \tag{7}$$

where where η is the magnetic diffusivity, $\Pi = P + |\mathbf{b}|^2/2$ the modified pressure, $\bar{\tau}$ is the Reynolds SFS stress tensor,

$$\bar{\tau}_{ij} = \overline{v_i v_j} - \bar{v}_i\bar{v}_j - (\overline{b_i b_j} - \bar{b}_i\bar{b}_j)\,, \tag{8}$$

and $\bar{\tau}^b$ is the electromotive-force SFS stress tensor,

$$\bar{\tau}^b_{ij} = \overline{b_i v_j} - \bar{b}_i\bar{v}_j - (\overline{v_i b_j} - \bar{v}_i\bar{b}_j)\,. \tag{9}$$

Note that the extension of eddy viscosity to eddy resistivity employs the usual form for $\bar{\tau}$ involving only the filtered velocity while the expression for $\bar{\tau}^b$ similarly only involves the filtered magnetic field [31]. Meanwhile, Eqs. (8) and (9) make it explicitly clear that interactions between the two fields *at subfilter scales* must be taken into account.

Another problem with extending eddy-viscosity concepts to MHD is that they can be related to a known power law of the energy spectrum [4]. This is inappropriate for MHD as neither kinetic nor magnetic energy is a conserved quantity and the general expression of the energy spectrum is not known at this time [19, 22, 9, 24, 23]. Additionally, MHD has been shown to have a greater amount of nonlocal interactions between large and small (superfilter and subfilter) scales (e.g., the large-scale magnetic field provides the restoring force for small-scale Alfvén waves) than Navier-Stokes [1]. This complex interaction is a challenge in general for MHD-LES, but may be an advantage for the Lagrangian-averaged approach as energy exchange with sub-α scales may disrupt rigid body formation. Some limited case MHD LES include the cross-helicity model [28] assuming alignment between the fields and the low magnetic *Re* LES [21, 29]. However, there are many regimes of MHD dependent on the ratios between the various conserved quantities and ν/η. As a result, there is not yet any generally applicable LES for MHD.

3.1 LAMHD$-\alpha$ and absence of rigid bodies

The Lagrangian-averaged MHD $\alpha-$model (LAMHD$-\alpha$) [17] is given by, where the velocity if filtered as before and $\bar{\mathbf{b}} = (1-\alpha^2\nabla^2)^{-1}\mathbf{b}$:

$$\begin{aligned} \partial_t \mathbf{v} + \boldsymbol{\omega} \times \bar{\mathbf{v}} &= \mathbf{j} \times \bar{\mathbf{b}} - \nabla\pi + \nu\nabla^2\mathbf{v} \\ \partial_t \bar{\mathbf{b}} &= \nabla \times (\bar{\mathbf{v}} \times \bar{\mathbf{b}}) + \eta\nabla^2\mathbf{b} \\ \nabla\cdot\mathbf{v} = \nabla\cdot\bar{\mathbf{v}} &= \nabla\cdot\mathbf{b} = \nabla\cdot\bar{\mathbf{b}} = 0 . \end{aligned} \tag{10}$$

The model preserves all the ideal invariants of MHD (in the H^1_α norm) as well as Alfvén's theorem for frozen-in field lines. Physically, it supports Alfvén waves at all scales while slowing and hyper-diffusively damping waves with wavelengths, λ, smaller than α [15]. In examinations of its SFS physical properties LAMHD$-\alpha$ (dashed lines) displays neither positive-exponent power-law scaling nor superfilter-scale spectral contamination (see Fig. 3). Under similar conditions LANS$-\alpha$ (not shown) displays these signs of rigid body formation. Further examinations with larger filters and higher *Re* were unable to unravel any sign that rigid bodies form for LAMHD$-\alpha$ [15].

3.2 LAMHD$-\alpha$ as a SFS model

Given that LAMHD$-\alpha$ does not display the same limitations as LANS$-\alpha$, we test it as a SFS model for large kinetic and magnetic Reynolds numbers, ≈ 3300. A DNS of MHD is computed at a resolution of 1024^3. The initial conditions for $\mathbf{v}$ and $\mathbf{b}$ are a super-position of ABC modes [6] with random phases and wavenumbers

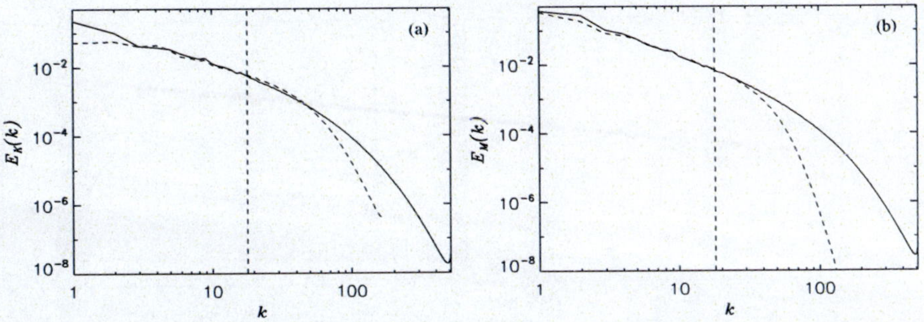

Fig. 3. Kinetic (a) and magnetic (b) energy spectra ($2\pi/\alpha$ vertical dashed line). The largest scales are affected by differences from the MHD DNS (solid lines) in initial conditions. LAMHD$-\alpha$ (dashed lines) exhibits neither the positive power-law nor the superfilter-scale spectral contamination associated with high *Re* LANS$-\alpha$.

$k \in [1,4]$. No external forcing is applied and the total energy is allowed to freely decay. LAMHD$-\alpha$ is computed for identical conditions at a resolution of $N^3 = 168^3$ with a filter size $\alpha = 6\Delta = 2\pi/28$. As a base-level comparison we also compute an under-resolved (or no-model) solution of the MHD equations at $N^3 = 168^3$. Time evolution of the total energies and enstrophy are shown in Fig. 4. In comparison with under-resolving MHD, LAMHD$-\alpha$ shows errors of approximately the same magnitude in these global quantities. Comparisons of energy spectra (Fig. 5), however, show an improvement in predictive quality for LAMHD$-\alpha$, especially for the magnetic energy spectrum. As turbulence develops, energy begins to pile up at small-scales and deplete at intermediate scales for 168^3 MHD. LAMHD$-\alpha$ improves the prediction of superfilter-scale spectra compared to no SFS model.

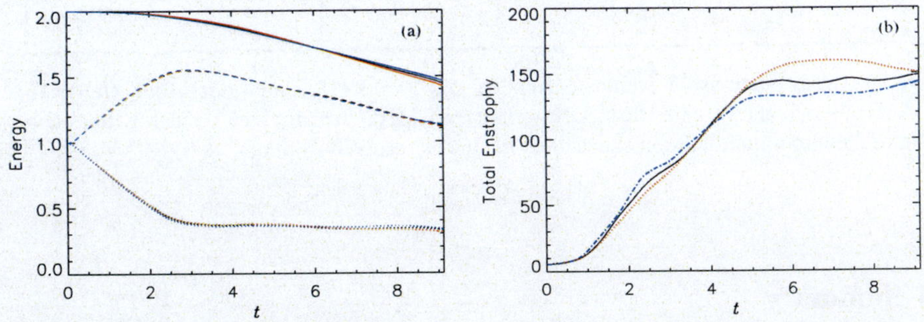

Fig. 4. Time evolution of total (solid), magnetic (dashed), and kinetic (dotted) energies (a) and total enstrophy (b) for 1024^3 DNS (black/solid), 168^3 LAMHD$-\alpha$ (blue/dash-dotted), and 168^3 no-model (red/dotted). LAMHD$-\alpha$ provides no improvement in prediction of global quantities.

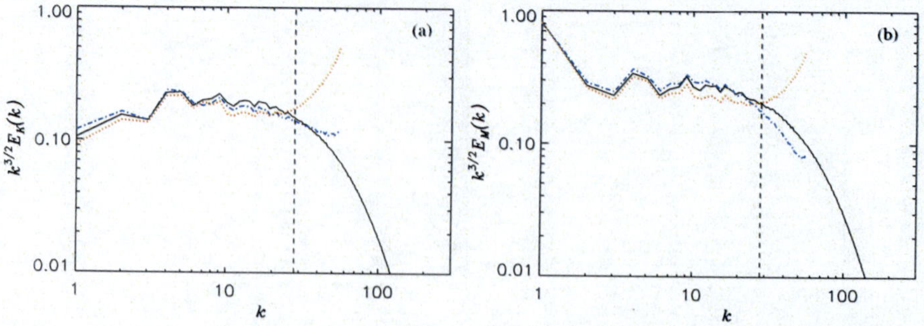

Fig. 5. Compensated kinetic (a) and magnetic (b) energy spectra for $t \in [8, 8.8]$. Labels are as in Fig. 4. Energy piles up at small scales in the no-model approach (under-resolved DNS) and LAMHD$-\alpha$ is seen to provide improved predictions of the superfilter-scale spectra, especially for the magnetic field.

Cross-sections of $|\mathbf{j}|^2$, shown in Fig. 6 at $t = 8.4$ indicate that LAMHD-α finds sharper and better defined, more intermittent current sheets than the under-resolved run compared to the DNS.

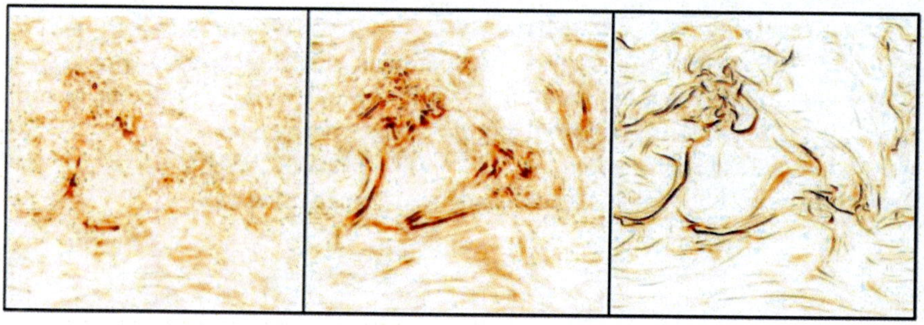

Fig. 6. Cross sections of square current, j^2, at $t = 8.4$ for no-model 168^3 (left), 168^3 LAMHD$-\alpha$ (center), and 1024^3 DNS (right). LAMHD$-\alpha$ provides a much better capturing of the intermittent current sheets than the under-resolved solution.

4 Summary

Incompressible LANS$-\alpha$, while it performed well at moderate Reynolds number is limited as a high *Re* SFS model. Due to its strong suppression of spectrally local interactions at subfilter-scales, and consistent with its conservation of small-scale circulation, LANS$-\alpha$ develops rigid bodies which contaminate the superfilter-scale energy spectrum. In contrast, Clark$-\alpha$ and Leray$-\alpha$, neither of which conserve small-scale circulation do not develop energy-spectrum contamination from rigid bodies. LANS$-\alpha$, however, best matches the intermittency properties of Navier-Stokes fluid turbulence.

In MHD, a mechanism for local small-scale transfer is the interaction of small-scale Alfvén waves. As LAMHD$-\alpha$ supports Alfvén waves at all scales while slowing and hyperdiffusively damping those with wavelength $\lambda < \alpha$, it more gently suppresses SFS local interactions than LANS$-\alpha$. This together with the greater nonlocality in MHD and the Lorentz-force source of small-scale circulation, inhibits the formation of rigid bodies in LAMHD$-\alpha$. For this reason, we find LAMHD$-\alpha$ to be a viable model at high *Re* in 3D. As LAMHD$-\alpha$ has been previously found to reproduce the difficult to model properties of MHD at high *Re* in 2D [26] and moderate *Re* in 3D [25], we believe it will prove to be a generally applicable MHD LES.

References

1. A. Alexakis, P. D. Mininni, and A. Pouquet. Shell-to-shell energy transfer in MHD. I. Steady state turbulence. *Phys. Rev. E*, 72(4):046301–+, 2005.
2. C. Cao, D. D. Holm, and E. S. Titi. On the Clark α model of turbulence: global regularity and long-time dynamics. *Journal of Turbulence*, 6:N20, 2005.
3. S. Chen, C. Foias, D. D. Holm, E. Olson, E. S. Titi, and S. Wynne. Camassa-Holm Equations as a Closure Model for Turbulent Channel and Pipe Flow. *Physical Review Letters*, 81:5338–5341, 1998.
4. J.-P. Chollet and M. Lesieur. Parameterization of Small Scales of Three-Dimensional Isotropic Turbulence Utilizing Spectral Closures. *Journal of Atmospheric Sciences*, 38:2747–2757, 1981.
5. C. Foias, D. D. Holm, and E. S. Titi. The Navier-Stokes-alpha model of fluid turbulence. *Physica D Nonlinear Phenomena*, 152-153:505–519, 2001.
6. D. Galloway and U. Frisch. Dynamo action in a family of flows with chaotic streamlines. *Geophysical and Astrophysical Fluid Dynamics*, 36:53–83, 1986.
7. B. J. Geurts and D. D. Holm. Regularization modeling for large-eddy simulation. *Physics of Fluids*, 15:L13–L16, 2003.
8. Bernard J. Geurts and Darryl D. Holm. Leray and LANS-α modelling of turbulent mixing. *Journal of Turbulence*, 7(10):1–33, 2006.
9. P. Goldreich and S. Sridhar. Toward a theory of interstellar turbulence. 2: Strong alfvenic turbulence. *ApJ*, 438:763–775, 1995.
10. D. O. Gómez, P. D. Mininni, and P. Dmitruk. MHD simulations and astrophysical applications. *Advances in Space Research*, 35:899–907, 2005.
11. D. O. Gómez, P. D. Mininni, and P. Dmitruk. Parallel Simulations in Turbulent MHD. *Physica Scripta Volume T*, 116:123–127, 2005.
12. J. Pietarila Graham, D. Holm, P. Mininni, and A. Pouquet. Highly turbulent solutions of the Lagrangian-averaged Navier-Stokes alpha model and their large-eddy-simulation potential. *Phys. Rev. E*, 76:056310–+, 2007.
13. J. Pietarila Graham, D. D. Holm, P. Mininni, and A. Pouquet. Inertial range scaling, Kármán-Howarth theorem, and intermittency for forced and decaying Lagrangian averaged MHD equations in 2D. *Physics of Fluids*, 18:045106, 2006.
14. J. Pietarila Graham, D. D. Holm, P. D. Mininni, and A. Pouquet. Three regularization models of the Navier-Stokes equations. *Physics of Fluids*, 20(3):035107–+, 2008.

15. J. Pietarila Graham, P. D. Mininni, and A. Pouquet. Lagrangian-averaged model for magnetohydrodynamic turbulence and the absence of bottlenecks. *Phys. Rev. E*, 80(1):016313–+, 2009.
16. J.-L. Guermond. On the use of the notion of suitable weak solutions in CFD. *International Journal for Numerical Methods in Fluids*, 57:1153–1170, 2008.
17. D. D. Holm. Lagrangian averages, averaged Lagrangians, and the mean effects of fluctuations in fluid dynamics. *Chaos*, 12:518–530, 2002.
18. D. D. Holm, J. E. Marsden, and T. S. Ratiu. The Euler-Poincaré Equations and Semidirect Products with Applications to Continuum Theories. *Adv. in Math.*, 137:1–81, 1998.
19. P. S. Iroshnikov. Turbulence of a Conducting Fluid in a Strong Magnetic Field. *Soviet Astronomy*, 7:566–+, 1964.
20. T.-Y. Kim, M. Cassiani, J. D. Albertson, J. E. Dolbow, E. Fried, and M. E. Gurtin. Impact of the inherent separation of scales in the Navier-Stokes-$\alpha\beta$ equations. *Phys. Rev. E*, 79(4):045307–+, 2009.
21. B. Knaepen and P. Moin. Large-eddy simulation of conductive flows at low magnetic Reynolds number. *Physics of Fluids*, 16:1255–+, 2004.
22. R. H. Kraichnan. Inertial-range spectrum of hydromagnetic turbulence. *Physics of Fluids*, 8:1385–1387, 1965.
23. E. Lee, M. E. Brachet, A. Pouquet, P. D. Mininni, and D. Rosenberg. On the lack of universality in decaying MHD turbulence. arXiv:0906.2506, 2009.
24. J. Mason, F. Cattaneo, and S. Boldyrev. Numerical measurements of the spectrum in magnetohydrodynamic turbulence. *Phys. Rev. E*, 77(3):036403–+, 2008.
25. P. D. Mininni, D. C. Montgomery, and A. Pouquet. Numerical solutions of the three-dimensional MHD α model. *Phys. Rev. E*, 71(4):046304–+, 2005.
26. P. D. Mininni, D. C. Montgomery, and A. G. Pouquet. A numerical study of the alpha model for two-dimensional magnetohydrodynamic turbulent flows. *Physics of Fluids*, 17(3):035112–+, 2005.
27. K. Mohseni, B. Kosović, S. Shkoller, and J. E. Marsden. Numerical simulations of the Lagrangian averaged Navier-Stokes equations for homogeneous isotropic turbulence. *Physics of Fluids*, 15:524–544, 2003.
28. W.-C. Müller and D. Carati. Dynamic gradient-diffusion subgrid models for incompressible MHD turbulence. *Physics of Plasmas*, 9:824–834, 2002.
29. Y. Ponty, H. Politano, and J.-F. Pinton. Simulation of Induction at Low Magnetic Prandtl Number. *Physical Review Letters*, 92(14):144503–+, 2004.
30. G. I. Taylor and A. E. Green. Mechanism of the Production of Small Eddies from Large Ones. *Proceedings of the Royal Society of London*, A158:499, 1937.
31. M. L. Theobald, P. A. Fox, and S. Sofia. A subgrid-scale resistivity for magnetohydrodynamics. *Physics of Plasmas*, 1:3016–3032, 1994.

When does eddy viscosity damp subfilter scales sufficiently?

Roel Verstappen

Johann Bernoulli Institute for Mathematics and Computer Science
University of Groningen
P.O. Box 407, 9700 AK Groningen, The Netherlands.
r.w.c.p.verstappen@rug.nl

Summary. The very essence of large-eddy simulation is that the computed flow field contains only eddies of size $\geq \Delta$. This restriction determines the eddy viscosity ν_e. Although the principle is quit apparent, it is not yet fully understood how to derive ν_e from that. Therefore we address two related questions, namely how much eddy viscosity is needed to damp any disturbances having a scale smaller than Δ and when does eddy viscosity stop the production (by means of vortex stretching) of scales $< \Delta$. From this we deduce that the ν_e has to depend on the invariants q and r of the filtered strain tensor. The simplest model is then given by $\nu_e = c^2 \Delta^2 r^+ / q$.

Key words: eddy viscosity, confinement of dynamics, invariants of strain tensor

1 Problem setting

The Navier-Stokes equations provide an appropriate model for turbulent flow. In the absence of compressibility ($\nabla \cdot u = 0$), the equations are

$$\partial_t u + (u \cdot \nabla) u + \nabla p = 2\nu \nabla \cdot S(u), \tag{1}$$

where u is the fluid velocity field, p stands for the pressure, ν denotes the viscosity, and the symmetric part of the velocity gradient (the rate of strain tensor) is defined by

$$S(u) = \tfrac{1}{2}\left(\nabla u + \nabla u^T\right).$$

Turbulent flow is often visualized as a cascade of kinetic energy from large to small scales of motion (eddies). The energy introduced at the large-scale components of the flow is transferred to smaller and smaller eddies - nearly without viscous dissipation - until eddies become sufficiently small to dissipate energy efficiently. The entire spectrum ranging from the wavenumbers where energy is injected to the dissipation range is to be resolved when turbulence is computed directly from the Navier-Stokes equations. This requirement limits direct numerical simulations

M.V. Salvetti et al. (eds.), *Quality and Reliability of Large-Eddy Simulations II*,
ERCOFTAC Series 16, DOI 10.1007/978-94-007-0231-8_38,

(DNS) to relatively low Reynolds numbers, since otherwise the dissipative scales are too small to be resolved numerically [1].

Large-eddy simulation (LES) aims at overcoming this problem by computing the evolution of the large scales only (see e.g. [2] and the references therein). To that end a spatial operator $u \mapsto \overline{u}$ is introduced that maps the full velocity field u to the velocity $\overline{u}$ of the large scales. Applying this filtering operator to the Navier-Stokes equations yields

$$\partial_t \overline{u} + (\overline{u} \cdot \nabla)\overline{u} + \nabla \overline{p} - 2\nu \nabla \cdot S(\overline{u}) = \overline{(u \cdot \nabla)u} - (\overline{u} \cdot \nabla)\overline{u} \tag{2}$$

provided $u \mapsto \overline{u}$ commutes with differentiation. The right-hand side of Eq. (2) depends on both u and $\overline{u}$ because the Navier-Stokes equations are nonlinear. To close the system (in terms of $\overline{u}$) the right-hand side of (2) is replaced by a approximate model. Usually these models are based on phenomenological arguments that cannot be derived formally from the Navier-Stokes equations. The most used model in LES is the eddy-viscosity model:

$$\partial_t v + (v \cdot \nabla)v + \nabla q = 2\nabla \cdot (\nu_t S(v)), \tag{3}$$

where the total viscosity is given by $\nu_t = \nu + \nu_e$. Here the variable name is changed from $\overline{u}$ to v to stress that the solution of Eq. (3) differs from that of Eq. (2), because the closure model is not exact.

The classical Smagorinsky model reads

$$\nu_e = C_S^2 \Delta^2 |S(v)| \tag{4}$$

where C_S is the Smagorinsky constant, Δ is the characteristic length scale set by the operator $u \mapsto \overline{u}$ (i.e., the width of the filter) and $|S(v)| = \sqrt{2\,\mathrm{tr}\,(S(v)^2)}$. It may be noticed that the precise definition of $\overline{u}$ is not of much significance in case the Smagorinsky model is used, since Eq. (4) depends only on the characteristic length scale Δ set by the map $u \mapsto \overline{u}$, and not on the details of the mapping.

Various value for the Smagorinsky constant have been proposed, mainly ranging from $C_S = 0.1$ to $C_S = 0.2$, see [1]-[2], e.g. Instead of adhering to a constant value, C_S is also treated as a model coefficient which is determined during the simulation, that is $\nu_e = \nu_e(v)$. In the well-known dynamical procedure the coefficient C_S is computed with the help of the Jacobi identity (in least-square sense) [3].

The very essence of LES is that the range of dynamically significant scales in the solution v of Eq. (3) is much smaller than the full range of scales in the Navier-Stokes solution u. This property enables us to solve (3) numerically when it is not feasible to compute the full solution of the Navier-Stokes equations (1). A simulation based on Eq. (3) differs from a DNS only in the use of a modified viscosity; hence, the desired effect thereof is a confinement of the dynamics, that is the smallest characteristic length-scale in v is bound to be Δ. Therefore, we view the eddy viscosity as a function v that is to be determined such that the smallest characteristic length-scale is (at least) equal to Δ. To get an idea about the required amount of eddy viscosity two questions are posed for discussion: how much eddy viscosity is needed to damp any disturbances having a length scale smaller than Δ (Section 2) and when does eddy viscosity stop the production of smaller scales (by means of vortex stretching) at scale Δ (Section 3). This leads us to the conclusion that ν_e has to depend on the invariants q and r of $S(v)$ (Section 4).

2 When does eddy viscosity damp any subfilter-scale disturbances?

The eddy viscosity must keep subfilter-scale disturbances from growing - stated otherwise, the dynamics governed by Eq. (3) should be stable with respect to these disturbances. To analyze the stability, we consider an arbitrary part D with volume $\mathcal{O}(\Delta^3)$ of the flow domain and superimpose an instantaneous, solenoidal, small-scale perturbation δv to v on D. Initially, say at time $t = t_0$, δv equals zero outside D and

$$\int_D \delta v(x,t) \cdot v(x,t) dx = 0 \tag{5}$$

At $t = t_0$ one may conceive v as being constant on D, whereas δv is any (non-constant) periodic function on D. In other words, δv and v are orthogonal at $t = t_0$, that is, $(\delta v, v) = 0$, where the innerproduct is defined in the usual way: $(a,b) = \int_D a(x,t) \cdot b(x,t)\, dx$. The disturbance δv can be caused externally, or by a numerical error, or perhaps the non-linear dynamics governed by Eq. (3) is not correctly closed and a unwanted small scale δv is produced at $t = t_0$. Anyway, the evolution of the perturbed velocity $v + \delta v$ is governed by Eq. (3), with v replaced by $v + \delta v$. Here we take the eddy viscosity independent of δv: $\nu_t = \nu_t(v)$. The dynamics of the disturbance is then given by

$$\partial_t \delta v + (v \cdot \nabla)\delta v + (\delta v \cdot \nabla) v + (\delta v \cdot \nabla)\delta v + \nabla \delta p = 2\nabla \cdot (\nu_t S(\delta v)), \tag{6}$$

where δv stands for a arbitrary vector field having divergence equal to zero. Eddies of size smaller than Δ ought not become significant; hence, their energy should not increase for all admissible disturbances δv at $t = t_0$,

$$\frac{\mathrm{d}}{\mathrm{dt}} \int_D \tfrac{1}{2} \delta v \cdot \delta v\, dx \overset{(6)}{=} -(\delta v, S(v)\delta v) - (S(\delta v), 2\nu_t S(\delta v)) \leq 0 \tag{7}$$

Notice that a number of terms vanish here, because the convective operator is skew-symmetric: $((a\cdot\nabla)b, c) = -(b, (a\cdot\nabla)c)$. Intuitively, inequality (7) can always be satisfied by taking the eddy viscosity sufficiently large. To obtain a lower-bound, we take the eddy viscosity constant on D. Then,

$$-(S(\delta v), 2\nu_t S(\delta v)) = -\nu_t (\nabla \delta v, \nabla \delta v) \leq -\nu_t\, \alpha_\Delta\, (\delta v, \delta v)$$

where the positive constant α_Δ denotes the smallest (non-zero) eigenvalue of the dissipative operator $-\nabla^2$ on D. The inverse of α_Δ is known as the Poincaré constant for the domain D. The Poincaré constant of any convex domain with diameter Δ is given by $(\Delta/\pi)^2$. So, in conclusion, Eq. (7) leads to the following restriction on the eddy viscosity

$$\nu_t \, \alpha_\Delta \geq -\frac{(\delta v, S(v)\delta v)}{(\delta v, \delta v)} \tag{8}$$

Rayleigh's principle states that

$$\max_{\delta v \neq 0} -\frac{(\delta v, S(v)\delta v)}{(\delta v, \delta v)} = -\lambda_1(v) \tag{9}$$

where $-\lambda_1(v)$ is the largest eigenvalue of the symmetric matrix $-S(v)$ on D, i.e., λ_1 is the smallest eigenvalue of S on D. We order the eigenvalues of S by $\lambda_1 \leq \lambda_2 \leq \lambda_3$. The sum of the eigenvalues is zero, because $\mathrm{tr}(S) = 0$. Hence, $\lambda_1 \leq 0$ and $\lambda_3 \geq 0$. The characteristic equation reads

$$\lambda^3 - q\lambda + r = 0, \tag{10}$$

where the invariants of the tensor $S(v)$ are given by

$$q = \tfrac{1}{2}\mathrm{tr}(S^2(v)) \qquad \text{and} \qquad r = -\tfrac{1}{3}\mathrm{tr}(S^3(v)) \tag{11}$$

The three roots of the cubic equation (10) can be computed analytically:

$$\lambda_1 = -\sqrt{\tfrac{1}{3}}|S(v)|\cos\left(\tfrac{\theta}{3}\right) \tag{12}$$

$$\lambda_2 = -\sqrt{\tfrac{1}{3}}|S(v)|\cos\left(\tfrac{\theta}{3} - \tfrac{2\pi}{3}\right) \tag{13}$$

$$\lambda_3 = -\sqrt{\tfrac{1}{3}}|S(v)|\cos\left(\tfrac{\theta}{3} + \tfrac{2\pi}{3}\right) \tag{14}$$

where

$$\theta = \arccos\left(\tfrac{1}{2}r / \tfrac{1}{3}\sqrt{q^3}\right) \tag{15}$$

It may be noted that the three roots of Eq. (10) must be real-valued because S is symmetric. Eq. (10) has three real roots if and only if $27r^2 - 4q^3 \leq 0$. Hence, $\theta \in [0, \pi]$.

Substituting (9)+(12) into (8) shows that any subfilter perturbations are damped if

$$\nu_t \geq \frac{\cos\left(\frac{\theta}{3}\right)}{\sqrt{3}\,\alpha_\Delta}|S(v)| \tag{16}$$

The above estimate resembles the Smagorinsky model if we take (a) $\alpha_\Delta = (\pi/\Delta)^2$; and (b) the smallest value of ν_t that satisfies (16). In this way we get

$$C_S^2 = \frac{\cos\left(\frac{\theta}{3}\right)}{\pi^2\sqrt{3}}. \tag{17}$$

The resulting Smagorinsky 'constant' C_S ranges from $\frac{1}{2\pi 3^{1/4}} \approx 0.12$ to $\frac{1}{\pi 3^{1/4}} \approx 0.24$, because $\cos\left(\frac{\theta}{3}\right) \in [\frac{1}{2}, 1]$. Most of the constant values of C_S that have been proposed fall in the same range. This holds for values resulting from an empirical fit to DNS results as well as for Kolmogorov-based values. A well-known example of the latter is Lilly's result, $C_S = 0.17$ [4].

In (9) we maximized the Rayleigh quotient without restricting δv to subfilter scales, i.e., without imposing the constraint (5). Without this constraint the maximum value is attained if δv is aligned with the eigenvector e_1 belonging to λ_1. Yet, if δv is constrained to a non-constant function in D it cannot be fully aligned with e_1, because e_1 depends on v and v is supposed to be constant in D (at $t = t_0$). Consequently, the lower-bound (16) on the eddy viscosity decreases if we impose the appropriate constraint. Then, $-\lambda_1(v)$ is to be replaced by the function

$$\mu(v) = \max_{(v,\delta v)=0} -\frac{(\delta v, S(v)\delta v)}{(\delta v, \delta v)} \tag{18}$$

where we can take $(\delta v, \delta v) = 1$ without loss of generality. Thus we get

$$\nu_t \geq \frac{\mu(v)}{\alpha_\Delta} = \frac{\Delta^2}{\pi^2}\mu(v) \tag{19}$$

The function $\mu(v)$ exists because it is the maximum of a continuous function on a compact set. According to Courant's minimax theorem, $-\lambda_2$ is the least of all possible values of $\mu(v)$ and $\mu(v)$ is minimized when v is aligned with e_2. Therefore, $-\lambda_2(v) \leq \mu(v) \leq -\lambda_1(v)$. This interval can also be interpreted as a possible range for the Smagorinsky constant C_S, that is instead of Eq. (17) we get $-\frac{1}{2} \leq \cos\left(\frac{\theta}{3} - \frac{2\pi}{3}\right) \leq \pi^2\sqrt{3}C_S^2 \leq \cos\left(\frac{\theta}{3}\right) \leq 1$. Furthermore, near a no-slip wall $y = 0$, Eq. (18) gives $\mu(v) \to 0$ if $y \to 0$.

The function μ depends on the invariants q and r of the tensor $S(v)$. Therefore μ is invariant under rotation of the coordinate axes. Unfortunately, we cannot obtain a useful, analytic expression for the solution μ of the minimization problem (18). However, μ can be determined from DNS-data. To that end the DNS solution u is to be mapped to $\overline{u}$. This gives an approximation of v and also $S(v)$. From that μ can be computed numerically.

It may be noted that Smagorinsky's model is formulated in terms of ν_e, whereas (7) yields a condition on $\nu_t = \nu + \nu_e$. This difference disappears, i.e., we get conditions on ν_e rather than on ν_t, if (7) is replaced by the slightly stronger condition

$$\frac{\mathrm{d}}{\mathrm{dt}}\int_D \tfrac{1}{2}||\delta v||^2\,dx \leq -\nu\int_D ||\nabla\delta v||^2\,dx$$

A noticeable difference between (19) and the Smagorinsky model (4) is that μ depends on both r and q, whereas the Smagorinsky model, $\nu_e = C_S^2\Delta^2\sqrt{4q}$, depends only on the invariant q if C_S is taken constant. This raises the fundamental question: is the invariant r important, or not?

A dimensional analysis shows that

$$\nu_e = \Delta^2\sqrt{q}\,\sum_\beta C_\beta\left(r/\sqrt{q^3}\right)^\beta \tag{20}$$

where C_β denotes a dimensionless constant. The standard Smagorinsky model corresponds to $\beta = 0$. Expressions based on the eigenvalues of S introduce a dependency on $r/\sqrt{q^3}$. See, for example, (15)-(17), or the function $\mu \geq -\lambda_2$. This lower-bound satisfies $|r/q| \leq |\lambda_2| \leq \frac{3}{2}|r/q|$ which indicates that in case of (19) we have to consider $\beta = 1$.

3 When does eddy viscosity stop vortex stretching from continuing at subfilter scales?

Again, we view the eddy viscosity as a function of the velocity v that is to be determined from the requirement that the dynamics stays confined to eddies of size $\geq \Delta$; in particularly, the smallest scales of motion that dissipate the energy are to be size Δ, or larger. To derive a condition on the eddy viscosity, we consider an arbitrary part Ω of size $\mathscr{O}(\ell^3)$ of the flow domain, with $\ell \geq \Delta$. We consider both the incompressible Navier-Stokes equations (1) and the LES-model given by Eq. (3) on the 3D box Ω, with periodic boundary conditions. At time $t = t_0$ the Navier-Stokes solution u is given by $u(x,t_0) = u_0(x)$, for all $x \in \Omega$. The initial condition for the LES-model reads $v(x,t_0) = \overline{u}_0(x)$. The initial conditions supply energy to the flow. This energy cannot escape from the box Ω, since we have applied periodic conditions. Hence, the energy is to be dissipated within Ω. In case of the Navier-Stokes equations the evolution of the energy $E(t) = \int_\Omega \frac{1}{2}||u||^2 dx$ is given by $dE/dt = -\varepsilon$ with $\varepsilon = \nu \int_\Omega ||\nabla u||^2 dx$. In the LES-model, the dissipation rate becomes $\int_\Omega (\nu + \nu_e)||\nabla v||^2 dx$. In the absence of eddy-viscosity, i.e., $\nu_e = 0$, this integral is much smaller than ε if $v \approx \overline{u}$. Indeed, the mapping $u \mapsto \overline{u}$ reduces the velocity gradient. Now suppose that the amount of eddy-viscosity is taken too little. Then, $||\nabla v||^2$ will have (a tendency) to increase, because the energy that is supplied to the flow has to be dissipated anyway. Since the norm of the velocity gradient $||\nabla v||$ provides a consistent characterization of the reciprocal of the time scale, an increase of $||\nabla v||$ implies that smaller time-scales are produced. Then, the eddies of scale ℓ in the velocity field v are unstable and break up, transferring their energy to smaller eddies. These smaller eddies undergo a similar break-up process, and transfer their energy to smaller eddies, and so on till the energy can be dissipated effectively. So, in conclusion, an increase of $\int_\Omega ||\nabla v||^2 dx$ indicates that scales with a length smaller than ℓ are produced. In a LES this causes no problem if $\ell > \Delta$. But, in order to confine the dynamics to scales $\geq \Delta$ this process has to stop at the scale set by the filter. Therefore, we determine the eddy viscosity from the requirement that

$$\frac{\mathrm{d}}{\mathrm{dt}} \int_{D=\mathscr{O}(\Delta^3)} ||\nabla v||^2 dx \leq 0 \tag{21}$$

This condition can also be interpreted in terms of the amplification of the vorticity vector $\omega = \nabla \times v$, since we have the equality

$$\int_\Omega ||\nabla v||^2 dx = \int_\Omega ||\omega||^2 dx \tag{22}$$

Thus we see that Eq. (21) states that the vortex-stretching mechanism stops from continuing if the volume Ω becomes $\mathscr{O}(\Delta^3)$.

To provide a more formal derivation of (21), we make use of Poincaré's inequality. This inequality states that there exists a constant C, depending only on Ω, such that for every function v in the Sobolev space $W^{1,2}(\Omega)$,

$$\int_\Omega ||v-\overline{v}_\Omega||^2\,dx \leq C\int_\Omega ||\nabla v||^2\,dx \tag{23}$$

where

$$\overline{v}_\Omega = \frac{1}{|\Omega|}\int_\Omega v\,dx$$

is the average value of v over Ω. In the present context, $||v-\overline{v}_\Omega||^2$ measures the subfilter component of v for a box filter with box size $|\Omega|$. Poincaré's inequality (23) demonstrates that the subfilter component of v is bounded by a constant (independent of v) times the $L^2(\Omega)$ norm of ∇v. By requiring that this upper-bound does not increase in time if $\Omega = \mathscr{O}(\Delta^3)$, that is by imposing condition (21), we keep the subfilter component of v under control. In this way, it can be shown that Eq. (3) is stable with respect to sub(box)filter disturbances; see Section 4.

To turn (21) into a lower-bound for the eddy viscosity, we take the L^2 innerproduct (3) with Δv, and integrate by parts to derive

$$\frac{\mathrm{d}}{\mathrm{dt}}\int_D \tfrac{1}{2}||\nabla v||^2 dx = \int_D \big(\,(v\cdot\nabla)v\cdot\Delta v - \nu_t||\Delta v||^2\,\big)\,dx \tag{24}$$

Note that the boundary terms that result from the integration by parts vanish because D is a periodic box. In [5] Dongho Chae shows that, for a 3D periodic box, the first term in the right-hand side of Eq. (24) can be written as:

$$\int_D (v\cdot\nabla)v\cdot\Delta v\,dx = \int_D \big(\,\tfrac{1}{4}\,\omega\cdot S\omega - \mathrm{tr}(S^3)\,\big)\,dx \tag{25}$$

By taking the curl of Eq. (3) we obtain the vorticity equation, and from that we get

$$\frac{\mathrm{d}}{\mathrm{dt}}\int_D \tfrac{1}{2}||\omega||^2 dx = \int_D \big(\,\omega\cdot S\omega - \nu_t||\nabla\omega||^2\,\big)\,dx \tag{26}$$

Now, since $\int_D ||\Delta v||^2 dx = \int_D ||\nabla\omega||^2 dx$, we deduce from Eqs. (22)-(26) that the vortex stretching term can be expressed in terms of the invariant r of S:

$$\int_D \omega\cdot S\omega\,dx = -\tfrac{4}{3}\int_D \mathrm{tr}(S^3)\,dx = 4\int_D r\,dx \tag{27}$$

Thus condition (21) becomes

$$\frac{\mathrm{d}}{\mathrm{dt}}\int_D \tfrac{1}{2}||\nabla v||^2\,dx = \int_D (4r - \nu_t||\nabla\omega||^2)\,dx \leq 0 \tag{28}$$

The equality $\int_D ||\nabla\omega||^2\,dx = \int_D ||\nabla\cdot S||^2\,dx$ shows once again (see Eq. (7)) that ν_t depends solely on S, that is on its invariants q and r (since ν_t has to be objective). To arrive at an explicit relation between ν_t and the invariants (q,r) we note that

$$\int_D ||\nabla\omega||^2\, dx = (c\Delta)^{-2}\int_D ||\omega||^2\, dx \geq \alpha_\Delta \int_D ||\omega||^2\, dx = \alpha_\Delta \int_D 4q\, dx \qquad (29)$$

The combination (28)+(29) establishes that the eddy viscosity stops the production of smaller scales by means of vortex stretching from continuing at subfilter scales if

$$\nu_t \int_D q\, dx \geq \alpha_\Delta \int_D r\, dx \qquad (30)$$

Finally it may be remarked that we may also sharpen condition (21) slightly:

$$\frac{\mathrm{d}}{\mathrm{dt}}\int_D \tfrac{1}{2}||\nabla v||^2 dx \leq -\nu \int_D ||\nabla\cdot\nabla v||^2 dx \qquad (31)$$

This has very little influence on our analysis: we just need to replace ν_t by ν_e in Eq. (24)-(30).

4 Eddy viscosity revisited

The very essence of LES is that the solution v of Eq. (3) contains only eddies of size $\geq \Delta$. This restriction determines the eddy viscosity. Although the principle is quit apparent, it is not yet fully understood how to determine the eddy viscosity from that. Therefore we have addressed two basic questions, namely how much eddy viscosity is needed to damp any disturbances having a length scale smaller than Δ (Section 2) and when does eddy viscosity stop the production of smaller scales (by means of vortex stretching) at scale Δ (Section 3).

Poincaré's inequality (23) relates these questions to each other. Indeed, for any sub(box)filter disturbance δv we have

$$\begin{aligned}\int_D \tfrac{1}{2}||\delta v||^2(x,t)\, dx &= \int_D \tfrac{1}{2}||v-\bar{v}_D||^2(x,t)\, dx \overset{(23)}{\leq} C\int_D \tfrac{1}{2}||\nabla v||^2(x,t)\, dx \\ &\overset{(31)}{\leq} Ce^{-\nu(c\Delta)^{-2}(t-t_0)}\int_D ||\nabla v||^2(x,t_0) dx \qquad (32)\end{aligned}$$

which shows that the energy of any sub(box)filter disturbance decreases eventually in time if (31) holds. In other words, Eq. (3) is stable (here meaning that any disturbances decreases eventually exponentially) with respect to subfilter-scale disturbances if (31) holds. Notice that the time-scale of the decay is given by $(c\Delta)^2/\nu$ with c according to (29).

The main conclusion of our analysis is that the eddy viscosity depends on the two invariants q and r of the strain tensor S. The most general form is then given by Eq. (20). The standard Smagorinsky model does not dependent on the invariant r of S. This is because the coefficient C_S is approximated using

$$< \overline{S}^3 > \approx < \overline{S}^2 >^{3/2}$$

see e.g. Lilly [4]. Note that we use the notation of Ref. [1] here. This approximation effectively eliminates r, and yields $C_S \approx 0.17$.

Eq. (27) shows that r is a measure for vortex stretching. The eddy viscosity has to prevent the intensification of vorticity at the scale Δ set by the map $u \mapsto \overline{u}$, otherwise smaller and smaller vortical structures are produced. From this point of view, it seems reasonable to assume that the eddy viscosity depends on r, and not just on q. The simplest form depending on both r and q is given by

$$\nu_t = \max\left\{c^2\Delta^2\,\frac{r}{q},\; \nu\right\} \tag{33}$$

where c^2 is a positive constant. This model follows directly from Eq. (30) if we take the smallest value of ν_t that satisfies (30). Again it may be remarked that taking (31) instead of (21) implies

$$\nu_e = c^2\Delta^2\,\frac{r^+}{q} \tag{34}$$

where $r^+ = \max\{r, 0\}$. Notice that the invariant q is by definition negative, whereas r can be both positive or negative. From $r = -\lambda_1\lambda_2\lambda_3$ it can be deduced that the sign of r and the sign of the middle eigenvalue λ_2 are the same. The molecular viscosity ν is certainly large enough to stop the vortex stretching at scale Δ in case $r \le 0$, i.e., Eq. (30) is trivially satisfied by taking $\nu_e = 0$ if $r \le 0$. Thus the eddy-viscosity model is to be turned off in the parts of the fluid domain where $r \le 0$. If this happens, we get a solution v depending on the molecular viscosity ν, i.e., on the Reynolds number Re.

According to Eq. (29) the value of c becomes $1/\pi$, provided the smallest (non-zero) eigenvalue of the dissipative operator $-\nabla^2$ on D is given by $\alpha_\Delta = (\pi/\Delta)^2$. In a numerical simulation one can also base c on the smallest (non-zero) eigenvalue of the discrete dissipative operator. Alternatively, c^2 can be viewed as a model parameter that is still to be determined, by a dynamical procedure, for instance.

Eq. (34) can also be displayed in the standard form (4) by taking the Smagorinsky constant C_S non-constant, that is by taking

$$C_S^2 = \frac{c^2 r^+}{2\sqrt{q^3}} \tag{35}$$

In terms of the Reynolds number Re the quotient of r and q scales like $r/q \propto Re^{3/2}/Re = Re^{1/2}$. Therefore we obtain that $\nu_t \to \nu$ if $\nu \propto Re^{-1} \propto \Delta^2 r/q \propto \Delta^2 Re^{1/2}$, that is if $\Delta \propto Re^{-3/4}$. This shows that the eddy viscosity given by Eq. (33)-(34) vanishes as Δ approaches the smallest scale in a turbulent flow.

The Smagorinsky model (4) predicts a nonvanishing eddy viscosity in regions where the flow is laminar. This shortcoming of the Smagorinsky model can be overcome by using a dynamic procedure for determining the model coefficient [3]. Vreman [6] developed a different approach. The principle constituent of his eddy-viscosity model is that the subfilter-scale dissipation vanishes for canonical laminar shear flows. Laminar flows can be characterized by $r = 0$ (no vortex stretching); hence (34) yields $\nu_e = 0$ in laminar flow. In fact this observation can be extended to any two-dimensional flow. Indeed in 2D flow, we have $\lambda_2 = 0$, which implies that

$r = 0$. Therefore Eq. (34) yields $\nu_e = 0$ for any two-dimensional flow. In this manner, Eq. (34) recognizes the salient differences of 2D and 3D turbulence.

Another shortcoming of the Smagorinsky model (4) is that the eddy viscosity does not vanish at no-slip walls. The near-wall behavior of the invariants r and q is given by $r \propto y^3$ and $q \propto y^0$, respectively, where y denotes the distance to the wall. Consequently, Eq. (34) results into an eddy viscosity that vanishes at the wall: $\nu_e \propto y^3$.

It goes without saying that the performance of (34) has to be investigated for many cases. As a first step it was tested (with $c = 1/\pi$) for turbulent channel flow (Re_τ = 180 and Re_τ = 395) using 32×32×16 and 64×64×32 grid points, respectively. Unlike the standard Smagorinsky model (with $C_S = 0.1$), the present model showed an appropriate behavior. In conclusion, it was observed to be as accurate as the dynamic Smagorinsky model, but also slightly less accurate than the regularization model C_4 in [7].

References

1. Pope SB (2000) Turbulent Flows. Cambridge University Press
2. Sagaut P (2001) Large Eddy Simulation for Incompressible Flows. Springer, Berlin Heidelberg New York
3. Germano M, Piomelli U, Moin P, Cabot WH (1991) Phys Fluids A3:1760–1765
4. Lilly DK (1967) The representation of small-scale turbulence in numerical simulation experiments. In: Goldstine H (ed) Proc. IBM Scientific Computing Symp. on Environmental Sciences, pp. 195–210. Yorktown Heights, IBM, New York
5. Dongho Chae (2005) Commun Math Phys 263:789–801
6. Vreman AW (2004) Phys Fluids 16(10):3670–3681
7. Verstappen R (2008) Comp & Fluids 37:887–897